SOLUTION PROCESSING OF INORGANIC MATERIALS

SOLUTION PROCESSING OF INORGANIC MATERIALS

Edited by

David B. Mitzi

WILEY

A JOHN WILEY & SONS, INC., PUBLICATION

Published by John Wiley & Sons, Inc., Hoboken, New Jersey
Published simultaneously in Canada

Library of Congress Cataloging-in-Publication Data:

Solution processing of inorganic materials / edited by David B. Mitzi.
p. cm.
Includes index.
ISBN 978-0-470-40665-6 (cloth)
1. Inorganic compounds. 2. Materials. 3. Solid state chemistry.
4. Solution (Chemistry) I. Mitzi, David B.
TA403.6.S62 2009
621.381—dc22

2008021430

Printed in the United States of America

10 9 8 7 6 5 4 3 2 1

CONTENTS

PREFACE

Rapid technological progress is transforming our world into one in which electronic capabilities integrate throughout all aspects of everyday life. Cell phones, laptops, digital assistants, and portable media players now provide unprecidented connectivity among people, information, and entertainment. Future advances promise to bring even more seamless integration, including flexible, wearable, and/or very large-area electronics with advanced functionality. Renewable forms of power generation (e.g., photovoltaics) will also hopefully replace carbon-based power sources in order to address both growing energy needs as well as environmental concerns. All of these changes require fundamental invention in the area of electronic materials processing to attain technological/economical viablity—including most notably in the area of low-cost deposition of high-quality functional films, which form the basis of modern electronics. Solution-based approaches for thin-film deposition are particularly desirable because of the low capital cost of the deposition equipment, relative simplicity of the processes, and potential compatibility with high-throughput (e.g., roll-to-roll) processing. Although most of the work toward this goal has focused on molecular and polymeric *organic* films, the search for solution-processible *inorganic* materials is at least as important, offering the potential for much higher performance and better thermal/mechanical stability than comparable organic-based systems. This book offers an exploration of the various means of overcoming technological barriers to the solution-deposition and patterning of inorganic electronic components. Throughout the text, emphasis is also placed on providing concrete examples of applications that employ the described solution-processed inorganic films (e.g., transistors, solar cells, and sensors).

In Chapter 1, the issues confronting the solution-based processing of inorganic films will be introduced, with a primary focus on two questions. The first question is why solution processing of inorganics (versus solution-processed organics and inorganics processed by other means) is important in the context of current trends and needs in the electronics and energy industries. The term "macroelectronics," or low-cost, potentially large-area and/or flexible, high-performance electronics (e.g., from flexible displays to solar panels and antenna arrays), will be introduced as an important future direction of technology evolution. The second question relates to what needs to be achieved, with respect to film performance, fabrication costs, and device characteristics, in order to have the desired impact on selected applications.

Chapter 2 introduces techniques for solution deposition (e.g., spin coating, spray coating, printing, and stamping) and discusses some of the basic concepts common to the solution deposition of inorganic materials, using chemical solution deposition (e.g., sol-gel, chelate, and metal-organic deposition) as an example. The concepts explored include the issues of substrate surface preparation, solution properties, film formation, crystal nucleation and growth, removal of reaction products during heating, and thermodynamic phase stability. Many of the issues introduced in this chapter reappear in later chapters.

After discussion of basic issues confronting solution-based film deposition, the next chapters address specific developments in the deposition of three important classes of electronic materials—chalcogenides (Chapter 3), oxides (Chapter 4), and silicon (Chapter 5)—with discussion of both the deposition techniques as well as the device results employing these films. The focus in these chapters is on how to get a relatively insoluble inorganic material into solution so that it can be deposited on a substrate, and the common theme is the formation of a soluble precursor that can be thermally and cleanly decomposed to yield the desired phase. Once in solution, spin coating is used as the primary means of delivering the inorganic material to the substrate. Relatively high-performance solar cells and thin-film transistors (TFTs) have been demonstrated using these approaches.

Beyond spin coating, Chapters 6–8 explore other means of delivering the inorganic material to a substrate. For example, spray pyrolysis and spray CVD (Chapter 6) represent a promising direction for high-throughput deposition. Particular emphasis is placed on the development of single-source precursors for use in the spray-based preparation of photovoltaic components. Chemical-bath-based techniques are described in Chapter 7. Rather than effecting a rapid chemical reaction of sprayed precursors on a heated substrate (as for spray deposition), the reaction between soluble metal salt and chalcogen source occurs more slowly in a chemical bath under more mild thermal conditions. Chemical bath deposition, electrodeposition, and electroless deposition are each explored as a means of depositing films for solar cells and superconductors. Successive-ionic-layer-adsorption-and-reaction (SILAR), ion layer gas reaction (ILGAR), and electrochemical atomic layer epitaxy (ECALE) deposition are described in Chapter 8 and rely on the sequential deposition of the cationic and anionic components of the desired inorganic materials. Film thickness is controlled in these techniques by the number of dipping cycles completed.

The next three chapters (Chapters 9–11) focus on the deposition of nanostructured or microstructured films and entities. Porous oxide thin films are, for example, of great interest due to potential application of these films as low-K dielectrics and in sensors, selective membranes, and photovoltaic applications. One of the key challenges in this area is the problem of controlling, ordering, and combining pore structure over different length scales. Chapter 9 provides an introduction and discussion of evaporation-induced self-assembly (EISA), a method that combines sol-gel synthesis with self-assembly and phase separation to produce films with a tailored pore structure. Chapter 10 describes how nanomaterials can be used as soluble precursors for the preparation of extended inorganic films. In this respect, nano-entities (e.g., nanoparticles, nanorods, nanowires, and nanotetrapods) provide an exciting pathway to tailor material properties through size/shape selection, compositional flexibility, and formation of core-shell structures. Chapter 11 focuses on how functional structures can be assembled from nanowire building blocks (i.e., nanowire electronics). In recent studies, fluids have been used to disperse nanowires, and these fluids are used to solution-deposit and orient these entities onto a substrate, yielding an array of nano- or micro-entities. A "holy grail" for this type of research would be to have the capability to chemically and physically functionalize the substrate and nano-entities such that they would self-assemble into relatively complex and predetermined patterns on a substrate.

In addition to depositing and characterizing blanket films, modern electronics relies on the ability to pattern and assemble the resulting insulating, semiconducting and metallic entities into functional devices. Many of the solution-based techniques offer a natural opportunity to achieve this goal during deposition. Chapter 12 gives an overview of patterning techniques that are unique to solution processing, such as ink-jet printing, flexography, gravure, and screen printing. All of these techniques rely on the ability to put the inorganic material or a precursor into solution (i.e., the formation of an ink) and there will therefore also be a discussion of what is needed to make an optimal ink for the different solution-based patterning techniques. Chapter 13 continues with the theme of printing by providing an introduction to rubber-stamping approaches, focusing on techniques that can be used to deposit and pattern single-crystalline micro- and nano-entities on a substrate (no heating of the substrate required). The versatile transfer printing process enables facile integration into heterogeneous 2D and 3D electronic devices and circuits. Although the approaches described in Chapter 13 are primarily "dry" techniques (no solutions required), they are included because they represent an exciting new direction in the low-cost processing of inorganic materials and involve related issues with respect to the stamping process.

Finally, the concluding chapter will discuss the potential implications of the above-described thin-film deposition techniques with respect to technology, highlighting common issues and imminent (or actual) applications of solution-processed inorganics. Although solution-based inorganic film technology is largely in its infancy, commercialization efforts are beginning to ramp up. It is hoped that the current book will give the reader not only a snapshot of the state-or-the-art and a primer on the basics of solution-processing of inorganic materials but also a view of critical areas that need to be addressed (from a materials point of view) before the new technologies can become a commercial reality, thereby giving a direction for future research in the field.

David B. Mitzi

Yorktown Heights, NY
August 2008

CONTRIBUTORS

Jong-Hyun Ahn, School of Advanced Materials Science and Engineering, SKKU Advanced Institute of Nanotechnology, Sungkyunkwan University, Suwon, Korea

Jeremy T. Anderson, Department of Chemistry, Oregon State University, Corvallis, OR

Alfred J. Baca, University of Illinois, Department of Chemistry, Materials Research Laboratory, Urbana, IL

Kulbinder K. Banger, University of Cambridge, Cavendish Laboratory, Optoelectronics, Department of Physics, Cambridge, UK

Raghu N. Bhattacharya, National Renewable Energy Laboratory, Golden, CO

Paul W. Brazis, Corporate Research, Underwriters Laboratories, Northbrook, IL

Babu R. Chalamala, Research Triangle Park, NC

John E. Dickman, Ypsilanti, MI

Dina Fattakhova-Rohlfing, Ludwig-Maximilians-Universität München, Department of Physical Chemistry and Biochemistry, Munich, Germany

Masahiro Furusawa, SEIKO EPSON Corporation, Frontier Device Research Center, Nagano-ken, Japan

Daniel Gamota, Motorola Physical Realization Research Center, Schaumburg, IL

David Ginley, National Renewable Energy Laboratory, Golden, CO

Jerry D. Harris, Northwest Nazarene University, Department of Chemistry, Nampa, ID

Aloysius F. Hepp, NASA Glenn Research Center, Cleveland, OH

Yu Huang, Department of Materials Science and Engineering, University of California, Los Angeles, CA

Michael H.-C. Jin, University of Texas at Arlington, Department of Materials Science and Engineering, Arlington, TX

Douglas A. Keszler, Department of Chemistry, Oregon State University, Corvallis, OR

Dahl-Young Khang, Yonsei University, Department of Materials Science and Engineering, Seoul, Korea

Hoon-Sik Kim, University of Illinois, Department of Materials Science and Engineering, Materials Research Laboratory, Urbana, IL

Markku Leskelä, University of Helsinki, Department of Chemistry, Helsinki, Finland

Seppo Lindroos, University of Helsinki, Department of Chemistry, Helsinki, Finland

Jeremiah S. McNatt, NASA Glenn Research Center, Cleveland, OH

Matthew A. Meitl, Semprius, Inc., Research Triangle Park, NC

Stephen T. Meyers, Department of Chemistry, Oregon State University, Corvallis, OR

David B. Mitzi, IBM T.J. Watson Research Center, Yorktown Heights, NY

Manoj Narayanan, Argonne National Laboratory, Energy Systems Division, Argonne, IL

Robert H. Reuss, Fountain Hills, AZ

John A. Rogers, University of Illinois at Urbana Champaign, Department of Materials Science and Engineering, Urbana, IL

Robert W. Schwartz, Missouri University of Science and Technology, Rolla, MO

Bernd Smarsly, Institute of Physical Chemistry, Justus-Liebig-University of Giessen, Giessen, Germany

John Szczech, Motorola Physical Realization Research Center, Schaumburg, IL

Dmitri V. Talapin, The University of Chicago, Department of Chemistry, Chicago, IL

Hideki Tanaka, SEIKO EPSON Corporation, Frontier Device Research Center, Nagano-ken, Japan

Maikel van Hest, National Renewable Energy Laboratory, Golden, CO

Jie Zhang, Motorola Physical Realization Research Center, Schaumburg, IL

CHAPTER 1

Introduction to Solution-Deposited Inorganic Electronics

ROBERT H. REUSS
Fountain Hills, AZ

BABU R. CHALAMALA
Research Triangle Park, NC

1.1 BACKGROUND AND MOTIVATION

1.1.1 Electronics Technologies

Two thrusts currently dominate efforts in electronics research. In both thrusts, the business opportunities stem from society's desire for a more pervasive and integrated electronics environment. However, the technical methods and strategies for achieving this goal are fundamentally different. The first and most widely discussed thrust focuses on what is popularly referred to as Moore's Law and the seemingly endless progression to smaller device feature sizes and the increasing numbers of transistors integrated onto a chip.[1] These chips (i.e., microelectronics) have enabled everything from laptop computers to cell phones, from smart cards to smart toys. The essence of the success of Moore's Law is that by creating technology to make devices smaller, the density and performance increases and the functionality goes up, whereas cost/function goes down. This amazingly successful thrust has created a technological revolution and has been an engine for worldwide economic growth because it results in faster and more compact products for computing and communications.

While over the last 40 years microelectronic integrated circuits based predominantly on silicon technology have made possible our current capabilities in everything from computers and phones to appliances and toys, even greater opportunities would exist if the circuits could be made more lightweight, flexible, and inexpensive. Everything from flexible displays, to radio frequency identification (RFID) tags that conform to a product's shape, to large and pliable "sensor sheets" that are integrated into airplanes, bridges, or even people to monitor and augment their physical condition, could become possible.[2] This concept is much newer, and the required technology is not nearly as mature. The distinguishing feature of this newer approach is that small device size is not a critical factor. Rather than fabrication of smaller devices and circuits, described below are two microelectronics-related electronics technologies that have become successful by fabricating modest-sized devices over larger and larger glass substrates (large-area electronics). This work is just now being extended to plastic substrates to provide reduced weight and novel form factors.

Given that microelectronics and large-area electronics are both electronics technologies, it might be assumed that the second is derived from and will evolve with the mainstream semiconductor industry. However, microelectronics is driven to produce smaller feature sizes and higher complexity chips. There are, however, many applications where microelectronics is not an appropriate technology, and in fact, it represents too complex or costly of a solution. Thus, the requirements and drivers are so different that few, if any, of the mainstream integrated circuit (IC) technologies are applicable to this second and newer thrust. While sharing many concepts with microelectronics, the second thrust is NOT for the most part derived from the IC industry and does not really benefit from its learning curve, but rather it originates from needs not adequately addressed by conventional microelectronics. It has different drivers, product attributes, and metrics and will be successful in its own product space, or by complementing conventional ICs to create solutions that neither could provide individually. Because of this distinction, varieties of names have been used to describe this non-microelectronic, large-area electronics technology. Because the device dimensions are generally large compared with microelectronics and product applications are physically large compared with microchips, one popular name for this form of electronics is "macroelectronics."[3]

1.1.2 Commercial Macroelectronic Technology

The most successful application of macroelectronics, the flat panel display (FPD) industry,[4,5] now rivals the microelectronics industry in revenue; yet, from an electronics perspective, it is based on nothing more than manufacture of modest-sized transistor switches distributed over glass substrates as large as a meter on a side.[6] Within 10 years, the FPD industry has almost reached the $100B mark (see Figure 1.1), whereas more traditional semiconductor industry growth has become relatively mature with slowing growth prospects. Another interesting aspect of the FPD story is that it has been accomplished while undergoing rapid changes in the manufacturing technologies. As shown in Figure 1.2, the size of the glass panels used have progressed rapidly through multiple manufacturing generations, which means that the panel size has

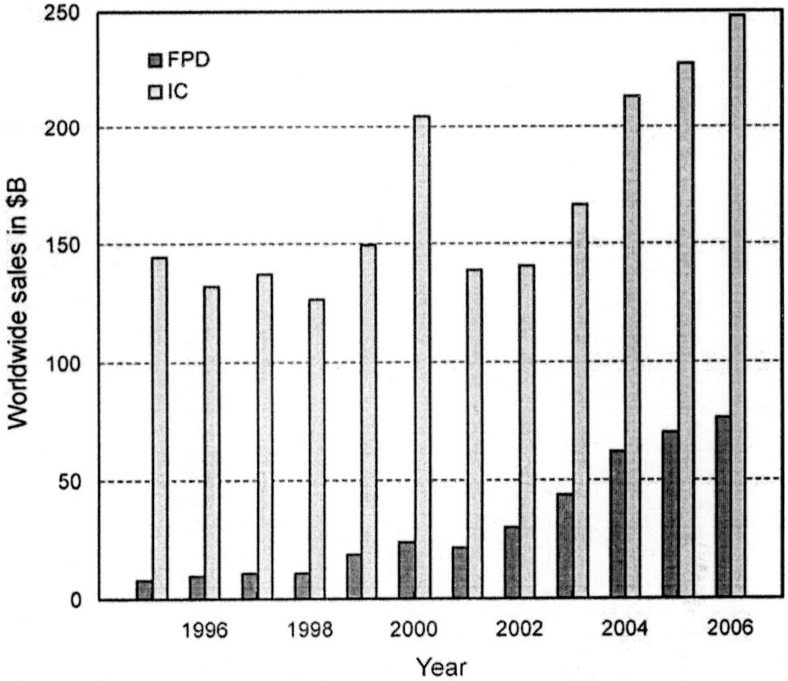

Figure 1.1. Growth of semiconductor and flat panel display industries. [Data Source: Semiconductor industry sales data from Semiconductor Industry Association (SIA) and flat panel display data from Displaysearch Corp, San Jose, CA.]

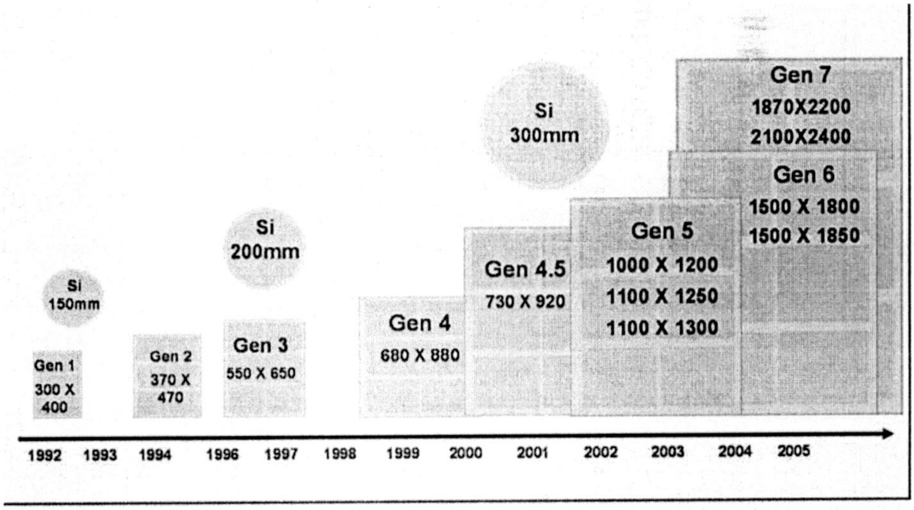

Figure 1.2. Substrate size comparison between Si wafers and glass substrates used in flat panel displays.

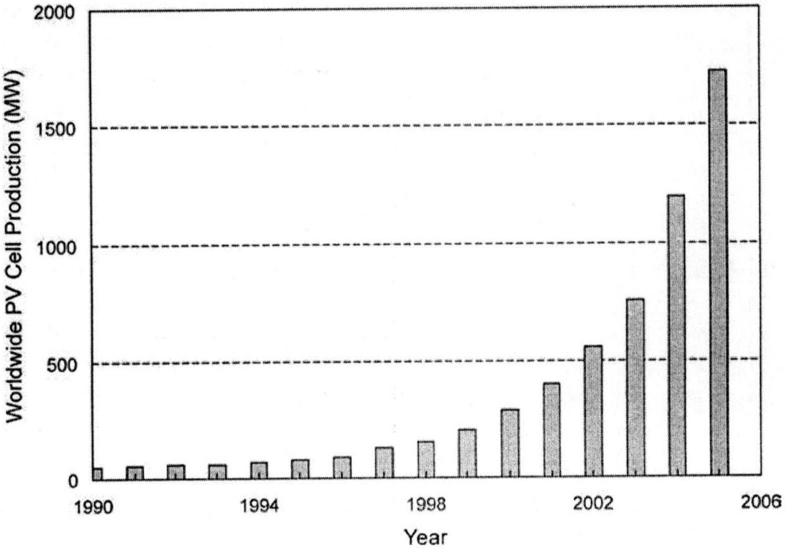

Figure 1.3. Worldwide production volume of photovoltaic modules. [Data source: European Photovoltaic Industry Association, 2006.]

increased by more than 30 times to more than 2 m × 2 m for the current eighth-generation production facilities. In contrast, most semiconductor fabrication has been limited to wafer sizes of 300 mm or smaller.

The second large-area electronics technology to reach high volume and revenue, photovoltaics (PV), which is based on "simple" diodes distributed over many hundreds of square centimeters, is also now poised to grow into a major industry as new energy sources become more essential.[7,8] With the declining stock of fossil fuels and worries about global climate change, solar energy using photovoltaics has become increasingly attractive. The cumulative installed capacity of PV systems has reached over 4 GW and is expected to double every year over the next three years. The annual production volume of PV modules reached the 1-GW milestone in 2004 (see Figure 1.3). With the commencement of a number of a high-volume PV manufacturing facilities over the last two years, production volumes have been growing rapidly ever since, with worldwide manufacturing capacity reaching more than 5 GW in 2007.[9] Most expansion in photovoltaics is based on crystalline and multicrystalline silicon materials.[10] With the rapid growth of the solar energy market, the availability of reasonably priced silicon feedstock has become a major barrier for future growth. Thin-film technologies offer efficient materials utilization and opportunities for large-area processing. Several companies are building thin-film silicon PV production lines based on large glass substrates (e.g., 4-m^2 glass sheets), which are similar to those employed by the FPD industry. The main barriers for thin-film PV technologies, which need to be overcome in order for thin-film PV to become pervasive, are improved conversion efficiencies (approaching that of crystalline Si devices) and lower capital cost of some production equipment, especially vacuum deposition tools.[11]

To overcome these barriers, the industry is aggressively pursuing both alternative materials and manufacturing methods. Although the major focus of the PV industry is on silicon-based devices, several companies have developed products based on thin-film chalcopyrite and cadmium telluride, as these materials offer opportunities for lower cost production through solution-based processing. Chalcopyrite, or more specifically the $Cu(In,Ga)(Se,S)_2$ family of compounds, has achieved the highest conversion efficiencies (~20%) of any polycrystalline thin-film material.[12] Several companies have commercialized the technology and are approaching volume production.[13] Several companies have started producing CdTe in volumes beyond pilot production. Small-area efficiencies of 16% together with simple production technologies make this material very attractive. State-of-the-art commercial CdTe PV modules that have efficiencies in the 9% range are in volume production.[14]

1.1.3 Macroelectronics Potential

The success of the FPD industry and the rapidly developing PV industry are testaments to the potential for large-area electronics for other system solutions. For example, conformal and flexible form factors are very desirable attributes to provide either portability and/or the ability to install large-area electronics in a variety of locations. These needs are receiving much attention, although fulfilling them is proving to be difficult. Moreover, despite the success of microelectronics, there are applications where it has not been good enough to meet all requirements. Specifically, applications where very low cost is the product driver, rather than performance, can prove challenging for conventional microelectronics. As much as microelectronics has reduced the cost/transistor, the costs are still not low enough to meet the few pennies/item targets for electronic applications that are intended to be disposable, such as RFID tags and product expiration sensors. Similarly, although the number of microelectronic transistors per square centimeters (areal density) has remarkably increased over the last 40 years, the ability to distribute even moderate numbers of transistors over large areas onto a variety of substrates is just beginning to be commercialized for applications such as flexible displays.[15] Transistors at low density can be fabricated over large pieces of glass, but at great sacrifice to performance characteristics compared with mainstream "Moore's Law" devices.

Although initial applications of large-area electronics have focused on displays and PV, future product opportunities are expected to include sensors, imagers, distributed lighting, electronics that are embedded into clothing or gear already carried (radios, computers), and health monitoring/control of vehicles and even people (Figures 1.4–1.7).[3,16,17] Figure 1.4 captures a concept long championed by many display manufacturers. It proposes that at some point in the future, the display manufacturing capability will be able to provide

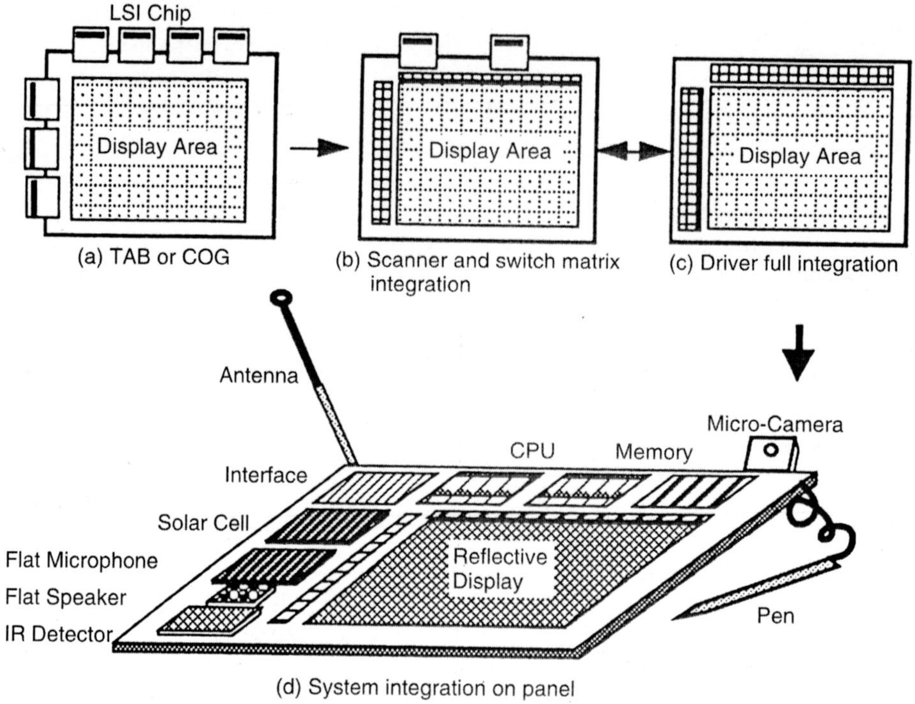

Figure 1.4. Schematic showing the evolution of displays toward a system on a flexible panel. (a) Direct chip on glass attachment technology, moving toward (b) partial display driver integration, (c) fully integrated drivers on glass, followed by (d) a fully integrated "system on a flexible panel," showing how high-performance thin-film transistors enable display drivers and other system components to be integrated on a flexible metal foil. [Schematic courtesy of Sharp Corp, Osaka, Japan.]

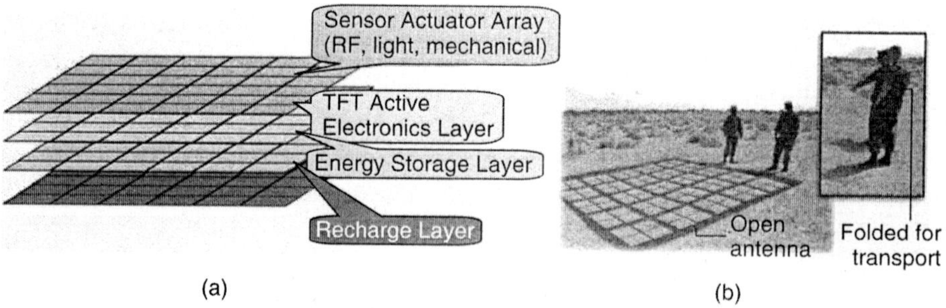

Figure 1.5. Schematic of a fully integrated macroelectronic system and an example of an application concept for macroelectronic systems. (a) Building blocks for a generic macrosystem. (b) Mockup of a large-area antenna array. [Figure courtesy of Sarnoff Corporation. Used with permission.]

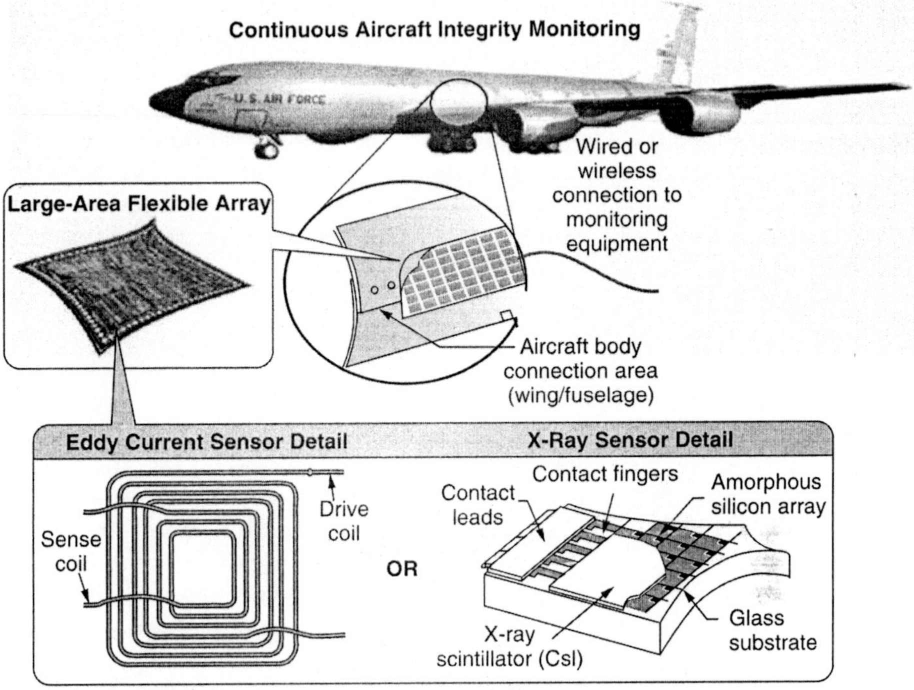

Figure 1.6. Application of large-area embedded flexible control electronics includes structural health monitoring of large objects such as airframes.

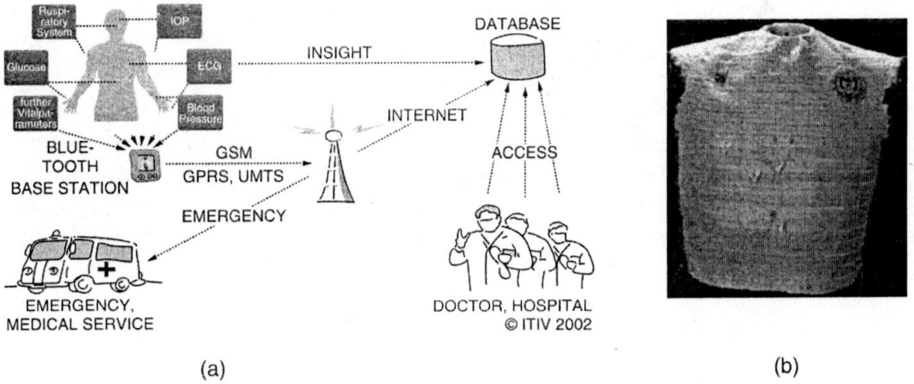

Figure 1.7. Large-area electronic fabrics and health monitoring systems for soldiers and personnel employed in high-risk field operations. (a) Operation of a personnel health monitoring system. (b) Example of a vest with integrated sensors for monitoring body temperature, respiration rate, and other bodily parameters.

not just a display, but also an entire wireless laptop tablet complete with camera, microphone, and solar cell for power. It is unclear when, or even if, such a vision will ever make technical (let alone economic) sense. Nevertheless, the point is that this is one example of a vision that is driving the creation of macroelectronics technology. Figure 1.5 shows yet another large-area concept. Here, the notion is a sensor system that comes integrated with control electronics, energy storage mechanism, and energy harvesting layer, all made via yet-to-be-determined manufacturing technology. Figure 1.6 takes this idea one step further and suggests how a conception suggested in Figure 1.5 might be implemented into an aircraft for structural health monitoring, active flight control and multifunctional applications such as load-bearing structural antennas. Finally, in Figure 1.7, the personal health status application involves a range of sensors, processors, and transmitters integrated within composite materials or mounted on/in human biomaterials (skin/tissues). This assembly is then able to sense and report faults to ensure proper drug usage, or to continuously sense, modify, and transmit physiological and cognitive status. The weight and material integration issues with conventional ICs make these applications impossible or unlikely with a purely microelectronics approach. The major challenge for macroelectronics technology is to enable applications beyond displays that involve large areas and applications that cannot be cost-effectively achieved through traditional packaged-chip fabrication followed by pick-and-place assembly. Nonetheless, these applications will still require sophisticated, high-functionality circuits. The large-scale applications envisioned give rise to the requirement for properties heretofore not associated with IC applications, including the thinness, ductility, and elasticity of electronic components, even during operation. Potentially, cofabrication of electronics and physical structures might be possible. This process would enable the electronics to be built directly onto or within the structure from which it controls, senses, or communicates. Ideally, the electronics would be synergistic with and inseparable from the system. A conceptual model might be the human nervous system. However, the opposite is true for traditional microelectronics, in which passive devices, packaged chips, boards, and boxes are each fabricated separately and only later integrated into the final structure. This difference in manufacturing approach creates major differences for the materials, electronic design, and fabrication methods for macroelectronics versus microelectronics.

1.2 IMPORTANCE OF SOLUTION PROCESSING

Researchers have many obstacles to overcome in the quest to make macroelectronics the “next big thing.” The keys to achieving the desired levels of functionality for a wide range of large-area electronic functions are advances in materials and processes and device structures that can get cost down to pennies (rather than dollars) per square centimeter. Tools and process methods that provide these devices and their interconnections, at adequate levels of integration and in high yield on a wide range of substrates, must be developed. Some of the required advances in processing and tools will be adopted from the display and photovoltaic industry. However, to achieve the device/circuit performance for more demanding electronic functions, significant improvement in materials and device characteristics must be achieved.

To manufacture flexible integrated circuits, it is not the transistors themselves that are inflexible; it is the relatively thick, bulk wafer on which that the transistors are manufactured. Thinning the wafer to harvest just the upper active circuits is possible but also time consuming, difficult, and expensive. Therefore, in the macroelectronics thrust, the focus is on developing techniques for depositing semiconductors very inexpensively and, in most cases, over a large area on a variety of substrates (to include even plastics and fabrics). The result is a different transistor structure known as a thin-film transistor (TFT; see Section 1.3.1).[2] The ideal method for fabrication of TFTs for macroelectronics requires that the materials used to create the devices be directly deposited on a thin (and ideally, flexible) substrate. In contrast to microelectronics, with TFTs for macroelectronics, the feature size and level of performance are not the primary drivers. Rather, the processing cost, compatibility with diverse substrates, and attributes of the end item (area, weight, bendability, durability/ruggedness) represent the critical factors. These challenges generally require mild processing conditions not significantly different from the ambient (in contrast to the temperatures and chemicals associated with microelectronics fabrication). Processing under such conditions is much more conducive to a variety of electronic substrates and to the integration of diverse functionality, including computational devices, sensors, photovoltaics, and displays. Therefore, fabrication technologies that promise lowest possible cost while delivering at least adequate electrical performance are of great interest.

Because cost/square centimeter is such a major driver for macroelectronic applications, established methods for low-cost manufacture are of great interest. Solution processing for all manner of printed products has a long history and well-developed infrastructure that addresses multiple applications with a wide range of inexpensive materials and patterning methods. Therefore, solution processing has received significant attention, because the essential steps of macroelectronic TFT circuit fabrication can (in principle) all be accomplished using the ordinary, relatively cheap, and widely available technologies used to print ink.[18–20] One method is a modification of ink-jet printing, and another adapts roll-to-roll processing, which is commonly used to print fabrics and newspapers (for a more detailed discussion, see Chapter 12). Unfortunately, to date, the problem with both of these approaches is the ability to produce transistors that can operate fast enough for potential applications of interest. Although adequate for displays, the TFTs produced easily with these printing methods are much too slow for many applications. Thus, macroelectronics research seeks to exploit this rich printing infrastructure, but with incorporation of materials required to fabricate higher performance TFT-based electronics.

One means that has been pursued to achieve low-cost, multi-material processing is based on organic semiconductors, because of the well-established potential for compatibility with printing technology.[21–24] However, to date, inorganic semiconductors have achieved the highest and most stable TFT performance.[25] Recent results have provided encouraging results based on solution deposition of inorganic materials rather than requiring the standard vacuum deposition methods. Because of the relatively mature theoretical understanding of inorganic semiconductor devices, and the difficulty of obtaining organic-based TFTs with adequate device characteristics, new ways to solution-deposit and fabricate inorganic semiconductors have received increasing attention, as will be explored in subsequent chapters.

Macroelectronics thus seeks to create a new fabrication methodology based on techniques that are currently alien to microelectronics processing. Ideally, roll-to-roll substrate handling will replace wafer batches, with material deposition via solution processing replacing vacuum evaporation, and material patterning by printing eliminating the need for etching. Given the diversity of materials, devices, and applications that may eventually encompass "large-area electronics technology," it may well be that no "standard process" and "standard equipment" will ever exist for macroelectronics, as it does for mainstream CMOS IC manufacturing. However, Figure 1.8 provides some idea of

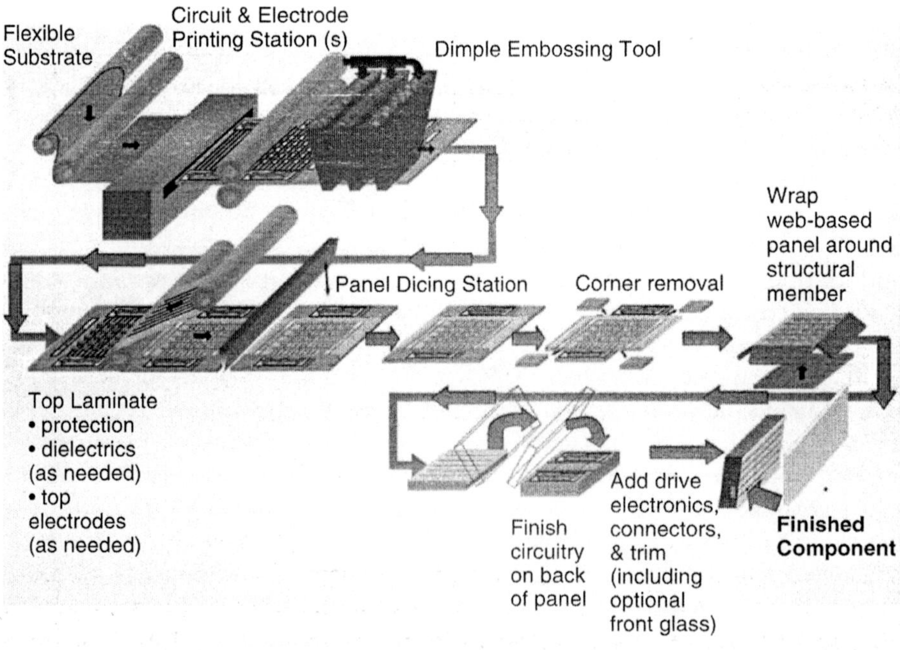

Figure 1.8. Conceptual roll-to-roll manufacturing process.

what a future "printed electronics" manufacturing line might look like. As much research in this field is still in its infancy, it is difficult to forecast what the technology will actually look like. Whatever the final details are, tools and methods will be chosen for their ability to handle large areas at a low cost/ square centimeter. The end result of this new manufacturing paradigm is expected to provide electronics solutions inexpensive enough to be "thrown away" or to cover a vast area (e.g., pervasive PV arrays).

This shift in the means of production can be appreciated as follows. The capital required to build a microelectronics state-of-the-art 300-mm wafer fabrication facility is more than $3B, whereas macroelectronics facilities are anticipated to cost at least an order of magnitude less. Furthermore, the cost per unit area of product is expected to drop from the order of $10,000 ft^{-2}$ for microelectronics to $100 ft^{-2}$ for macroelectronics.[16] Of course, the cost of making electronics with a roll-to-roll process is not expected to be as inexpensive as conventional ink printing. However, by eliminating the high-temperature, high vacuum steps used in the conventional chip manufacturing, a significant reduction in the manufacturing cost for macroelectronic circuits is expected. Perhaps of more significance, less demanding, lower cost macroelectronics fabrication technology may lead to its integration into the manufacture of a wide range of products, which would provide additional functionality not possible today.

1.3 APPLICATION CHALLENGES: TFT DEVICES AND CIRCUITS

Macroelectronics is a key enabler for several platform technologies.[3] Now, after a decade of fundamental materials and device research, the area is witnessing the first major push toward commercialization for products other than PV and displays. However, productization requires device architectures and materials processes that are reliable and well qualified for high-volume manufacturing (i.e., additional maturization is needed in these areas). Although there has been tremendous progress in device development, future advances need to occur not only at the device level, but also at the circuit and systems level. With continued progress, the technology could provide a practical means for delivering unique application solutions that are otherwise not possible. Some key applications that are expected to benefit and derive enhanced functionality through printed large-area electronics include flexible displays, new imaging devices, transparent electronics, RF tags, photovoltaics, solid state lighting, electronic textiles, robotics, and several unique applications in distributed electronics, including large antenna arrays and structural health monitoring of buildings, ships, and airframes.[26]

Unfortunately, TFT circuit performance has been limited by relatively poor device characteristics compared with bulk Si. Existing amorphous Si and organic TFT devices are constrained by materials and/or substrate process limitations and result in TFTs with low mobility (less than $1 cm^2/V\text{-}s$). Thus, applications that require even modest computation, control, or communication functions cannot be addressed by today's TFT technology. To achieve the desired ability to implement diverse electronic functions, it will be necessary to develop a transistor technology that can switch at least millions of times a second, rather than only thousands of times a second as is possible with today's technology. Doing so requires balancing process parameters and resulting electrical parameters suitable for intended applications.

1.3.1 TFT Device Fundamentals

A detailed discussion of TFT device operation is outside the scope of this chapter. Comprehensive review of TFT design parameters and related electrical characteristics is available in the published literature.[27,28] The actual performance of a TFT is determined by a wide range of interacting characteristics that depend on material parameters and dimensions of the device. Here we will focus on the critical first-order parameters that must be significantly improved as the minimum first step in achieving enhanced performance. Although the detailed description is complex, the fundamental features that dominate macroelectronic device performance are simple as shown in Figure 1.9. To first order, the higher the mobility (μ) and the shorter the channel (L), the higher is the transit frequency (f_T, a measure of the speed of the device). f_T is a key parameter that determines maximum frequency performance of a device in both digital and RF applications. The mobility is a measure of the ease of movement of charge carriers from the source across the channel to the drain. A low number indicates that there are many impediments to charge transport. The channel length is simply the distance that separates the source

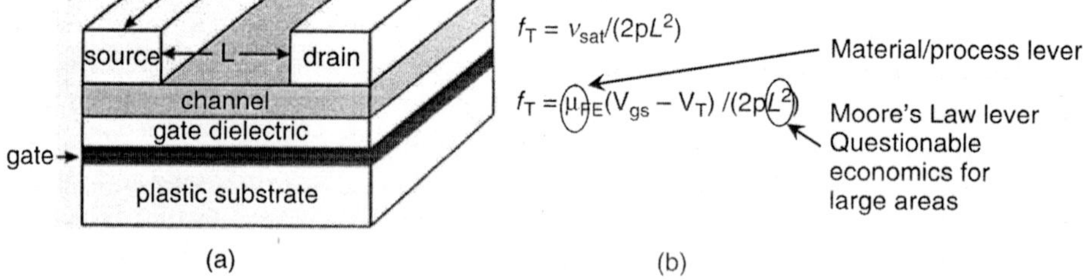

Figure 1.9. Device, materials, and process parameters to achieve high-performance thin-film transistors. (a) Thin-film transistor structure. TFTs consist of a substrate (in this case, a plastic sheet), a conducting "gate"; a semiconducting "channel" layer; a "gate dielectric," which separates the gate from the semiconductor; and source/drain contacts. The device operates by modulating the current flow in the channel (between source and drain) by applying a voltage to the gate. (b) Equations governing the performance (i.e., speed) of a TFT.

of carriers (source) from the collection point (drain). Although shorter channels are an obvious means for faster device operation, they also can be important because the shorter distance means that the charge carriers will interact with fewer impediments when crossing the channel.

One reason macroelectronic TFTs have significantly degraded performance is because, in microelectronic crystalline semiconductor TFTs, there are relatively few impediments and the device mobilities are in the range of hundreds to thousands of centimeters squared/volt-seconds. In comparison, macroelectronic TFTs, which comprise amorphous or polycrystalline materials, have many impediments, particularly at grain boundaries or surface discontinuities. Thus, these devices have mobilities of about 1 cm^2/V-s or even less. Therefore, for a given feature size, a macroelectronics TFT could have performance degraded by two or even three orders of magnitude compared with conventional crystalline semiconductors, simply as a result of intrinsic materials properties. Therefore, as will be described in detail in subsequent chapters (e.g., see Chapters 3 and 5), a major macroelectronics device development emphasis has been on finding materials and processing techniques that offer significant improvement in mobility, with little or no cost impact compared with amorphous silicon. Organic semiconductors devices (OFETs) tend to be consistent with the cost objectives, but they have mobilities about the same (or worse) than amorphous silicon. As noted, polycrystalline, rather than amorphous, silicon provides a significant performance improvement. But its increased processing cost becomes an important issue.

Since f_T increases exponentially with shorter channel length, all other factors being equal, the best way to achieve higher performance is to make devices with shorter channels. This is essentially responsible for the major advances in semiconductor microelectronics technology for the last several decades. Channel lengths are now in the 100 nm or less regime, which results in transistors with operating frequencies of >100 GHz and digital circuits that operate at 1 GHz or more. However, typical printing technology aimed at macroelectronics provides feature sizes of 100 μm or larger, with advanced methods capable of about 20 μm. Thus, for the same materials, a microelectronics device (channel = 100 nm) would have a f_T 40,000 [$(20/0.1)^2$] times larger than a macroelectronic device (channel = 20 μm). The difference in potential performance between the two types of devices can be appreciated as follows. Even if a macroelectronics TFT is fabricated with a channel length of around 1 μm (very difficult/expensive in a commercial process), it could be as much as 100 times slower than the corresponding deep submicron microelectronics transistor (channel < 100 nm). Obviously, because shorter channels and overall smaller devices can provide so much performance improvement, many printing methods capable of smaller and smaller features are being explored and developed for adaptation to macroelectronics. The scope and limitations of several of these will be described in detail in later chapters.

Although the above discussion addresses the major concerns associated with the semiconductor component of high-performance macroelectronic

TFTs, other issues must also be satisfactorily addressed. The gate dielectric (see Figure 1.9), for example, must be deposited via a method that results in no pinholes and ensures the integrity of the dielectric layer between gate and channel. Ideally, the deposition process will also be able to provide a thin dielectric film so that the gate voltage that must be applied to start current flow is not too high for practical circuits. In addition, the deposited dielectric must have very little trapped charge or other defects, especially near the semiconductor interface. Such defects are another source of degraded mobility, as they also impede the smooth flow of charge carriers across the channel. Besides the dielectric and semiconductor components, other device features can also have significant impact. For example, low-resistance, ohmic contacts are also essential to minimize the effects of parasitics on device performance.[29]

The capability to fabricate an individual high-quality thin-film transistor with attributes as described above is certainly important. However, from a circuit perspective, high device yield and uniformity also must be achieved. Furthermore, the device characteristics must remain reasonably constant and metal interconnects must provide both low resistance and durable, rugged performance when the substrate is flexed.[30–33] Promising technology capable of yielding resistivity comparable with bulk metals, but rugged enough for thin, flexible substrates has been achieved using directed self-assembly[34] and thermal spray deposition.[35,36] Both methods provide a means for room temperature deposition of a variety of metals and insulators for fabrication of passive electronic components. The directed self-assembly technique is adapted from the well-known laboratory technique, but it has been scaled to process large glass plates and fabrics. Thermal spray is somewhat similar to a variety of vacuum deposition techniques except that it operates in an ambient atmosphere and the vaporized material can be shaped to provide a beam size of about 100 μm.

In summary then, current macroelectronic device technology results in devices with mobility of $<10\,cm^2/V\text{-}s$ and channel lengths of $>10\,\mu m$. Research is being aggressively pursued to improve both of these key metrics significantly. However, it needs to be understood that unless low-cost methods are developed to provide micrometer-sized channel length devices and semiconductor materials with mobility more than $100\,cm^2/V\text{-}s$, macroelectronic transistors will suffer a significant performance penalty and limitations on the applications that can be addressed. System designers will need to deal with this issue to increase overall functionality without the benefit of the best device performance.

1.3.2 Next-Generation TFTs

If amorphous Si TFTs and OFETs are adequate only for lower performance, lower cost applications, what are the options for more advanced applications? To achieve the higher performance that is desired while keeping process technology consistent with the goals of macroelectronics, several different fabrication methods and materials are being explored. Primarily, work in this area has focused on crystallization of vacuum-deposited amorphous silicon to create polysilicon (known as LTPS) device islands at low processing temperatures, which are compatible with glass and even plastics. This advance has resulted in devices with mobilities of $100\,cm^2/V$-s and even higher.[37–39] Devices and circuits have even been produced on polyimide with operation at 100 MHz, which shows the potential for good device performance even on flexible substrates.[40] Significant literature is already available that describes efforts at improving the device/circuit characteristics of conventional polysilicon-based TFTs. The interested reader is referred to several review articles[27,37,39] and to several excellent reference volumes[41] that describe the efforts to circumvent process temperature issues by fabrication of LTPS devices via laser-induced crystallization. We note that display manufacturers are now introducing products with on-display LTPS driver and demux circuits, with built-in memory expected in the near future.[6,42]

Rather than laser processing, another approach for solving process temperature limitations is to look for new substrates that are less sensitive to heat. In particular, plastics made from liquid crystals or silicone resins are showing some promise.[43–45] For applications that do not need transparent substrates, thin metal foils can replace the plastic.[39] LTPS circuits on stainless steel foil that switch at rates of several hundred megahertz up to a gigahertz have been fabricated. Devices fabricated with this approach also provide performance that is satisfactory for operation of analog circuits.[46] A key issue here is the surface roughness of the foil. Planarization is critical for avoiding defects caused by pinholes in the over-layers and resulting low yields of LTPS circuits because of shorting to the metal substrate.

Although LTPS technology has made great strides for display applications as noted above, deposition and crystallization of semiconductors on flexible, low-temperature substrates remains a significant problem, with currently only relatively expensive solutions. Therefore, researchers have turned to solution deposition as a means of providing high-mobility semiconductors at relatively low cost. Promising results in the range of $10–100\,cm^2/V$-s have been reported.[47,48] These approaches include TFTs made from a variety of polycrystalline materials. The end objective is the capability to provide polycrystalline semiconductor materials by a variety of printing/deposition methods that do not require vacuum deposition and/or high-temperature thermal processes to achieve high-quality TFTs. Note also that a variety of semiconductors can be used such that optimum material for a given function can be employed. Subsequent chapters will describe the range of materials and processes being developed (see Chapters 3–5 and 7).

1.3.3 Technology for RF TFTs

As summarized, significant advances have been made with LTPS from a performance and cost perspective. In addition, lower cost solution-based processing of inorganics as an alternative to LTPS is progressing. However, even these advances have limitations, and therefore, one of the most important and most challenging applications for the proposed macroelectronics technology remains unsolved. TFT technologies able to reach the level of RF Performance needed to support sensor-based consumer electronics and electronically scanned antennas for government communication and radar systems, must be found. Many next-generation radar systems require very large, electrically scanned antenna arrays, which are flexible and conformal.[49] Antenna sizes of up to 5000 square feet or more are needed for some applications. A extremely low weight is critical for airborne or space-borne applications, and a low cost is required to attain system affordability goals. The goal is RF circuit performance at over 500 MHz or even several gigahertz if possible. Flexible plastic antennas incorporating active circuitry at each element of the array (to include low-noise amplifiers, RF switches, active RF combiners, and digital control circuits) are ideal candidates to meet these requirements. The challenges here include not only gigahertz-type switching, but also low noise figure operation for adequate amplifier operation. In addition, an ideal solution is a transceiver, which requires that the TFT technology provide at least milliwatt levels of output power at an efficiency of more than 10% to achieve practical solutions.

The challenges for application of macroelectronic TFTs for RF are significant and the available technical solutions are limited.[3] As described in Section 1.3.1, high f_T devices required for RF can be most effectively achieved "simply" by reduction of the channel length. Unfortunately, production of such short channels is both expensive and difficult, as is well known from microelectronics experience. This is even more so for macroelectronic devices because it is difficult to generate such small features reproducibly over the large substrates desired for production. Furthermore, the large substrates may be glass or various plastics, which tend to have rougher and undulating surfaces (compared with semiconductor wafers), making it even more difficult to control small feature sizes.

However, the benefit of smaller geometries is significant. Thus, if all other factors were the same, a lithographically defined device at 1 μm would have a theoretical f_T 400 times higher than the corresponding ink-jet-fabricated device with a 20 μm channel. For high-performance TFTs, an improvement of several orders of magnitude over existing technology can be expected by simultaneously increasing mobility and decreasing device dimensions. For example, using values consistent with modest semiconductor technology, for mobility (>100 cm^2/V-s) and channel length (1–2 μm), Figure 1.9 indicates that the theoretical maximum f_T is in the 5–15 GHz range. Real devices, of course, have parasitic elements that degrade the performance and practical circuits do not operate at the peak performance of an individual device. Therefore, a realistic expectation for a macroelectronics TFT RF circuit is probably about 1 GHz, possibly as high as 3 GHz. This is, obviously, very mediocre when compared with hundreds of gigahertz achieved by microelectronic devices and circuits.

But, it is substantially better than the tens of kilohertz that classic amorphous silicon or OFETs can achieve. It is also in frequency ranges of interest for many mobile communication systems and radar detection bands and, therefore, would provide sufficient capability for a range of RF opportunities.

High-resolution lithography is usually associated with microelectronics. However, a promising solution to the production of small feature sizes for TFT fabrication even on flexible substrates is available, but it has not yet received widespread use. It is based on scanning laser projection lithography with resolution and alignment capability in the 1–2-micrometer range.[50] If this approach can be successfully integrated into a solution-based fabrication approach, it should provide a significant advantage over available printing methods. Microcontact printing (see Chapters 12 and 13) is expected to also play a major role here in the future.[51]

Although significant reduction in feature size is essential, other advances will also likely be required. In the next section, we describe results that promise to improve the semiconductor materials significantly and to provide dramatic improvement over existing LTPS TFT technology. Of course, achieving these material and lithography advances in a cost-effective manner is yet an additional challenge that must be overcome.

1.3.4 Exploratory TFT Concepts

Because of the desire to address increasingly more demanding electronic functions, approaches that employ crystalline semiconductors may be required. However, a successful solution will also require that low-cost manufacturing, potentially available via printing methods, be achieved. Recently, approaches that use micro/nanoscale objects—e.g., nanotubes, nanowires, ribbons, disks, and platelets—of high-quality, single-crystal semiconductors have been reported.[52] A collection of these building blocks constitutes a type of material that can be deposited and patterned, by dry transfer printing or solution casting, onto plastic substrates, thereby forming an effective semiconductor layer to yield mechanically flexible thin-film transistors that have excellent electrical properties.[52–54] Because this approach separates the semiconductor growth process from the device substrate, it is independent of traditional requirements for epitaxy, thermal budgets for processing, and other considerations. As a result, it is well matched not only to flexible electronic systems on plastic substrates but also to devices that require heterogeneous or three-dimensional integration (see Chapter 13 for more details). This type of approach has been successfully demonstrated for highly bendable device arrays of modest size and should be scalable to large areas.[52,53,55] Devices fabricated with microstructured GaAs and 2-μm channels, with an f_T of almost 2 GHz, have been reported.[54] An interesting alternative technical approach here may be the application of direct laser transfer to move desired materials or devices from a temporary carrier transfer sheet onto the host substrate.[56]

Another strategy to integrate thin, high-quality semiconductor materials involves printing nanowires onto a low-temperature substrate.[57,58] In this approach, the nanowires, which may possess a core-shell structure, including a thin outer layer of high-quality silicon dioxide, are grown in an offline reactor. The nanowires are then suspended in a solution and printed onto the substrate. Work to date indicates that performance approaching that of bulk Si devices can be achieved with hole mobilities of >200 cm^2/V-s demonstrated. Non-RF optimized devices fabricated with the nanoribbon/wire approach with gate lengths of 2 μm have shown f_T of over 100 MHz. Perhaps of greater interest, not only Si nanowire devices can be fabricated. Other semiconductor nanowire devices, including those fabricated from GaAs, InP, and CdS via the same approach, have been reported.[58] A detailed discussion of nanowire structures and synthesis, as well as their use in fabrication of electronic devices, will be given in Chapters 10 and 11.

Although both nanostructured and nanowire approaches can avoid the need for high-temperature processing of the gate insulator to facilitate fabrication on low-temperature substrates, these methods may not always be applicable. Another promising approach to achieve low-temperature, yet high-quality, gate oxides is the development of solution-deposited, self-assembled nanodielectrics.[59,60] These dielectrics can be solution-deposited at thicknesses of 15 nm or less. The resultant films have leakage currents of less than 10^{-8} A/cm^2 and breakdown voltages of 6–7 MV/cm. Device mobility is several times higher and on-current is about 10 times higher compared with values achieved from devices with a conventionally deposited SiO_2 dielectric. A more complete description of solution-deposited dielectrics is presented in Chapter 4.

Although not a classic semiconductor material, single-wall carbon nanotube (SWNT) based TFTs have also received significant attention for microelectronic applications recently. Because of their unique properties, including potential mobilities of 10,000 cm^2/V-s or higher, as well as the ease and low expense of producing them in large quantities, they provide significant promise for a variety of device applications. These properties have also made them attractive for large-area electronics because of the possibility of achieving extremely high field-effect mobilities via printing a nanotube suspension onto substrates.[61,62] Because of the extremely high intrinsic per-tube mobilities, it is expected that very high device mobilities will be possible by increasing the fill factor of tubes from the current relatively low values of 1–2%.

Several major challenges must be overcome in order to take full advantage of carbon nanotubes as an electronic material. A major barrier to very low-temperature processing is the gate dielectric; it is difficult to form low-temperature deposited dielectrics with high dielectric integrity and with low hysteresis for the SWNT devices. A promising solution for this problem may also be the nanodielectrics mentioned previously.[63] Perhaps a bigger issue is that as-grown carbon nanotubes can be either semiconducting or metallic. Therefore, a means must be found either to remove the metallic tubes selectively or to render them insulating; otherwise, the metallic tubes will act as electrical shorts. Laboratory methods to achieve this have been reported,[62] but, to date, no method adequate for manufacturing has been reported. Finally, for maximum device performance, it is desirable to have a large density of nanotubes aligned in parallel between source and drain. Just as in the case of semiconductor nanowires, a reliable means must be developed to orient and position the nanotubes (see Chapter 11). Here again, not only would a successful approach provide a means to high-performance devices for large-area electronics, but also as with the nanowire-type approach, "printing" of carbon nanotubes may also be a viable route to three-dimensional (3D) and heterogeneous integration.

1.3.5 Technology Computer Aided Design for TFTs

For either conventional polycrystalline semiconductors or nanotubes and nanowires to be successful, the development of model and simulation tools that can be used for device and circuit design as well as for predictive engineering must be available. Since these devices are not necessarily based on single "wires" or single crystals, but rather on an ensemble of particles, the aggregate behavior must be considered. Initial efforts to provide the necessary physical understanding and device models using percolation theory have been reported.[64,65]

Although computer-aided design technology is highly developed for microelectronics, there is much that that needs to be modified, extended, or developed for macroelectronic devices. Design requirements, tools, and techniques have not been evaluated, optimized, or even defined for many of the materials, structures, and applications in the frequency ranges of interest. For more demanding applications like RF devices, advanced modeling capabilities will be required to simulate RF circuits accurately. Ideally, two-dimensional models should be available to provide a higher level of model fidelity and assistance in improving TFT reliability.[66]

1.4 APPLICATION CHALLENGES: OPTOELECTRONICS

Although high-performance TFTs are needed for several electronic applications, the potential for printed, inorganic electronics encompasses other devices and applications. A major opportunity is in optoelectronic applications, which impose different requirements, challenges, and opportunities (see Chapters 6, 7, 9, and 11 for discussion of solution-processed solar cells and other printed optical devices).

1.4.1 Photovoltaics

As discussed previously (e.g., see section 1.1.2), photovoltaics is already a commercially successful application of macroelectronics and appears poised to expand even further with solution-based inorganic materials and large-area, roll-to-roll processing reaching maturity. Just as TFT-based flat panel displays have become successful based on a relatively simple electronic switch, PV's success is also based on a relatively simple device, i.e., a pn junction photodiode. Comprehensive descriptions of the specifics of junction diodes and their use in PV technology are available elsewhere.[67,68] Here, we will focus on the key challenges to the development of a successful solution-based manufacturing technology. From this perspective, the focus is on the following material properties: energy band gap, quantum efficiency, photo-efficiency, and suitability for solution-based processing.

The band gap is important because it determines the energy of the incident photon required to create an electron in the semiconductor diode. When light is absorbed, an electron–hole pair is created and the charge carriers are swept away and collected at the respective diode electrodes. If the band gap of the material is small, a large percentage of the incident photons can generate an electron. However, because the band gap (difference in energy between valence and conduction bands) is small, the electrons that are produced are relatively low voltage and of limited utility in an external circuit. If the band gap is large, then the electrons have higher and more useful voltages. Unfortunately, the percentage of incident photons that have high enough energy to create an electron falls off at shorter wavelength/higher energy. Therefore, selection of the material with the optimum band gap for the intended PV application must be done carefully.

The next challenge is quantum efficiency. This is a measure of what percentages of absorbed photons are converted to electrons. Energy from absorbed photons that is not converted to electrons is dissipated as heat. Such nonradiative processes are more of a problem for solution-processed inorganic materials because the noncrystalline or polycrystalline nature of the solution-deposited film tends to create more locations where electron-hole pairs can relax back to the ground state without production of an electron, thus reducing the efficiency of the cell. Here, as in TFTs, defects can trap electrons and prevent efficient collection of the generated carriers (electrons/holes), thus degrading the efficiency of the PV device. The problem that solution processing must successfully address is, therefore, to find deposition and post-processing conditions that produce devices with quantum and overall efficiency comparable with those fabricated with vacuum deposition technology (i.e., to achieve comparable film quality for the active layer).

Finally, the materials must be readily suited for solution-based deposition on large areas, either on large glass plates or on flexible metal foils. Although solution-based processing is potentially economical in the use of materials, achieving throughput comparable with large vacuum deposition processes is a major challenge.

Although this discussion has focused on challenges for solution-based processing for PV, the fundamentals apply to other optoelectronic applications to be discussed in the following sections. First, the material characteristics must be achieved to maximize absorption (or transmission depending on the application). Next, the efficiency of the photo process must be maximized so that the absorption energy is not dissipated as heat, and finally, the electrons created by the photo process must be efficiently collected at the electrodes (or vice versa for emissive devices).

1.4.2 Transparent Conductive Oxides

Transparent conductive oxides (TCOs) of various types have been used in a wide range of applications, such as conducting electrodes for displays, solar panels, and electroluminescent lamps; as materials for electromagnetic interference (EMI) / radiofrequency interference (RFI) shielding applications; and in architectural and automotive window glasses.[69] The major considerations and development activity has been focused on maximizing both conductivity and transmissivity. Although there is a large class of TCO materials, such as CdO, Cd_2SnO_4, $CuAlO_2$, $CuGaO_2$, $BaCu_2S_2$ and $SrCu_2O_2$, the most commonly used transparent conductors are simple binary oxides, such as SnO_2, In_2O_3, and ZnO, or ternary alloys, such as $In_2O_3SnO_2$.[70]

For more than half a century, indium tin oxide, commonly referred to as ITO, has been the transparent conductor of choice for display applications, whereas SnO_2 and doped ZnO have been the most widely used materials in solar cells, architectural glass, and RFI/EMI applications. Although several other transparent oxides and their respective alloys have been explored as possible alternatives, the ease of processing for large-area applications and superior electrical and optical properties of ITO, SnO_2, and ZnO have made them the most widely used. These materials are highly transparent within the near-ultraviolet (UV)and visible range. Appropriate electrical conductivity values are achieved through the proper selection of material processing and the introduction of native (controlling the stoichiometry within the film) or substitutional dopants. The obvious challenge here for solution-processed inorganic systems is to achieve comparable performance to these well-established vacuum-deposited materials.

1.4.3 Transparent Transistors

Although transparent diodes (of interest for PV and "smart windows" applications) were fabricated over a decade ago, successful demonstration of transparent TFTs using various group IV oxide semiconductors are also beginning to show promise. Although these device demonstrations are scientifically interesting, the technology is far from offering significant performance improvement over existing TFT technology. Researchers have reported fabricating TFTs using sputtered ZnO on Si substrates but with field-effect mobilities of only ~–2 $cm^2/V\text{-}s$.[71,72] Improved device performance using epitaxial $InGaO_3(ZnO)_5$ has been reported.[73] However, neither of these material systems is desirable for commercial applications, especially when considered from a solution-based deposition perspective.

1.4.4 Light-Emitting Diodes

Optoelectronic applications such as blue and near-UV light-emitting diodes (LEDs) and detectors are well established in microelectronics. Equivalent macroelectronic devices may become feasible, if suitable p-type materials become available. The most fundamental problem is in finding p-type TCOs that can be used to form efficient pn junctions. P-type TCO materials such as $CuAlO_2$, $CuGaO_2$, and $SrCu_2O_2$ are in early development. What is really needed is a ZnO-based p-type semiconductor to create a transparent analog of a Si pn diode. Additional discussion on low-temperature solution processing of oxide electronic materials is presented in Chapter 4 (although in this case focusing on dielectric applications). Here, we simply note that the initial challenges are the fundamental electrical and material properties required for adequate electrical characteristics. After this will come the challenge of maintaining adequate optoelectronic performance using solution-deposited materials and methods.

1.4.5 Solid-State Lighting

An area of significant technological interest is solid-state lighting and its application to distributed light sources. Although general lighting technology options may seem straightforward, the cost of ownership is a significant issue that adds considerable complexity. A 100-W incandescent light bulb costs less than \$1. With an efficiency of 15 lm/W, this represents a lamp cost of less than \$1 per kilolumens. The cost for fluorescent lighting is approximately a factor of two higher. However, the average lifetime of an incandescent bulb is about 1000 hours, whereas that of a fluorescent tube is approximately 20,000 hours. Therefore, the cost of ownership calculated over the life of the lamp is much lower for a fluorescent lamp compared with an incandescent lamp. Thus, cost of ownership is particularly important in understanding market penetration of alternative technologies in the commercial and industrial markets.[74]

Although there has been tremendous progress in conventional microelectronic III–V-based LED technology, the major commercialization challenge for penetration into the general lighting market is achieving the cost targets that will be required to compete with existing incandescent and fluorescent lamps.[75] As these LEDs are manufactured on small (2–3″ typical), expensive substrates and are point sources, system integration costs, including backend packaging, light, and thermal management issues, pose a significant barrier to cost reduction over the next decade. Healthy penetration in automotive exterior lighting as well as niche architectural lighting applications are expected[76] and will continue to push down the cost of this technology. Since manufacturing cost is such a significant issue, a printed electronics-based solution could be attractive. Unfortunately, currently, the technical challenges coupled with the highly competitive cost structure seem to be too difficult for a viable point source lighting solution based on solution processing.

In contrast, extended (distributed) light sources, by their nature, circumvent such issues and are more consistent with printed electronics technology. Organic LEDs (OLEDs) represent another technology platform suitable for large-area solution processing.[77] However, the improvements in the reliability of organic semiconductors required to meet general lighting requirements continues to be very challenging. More environmentally stable thick-film inorganic electroluminescent devices have been used to fabricate flexible and thin illumination sources for LCD backlights, some automotive applications and novelty applications such as night lights. Even though this technology is well suited for solution-based large-area processing, efficiency, and lifetime remain as major problems. Because of a short lifetime, low luminance, and high operating voltage, this technology has found only limited applications in general illumination or displays. However, the attractive features of LEDs for lighting and the solution processability of OLEDs and inorganic electroluminescent materials, coupled with the importance of lighting in overall energy consumption, suggests that this may be an active area for future research.

1.4.6 Si-Based Integrated Emitters

Nanocrystalline silicon presents opportunities for integrating optoelectronic functionality on large-area substrates. Recent reports of light emission from silicon represent an exciting and innovative area of research that combines advances in nanoscale silicon and silicon oxides and rare earth materials.[78] Adoption of methods that have been developed in the flat panel display industry for silicon-based solid-state lighting provide the advantage of having manufacturing processes that are compatible with CMOS and flat panel display processing. Therefore, this technology can be integrated with existing silicon devices, or be fabricated on large-area glass substrates (similar to those used in FPD fabs) with great potential for cost improvement as compared with III–V-based LEDs. In addition, this approach offers the possibility for developing large-area light sources.

One of the most exciting and promising recent areas of research has been the development of silicon nanocrystals embedded in a silicon oxide matrix. This activity initially focused on the development of high-density-embedded memories.[79] More recently, it has led to interesting applications in silicon microphotonics. Although much of the current effort is on synthesis of nanocrystals with proper size and passivation, in order to achieve saturated emission color and high luminescent efficiency, several other technological advances need to occur. These include the development of efficient and low-cost methods to insert and activate rare earth ions. Downstream use of nanocrystals for fabrication of efficient electrically pumped devices is in the early stages of development. As the technology matures, adaptation of printing and solution deposition techniques would allow exploitation for novel large-area applications.

Innovative technologies that consist of highly efficient emitters, compatible with CMOS production processes and inherently large area, so as to be manufacturable on flat glass panel substrates, present significant potential as they cut across several markets (e.g., display, telecom, and lighting) and offer a paradigm shift in manufacturing cost. Furthermore, the increase in substrate size that can be used for product manufacture represents a major factor in reducing production cost even more dramatically, especially if FPD infrastructure can be leveraged. What has been demonstrated is based on traditional semiconducting processing methods; adoption of methods similar to what is discussed in Chapter 5 (Liquid Silicon Materials) to silicon nanophotonics would offer exciting possibilities for large-area silicon-based optoelectronics.

1.5 APPLICATION CHALLENGES: POWER SOURCES, SENSORS, AND ACTUATORS

Although a wide range of potential electronic and optoelectronic applications have been described, at least one more component must be considered to achieve a truly integrated macroelectronic system. Perhaps the most difficult of the major challenges for macroelectronics is providing adequate power (and power storage), especially for systems that are field portable such as large-area PV and antenna arrays. Although existing lithium ion batteries are well suited for conventional microelectronic portable device applications, thin-film batteries offer possibilities for device level integration into macroelectronic systems. What would be truly innovative would be the integration of thin-film microbatteries as power sources for macroelectronics (flexible displays, large-area electronics, and integrated flexible photovoltaic modules). This could lead to new applications in autonomous sensor units, miniature remotely piloted vehicles, self-powered 3D circuits, and power modules with integrated batteries and photovoltaic layers. Unlike commercial lithium ion batteries, which are fabricated using solution-based processing methods, existing thin-film battery technology is based on vacuum deposition methods.[80] Even though large-area, thin-film batteries on flexible Kapton films have been commercialized,[81] for them to become primary sources of energy for macroelectronics, energy capacity needs to be increased significantly along with methods to fabricate these devices at lower temperatures to facilitate integration with other components and multiple substrate types. Rather than lithium-based technology, much older electrochemical cell technology has been successfully converted to roll-to-roll manufacturing for integration into various large-area applications. Zinc/MnO_2 is a well-known and well-understood technology. Several companies are now making it available in thin-film format (Figure 1.10).[82,83]

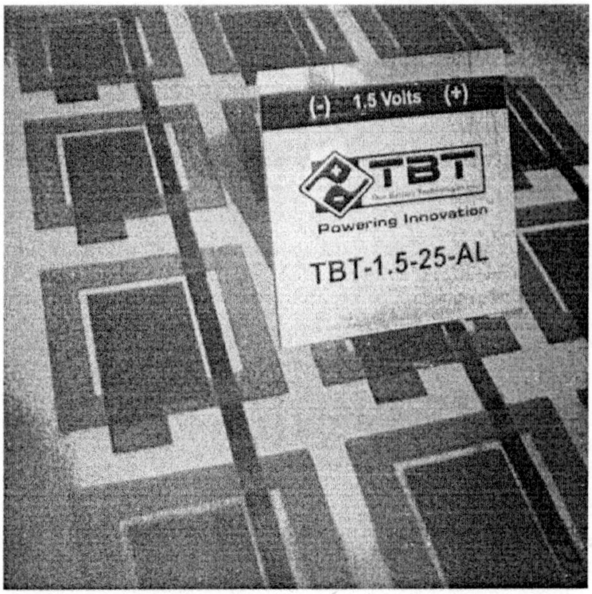

Figure 1.10. Example of fully integrated battery and electronics. [Figure courtesy of Thin Battery Technologies, Inc. Used with permission.]

Of course energy storage is just part of the power source problem. If the battery cannot be replaced or recharged readily, then energy harvesting must be considered. Even though the technology for stand-alone PV is well established and progressing at a rapid pace, integration of PV cells with other electronic functions as suggested earlier is still some years and several research breakthroughs away. Currently, printed organic-based PV technology has shown promise for this type of application.[84] However, solution-deposited inorganics (see Chapters 6 and 7) are a more attractive solution because of the demonstrated higher conversion efficiency and greater stability to oxygen, water, and light. The same can also be said for various other energy scavenging mechanisms, which are reasonably well understood as a discrete component, but will likely need considerable effort to integrate into a complete macroelectronics system.

There are more issues and complexity to be considered if various microelectromechanical (MEMS)-type devices are included in the macroelectronics tool kit. As described previously, the materials and devices required for TFTs and circuits can provide adequate electromagnetic (visible and RF) sensitivity for many image-type applications. These materials may also provide satisfactory performance in pressure and strain sensors. Nanotube/nanowire-based devices look promising for various chem-bio sensors.[85] However, there is little that is known about the ability to integrate printed microfluidic devices (and other such devices with moving parts) into a roll-to-roll-type process.

Questions regarding applicability of particular technologies to particular applications must, of course, be resolved eventually. However, mainstream focus has been on achieving adequate TFT device and circuit performance to support the control, computation, and communication functions for a wide range of potential applications. As satisfactory electrical circuit performance is attained for additional functionality, questions that relate to the quality of other device types, their integration into an overall process flow, and perhaps most importantly, the robustness and stability of the final product, will receive much more attention.[86–88]

1.6 CONCLUSIONS

In this introductory chapter, the intent has been to identify and describe the two major electronics technologies, with an emphasis on why and how Macroelectronics is related to, but yet very different from microelectronics. The differences in intended product applications drive macroelectronics to have very different requirements in terms of manufacturing technology and required electrical performance. In contrast to microelectronics, the main objectives for macroelectronics are very low cost and processing on large-area, preferably flexible substrates. These attributes, when achieved, will allow a wide range of interesting and novel applications ranging from surface conformal electronics to rollable/foldable multi-meter-area distributed electronics.

The display and PV industries are the first examples of commercially successful macroelectronics. Research is being aggressively pursued to extend these initial successes into a wide range of new areas that require lower cost, higher performance, and compatibility with a variety of substrates. Existing solutions based on amorphous silicon, LTPS and organic TFTs have not been successful because of either cost and/or performance limitations.

To achieve more aggressive cost and performance objectives, new materials, devices and manufacturing methods must be developed and combined into an integrated, cost-effective macroelectronics solution. One of the most promising approaches for this next generation of macroelectronics technology is solution deposition of inorganic materials. This class of approaches should provide the means to deposit the key electronic materials—metals, semiconductors, and insulators. Furthermore, the methodology is scalable from wafer-sized, to sheet-sized, to roll-sized substrates. It also offers the potential to enable high-performance devices, with mobilities of more than 50 cm^2/V-s or even as high as several hundred, approaching that of crystalline silicon used in current microelectronic devices. Just as critical, numerous inorganic materials can be solution deposited to facilitate high-throughput, low-cost processing and multifunction systems. It can be accomplished at near-ambient temperature so that many substrate types can potentially be used. Finally, inorganic materials tend to be stable to light, oxygen, and water, all of which create problems for the TFTs currently used in manufacturing.

Another critical challenge is combining the materials and device structures with a patterning technique that is consistent with the low-cost objectives, but still providing resolution and alignment features in the less than 10 μm range (preferably in the 1-μm range). Furthermore, this capability must be achievable over substrate sizes of meters and/or in a roll-to-roll format. Achieving both good materials characteristics and small feature sizes will be necessary if significant RF applications (at least several gigahertz) are to be addressed.

Although adequate materials and devices are essential, successful manufacturing will require other capabilities as well. First, the process must have high yield, which implies low variability, and provide robust stability to environmental factors. To produce the envisioned products, there must be readily available electronic design tools that can adequately simulate both device and circuit performance. Although some of these computer-aided design tools are available from microelectronics technology, others must either be modified, because of the differences in the thin-film devices, or created anew because the devices have no equivalent (nanowires and nanotubes).

Since macroelectronics will be most successful if it can integrate diverse materials and devices, the above statements also apply to the creation of technology for solution-deposited inorganic optoelectronic devices, batteries, and sensors, deposited on a wide range of substrate types (including those that are flexible). The fabrication process must be nimble enough to adjust for many product types and many manufacturing variables. As the remaining chapters describe, there is good reason to believe that solution-based deposition of inorganic materials will ultimately provide the necessary device performance and manufacturing capability.

REFERENCES

1. Rajeshwar, K.; Roozeboom, F. 2007. Gordon Moore and his legacy: four decades and counting. *Electrochem. Soc. Interface* Spring:11–14.
2. Chalamala, B.; Temple, D. 2005. Big and bendable. *IEEE Spectrum* 42:50–56.
3. Reuss, R. H.; Chalamala, B.; Moussessian, A.; Kane, M. K.; Kumar, A.; Zhang, D. C.; Rogers, J. A.; Hatalis, M.; Temple, D.; Moddel, G.; Elissson, B. J.; Estes, M. J.; Kunze, J.; Handy, E. S.; Harmon, E. S.; Salzman, D. B.; Woodall, J. M.; Alam, A.; Murthy, J. Y.; Jacobsen, S. C.; Olivier, M.; Markus, D.; Campbell, P. M.; Snow, E. 2005. Macroelectronics: Perspectives on technology and applications. *Proc. IEEE* 93:1239–1256.
4. Wener, K. 1997. The flowering of flat displays. *IEEE Spectrum* 34:40–49.
5. Mentley, D. E. 2002. State of flat-panel display technology and future trends. *Proc. IEEE* 90:453–459.
6. Jin, H. C.; Kang, I. B.; Jang, E. S.; Moon, H. M.; Oh, C. H.; Lee, S. H.; Yeo, D. 2007. Development of 100-in. TFT-LCDs for HDTV and public-information-display applications. *J. Soc. Information Display* 15:277–280.

7. Slaoui, A.; Collins, R. T. 2007. Advanced inorganic materials for photovoltaics. *MRS Bull.* 32:211–214.
8. For details on the status of the world wide PV industry, refer to International Energy Agency's photovoltaic energy data base at www.iea-pvps.org.
9. International Energy Agency (IEA). 2006. Trends in photovoltaic applications. Report: IEA-PVPS T1-15.
10. Schropp, R. E. I.; Carius, R.; Beaucarne, G. 2007. Amorphous silicon, microcrystalline and thin-film polycrystalline silicon solar cells. *MRS Bull.* 32:219–223.
11. Jager-Waldau, A. (Editor). 2007. *2nd International Workshop on Thin Films in the Photovoltaic Industry*, EUR 22630. (*Ispra, Italy; Nov. 9–10, 2006*).
12. Beach, J. D.; McCandless, B. E. 2007. Materials challenges for CdTe and $CuInSe_2$ photovoltaics. *MRS Bull.* 32:225–229.
13. Nanosolar, Inc., Palo Alto, CA. Accessed: www.nanosolar.com.
14. von Roedern, B.; Ullal, H. S.; Zweibel, K. 2006. Polycrystalline thin-film photovoltaics: From the laboratory to solar fields. *Proc. of the IEEE 4th World Conference on Photovoltaic Energy Conversion* (*Waikoloa, HI; May 7–12, 2006*). Accessed: http://www.nrel.gov/pv/thin_film/docs/.
15. Mokhoff, N. 2007. E-paper, flexible displays creep closer to market. *EE Times*. May 28. Accessed: www.eetimes.com/issue/na/showArticle.jhtml?articleID=199701485.
16. Reuss, R. H.; Hopper, D. G.; Park, J.-G. 2006. Macroelectronics. *MRS Bull.* 31: 447–450.
17. Allen, K. J. 2005. Reel to real: Prospects for flexible displays. *Proc. IEEE* 93: 1394–1399.
18. Chabinyc, M. L.; Wong, W. S.; Arias, A. C.; Ready, S.; Lujan, R. A.; Daniel, J. H.; Krusor, B.; Apte, R. B.; Salleo, A.; Street, R. A. 2005. Printing methods and materials for large-area electronic devices. *Proc. IEEE* 93:1491–1499.
19. Parashkov, R.; Becker, E.; Riedl, T.; Johannes, H.-H.; Kowalsky, W. 2005. Large-area electronics using printing methods. *Proc. IEEE* 93:1321–1329.
20. Chason, M.; Gamota, D. R.; Brazis, P. W.; Kalyanasundaram, K.; Zhang, J.; Lian, K. K.; Croswell, R. 2006. Toward manufacturing low-cost, large-area electronics. *MRS Bull.* 31:471–475.
21. Zyung, T.; Kim, S. H.; Chu, H. Y.; Lee, J. H.; Lim, S. C.; Lee, J.-I.; Oh, J. 2005. Flexible organic LED and organic thin-film transistors. *Proc. IEEE.* 93:1265–1272.
22. Chason, M.; Brazis, P. W.; Zhang, J.; Kalyanasundaram, K.; Gamota, D. R. 2005. Printed organic semiconducting devices. *Proc. IEEE* 93:1348–1356.
23. Bock, K. 2005. Polymer electronics systems—polytronics. *Proc. IEEE* 93: 1400–1406.
24. Lee, S.; Koo, B.; Park, J.-G.; Moon, H.; Hahn, J.; Kim, J. M. 2006. Development of high-performance organic thin-film transistors for large-area displays. *MRS Bull.* 31:455–459.
25. Sazonov, A.; Striakhilev, D.; Lee, C.-H.; Nathan, A. 2005. Low-temperature materials and thin-film transistors for flexible electronics. *Proc. IEEE* 93: 1420–1428.

26. Qing, X. P.; Chan, H.-L.; Beard, S. J.; Kumar, A. 2006. An active diagnostic system for structural health monitoring of rocket engines. *J. Intelligent Material Systems and Struct.* 17:619–628.

27. Kagan, C. R.; Andry, P. (Editors). 2006. *Thin-Film Transistors.* CRC Press, Boca Raton, FL.

28. Street, R. A. (Editor). 1999. *Technology and Applications of Amorphous Silicon.* Springer Verlag, Berlin.

29. Hamadani, B. H.; Natelson, D. 2005. Extracting contact effects in organic FETs. *Proc. IEEE* 93:1306–1311.

30. Lacour, S. P.; Jones, J.; Wagner, S.; Suo, Z. 2005. Stretchable interconnects for elastic electronic surfaces. *Proc. IEEE* 93:1459–1467.

31. Lewis, J.; Grego, S.; Chalamala, B.; Vick, E.; Temple, D. 2004. Electromechanics of a highly flexible transparent conductor for display applications. *Society of Vacuum Coaters 47th Annual Techn. Conf. Dig.* 47:129–132.

32. Ahn, J.-H.; Kim, H.-S.; Menard, E.; Lee, K. J.; Zhu, Z.; Kim, D.-H.; Nuzzo, R. G.; Rogers, J. A. 2007. Bendable integrated circuits on plastic substrates by use of printed ribbons of single-crystal silicon. *Appl. Phys. Lett.* 90:213501–213509.

33. Servati, P.; Nathan, A. 2005. Functional pixel circuits for elastic AMOLED displays. *Proc. IEEE* 93:1257–1264.

34. Lalli, J. H.; Hill, A.; Demerci, N.; Homer, M.; Claus, R. O. 2007. Metal rubber™ sheet and fabric materials. *Proc. SPIE Smart Materials and Structures Conference. (San Diego, CA)*, pp. 6525–6563.

35. Pique, A.; Chrisey, D. B. 2001. Direct-write electronics for rapid prototyping applications. In *Sensors, Electronics and Integrated Power Sources.* Academic Press, New York.

36. Chrisey, D. B.; Gamota, D. R.; Helvajian, H.; Taylor, D. P. 2000. Materials Development for Direct Write Technologies. *MRS Proceedings*, Vol. 624. Materials Research Society, Warreudale, PA.

37. Han, S.-M.; Lee, M.-C.; Shin, M.-Y.; Park, J.-H.; Han, M.-K. 2005. Poly-Si TFT fabricated at 150C using ICP-CVD and excimer laser annealing. *Proc. IEEE* 93:1297–1305.

38. Vikas, R.; Ishihara, R.; Horoshima, Y.; Abe, D.; Inoue, S.; Shimoda, T.; Metselaar, J. W.; Beenakker, C. I. M. 2005. High performance single grain Si TFTs inside a location-controlled grain by micro-Czochralski process with capping layer. *IEEE Intern. Electron Dev. Meeting Tech. Digest* 2005:941–944.

39. Van der Wilt, P. C.; Kane, M. G.; Limanov, A. B., Firester, A. H.; Goodman, L.; Lee, J.; Abelson, J. R.; Chitu, A. M.; Im, J. S. 2006. Low-temperature polycrystalline silicon thin-film transistors and circuits on flexible substrates. *MRS Bull.* 31:461–465.

40. Kane, M. G.; Goodman, L.; Firester, A. H.; van der Wilt, P. C.; Limanov, A. B.; Im, J. S. 2005. 100 MHz CMOS circuits using sequential laterally solidified silicon thin-film transistors on plastic. *IEEE International Electron Devices Meeting Tech. Digest* 2005:1087–1089.

41. Kuo, Y. (Editor). 2004. *Thin-Film Transistors: Materials and Processes, Amorphous Silicon Thin-Film Transistors, Polycrystalline Silicon Thin Transistors.* Kluwer, New York.

42. Hilkes, R. 2007. LTPS display architectures. *EE Times*. April 16:44.

43. Yan, M.; Kim, T. W.; Erlat, A. G.; Pellow, M.; Foust, D. F.; Liu, J.; Schaepkens, M.; Heller, C. M.; McConnelee, P. A.; Feist, T. P.; Duggal, A. R. 2005. A transparent, high barrier, and high heat substrate for organic electronics. *Proc. IEEE* 93:1239–1256.

44. Cerny, G. 2007. SiC:H alloy barrier films. *2007 Flexible Displays and Microelectronics Conference (Phoenix, AZ; Feb. 5–8, 2007)*, p. 14.3.

45. MacDonald, B. 2007. Developments in polyester film for flexible electronics. *2007 Flexible Displays and Microelectronics Conference (Phoenix, AZ; Feb. 5–8, 2007)*, p. 15.1.

46. Afentakis, T.; Hatalis, M.; Voutsas, A. T.; Hartzell, J. 2006. Design and fabrication of high-performance polycrystalline silicon thin-film transistor circuits on flexible steel foils. *IEEE Trans. Electron Dev.* 53:815–822.

47. Mitzi, D. B.; Milliron, D. J.; Copel, M.; Murray, C.; Kosbar, L. 2005. High mobility solution-deposited chalcogenide films for flexible applications. *IEEE VLSI-TSA International Symp. on VLSI Technol. (April)*, pp. 41–44.

48. Shimoda, T.; Matsuki, Y.; Furusawa, M.; Aoki, T.; Yudasaka, I.; Tanaka, H.; Iwasawa, H.; Wang, D.; Miyasaka, M.; Takeuchi, M. 2006. Solution-processed silicon films and transistors. *Nature* 440:783–786.

49. Fulghum, D. B.; Barrie, D.; Wall, R. 2006. Small aircraft may become large sensors. *Aviation Week and Space Technology* 127 (October 28, 2006). The article can be accessed at: http://www.aviationweek.com/aw/generic/story_channel.jsp?channel=defense&id=news/aw103006p1.xml

50. Jain, K.; Klosner, M.; Zemel, M.; Raghunandan, S. 2005. Flexible electronics and displays: high resolution, roll-to-roll, projection lithography and photoablation processing technologies for high-throughput production. *Proc. IEEE* 93:1500–1510.

51. Michel, B.; Bernard, A.; Bietsch, A.; Delamarche, E.; Geissler, M.; Juncker, D.; Kind, H.; Renault, J. P.; Rothuizen, H.; Schmid, H.; Schmidt-Winkel, P.; Stutz, R.; Wolf, H. 2001. Printing meets lithography: Soft approaches to high-resolution patterning. *IBM J. Res. Develop.* 45:697–719.

52. Ko, H. C.; Baca, A. J.; Rogers, J. A. 2006. Bulk quantities of single-crystal silicon micro-/nanoribbons generated from bulk wafers. *Nano Lett.* 6:2318–2324.

53. Lee, K. J.; Meitl, M. A.; Ahn, J.-H.; Rogers, J. A.; Nuzzo, R. G.; Kumar, V.; Adesida, I. 2006. Bendable GaN high Electron mobility transistors On plastic substrates. *J. Appl. Phys.* 100:124507–124510.

54. Sun, Y.; Menard, E.; Rogers, J. A.; Kim, H.-S.; Kim, S.; Chen, G.; Adesida, I.; Dettmer, R.; Cortez, R.; Tewksbury, A. 2006. Gigahertz operation in flexible transistors on plastic substrates. *Appl. Phys. Lett.* 88:183509–183511.

55. Lagally, M. 2007. Silicon nanomembranes. *MRS Bull.* 32:57–63.

56. Arnold, C. B.; Serra, P.; Pique, A. 2007. Laser direct-write techniques for printing of complex materials. *MRS Bull.* 32:23–32.

57. Lieber, C. M.; Wang, Z. L. 2007. Functional nanowires. *MRS Bull.* 32:99–104.

58. Duan, X. 2007. Assembled semiconductor nanowire thin-films for high-performance flexible macroelectronics. *MRS Bull.* 32:134–141.

59. Ju, S.; Lee, K.; Yoon, M.-H.; Facchetti, A.; Marks, T. J.; Janes, D. B. 2007. High performance ZnO nanowire field effect transistors with organic gate nanodielec-

trics: effects of metal contacts and ozone treatment. *Nanotechnol.* 18: 155201–155208.

60. Facchetti, A.; Yoon, M.-H.; Marks, T. J. 2005. Gate dielectrics for organic field-effect transistors: New opportunities for organic electronics. *Adv. Mater.* 17:1705–1725.
61. Kang, S. J.; Kocabas, C.; Ozel, T.; Shim, M.; Pimparkar, N.; Muhammad, M. A.; Rotkin, S. V.; Rogers, J. A. 2007. High-performance electronics using dense, perfectly aligned arrays of single-walled carbon nanotubes. *Nature Nanotechnol.* 2:230–236.
62. Gruner, G. 2007. Carbon nanonets spark new electronics. *Sci. Amer.* 296:76–83.
63. Hur, S.-H.; Yoon, M.-H.; Gaur, A.; Shim, M.; Facchetti, A.; Marks, T. J.; Rogers, J. A. 2005. Organic nanodielectrics for low voltage carbon nanotube thin-film transistors and complementary logic gates. *J. Am. Chem. Soc.* 127: 13808–13809.
64. Alam, M. A.; Pimparkar, N.; Kumar, S.; Murthy, J. 2006. Theory of nanocomposite network transistors for macroelectronics applications. *MRS Bull.* 31:466–470.
65. Kang, S. J.; Kocabas, C.; Ozel, T.; Shim, M.; Pimparkar, N.; Alam, M. A.; Rotkin, S. V.; Rogers, J. A. 2007. High-performance electronics using dense, perfectly aligned arrays of single-walled carbon nanotubes. *Nature Nanotechnol.* 2:230–236.
66. Iniguez, B.; Picos, R.; Veksler, D.; Koudymov, A.; Shur, M. S.; Ytterdal, T.; Jackson, W. 2007. Universal compact model for long- and short-channel thin-film transistors. *Proc. of the International TFT Conference* (*Jan 7*).
67. Madan, A.; Shaw, M. P. 1988. *The Physics and Applications of Amorphous Semiconductors*. Academic Press, Boston, MA.
68. Green, M. A. 2003. *Third Generation Photovoltaics: Advanced Solar Energy Conversion*. Springer, Berlin.
69. Fortunato, E.; Ginley, D.; Hosano, H.; Paine, D. C. 2007. Transparent conducting oxides for photovoltaics. *MRS Bull.* 32:242–247.
70. For review of transparent conductive oxides, refer to the special issue of *MRS Bull.* 25(8), 2000 and the *Proceedings of the MRS Workshop on Transparent Conducting Oxides* (Denver, CO, June 19–20, 2000).
71. Carcia, P. F.; McLean, R. S.; Reilly, M. H.; Nunes, Jr., G. 2003. Transparent ZnO thin-film transistor fabricated by rf magnetron sputtering. *Appl. Phys. Lett.* 82: 1117–1119.
72. Hoffman, R. L.; Norris, B. J.; Wager, J. F. 2003. ZnO-based transparent thin-film transistors. *Appl. Phys. Lett.* 82:722–735.
73. Nomura, K.; Ohta, H.; Ueda, K.; Kamiya, T.; Hirano, M.; Hosono, H. 2003. Thin-film transistor fabricated in single-crystalline transparent oxide semiconductor. *Science* 300:1269–1273.
74. For an overall review of solid state lighting, refer to a number of recent U.S. Department of Energy Solid State Lighting Workshop reports. Many of these documents are accessible online at: http://www.netl.doe.gov/ssl/.
75. Navigant Consulting. 2003. *Energy Savings Estimates of Light-Emitting Diodes in Niche Lighting Applications*. US Department of Energy, Washington, DC. Accessed: http://www.netl.doe.gov/ssl/PDFs/Niche%20Final%20Report.pdf.

76. Krames, M. R.; Shchekin, O. B.; Mueller-Mach, R.; Mueller, G. O.; Zhou, L.; Harbers, G.; Craford, M. G. 2007. Status and future *of* high-power light-emitting diodes for solid-state lighting. *IEEE/OSA J. Display Technol.* 3:160–175.
77. Duggal, A. R.; Heller, C. M.; Shiang, J. J.; Liu, J.; Lewis, L. N. 2007. Solution-processed organic light-emitting diodes for lighting. *IEEE/OSA J. Display Technol.* 3:184–192.
78. Iaconaa, F.; Pacifici, D.; Irrera, A.; Miritello, M.; Franzo, G.; Priolo, F.; Sanfilippo, D.; Di Stefano, G.; Fallica, P. G. 2002. Electroluminescence at 1.54 μm in Er-doped Si nanocluster-based devices. *Appl. Phys. Lett.* 81:3242–3244.
79. Hannafi, H. I.; Tiwari, S.; Khan, I. 1996. Fast and long-retention-time nanocrystal memory. *IEEE Trans. Electron Devices.* ED-43:1553–1558.
80. Bates, J. B.; Dudney, N. J.; Neudecker, B.; Ueda, A.; Evans, C. D. 2000. Thin-film lithium and lithium-ion batteries. *Solid State Ionics.* 135:33–45.
81. Excellatron Corp, Atlanta, GA. Accessed: www.excellatron.com.
82. Ghiurcan, G. A.; Liu, C.-C.; Webber, A.; Feddrix, F. H. 2003. Development and characterization of a thick-film printed zinc-alkaline battery. *J. Electrochem. Soc.* 150:A922–A927.
83. Power Paper Ltd., Kiryat Arye, Petah Tikva, Israel. Accessed: www.powerpaper.com.
84. Dennler, G.; Sariciftci, N. S. 2005. Flexible conjugated polymer-based plastic solar cells: from basics to applications. *Proc. IEEE.* 93:1429–1439.
85. Patolsky, F.; Timko, B. P.; Zheng, G.; Lieber, C. M. 2007. Nanowire-based nano-electronic devices in the life sciences. *MRS Bull.* 32:142–149.
86. Cairns, D. K.; Crawford, G. P. 2005. Electromechanical properties of transparent conducting substrates for flexible electronic displays. *Proc. IEEE.* 93:1451–1458.
87. Servati, P.; Nathan, A. 2005. Functional pixel circuits for elastic AMOLED displays. *Proc. IEEE.* 93:1257–1264.
88. Bhattacharya, R.; Wagner, S.; Tung, Y.-J.; Esler, J. R.; Hack, M. 2005. Organic LED pixel array on a dome. *Proc. IEEE.* 93:1273–1279.

CHAPTER 2

Chemical Solution Deposition—Basic Principles

ROBERT W. SCHWARTZ

Missouri University of Science and Technology, Rolla, MO

MANOJ NARAYANAN

Argonne National Laboratory, Energy Systems Division, Argonne, IL

2.1 INTRODUCTION

Chemical solution deposition (CSD) procedures have been widely used for the production of both amorphous and crystalline thin films for more than 20 years.[1–4] Both colloidal (particulate) and polymeric-based processes have been developed. Numerous advances have been demonstrated in understanding solution chemistry, film formation behavior, and for crystalline films, phase transformation mechanisms during thermal processing. Several excellent review articles regarding CSD have been published, and the reader is referred to Refs. 5–12 for additional information on the topic. Recently, modeling of phase transformation behavior for control of thin-film microstructure has also been considered, as manipulation of film orientation and microstructure for various applications has grown in interest.[13–15]

In this chapter, some basic principles of CSD are discussed and remaining challenges in the field are noted. The chapter focuses on polymeric-based solution processing and is organized according to the typical steps of the CSD process, which are highlighted in Fig. 2.1.[16] The overall approach involves several steps, beginning with solution synthesis and typically ending with heat treatment. During each step, a variety of parameters must be controlled. These parameters are discussed below in the individual sections. The large number of variables and codependency among the variables can make identification (and understanding) of cause–effect relationships difficult, although there has been significant progress in understanding these relationships since

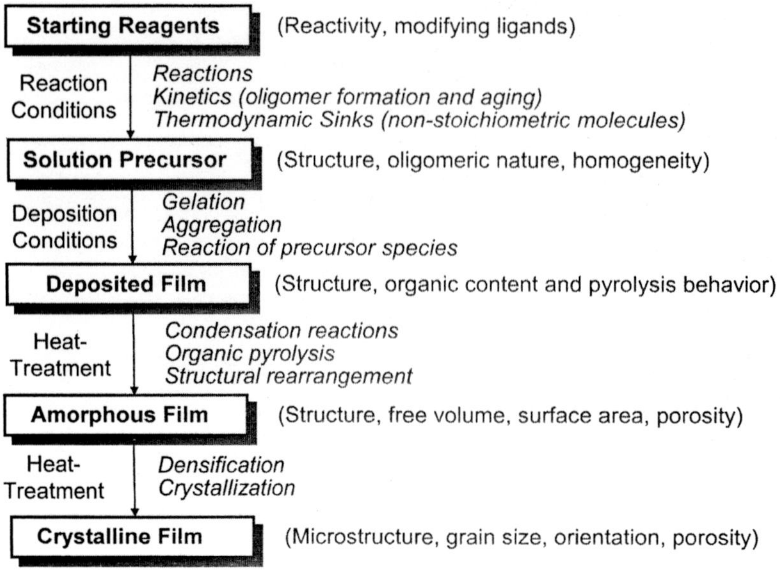

Figure 2.1. Processing stages in chemical solution deposition of thin films. Controllable parameters are shown on the left; dependent processes are shown in italics. [Reprinted from Ref. 16 with the permission of the publisher, Taylor & Francis, Ltd.: R. W. Schwartz et al., Sol-gel processing of PZT thin films: a review of the state-of-the-art and process optimization strategies, *Int. Ferro.*, **7**, 259, (1995).]

solution-based film deposition processes first became widely used. The technique also remains widely used for investigation of the properties of new material systems. Although the chapter will specifically focus on the CSD approach, many of the principles discussed apply to other solution-based deposition approaches.

2.2 SUBSTRATE SURFACE PREPARATION

Before the stock solution is prepared and used for coating, the coating surface of the substrate has to be cleaned, prepared, and dried appropriately to improve the adhesion and quality of the coating. General contaminants that are commonly found on the substrate surface include dust particles, oils, grease, fingerprints, hair, and fibers from paper products and application-specific contaminants include ions, salts, fluxes, oxide scales, and metal particles. These contaminants result in poor film adhesion to the substrate, formation of pores, bubbles, crack initiation sites in the film during subsequent heat treatments, and premature electrical failure caused by shorting. Therefore, it is absolutely necessary to clean and prepare the surface of the substrates using various chemicals and techniques prior to coating. The cleaning and treatment procedures depend on the kind of contaminants, solvents, and types of substrate. Solvents typically used are grouped as hydrophilic (polar), hydrophobic (nonpolar), or azeotropic. Hydrophilic solvents include organic solvents like acetone, methanol, ethanol, and isopropanol; water; and some detergent solutions. Hydrophobic solvents include organic solvents (xylenes, toluene, napthas), fluorocarbons, and chlorinated hydrocarbons (trichloroethylene). Polar solvents like alchohols, ketones, and aldehydes are commonly used to clean oil and grease on the surface. Azeotropic solvents include a mixture of both hydrophobic and hydrophilic solvents (e.g., 50:50 mixture of naptha and alcohol). Azeotropic solvents are more efficient in removing a wide range of contaminants in a single step, because they contain both polar and nonpolar solvents. Typically the substrates are dipped in different solvent baths for a certain period of time to remove the contaminants. The different mechanisms used to clean the substrates using these solvents, depending on the nature of the contaminants, are discussed below.[17]

The five different cleaning mechanisms used to remove the surface contaminants are as follows: *dissolution, chemical reaction, physical, plasma exposure*, and *vacuum baking. Dissolution* involves the process of dissolving the contaminants into excess solvents. A typical example includes the dissolution of oils and greases using organic solvents like acetone and isopropyl alcohol or some common salts using water. In a *chemical reaction*, the contaminant is converted to a soluble compound that is later cleaned by dissolution. Examples include the removal of metal oxides and scales using acids or alkalis. In *physical cleaning*, the surface contaminants are removed by blowing pressurized air or inert gas, scrubbing, brushing, swabbing, or ultrasonic energy. Typically dusts, fibers from paper products, and other general contaminants are removed from the surface using these physical methods. In *plasma cleaning*, the surface of the substrate is exposed to oxygen or argon plasma where the contaminants (organic residues) are dislodged by bombarding it with energized atoms. Typically, organic residues are cleaned (i.e., dislodged and oxidized) by this technique. *Vacuum baking* is a process of heating the substrate to 70–250 °C in a vacuum chamber. This process removes the volatile residues like solvents, chemicals, and moisture on the substrate surface.[17] Substrates are generally exposed to a combination of the above techniques to improve the efficiency of the cleaning process. For example, metal substrates are typically ultrasonicated in acetone and methanol to remove residues and grease before film deposition.[18]

After the cleaning process, other techniques are used to prepare the surface of the substrate for coating. Some techniques include drying, surface etching, and chemical surface preparation. Examples of chemical surface preparation include the formation of an oxide layer or the monolayer assembly of an adhesion promoter on the surface. These processes modify the surface of the substrates so as to facilitate the subsequent deposition process. In surface preparation, frequently, the hydrophilic/hydrophobic character of the surface is controlled to match the coating solution properties. For example, Van Driessche et al.[19] reported on improving the wettability of Ni-4at%W tapes by chemically cleaning them in a hot mixture of hydrogen peroxide and formic acid followed by ultrasonic cleaning in methanol. These tapes were degreased in trichloroethylene, acetone, and methanol prior to the chemical cleaning process for deposition of CeO_2 and $La_2Zr_2O_7$ buffer layers.[20]

2.3 STARTING REAGENTS AND SOLVENTS

2.3.1 Background

The first step in the CSD process is solution preparation, which involves reagent selection (chemical precursors) and solvent choice.[1,5–12,16] During solution preparation, other chemical modifiers may also be added to the solution to facilitate or limit chemical reactivity. Also during this stage of the process, identification of appropriate reaction conditions to promote other desired changes in precursor nature or solution characteristics is also considered. The goal for solution preparation is to develop a homogeneous solution of the necessary cation species that may later be applied to a substrate.

Choices of precursor(s) may be dictated by solubility, reactivity, or other property. For multicomponent systems, mutual solubility is another factor that must be considered. For such solutions, the solvent selected must facilitate dissolution of all precursors.

A final factor in precursor selection is the general synthetic route to be followed. Historically, these routes have been categorized as Pechini,[21] nitrate,[22] metallo-organic decomposition,[23–29] chelate,[16,30–32] or sol-gel.[1,3,4] Species selection and choice of reaction conditions serve to define the nature of the CSD process, including factors such as level of achievable film processing control, simplicity, and solution shelf life. Other solution characteristics that may be closely controlled are the concentration and viscosity of the solution, which will impact film formation and drying behavior. Classes of starting reagents and principal types of CSD routes are discussed below.

2.3.2 Starting Reagents

For polymeric CSD processes, three classes of metal organic (metallo-organic) compounds are used most often as starting reagents: metal alkoxides, metal carboxylates, and metal beta-diketonates. These species differ in their solubility and reactivity, as well as their tendency to react with one another, all of which are factors that may influence starting reagent selection. Representative structures of these classes of precursors are illustrated in Fig. 2.2.[8]

Metal alkoxide compounds, frequently represented as $M(OR)_x$, where M is a metal and R is an alkyl group, are the most common precursors in sol-gel CSD processes and are also frequently used in chelate processes.[3–12,30–34] Groups such as OR, which are bound to a metal center, are frequently referred to as *ligands*. Alkoxide compounds, including commonly used alkoxides such as

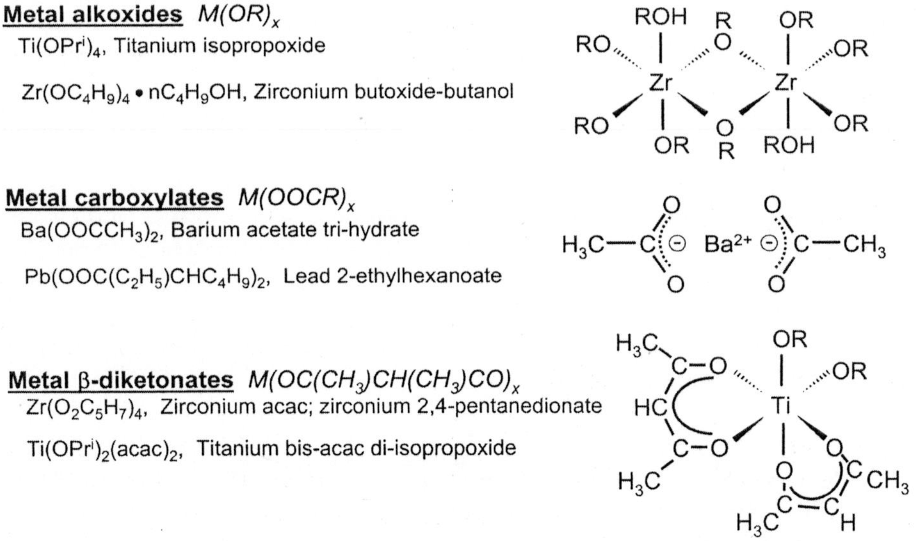

Figure 2.2. Classes and examples of common metallo-organic starting compounds used in chemical solution deposition processing. A common feature for all compounds is M–O–C bonding. Associated structures are also shown. [Adapted with permission from Ref. 8: R. W. Schwartz *et al.*, *C. R. Chemie*, **7**, 433 (2004).]

tetraethylorthosilicate (TEOS) [e.g., $Si(OCH_2CH_3)_4$] and titanium isopropoxide (TIPT) [e.g., $Ti(OCH(CH_3)_2)_4$] may be thought of as derivatives of alcohols.[34] Both of these compounds are liquid at room temperature.

The specific synthesis reactions for alkoxides depend on both the nature of the metal and the intrinsic reactivity, as well as on the parent alcohol. A typical reaction for a highly reactive metal would be represented by

$$M + ROH \rightarrow MOR + \frac{1}{2}H_2\ (g) \tag{1}$$

For less reactive metals, other synthesis schemes based on halide precursors are frequently employed. Bradley, Mehrotra, and Gaur have discussed alkoxide synthesis in detail.[34]

Alcohols (ROH) are often employed as the solvent medium for alkoxides compounds. In this situation, the alcohol also fills the role of reactant, in addition to its roles as solvent and common medium. Alcohol exchange reactions are well known in many of the more common systems that have been investigated. In simple terms, the reaction may be shown as follows:

$$M(OR)_x + xR'OH \rightarrow M(OR')_x + xROH \tag{2}$$

The exchange reaction may result in the formation of a less reactive precursor, which may be isolated for use, or used in the alcohol mixture.[35] One such common reaction in the preparation of lead zirconate titanate is the formation of titanium methoxyethoxide, $Ti(OCH_2CH_2OCH_3)_4$, from TIPT and 2-methoxyethanol.[35] The methoxyethoxide demonstrates lower sensitivity toward hydrolysis than the TIPT compound and affords more control during film preparation, as well as higher solution stability.

Compared with the other precursor species discussed below, alkoxides are typically more reactive to alcohols, other starting reagents added to the solution, or water. Control of the reactivity of these species may be achieved by control of the length of the alkyl group, which through steric hindrance effects, may limit the reactivity of the metal center. Additionally, the number of bonds between the metal center and the ligand may also be controlled (or selected) to influence reactivity. In TEOS, the ethoxy groups are *unidentate* in nature, meaning that they are bonded via one oxygen to the Si metal center. Bidentate and tridentate (bonding of the alkyl group to the metal center through two and three oxygens, respectively) ligands have also been employed in CSD processing to limit the reactivity of the precursor.[35] Reaction chemistry is discussed further below, and additional information may be found in Ref. 1. Choice or manipulation (via reaction) of such characteristics leads to control of properties ranging from sensitivity toward hydrolysis to film thickness during deposition.

Metal carboxylates, $M(OOCR)_x$, may be considered as derivatives of carboxylic acids. A representative reaction for the synthesis of sodium acetate is shown below[36]:

$$CH_3COOH + NaOH \rightarrow CH_3COONa + H_2O \quad (3)$$

Carboxylate chain length may be tailored to improve solubility, which also impacts precursor reactivity.[23–25] Short chain carboxylates, such as acetates,[37,38] are generally less soluble in organic solvents than longer chain carboxylates that possess a higher organic fraction. Carboxylate groups may be unidentate or bidentate and may also be bridging, with the nature of the species being detectable using infrared spectroscopy.[39] The carboxylate nature and bonding environment are also known to impact precursor reactivity. A schematic illustrating the carboxylate structure is shown in Fig. 2.2.[8] A variety of oxo-carboxylate compounds have also been synthesized for CSD purposes.

Metal β-diketonates, $M(CH_3COCHCOCH_3)_x$, are typically monomeric species. A schematic of the structure is shown in Fig. 2.2.[8] Of the three classes of compounds, β-diketonates typically demonstrate high solubility, but less tendency toward reaction than either alkoxides or carboxylates. The monomeric nature of the species also represents a significant difference compared with the other classes of starting reagents, which typically demonstrate polymerization during solution synthesis. This oligomerization can be beneficial during film formation.

Although the above classes of raw materials are the most widely used, depending on the precursor chemical approach and desire for a specific level of reactivity, other families of compounds are also used. These include nitrates, citrates, oxo-alkoxides, mixed ligand reagents, such as alkoxy-carboxylates, and alkoxy-β-diketonates. It is also worth noting that, during the past several years, the variety of commercially available metal organic reagents has greatly increased, giving the film fabricator a choice of options with regard to, among other properties, precursor reactivity and organic burnout characteristics. Another option is in-house synthesis of precursor compounds. For this option, numerous investigators have pursued the synthesis of mixed metal compounds of the desired stoichiometry to ensure material composition and reduce crystallization temperatures.[40–42] A review of the literature for the material system of interest will likely provide adequate information for the identification of suitable precursors and the development of an acceptable processing route. Additional discussion of precursor characteristics may be found below.

2.3.3 Solvents

The primary function of the solvent is to provide a medium for dissolution (and, in many instances, reaction) of the starting reagents. The importance of the solvent in organic chemistry is discussed in detail by Morrison and Boyd.[36] Reaction rate (over several orders of magnitude), whether or not a reaction will occur, and reaction pathway are all defined by the solvent, as are the characteristics of the solvated reactant, product, and transition-state species. Solvation characteristics, that is, the surrounding or clustering of the solvent around the particles or molecules of the starting reagent,[36,43] can influence behavior, which ranges from dissolution, to reaction occurrence and mechanism, to striation formation during film deposition.[44,45] This later characteristic is influenced through solvent quality effects, which describe the tendency of the solvent to prefer coordination to the molecular starting reagent species versus itself. Such traits lead to either uniform solutions, or potentially a tendency toward phase separation during the drying phase of film formation.[43–45]

The two key characteristics of the solvent that may be controlled to optimize solubility, as well as subsequent processing characteristics, are the polarity of the solvent and whether the solvent is protic or aprotic.[1,36] Protic solvents (e.g., H_2O and ROH) contain a hydrogen attached to either oxygen or nitrogen and are sufficiently acidic to form hydrogen bonds[36]; aprotic solvents do not form hydrogen bonds. The polarity of the solvent may be controlled, and the organic moieties on the starting reagents may be selected, to optimize solubility, reaction pathway, and so on. For dissolution, the general rule of thumb is "like dissolves like," or stated more specifically, polar solvents dissolve polar molecules and nonpolar solvents dissolve nonpolar molecules. The types of bonds that are broken and formed during dissolution obviously play a key role in defining reaction chemistry, but they are beyond the scope of this chapter.[36]

In polymeric CSD processes, the most commonly used solvents are alcohols. Frequently, for sol-gel processes, the solvent plays a role that goes beyond simply dissolution of the starting reagents. Alcohol exchange reactions,[35] as noted above, may occur, and solvent characteristics are often controlled to dictate hydrolysis sensitivity and reaction pathway. As will be discussed, these synthetic conditions may be controlled to such an extent that control of the nature of the resulting precursor molecules in solution may be exerted.[9,46–48] This may have very desirable benefits in controlling the subsequent processing behavior of the film. For CSD routes in which reactions between the solvent and the starting reagents do not occur, the solvent acts simply as a medium for dissolution of the starting reagents and other additives. Clustering of the solvent molecules around the starting reagent species will still occur,[36,43] assuming solvent quality issues have been addressed, but controlled modification of the nature of the precursor species is usually not achieved, nor is this a goal, in this case.

In addition to alcohols, which are most often used in traditional sol-gel and chelate routes, other solvents that have been used include xylene, toluene, dimethyl formamide (DMF), pyridine, tetrahydrofuran (THF), water, and dissolution agents, such as acetic acid. Brinker and Scherer have summarized solvent properties and have discussed the role of the solvent in CSD processing.[1] When processing routes with different types of starting reagents (for example, carboxylate and alkoxide reagents) are used, because of the different polarity and ionicity/covalency of reagents, careful selection of the solvent may be required. In such instances, a solvent with mixed polar and nonpolar character, such as 2-methoxyethanol ($CH_3OCH_2CH_2OH$), may be used.[1] Alternatively, as will be seen for chelate processes, mixed solvent systems (e.g., alcohol, acetic acid and water) are frequently used to ensure the desired dissolution and reaction behavior.[16,30–33,37,38]

Because of their beneficial solubilizing characteristics and role in promoting desired reactions, solvents such as 2-methoxyethanol have become widely employed in CSD processes. It should be noted, however, that such compounds frequently demonstrate adverse effects. For example, 2-methoxyethanol is a known teratogen. The reader is encouraged to always review the Material Safety Data Sheet (MSDS) for the selected solvent and starting reagents. These sheets are typically provided by the commercial vendor at the time of purchase.

In addition to polarity and protic/aprotic nature, other solvent characteristics, including molecular weight, boiling point, density, viscosity, and dipole moment, should be considered in the development of the solution processing approach. The influence of such characteristics is wide ranging, spanning solution reaction effects to film deposition characteristics and drying behavior.[49] Some of these effects have been investigated in greater detail than others, and often a more empirical approach is followed for solvent selection and specification of reaction conditions.

2.4 PRECURSOR SOLUTION PREPARATION AND CHARACTERISTICS

2.4.1 Background

Reagent selection, solvent, and choice of solution reaction conditions serve to define the nature of the solution precursor species that are formed and that will be used for film deposition. The characteristics of these solution species serve to define film processing characteristics, including aspects such as cracking tendency,[50] organic burn-out temperature,[48] and for crystalline films, crystallization temperature.[49,51]

The framework around which the cause–effect relationships in CSD have been studied is generally referred to as "structural evolution." This name has been used, in part, because structural rearrangement at various length scales typically occurs during the transitions from solution species to the final desired film, as outlined in Fig. 2.1.[16]

One key in defining structural evolution, and thus, the resulting characteristics of the final film, is the chemical reactions that occur (intended or otherwise) during solution preparation. These reactions have been investigated in great detail for a variety of material systems, and the basic reaction chemistry for the more common processes is well understood. This chemistry lends itself to categorization into three divisions: sol-gel, chelate, and metallo-organic decomposition (MOD) processes. These processes and their associated reaction chemistries are discussed below, prior to discussion of the role of solution species nature on structural evolution.

2.4.2 Sol-Gel Processes

The sol-gel process is typically based on the use of alkoxide starting reagents and alcohol solvents.[1,3,4,35,52–56] Depending on the material system and starting reagent reactivity, relatively common alcohols, such as methanol and ethanol, may be used. However, for more reactive alkoxide starting reagents, less common alcohols such as 2-methoxyethanol[3,4,35] and 1–3 propanediol[49,57–62] have also found widespread utilization. Considerations in selection of reagent(s) and solvent are to exert control over the hydrolysis and condensation reactions that lead to oligomerization, i.e., the development of short polymeric species.

The reactions in the sol-gel process that lead to the formation of oligomeric species with M—O—M bonds are as follows:

Hydrolysis

$$M(OR)_x + H_2O \rightarrow M(OR)_{x-1}(OH) + ROH \quad (4)$$

Condensation (water elimination)

$$2M(OR)_{x-1}(OH) \rightarrow M_2O(OR)_{2x-2} + H_2O \quad (5)$$

Condensation (alcohol elimination)

$$2M(OR)_{x-1}(OH) \rightarrow M_2O(OR)_{2x-3}(OH) + ROH \tag{6}$$

These reactions have been studied in detail for materials such as silica, and understanding of reaction mechanisms, as well as of the role of the precursor and catalyst (acid or base), has been well documented.[63,64] Similar studies have been carried out in other material systems, most notably, lead zirconate titanate [$Pb(Zr,Ti)O_3$; PZT].[52,65–68] For multicomponent (mixed-metal) systems such as those noted, prehydrolysis of less reactive alkoxides is sometimes employed to improve solution compositional uniformity. Other synthetic strategies to achieve molecular level mixing of reagents have also been employed. Here, synthesis of mixed-metal alkoxides has been a focus of investigators.[40–42] A key point is to restrict the amount of water and to control how it is added to form solubalizable precursor species, rather than to induce precipitation.[1,52,69,70]

The desire to control the characteristics of the solution precursor species in sol-gel processes originates from the influence of the precursor on subsequent film processing characteristics. Compared with other solution-based processes, such as chelate and metallo-organic decomposition, sol-gel processes offer the greatest level of control of precursor nature. Some precursor properties that might be controlled include molecular shape, extent of oligomerization, composition, and organic fraction. These properties are controlled by varying factors, including the alkoxide reagent, solvent, alkoxide concentration, water concentration, method used for water addition, catalyst, and solution preparation/reaction conditions.[1,52,70,71]

Manipulation of the factors noted above may be used to control the characteristics of the hydrolysis and condensation reactions, which will dictate precursor nature. These reactions have been studied extensively for silica,[1,63] and some similarity in chemical behavior to electronic material systems (PZT, etc.) has been noted because of the alkoxide reagents used in the synthesis of these material systems.[52,70,71]

For silica, much of the work in this area was carried out in the 1980s and the reader is referred to the series of proceedings volumes from the Materials Research Society entitled "Better Ceramics through Chemistry."[63] To illustrate typical precursor structures, and the role of reaction chemistry on precursor structure, Fig. 2.3 presents capillary gas chromatography results for the hydrolysis of tetramethoxysilane [TMOS: $Si(OCH_3)_4$].[72] Typical polymeric species formed (Fig. 2.3a) include linear and cyclic structures.

Because of the role of precursor structure on film processing behavior (consolidation, densification, crystallization behavior), the reaction pathways are typically biased through the use of the catalyst, which is simply an acid or a base. This steers the reaction toward an electrophilic or nucleophilic attack of the M—OR bond.[1,63] Hydrolysis sensitivity of singly or multiply hydrolyzed silicon alkoxides is also influenced by the catalyst, which contributes to the observed variations in oligomer length and structure. Figure 2.3b illustrates

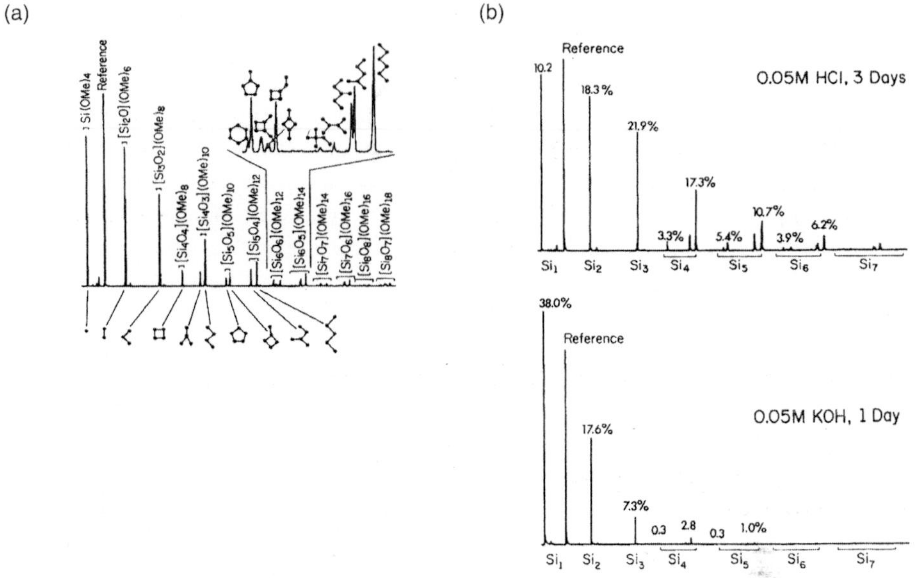

Figure 2.3. Capillary gas chromatogram of $Si(OCH_3)_4$ (3M): (a) after hydrolysis/condensation with H_2O (1.8M) and HCl (0.05M) showing assignments of molecular formulas and structures; and (b) with H_2O (1.5M) and catalysis conditions shown. For (a), linear and cyclical structures are indicated along the x-axis. All plots illustrate relative concentrations of species (y-axis) as a function of GC column retention time (x-axis). Higher mass species (e.g., hexamers (Si_6) and pentamers (Si_5)) demonstrate longer retention times. [Reprinted from Ref. 72, with permission.]

the role of the catalyst on the polymer distribution. The subscript shown indicates the number of Si atoms in the oligomer (e.g., Si_1—monomeric species or Si_2—dimer or two Si atoms, etc.). Acidic catalysis conditions (shown at the top of Fig. 2.3b) result in oligomeric species with more Si atoms (longer chains or cyclical structures).

Generally speaking, the susceptibility of the M–OR bond to hydrolysis is dependent on the polarity of the bond.[1,34,63] Bond polarity results when the sharing of the electron pair involved in the covalent bond is unequal. Stated otherwise, the electron cloud is denser around one nucleus and the resulting bond has a positive and a negative pole. Generally, as bond polarity increases, so does susceptibility to hydrolysis. This has several effects on reaction behavior, and unless properly controlled, the processing window (e.g., factors such as water/alkoxide ratio and precursor concentration) to form soluble oligomers and control gelation becomes smaller, whereas the tendency for precipitate formation is increased.[52,69–71] Reagents such as TEOS have relatively nonpolar bonds and demonstrate low sensitivity to hydrolysis. Materials like TIPT have bonds that exhibit much greater polarity and are highly susceptible to hydrolysis. Because of these differing characteristics, TEOS precursor solutions are frequently prepared under ambient laboratory conditions, whereas the initial phases of the preparation of TIPT-based solutions are nearly always carried out in an inert atmosphere box under low pO_2 and pH_2O conditions.

To address the hydrolysis sensitivity issue for materials with highly polar bonds, one of two strategies is typically employed: (1) alcohol exchange or (2) chelation (reaction) of the starting alkoxide with a ligand that is less susceptible to attack by water. Chelation processes are discussed in detail in the next section.

The alcohol exchange reaction was shown above in equation (2). The reactive alkoxy group (OR) is replaced by an alkoxy group that has less hydrolysis sensitivity (OR′). A representative example here is the use of reagents such as zirconium *n*-propoxide and titanium *i*-propoxide, both of which possess polar bonds, for the production of lead zirconate titanate films. Commonly in these processes, R′OH is 2-methoxyethanol ($CH_3OCH_2CH_2OH$), which is generally present as a bidentate ligand.[35]

As noted, sol-gel syntheses for materials based on water-sensitive reagents, such as titanium isoproproxide, must be carried out under low pH_2O conditions, which are typically achieved using either an inert atmosphere box (glove box) or under laboratory conditions using a Schlenk-line.[73] The Schlenk-line, shown in Fig. 2.4,[71] consists of a dual manifold arrangement with one line

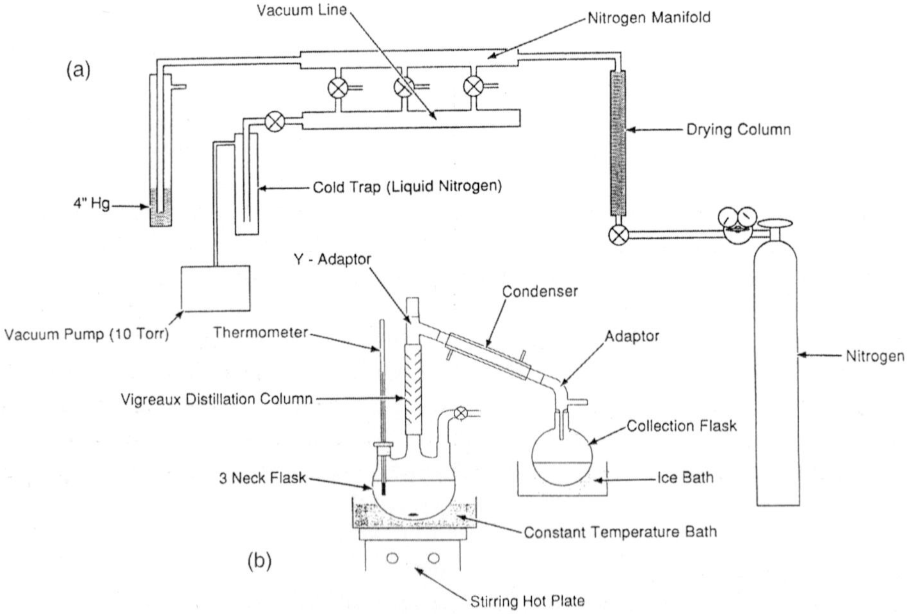

Figure 2.4. (a) Typical Schlenk-line and (b) distillation/reflux apparatus used in sol-gel synthesis of film precursor solutions. [Reprinted from Ref. 71, with permission.]

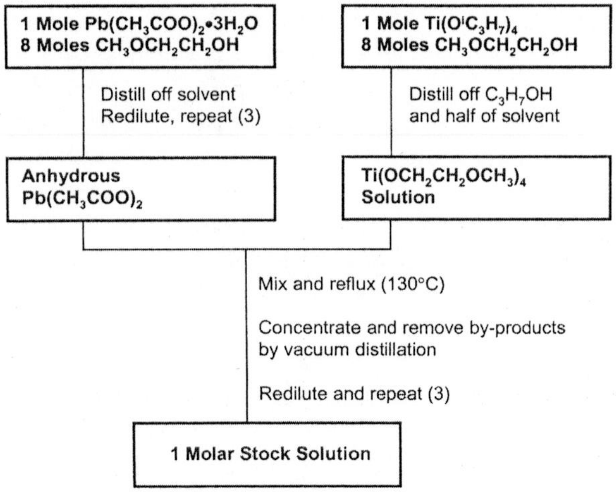

Figure 2.5. Flow diagram for typical sol-gel solution synthesis procedure. [Reprinted from Ref. 71, with permission.]

providing vacuum and the second line providing inert gas, typically N_2. Vacuum tubing may be used to connect the Schlenk-line to reaction vessels for refluxing and distillation under inert atmosphere. Typically, the reagents are transferred into the reaction vessels in the glove box and then are carefully connected to the Schlenk-line to prohibit contamination with air.

A typical sol-gel process for the synthesis of a lead titanate precursor solution is shown in Fig. 2.5.[71] Compared with chelate and MOD processes, sol-gel processes are generally more complicated from the perspective of the chemistry involved. This complexity, however, provides unique opportunities to control precursor structure compared with other CSD processes, which affords numerous other advantages. In addition, despite the more complex chemistry, many studies of reaction chemistry have now been carried out.[35,52,63,70,71,74–76] These studies provide significant guidance to the researcher looking to develop a process for a broad range of electronic ceramic thin films, while maintaining the ability to use control of a precursor structure to advantage in thin film processing (for discussion of precursor structure effects on processing, see Section 2.6.3.2). The most extensively investigated electronic materials have been lead zirconate titanate, barium titanate, strontium titanate, and strontium bismuth tantalate. Numerous relaxor material systems have also been synthesized and characterized, including lead magnesium niobate and lead zinc niobate.

2.4.3 Chelate Processes

A second category among widely used solution synthesis approaches is referred to as the "hybrid" or "chelate" process.[16,23–30,37,38,46–48] These processes are usually characterized by the use of multiple metal-organic precursors, such as carboxylate and alkoxide compounds, and rely on molecular modification[37,38,46–48] of the alkoxide compounds through reactions with other reagents that alter the chemical nature and structure of the precursor. The general goal of these processes is to reduce the hydrolysis sensitivity of the precursor species, reducing the tendency to excessive polymerization or precipitation.

Most often in these processes, compounds such as acetic acid, acetylacetone (acac, 2,4-pentanedione), or amine compounds are employed, since these compounds readily react with alkoxides.[46–48,77] A typical reaction involves the formation of new chemical species that (hopefully) possess physical and chemical characteristics that are more attractive in terms of solution stability and film formation behavior. An example reaction between a metal alkoxide and acetic acid is shown below:

$$M(OR)_n + xCH_3COOH \rightarrow M(OR)_{n-x}(OOCCH_3)_x + xROH \qquad (7)$$

In this instance, the resulting species possess both alkoxide and acetate ligands. Because the acetate species may be (and typically are) bidentate in nature,[77] as well as sterically larger than short chain alkoxy groups, they are less susceptible to attack by water, and the hydrolysis and condensation reactions are slowed. Because of the reduced sensitivity toward hydrolysis, the nature of the precursor species resulting from these processes is desirable for film formation (to be discussed in Sections 2.5 and 2.6).

Because of these precursor modification reactions, the process chemistry of chelate processes is as complex, or more so, than that involved in sol-gel processes.[78] However, it is typical for chelate processes that some control of process chemistry is sacrificed in return for more expedient solution preparation. For example, the hour-long (or longer) reflux processes that have been historically used in 2-methoxyethanol based sol-gel processing of ferroelectric films are not used. Rather, the entire solution preparation procedure is generally completed within one hour, with only the initial phase of the procedure being carried out under dry box and inert atmosphere conditions. Once the chelation reaction(s) has occurred, the hydrolysis sensitivity of the precursor solution is reduced to the point where the remaining process chemistry may be carried out under ambient conditions.[46]

A typical flowchart for a chelate process is shown in Fig. 2.6.[46] In this process, titanium isoproproxide is first added to zirconium butoxide. Acetic acid is then added to the precursor mixture for suppression of hydrolysis. Finally, lead acetate is then added to the solution, followed by the addition of alcohol and water for control of solution viscosity, solution stability, and oligomer formation.[46]

As for sol-gel processes, the product of the reactions in chelate processes is small oligomeric species that generate either chemical or physical gelation during film preparation. Although the nature of the species has not been

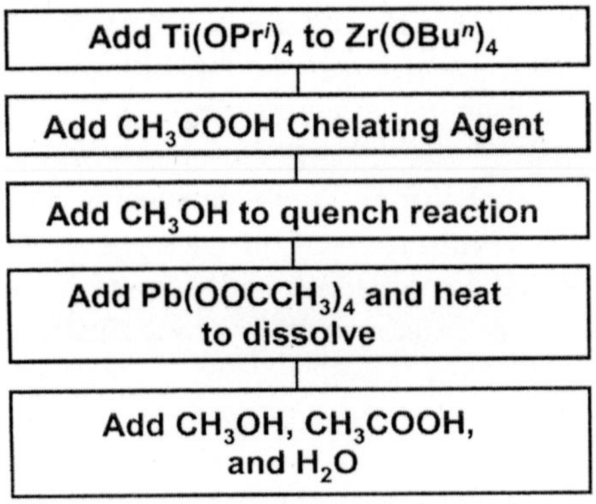

Figure 2.6. Flow diagram for typical chelate solution synthesis procedure. [Reprinted with permission from Ref. 46. Copyright 1993 American Chemical Society.]

investigated as thoroughly as in silica sol-gel processes, several studies have been completed. Some of these studies have focused on the growth of crystalline precursors of known stoichiometry and structure,[79–84] whereas others have addressed the general nature of the species and the reaction chemistry involved in the process.[46] Key reactions during solution preparation and aging were determined to be chelation, esterification, hydrolysis, and condensation. A variety of analytical and spectroscopic methods (discussed below) have been used in the characterization of reaction chemistry, and this work has effectively documented the reaction complexity associated with this type of acetate chemistry. Although the solution preparation is straightforward, the reaction chemistry is complex and solution composition varies with solution age.[78] These issues may complicate film deposition depending on the miscibility of the various phases present in the solution.

2.4.4 MOD Solution Synthesis

The third general classification of solution synthesis approaches used for inorganic electronic thin film fabrication is referred to as metallo-organic decomposition, or MOD for short.[23–29,37,38,85] Historically long-chain carboxylate compounds, such as lead 2-ethylhexanoate, zirconium neodecanoate, and titanium di-methoxy di-neodecanoate have been used.[23–29,85] Both commercially available precursors and in-house synthesized starting reagents have been used.

The approach typically consists of dissolving the metallo-organic compounds in a common solvent, typically xylene, in the desired stoichiometry. The long-chain length associated with the organic moieties of these compounds increases their dissolution tendency and concomitantly limits their hydrolysis tendency. Therefore, compared with the sol-gel and chelate processes, the nature of the precursor species in solution are believed to be little changed compared with the starting reagents and the solutions are essentially simple mixtures of the starting reagents. Process chemistry is, thus, significantly less complex than in the sol-gel and chelate routes, and the ability to manipulate the structure of the precursor species is diminished significantly. This could be viewed as a limitation of the process, although the fact that the chemistry is straightforward represents a significant advantage for an individual without an extensive chemistry background. This same characteristic also results in tremendous solution stability compared with solutions prepared by the chelate process. Solution synthesis is straightforward, and rapid compositional mapping of material systems is possible. Some material systems that have been prepared by this method include lead zirconate titanate, lead lanthanum zirconate titanate, barium zirconate titanate, and barium strontium zirconate titanate.[23–29,37,38,85]

2.4.5 Solution Preparation Summary

An overview of the precursors, process chemistry, and relative advantages and disadvantages of the three principal methods of inorganic electronic thin film preparation is shown in Table 2.1. Generally, sol-gel methods offer the greatest control over the nature of the solution precursor species, but they have involved

TABLE 2.1. Summary of the Principal Chemical Solution Deposition Methods

Method and Traits	Control (Incr.)	Simplicity (Incr.)
Sol-Gel	↑	↓
Metal alkoxides in alcohol solvent system		
H_2O and catalyst additions to promote polymer formation		
Control polymer structure by hydrolysis ratio (H_2O:alkoxide molar ratio) and catalyst type (acid or base)		
Typically forms true chemical gel during film formation		
Involved solution synthesis with extensive control of precursor properties		
Chelate		
Metal carboxylate, alkoxide, and/or β-diketonate precursors		
If alkoxides are employed, they usually reduce hydrolysis sensitivity through modifying/chelating agents (e.g., acetic acid and acetylacetonate)		
Metallo-Organic Decomposition (MOD)		
Typically, long-chain metal carboxylates and inert solvents, such as xylene		
Rapid solution synthesis with minimal reactivity between precursor compounds		

reaction chemistry. Solution preparation is typically time-consuming and carried out under inert atmosphere conditions, although control of the nature of the precursor species can be exercised in this method. Cation stoichiometry may also be controlled under certain conditions to yield a mixed-metal precursor. Chelate processes offer rapid solution preparation, but inherently, they have complex reaction chemistry. Control of the nature of the solution precursor species is generally less extensive than for sol-gel processes, although the general solution characteristics have also been found to be amenable to the preparation of thin films with acceptable electrical properties. Metal-organic decomposition methods typically have simpler chemistry but at the cost of limited control over precursor nature. The result of this aspect of the solution preparation process is that control of film properties must be gained exclusively through control of deposition and heat treatment conditions.

2.4.6 Other Processing Routes

The above methods represent the most commonly employed methods for inorganic electronic thin film preparation. A variety of other methods, including Pechini,[21] citrate,[86] nitrate,[23] and aqueous processes[87] have also been used. For a discussion of these methods, the reader is referred to Refs. 5 through 12, which highlight these methods for the preparation of various electronic ceramic materials.

2.5 FILM FORMATION BEHAVIOR

2.5.1 Background

Precursor structure and solution characteristics can have a significant impact on film formation behavior. In this section, film formation methods are first discussed, and then, in the subsequent section, the role of precursor structure on film formation and structural evolution into the desired crystalline state is reviewed.

After the precursor stock solution is prepared, various techniques can be used to coat the substrate, depending on the solution viscosity, required film thickness and coverage. The most common methods in the semiconductor industry are spin- and dip-coating. Other processes that are used for deposition include spray coating and stamping. A summary of the uses, limitations, and advantages of the various thin film deposition methods is reported in Table 2.2.

2.5.2 Spin Coating

Spin coating has been used to deposit ultrathin to relatively thick coatings on flat substrates for several decades. Various materials, including resin, epoxy,

TABLE 2.2. Summary of Film Deposition Methods

Technique	Thickness	Advantages	Limitations	Applications
Spin	10 nm–5 µm	Uniformity, reproducibility, excellent thickness control, low cost	Requires flat substrate, high material loss	Photoresists, dielectric layers, flat panel displays
Dip	20 nm–50 µm	Complex and irregular shapes can be coated, can be a continuous process, very simple	Thickness variations, voids, and pin holes	Insulating polymer layers, electronic parts, and assemblies
Spray	2 nm–20 µm	Fast and adaptable to complex shapes and sizes, conformal step coverage, high efficiency	Expensive, low-viscosity coating solution	Electrical insulation, circuit board housing
Stamping; microcontact printing	50 nm–1.5 µm	Direct development of patterned film	Lateral resolution	Waveguides, capacitors, diffraction gratings, lab-on-a-chip structures

polymers, and sol-gel stock solutions, have been successfully coated on metal, glass, ceramic, plastic, paper, and semiconductor substrates in different industries. This technique normally uses the material to be coated in its liquid form or dissolved in a liquid solvent. Typically the substrate to be coated is held in place using a motor-driven vacuum chuck, as shown in Fig. 2.7a,[8] and the coating solution is dispensed on the substrate either manually or by an automated robotic arm. The substrate is then accelerated to very high angular velocities (~300 to 10,000 rpm) during which the excess liquid is spun off from the substrate leaving a thin uniform coating. Thicknesses of less than 30 nm to a few microns per layer can be easily achieved.

The theory behind this coating technique involves the equilibrium between the centrifugal forces created by the rapid spinning and the viscous forces determined by the viscosity of the liquid.[1,88] The film thickness can be varied by controlling the spin speed and time, as well as the viscosity of the solution.

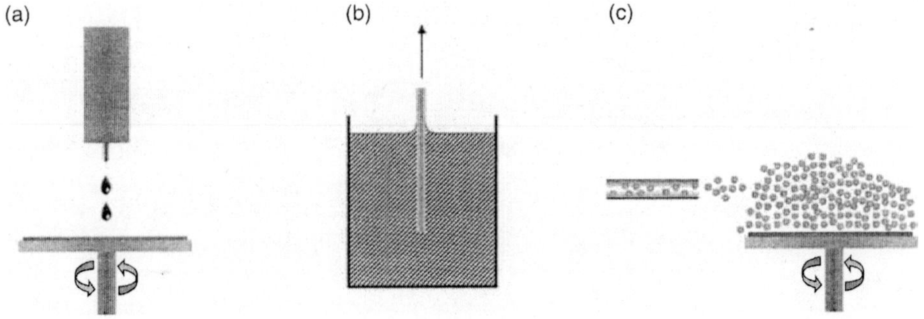

Figure 2.7. Schematic diagrams of (a) spin, (b) dip, and (c) spray coating. [Adapted with permission from Ref. 8: R. W. Schwartz et al., *C. R. Chemie*, **7**, 433 (2004).]

Empirically, it has been observed that the film thickness (t) is inversely proportional to the square root of the spin speed (ω: angular velocity).[88]

$$t \propto \frac{1}{\sqrt{\omega}} \tag{8}$$

Extensive work has been reported by Birnie and coworkers on the common striation and cellular-type defects created during spin coating and the means to avoid them by proper solvent selection.[44,45] The major advantages of spin coating are reproducibility, uniformity, simplicity, ease of integration, ability to use different substrate materials, and low cost. The main disadvantage of the method is that a smooth, flat substrate should be used. Step coverage can prove difficult, and complex structures cannot be coated using this technique.

2.5.3 Dip Coating

Dip coating is another low-cost solution deposition technique that is commonly used in the semiconductor industry, mainly to coat irregular and complex shapes. In this process (Fig. 2.7b),[8] the complex part or substrate is immersed into the coating solution and the film is formed by either removing the substrate vertically from the solution or draining the solution from the reservoir at a constant rate. Dip coating can be performed employing either continuous or batch processing but entails the same key stages: immersion, withdrawal, film formation, solvent evaporation, and continued drainage of the excess solution. The physics behind film formation in the dip-coating process involves a balance among the viscous drag, gravitational forces, and surface tension in the concavely shaped meniscus.[89,90] Various factors affecting the thickness of the film formed are reviewed by Brinker and Scherer.[1] According to Brinker and Hurd,[89] the thickness of the film formed is determined by the viscosity (η) and density (ρ) of the solution and by the withdrawal speed or rate (U):[89]

$$t = 0.94 \frac{(\eta U)^{2/3}}{\gamma^{1/6}(\rho g)^{1/2}} \tag{9}$$

where γ and g are the surface tension and acceleration caused by gravity, respectively.

Common defects that are observed in this coating technique are voids, pin holes, thickness variations, and wavy surfaces. Voids and pinholes are caused by air entrapment in the film, surface contamination, or dust particles.

2.5.4 Spray Coating

The spray-coating technique[91–93] is mainly used for its ability to coat nonplanar structures like steps, trenches, and stacks on semiconductor chips. In this process, very fine droplets are formed from the coating solution using atomizers or nebulizers as shown in Fig. 2.7c.[8] Compared with spin and dip coating, the solution viscosity has to be greatly reduced to facilitate the nebulization process to form fine droplets. These fine droplets are then carried into the coating chamber with a carrier gas and deposited on the substrate by gravity or with an electrostatic field, where the negatively charged droplets are attracted to the electrically grounded substrate. The advantages of electrostatic spray coating over traditional gravitational spray coating are high deposition rates and low material loss. Materials savings of 25% to 50% over traditional spraying techniques have been reported.[17] The step coverage and conformal nature of the films deposited on nonplanar surfaces by spray coating have also been studied. A thickness variation of <20% was observed on 400-nm trenches with 200-nm thin films.[93] The conformal coating quality is determined by the size of the droplet, which can be reduced by decreasing the viscosity of the solution and by increasing the atomizing pressure or using a venturi nozzle.

2.5.5 Stamping and Microcontact Printing

Stamping and microcontact printing processes involve transferring a pattern to a thin film using a mold or patterned stamp.[94–98] In microcontact printing (μCP), self-assembled monolayers (SAMs), such as alkanethiols and organosilanes that are 6–30 Å in thickness, are used in conjunction with a stamp to develop a patterned film on a substrate.[94–97] The stamp is typically prepared from polydimethylsiloxane (PDMS) and may be fabricated through typical photolithography procedures. After fabrication, the stamp is coated (inked) with a solution containing the alkanethiol or organosilane SAMs. After inking, the stamp is brought into contact with the substrate, which may be flat or

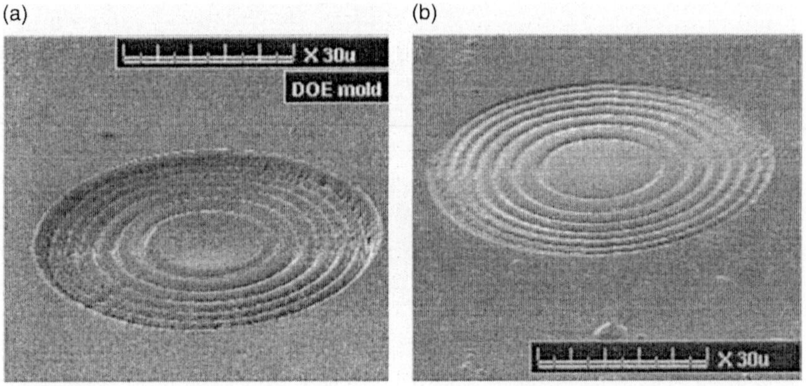

Figure 2.8. Scanning electron microscopy (SEM) photomicrographs of (a) a micro-diffractive optical element mold fabricated by a focused ion beam (FIB) and (b) the transferred optical element on a sol-gel film. [Reprinted with permission from Ref. 98.]

curved, to transfer the SAM pattern. The pattern is then overcoated with the desired material by a technique such as spin casting, and during heat treatment or via other processes, those sections of the coated film in contact with the SAMs may be easily removed, leaving a patterned thin film.

Although μCP has been widely investigated, other stamping processes to create patterned films have also been developed. An example of a typical stamping process is reported by Fu, and Bryan where a diffractive structure is transferred from a designed pattern to a replica using a hybrid sol-gel film.[98] The pattern (master), shown in Fig. 2.8a,[98] was created in a silicon substrate by focused ion beam milling (FIBM). A 1.5-μm-thick organosilicate–titanate layer was then spin coated at 1000 rpm onto a fused silica substrate by a typical sol-gel processing route. After prebaking the film at 50 °C to remove solvent, the pattern from the mold was then be transferred to the film by stamping. The resultant structure is shown in Fig. 2.8b. More details on these processes, and additional applications, are summarized in Table 2.2.[94–98]

2.6 STRUCTURAL EVOLUTION: FILM FORMATION, DENSIFICATION, AND CRYSTALLIZATION

2.6.1 Background

Over the course of the past 20–25 years, hundreds of studies on the fabrication of electronic thin films by chemical solution deposition methods have been reported. Because previous review articles have focused on summarizing the results of these investigations, they are not considered in detail here, but rather a more general approach to film formation and subsequent structural evolution during heat treatment is presented. The reader is referred to Refs. 5–12 and the references contained therein for more detailed information.

2.6.2 Film Formation

The precursor structure can significantly impact film formation characteristics. Affected properties during film formation may include sensitivity to hydrolysis, phase segregation, and tendencies toward dewetting and defect formation. Some solvent/solution properties that are believed to dictate film formation qualities are shown in Table 2.3. One of the key solvent characteristics is believed to be solvent quality, which when improperly controlled, may lead to phase segregation into polymer-rich and polymer-deficient regions.[43] This behavior can result in the formation of striations,[44,45] which are regions of different thicknesses in the films. Typically for sol-gel processed films, striations are readily apparent to the naked eye as film nonuniformities. Given the use of solution-derived films in capacitor and ferroelectric nonvolatile memory applications, such thickness variations may result in inconsistent capacitance or coercive field values, respectively.

Another key property noted in Table 2.3 is solvent vapor pressure. Significant work has been carried out to prepare films that range in thickness from tens of nanometers[99] to 10 μm.[59–62,100] Short-chain alcohols are commonly used as solvents for the former, whereas diol compounds, such as 1,3-propanediol[49,57–61] have been used to fabricate thicker layers. The short-chain alcohols have higher vapor pressures, which lead to capillary forces that can drive precursor species into greater proximity during film drying and thus lead to a chemical reaction between the precursor species. This behavior can lead to a film that possesses a higher crosslink density (between precursor species) and that may be more resistant to densification caused by the presence of M–O–M bonds. In contrast, low volatility solvents can inhibit reactivity between precursor species. This may lead to dewetting of the substrate.

Solution concentration and viscosity represent additional important factors in film processing. The greatest impact of these properties is on film thickness

TABLE 2.3. Solvent/Solution Characteristics that May Define Film Formation and Properties

Solvent/Solution Property	Film Property Affected
Solvent quality	Phase separation and tendency toward striation formation. Interactions with oligomers, solubility.
Solvent vapor pressure	Film formation, dewetting tendency, drying, and consolidation behavior.
Concentration and viscosity	Film thickness and uniformity.
Reaction tendencies and by-products	Film wetting behavior, film thickness, and microstructural evolution.

and the tendency toward cracking. Researchers have explored "critical thickness" limits above which film cracking will be observed. The critical thickness is described by[50]

$$h = (K_{IC}/\sigma\Omega_c(\Sigma))^2 \tag{10}$$

where h is the critical thickness, K_{IC} is the critical stress intensity factor, σ is the tensile stress in the film, and $\Omega_c(\Sigma)$ is a dimensionless quantity that is a function of the ratio of the Young's modulus of the film to that of the substrate. Tensile stresses are generated during removal of the solvent and organic moieties of the precursor species. As tensile stress and the Young's modulus of the film increase, the critical thickness decreases. This effect has led to a variety of multilayering approaches to achieve thicker films and to the use of lower vapor pressure solvents.[49,57–61]

2.6.3 Densification and Crystallization

2.6.3.1 General Aspects and Heat Treatment Effects. After film formation, for most of the films aimed at electronic applications (other than amorphous oxides, such as SiO_2), the films are subjected to a heat treatment process for removal of residual organic species (entrapped solvent as well as the organic constituents associated with the precursor species), densification (elimination of residual porosity and structural free volume in the film), and crystallization.

From a practical perspective, the organic removal step is typically carried out at temperatures of between 200 °C and 400 °C, and frequently a hot plate is used. The hot plate may be a standalone piece of equipment, as might be encountered in a small research setting or, in production environments, may be part of a spin-track system. Densification and crystallization have been carried out using box or tube furnaces, although rapid thermal annealing has also been used in many instances.[101–103] Control of ramp rate can be used to impact densification behavior and crystallization temperature. Heat treatment schedules that use a single thermal process for all three processing aspects have been used, as have schedules that use separate pyrolysis and densification/crystallization steps. The details of these different approaches have been discussed previously.[5–12]

Typical physical processes that occur after film deposition and associated temperature ranges are shown in Fig. 2.9.[1,51] As solvent is removed from the film (which may begin during the spin-off stage of the deposition process and continue during drying, and then during low-temperature heat treatment), fine-scale (nm size) porosity may be created. At the interface with the solvent remaining in the pores and pore channels, significant pressures can develop.[1] These pressures contribute to capillary contraction, which can be a substantial driving force for the structural collapse of the film, which leads to densification. As the film collapses, reactive groups on different precursor molecules

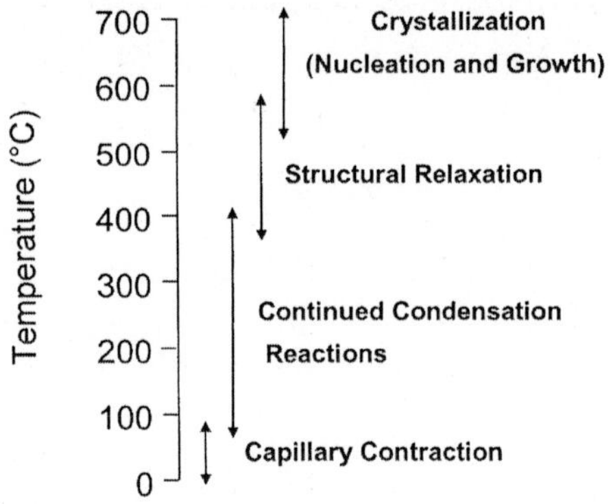

Figure 2.9. Approximate temperature ranges for densification processes contributing to structural evolution (non-isothermal processing with moderate ramp rate).

within the film come into closer proximity, which causes additional condensation reactions. This densification mechanism is accelerated as temperature is increased and organic groups from the precursor species are removed. Once the bulk of the organic materials have been removed from the film, other characteristics of the film will be altered, including the structural free volume associated with the inorganic amorphous state.[1,67,68,104,105] This "structural relaxation" leads to continued densification of the film. Finally, for crystalline materials such as high permittivity dielectrics and ferroelectric materials, crystallization occurs via a nucleation and growth process.[5–12,106–108]

The specific temperature ranges over which particular processes occur are highly dependent on the nature of the precursor species and on the heating rate employed. For the preparation of the highest density materials without residual porosity, it is advantageous to control the heat treatment process so that densification is completed prior to the onset of nucleation and growth.[51] One strategy employed to facilitate the achievement of high film densities is rapid thermal processing (RTP or rapid thermal annealing, RTA), which can facilitate densification prior to the onset of crystallization.[101–103] Representative results for zirconia thin films are shown in Fig. 2.10, where refractive index is used as a measure of film density. The refractive index values reported were obtained by ellipsometry, for films heated to the temperature indicated at three different ramp rates using a rapid thermal annealing furnace, and then quenched. For the films heated at the slowest ramp rate, pyrolysis of organic materials is facilitated at lower temperatures, which allows crystallization to occur prior to the completion of densification. Grain-to-grain contact and a change in densification mechanism from those reported in Fig. 2.9[51] to a solid-state diffusion mechanism limits densification and final film density; residual porosity is observed. For films heated at higher ramp rates, organic pyrolysis is delayed to higher temperatures, which impedes the onset of crystallization and facilitates the densification processes shown in Fig. 2.9.[51] Crystalline films of higher density may more readily be obtained under these conditions.

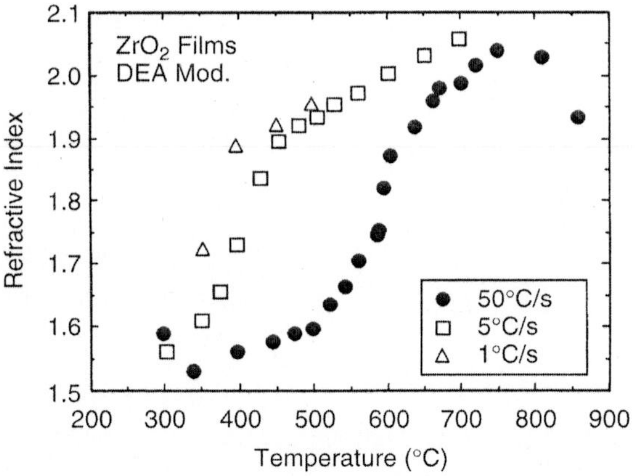

Figure 2.10. Variations in refractive indices of zirconia thin films prepared with the drying control agent, diethanolamine (DEA), and using three different heating ramp rates.

2.6.3.2 Precursor Structure Effects. The precursor structure can impact a broad range of properties, including crystallization temperature, the formation of intermediate phases during thermal treatment and film density, among other properties. Table 2.4 reports some of the key precursor properties that may affect densification and crystallization behavior, as well as the final film microstructure.

Numerous investigators have attempted to control the precursor structure and related solution chemistry effects with varying degrees of success, to influence subsequent processing behavior, such as crystallization temperature.[40–42,78,109,110] Particular attention has been given to precursor characteristics such as structural similarity to the desired product and the chemical homogeneity of the precursor species. For multicomponent films, this latter factor is believed to influence the interdiffusional distances associated with the formation of complex crystal structures, such as perovskite compounds. Synthetic approaches have been geared toward the preparation of multimetal species with cation stoichiometry identical to that of the desired crystalline phase.[40–42,83,84]

In some instances, this approach has proven successful, with comparatively low crystallization temperatures being observed. For example, Eichorst and Payne in the synthesis of $LiNbO_3$ noted crystallization temperatures of 400–500 °C for a mixed-metal alkoxide precursor.[111] In other instances, these attempts have proven less successful. Numerous attempts have been made to synthesize Pb-Zr and Pb-Ti precursors, each with the 1:1 cation stoichiometry of the desired $PbZrO_3$ and $PbTiO_3$ compounds.[83,84] Unfortunately, 1:1 stoichiometric ratio compounds have not always been obtained, with crystalline compounds of other stoichiometries precipitating from the solution, as illustrated in Fig. 2.11.[83] This figure shows the crystal structure of $PbTi_2[\mu(4)$–

TABLE 2.4. Precursor Properties that May Affect Film Characteristics

Precursor Property	Film Property Affected
Chemical homogeneity (M, C, H, O)	Crystallization temperature; interdiffusion distances.
Solids and organic content	Cracking tendency, crystallization temperature, *in situ* atmosphere, active nucleation sites.
Molecular size and reactivity	Physical characteristics (surface area, pore size, and pore volume), transformation driving force, consolidation behavior.
Molecular shape and other structural aspects	Oligomer alignment during deposition, thickness, and porosity. Structural similarity to crystalline phase can affect crystallization temperature.

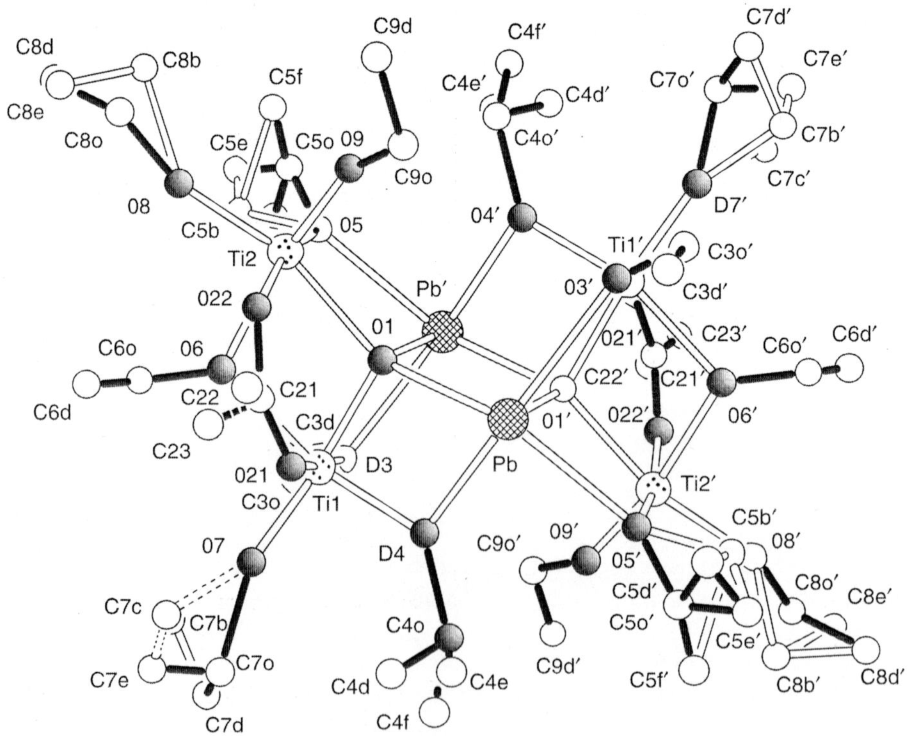

Figure 2.11. Crystal structure of $PbTi_2[\mu(4)O(OOCCH_3)(OCH_2CH_3)_7]_2$. [Reprinted with permission from Ref. 83. Copyright 1994 American Chemical Society.]

$O(OOCCH_3)(OCH_2CH_3)_7]_2$, a compound originally designed to serve as a $PbTiO_3$ precursor. The crystal structure shows two central lead atoms (Pb and Pb′), oxygen species that bridge the Pb and Ti species (including O1 and O1′; these bridge four cations, hence the μ(4) notation), and other terminal and bridging metal organic groups. Despite the starting solution used for precursor synthesis possessing the desired 1:1 Pb:Ti stoichiometry, this stoichiometry was not achieved for the crystalline precursor. In this case, thermodynamics effects dictated the formation of a 1:2 stoichiometry phase. Such compounds are often described as "thermodynamic sinks," and their stability can limit, or prevent, synthesis of precursors of the desired stoichiometry.[80–82] Thus, although the structural characterization of the precursor species is informative, the stoichiometry does not reflect that desired for synthesis of the ferroelectric material, $PbTiO_3$.

In the analysis of precursor effects, a variety of characterization methods have been used (e.g., see Table 2.5).[46,112,113] The most commonly used methods

TABLE 2.5. Characterization Methods Used in Chemical Solution Deposition Processing

Characterization Tool and Process State	Property Studied
Precursor and Solution	
Multinuclear NMR (C,H,O); Liquid and solid	Structure, bonding environment.
Elemental analysis	Chemical composition and stoichiometry.
Gas chromatography	Organic species present in solution. By-products and concentrations.
Capillary rheometry	Molecular weight determination and distribution.
Small angle X-ray analysis	Oligomer size and shape.
pH	Solution reactions.
As-Deposited and Heat-Treated Films	
Ellipsometry and profilometry	Thickness, refractive index, and consolidation behavior during drying and crystallization. Thickness uniformity.
Secondary ion mass spectrometry (SIMS), Rutherford backscattering (RBS), auger electron spectroscopy	Composition vs. depth and compositional uniformity.
Extended X-ray absorption fine structure (EXAFS)	Amorphous film structure and local order.
Surface acoustic wave (SAW)	Film surface area and porosity.
Film dissolution behavior	Cross-link density; gel nature: chemical vs. physical.
Thermal analysis	Weight loss and crystallization temperature.

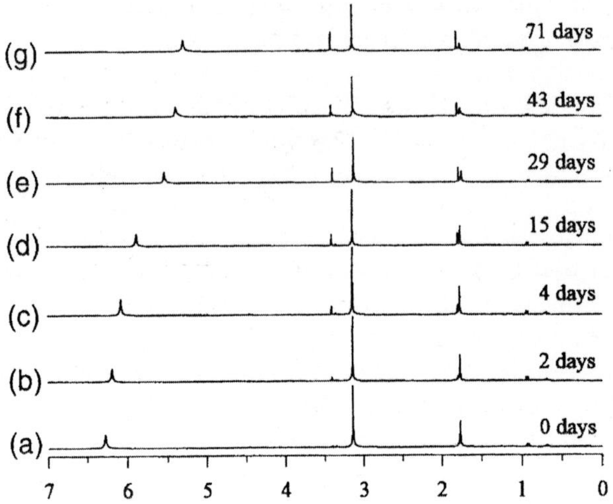

Figure 2.12. ^{1}H NMR spectra as a function of aging for a PZT precursor solution prepared by a chelate process: (a) 0, (b) 2, (c) 4, (d) 15, (e) 29, (f) 43, and (g) 71 days. [Reprinted with permission from Ref. 78.]

have been multi-nuclear magnetic resonance (NMR) spectroscopy, infrared spectroscopy (FTIR), thermal analysis, ellipsometry, x-ray diffraction (XRD), and electron microscopy.

Reaction chemistry and the general (organic) nature of the precursor species have been studied by both ^{1}H and ^{13}C NMR.[46] Particularly for chelate processes, this method has proven useful in identifying the reactions occurring during solution preparation and the organic ligands associated with the precursor species. Both liquid and solid-state NMR methods have been employed. Some key investigations in this area are summarized in Refs. 5–12, and Refs. 112 and 113 also contain detailed information on the analysis of structural evolution processes. The reader is referred to these papers for additional information.

Other authors have used NMR and pH measurements of the precursor solution during aging to understand variations in thin film thickness, microstructure, ferroelectric, and optical scattering properties.[48,78] ^{1}H NMR has proven to be particularly insightful, as shown in Fig. 2.12,[78] which demonstrates the use of this technique to study continued reactivity and by-product formation after solution preparation. Resonances at chemical shifts between 1.5 and 2.0 ppm and 3.0 and 3.5 ppm are related to the continued formation of ester compounds in the chelate solution. As these changes in solution composition occur, corresponding increases in optical scattering losses and film thickness were observed. Understanding the variations in the chemical composition of the solution can allow for other chemical additions to the solution to compensate for these changes.

The extent of control over film processing behavior that may be achieved through manipulation of precursor structure/nature is extensive, although many of the cause–effect relationships still need to be identified. This is because of the plethora of interrelationships inherent in the process; i.e., when one factor is changed, for example, organic content, numerous processing factors are simultaneously impacted. For the example cited, such dependent factors might include precursor interaction and organic entrapment during deposition, organic pyrolysis temperature, residual porosity after organic removal, densification pathways, and crystallization temperature, all of which may impact thin film microstructure. Some microstructural variations associated with differences in precursor nature may be eliminated during processing, depending on factors such as, for example, heating rate and hold time below the crystallization temperature.[105] But many variations persist, and as noted, the effects of solution chemistry on the film microstructure have been widely reported.

Because of the complexities involved in understanding cause–effect relationships, an alternative approach to control the thin film microstructure has been pursued by some investigators—the use of statistically designed experiments to identify key processing parameters.[114,115] In these approaches, as illustrated in Table 2.6 for a Plackett–Burman screening study,[114] limiting values for various experimental parameters are chosen. Films are then prepared from solutions synthesized under these conditions, and resulting film

TABLE 2.6. Typical Approach and Results for Statistical Design of Experiments Approach

Variable	Screening Values
Ti/Zr reaction time	0 or 15 minutes
Acetic acid amount	0.25 or 2.0x standard
Acetic acid reaction time	0 or 15 minutes
Solution Concentration	*0.4 or 0.8 M*
Pb dissolution temperature	75 or 100 °C
Refluxing	Yes or No
Additional Acetic Acid Added Post-Dissolution	*Yes or No*
Water volume	*0.5 or 5.0 ml*

	Acetic Acid (ml)	Sol'n Conc. (M)	Water (ml)	P_r* ($\mu C/cm^2$)	E_c* (kV/cm)	Microstruc. Quality	Hyst. Loop Quality
Nominal	10	0.40	2	27.0	43.0	3	40.0
Sweet spot 1	0	0.5	5	26.9	23.9	4	105.2
Sweet spot 2A	10	0.56	10	25.0	33.2	4	50.0
Sweet spot 2B	5	0.56	10	29.6	33.0	3	41.4

*P_r: remanent polarization; E_c: coercive field. [Reprinted with permission from Ref. 114.]

properties are studied and compared with the desired values. These values may be readily measured quantities, such as the coercive field of a ferroelectric film or the dielectric constant of a film intended for capacitor applications. Alternatively, they may be more subjective measures, such as microstructural quality, which may be quantified by the assignment of a numerical scale. As shown by the italicized solution preparation parameters in Table 2.6, the key processing parameters can be determined. These significant factors may then be subsequently analyzed by a variety of approaches. In the present case, a uniform shell design (see Figure 9 in Ref. 114) was used to determine "sweet spots" in processing space that produce films with optimum properties and that may later be used routinely for solution synthesis and film fabrication.

These statistical methods, although providing less fundamental insight into the cause–effect relationships associated with film processing, can prove powerful. The lower half of Table 2.6, which is reproduced from the work of Lockwood et al.,[114] clearly demonstrates the utility of the method. The coercive field was reduced at the sweet spot condition by 40% compared with the value obtained under the standard/original processing conditions. Given the large number of variables inherent in the process, it is perhaps surprising that the basic statistical design of experiments approach has not been employed more widely.

Researchers who have focused more on understanding cause–effect relationships in solution processing have given attention to film drying and pyrolysis behavior, densification processes, and nucleation and growth into the desired crystalline state. Both thermodynamic and kinetic factors associated with the phase transformation from the amorphous state to the crystalline state have been considered.[116–119] Control of these factors can lead to improvements in the ability to influence the microstructure. It is noted that in the previous sentence, "influence" has been carefully chosen, since the ability to manipulate the factors that govern the nature of the phase transformation to the extent that full control of the microstructure is possible remains to be demonstrated. However, trends in characteristics such as film orientation and columnar versus uniaxial grains have certainly already been achieved.[120]

A key area of focus for understanding the thermodynamic aspects of the phase transformation has been on consideration of thermodynamic driving forces associated with the phase transformation from the amorphous to the crystalline state.[106,118,119] As sol-gel processes were developed in the late 1970s to late 1980s, there was significant attention given to the nature of chemically derived amorphous materials and whether they were truly different from amorphous materials prepared from oxide precursors.[1,63,105] Distinguishing characteristics of chemically derived materials, which may be reduced or eliminated compared with the oxide-derived glass based on heat treatment conditions, include residual hydroxyl content, surface area associated with nanoporosity in the material, and skeletal density.[1,63,67,71,105] The role of these factors as a thermodynamic driving force are shown in Fig. 2.13.[64,121] This figure suggests that the driving force associated with the nucleation of the crystalline

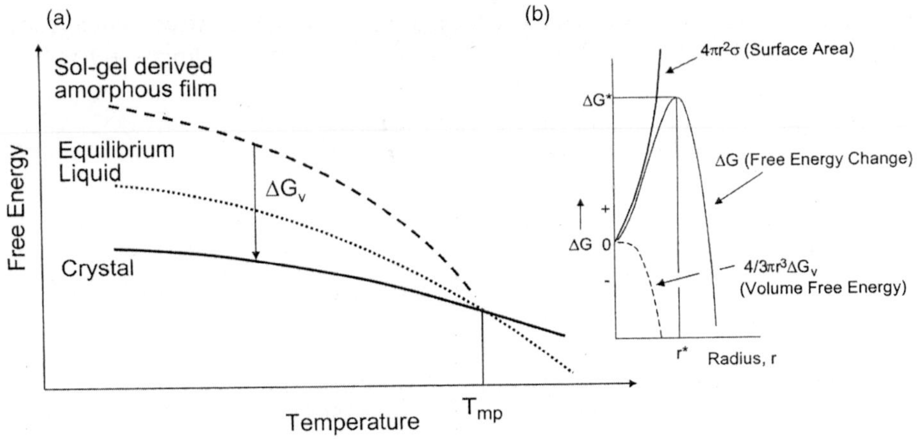

Figure 2.13. (a) Thermodynamic driving force (ΔG_v) associated with the phase transformation from the amorphous to the crystalline state and (b) classic picture of the energy barrier associated with nucleation resulting from the balance of the surface energy and volume free energy terms. [(a) Adapted from Ref. 64: C. J. Brinker and G. W. Scherer, Relationships between the sol-to-gel and gel-to-glass conversions, In *Ultrastructure Processing of Ceramics, Glasses, and Composites*, edited by L. L. Hench, D. R. Ulrich, pp. 43–59, 1984. Reprinted with permission of John Wiley & Sons, Inc.]

phase is dependent on both the crystallization temperature and the free energy of the amorphous state.

Using this thermodynamic picture, classic nucleation and growth theory was used to describe the phase transformation that occurs in these materials, despite the relatively unique synthesis method that is employed. The governing equation for homogeneous nucleation that describes the change in free energy associated with the formation of a spherical crystalline nucleus in an amorphous host is as follows:

$$\Delta G = -\frac{4}{3}\pi r^3 \Delta G_v + 4\pi r^2 \sigma \qquad (11)$$

where ΔG is the change in free energy, r is the radius of a spherical nucleus, ΔG_v is the driving force for the transformation, and σ is the surface energy. Taking the derivative of Eq. 11 with respect to radius allows for calculation of the energy barrier ΔG^* associated with nucleation:

$$\Delta G^* = \frac{16\pi\sigma^3}{3(\Delta G_v)^2} \qquad (12)$$

Surface energetics and thermodynamic driving force thus define the energy barrier for nucleation as shown in Fig. 2.13b. For heterogeneous nucleation,

Eq. 12 is modified by a contact angle term that is dependent on the surface energies of the interfaces among the substrate, crystal, and amorphous phases:

$$\Delta G^*_{hetero} = \Delta G^* \cdot f(\theta) \tag{13}$$

where

$$f(\theta) = \frac{2 - 3\cos\theta + \cos^3\theta}{4} \tag{14}$$

and θ is the contact angle measured between crystal and the solid surface on which nucleation occurs. Energetically, heterogeneous nucleation will always be preferred, confirming the importance of the substrate in the phase transformation of solution deposited films.

Brinker and Scherer[1,64] have carried out a detailed analysis of the impact that the physical factors noted above can have on the free energy of the amorphous material. This analysis leads to the ability to predict the driving force associated with the phase transformation into the crystalline state. A typical approach to this analysis is shown below for estimating the contribution of residual hydroxyl species.

If the chemical synthesis of SiO_2 is chosen for consideration, the role of hydroxyl species on free energy may be estimated from

$$Si(OH)_4 \rightarrow SiO_2 + 2H_2O; \quad \Delta G = -14.9\,kJ/mol$$

Therefore, as more hydroxyls are retained in the film, the free energy of the film is increased. Assuming that condensation is not complete (frequently the case) and that 0.33 to 1.48 OH species per Si are retained, the OH contribution to free energy may be estimated at 1.2–5.5 kJ/mol. Using reported values of surface area and skeletal density for solution-derived thin films leads to values of ~9–30 kJ/mol and ~6 kJ/mol for these contributions to the free energy of the amorphous material, respectively. Considering that the typical driving force for the transformation (ΔG_v) is in the range of 50–80 kJ/mol,[15] it may be seen that the physical characteristics of the amorphous state can make substantial contributions to the driving force for the transformation. Provided the variations in physical characteristics persist to the crystallization temperature, this can, in turn, readily impact the preferred nucleation sites (interface, surface, and/or bulk), which may thus dictate thin-film microstructure. Inadequate investigations have been carried out to characterize the role of these properties on the thin film microstructure.

In addition to the role of the physical characteristics of the amorphous state on nucleation and growth, as indicated by Eq. 11, surface energies also play a role. The role of these properties on film orientation and microstruc-

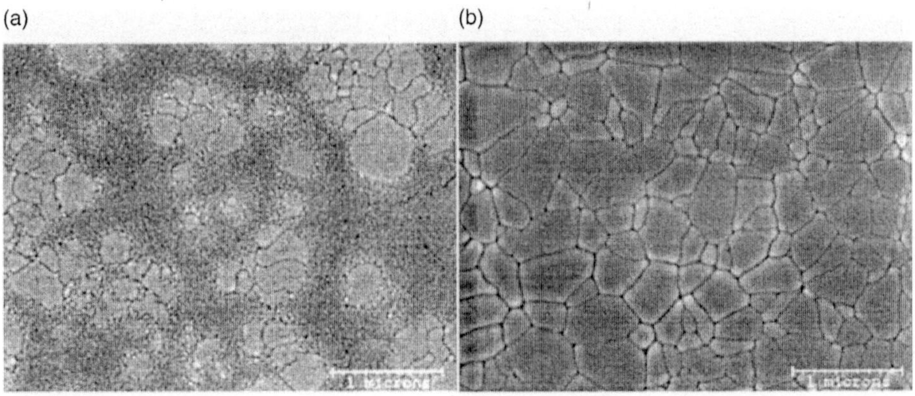

Figure 2.14. SEM photomicrographs of (a) poorly and (b) effectively processed 0.95 $PbZrO_3$–0.05 $PbTiO_3$ (PZT 95/5) thin films. [Reprinted with permission from Ref. 9. Copyright 1997 American Chemical Society.]

ture have been investigated by Norga and co-workers.[118,119] Control of the microstructure through substrate selection and the use of seed layers via homo- and heteroepitaxial nucleation has been also been widely (and successfully) explored.

Although solution chemistry may be complex and many physical processes occur during the transformation from the as-deposited to the crystalline state, films with high-quality ceramic microstructures are today routinely obtained. Figure 2.14 highlights the typical results for an antiferroelectric PZT 95/5 (0.95 $PbZrO_3$:0.05 $PbTiO_3$) thin film.[9] Shown on the left is a diphasic microstructure for this composition for a film prepared under improper processing conditions. This film consists of the desired perovskite phase as well as a detrimental pyrochlore/fluorite phase.[9,107,108] Here, the kinetic competition between the progress of the perovskite growth front that nucleates at the substrate and PbO volatilization were not properly balanced. The film shown in Fig. 2.14b,[9] on the other hand, is a single-phase perovskite film that contains no pyrochlore/fluorite and demonstrates excellent antiferroelectric response. Here, PbO loss has been properly compensated during the heat treatment process. These results are highlighted to demonstrate the ability to manipulate processing conditions to prepare thin films with electrical properties and microstructures that rival traditionally prepared bulk materials.

2.7 SUMMARY

Great progress has been demonstrated in the last 20 years in the development of wet chemical methods for the fabrication of inorganic films for electronic applications. A variety of solution chemistries and film fabrication strategies have been developed that may be used to fabricate thin films with excellent

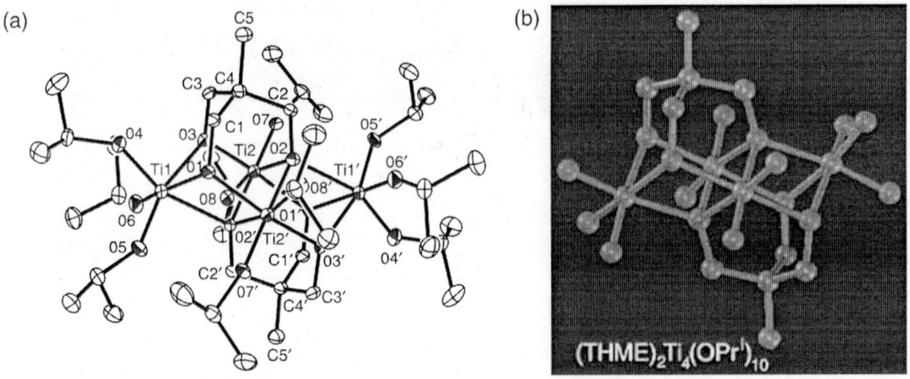

Figure 2.15. (a) ORTEP structure of TiO_2 precursor [(THME)$_2$Ti$_4$(OPr^i)$_{10}$] synthesized for use in thin-film preparation and (b) molecular dynamics simulation of the same precursor. [Reprinted with permission from Ref. 82. Copyright 1995 American Chemical Society.] (*See color insert.*)

electrical and ferroelectric properties. Through the large number of investigations that have been carried out, the general principles of the processes are, today, well understood. Investigations that were focused initially on one or two material systems of interest have become more encompassing, with a range of linear and nonlinear dielectric materials having been synthesized in thin-film form, over a range of film thicknesses, on a variety of substrates.[122–124] These technical achievements have involved the interdisciplinary efforts of chemists, electrical engineers, materials engineers, and physicists.

The use and study of CSD methods for fabrication of electronic films will undoubtedly continue in the future. It should be noted, however, that not all studies will be carried out solely by experimental methods. Simulation techniques, including molecular dynamics, may be used in conjunction with materials synthesis to elucidate the role of precursor structure on processing. Figure 2.15a shows the crystalline structure of the tetra-nuclear TiO_2 precursor,[82] (THME)$_2$Ti$_4$(OCHMe$_2$)$_{10}$, with the corresponding molecular dynamics simulation of the molecule on the right (Fig. 2.15b). The THME group is a trishydroxymethylethane derivative (shown in green in Fig. 2.15b without the protons for clarity), and Me is the methyl group. Protons and terminal isopropoxy groups are not shown in the simulation for clarity. The simulated structure effectively captures key aspects of the synthesized compound. Additional simulations of this structure and related structures might be used to probe processing aspects such as solvent quality to identify optimal solvents for minimization of striation tendencies. Other simulations could be carried out to investigate precursor thermal stability and decomposition pathways. Such studies can provide insight into densification and crystallization behavior.

Other computational techniques can be used to probe crystallization behavior, as recently used by Dobberstein and coworkers (Fig. 2.16).[13–15] Using an

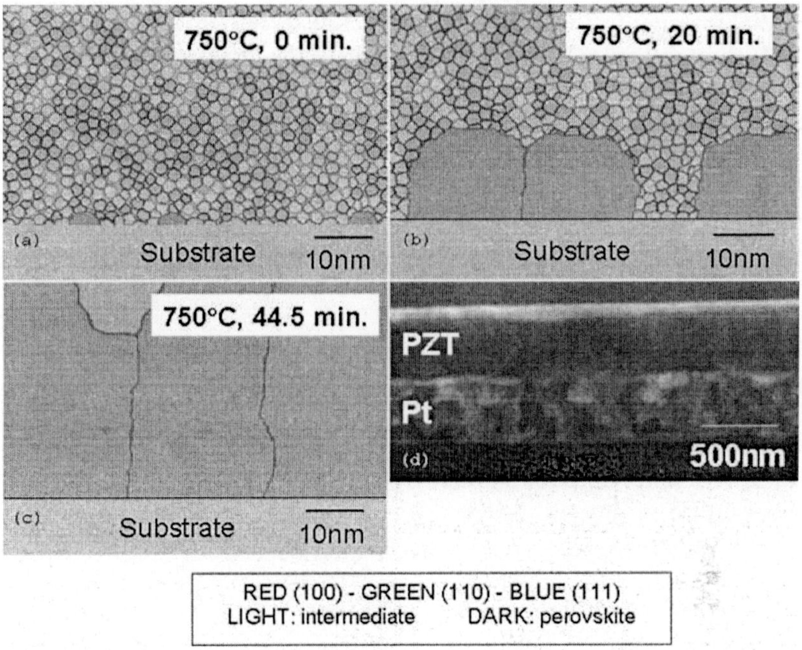

Figure 2.16. (a–c) Simulations of film structural evolution for PZT thin films at various times during heat treatment.[15] (d) A representative SEM photomicrograph illustrating the columnar microstructure of PZT.[48] The lower layer is the lower Pt electrode, the middle layer is the PZT, and the upper layer is the top Pt electrode. [(a)–(c) Reprinted with permission from Ref. 15. (d) Reprinted with permission from Ref. 9. Copyright 1997 American Chemical Society.] (*See color insert.*)

approach based on classic nucleation theory, the kinetic rate of crystal growth and bulk material properties, the authors developed a pixel-by-pixel approach to study the transformation behavior of a lead zirconate titanate thin film. Included in the model was the formation of a transitory fluorite phase. To simplify the simulation, crystallographic orientation was restricted to three principal directions: (100), (110), and (111). The simulation of the transformation presents "snapshots" in time of the transformation process, and it shows first the formation of the transitory fluorite phase, which is subsequently consumed by a perovskite phase that nucleates at the substrate surface and grows toward the surface of the film. A columnar microstructure is predicted. For comparison, a CSD-prepared PZT thin film microstructure is shown in Fig. 2.16d.[48] Although the predicted and actual grain sizes are different by approximately a factor of five, general aspects of the microstructure are effectively captured. In the future, the use of such methods might facilitate an improved understanding of the role of processing conditions on thin film microstructure.

The achievements that have been demonstrated in this field are noteworthy, although opportunities remain, both in the development of new material systems by these methods, as well as in future improvements in fundamental understanding. For example, although the general principles of thermodynamic driving force are well understood, systematic studies of the impact of this parameter, and properties such as surface energy, on thin film microstructure remain to be carried out. Systematic studies of the ability to control the nature of the amorphous film, including physical properties, in order to influence microstructure also remain to be completed. The focus of these efforts on model material systems that do not demonstrate transitory phases during crystallization might simplify the necessary analysis to understand these effects. There are also opportunities to "decouple" some factors, such as precursor organic content versus structure on processing behavior and microstructural evolution. More studies in these areas will only add to an improved fundamental understanding of the technique, aiding its commercial implementation.

REFERENCES

1. Brinker, C. J.; Scherer, G. W. 1990. *Sol-Gel Science*. Academic Press, Boston, MA.
2. Fukushima, J.; Kodaira, K.; Matsushita, T. 1984. Preparation of ferroelectric PZT films by thermal decomposition of organometallic compounds. *J. Mat. Sci.* 19:595–598.
3. Budd, K. D.; Dey, S. K.; Payne, D. A. 1985. Sol-gel processing of $PbTiO_3$, $PbZrO_3$, PZT, and PLZT thin films. *Brit. Ceram. Soc. Proc.* 36:107–121.
4. Dey, S. K.; Budd, K. D.; Payne, D. A. 1988. Thin-film ferroelectrics of PZT by sol-gel processing. *IEEE Trans. UFFC.* 35:80–81.
5. Schneller, T.; Waser, R. 2002. Chemical solution deposition of ferroelectric thin films—state of the art and recent trends. *Ferro.* 267:293–301.
6. Waser, R.; Schneller, T.; Hoffmann-Eifert, S.; Ehrhart, P. 2001. Advanced chemical deposition techniques—from research to production. *Int. Ferro.* 36:3–20.
7. Waser, R.; Schneller, T.; Ehrhart, P.; Hoffmann-Eifert, P. 2001. Chemical deposition methods for ferroelectric thin films. *Ferro.* 259(1–4):205–214.
8. Schwartz, R. W.; Schneller, T.; Waser, R. 2004. Chemical solution deposition of electronic oxide thin films. *C R Chemie* 7:433–461.
9. Schwartz, R. W. 1997. Chemical solution deposition of perovskite thin films. *Chem. Mat.* 9:2325–2340.
10. Tuttle, B. A.; Schwartz, R. W. 1996. Solution deposition of ferroelectric thin films. *Mat. Res. Bull.* 21:49–54.
11. Lakeman, C. D. E.; Payne, D. A. 1994. Sol-gel processing of electrical and magnetic ceramics. *Chem. Phys. Mat.* 38:305–324.
12. Haertling, G. H. 1992. Current status of thin/thick film ferroelectrics. *Ceram. Trans.* 25:1–18.

13. Dobberstein, H.; Schwartz, R. W. 2002. Modeling the nucleation and growth behavior of solution derived thin films. *Proc. 1st Symp. Adv. Mat. Next Generation—Prelude to Func.-Int. Mat. (AIST Chubu, Nagoya, Japan)*.

14. Schwartz, R. W.; Dobberstein, H. 2003. Modeling structural evolution in ferroelectric thin films. *Proc. 11th US-Japan Seminar on Diel. Piezo. Ceram.* (Sapporo, Japan), pp. 215–218.

15. Dobberstein, H. 2002. A thermodynamic and kinetic model for nucleation and growth in solution derived thin films. PhD dissertation. Clemson University, Clemson, SC.

16. Schwartz, R. W.; Boyle, T. J.; Lockwood, S. J.; Sinclair, M. B.; Dimos, D.; Buchheit, C. D. 1995. Sol-gel processing of PZT thin films: A review of the state-of-the-art and process optimization strategies. *Int. Ferro.* 7:259–277.

17. Licari, J. J. 2003. *Coating Materials for Electronic Applications: Polymers, Processes, Reliability, Testing*. Noyes Publications, William Andrew Publishing, New York.

18. Ma, B.; Balachandran, U.; Kaufman, D. Y.; Uprety, K. K. 2007. Strategy for the integration of PLZT thin films on base-metal foils for high voltage embedded passives. In *Heterogeneous Integration of Materials for Passive Components and Smart Systems*, edited by Nino, J. C.; Roozeboom, F.; Muralt, P.; Trolier-McKinstry, S.; LaVan, D. *Mat. Res. Soc. Symp. Proc.* 969:0969-W03–W07.

19. Van Driessche, I.; Penneman, G.; Abell, J. S.; Bruneel, E.; Hoste, S. 2003. Chemical approach to the deposition of textured CeO_2 buffer layers based on sol gel dip coating. *Mat. Sci. Forum.* 426:3517–3522.

20. Kotzyba, G.; Obst, B.; Nast, R.; Goldacker, W.; Holzapfel, B. 2006. Chemical solution deposition (CSD) of CeO_2 and $La_2Zr_2O_7$ buffer layers on cube textured NiW substrates. *J. Phys: Conf. Ser.* 43:345–348.

21. Pechini, M. 1967. Method of preparing lead and alkaline earth titanates and niobates and coating method using the same to form a capacitor. *U.S. Patent 3,330,697.*

22. Ng, M.-F.; Cima, M. J. 1997. Heteroepitaxial growth of lanthanum aluminate films derived from mixed metal nitrates. *J. Mat. Res.* 12:1306–1314.

23. Vest, R. W.; Xu, J. 1988. $PbTiO_3$ films from metallo-organic precursors. *IEEE Trans. UFFC.* 35:711–717.

24. Gross, M. E.; Hong, M.; Liou, S. H.; Gallager, P. K.; Kwo, J. 1988. Versatile new metalorganic process for preparing superconducting thin films. *Appl. Phys. Lett.* 52:160–162.

25. Klee, M.; Eusemann, R.; Waser, R.; Brand, W.; van Hal, H. 1992. Processing and electrical properties of $Pb(Zr_xTi_{1-x})O_3$ (x-0.2–0.75) films: Comparison of metallo-organic decomposition and sol-gel processes. *J. Appl. Phys.* 72:1566–1576.

26. Cui, T.; Markus, D.; Zurn, S.; Polla, D. L. 2004. Piezoelectric thin films formed by MOD on cantilever beams for microsensors and actuators. *Microsystem Tech.* 10:1432–1858.

27. Ousi-Benomar, W.; Xue, S. S.; Lessard, R. A.; Singh, A.; Wu, Z. L.; Kuo, P. K. 1994. Structural and optical characterization of $BaTiO_3$ thin films prepared by metal-organic deposition from barium 2-ethylhexanoate and titanium dimethoxy dineodecanoate. *J. Mat. Res.* 9:970–979.

28. Hasenkox, U.; Mitze, C.; Waser, R. 1997. Metal propionate synthesis of magnetoresistive $La_{1-x}(Ca,Sr)_xMnO_3$ thin films. *J. Am. Ceram. Soc.* 80:2709–2713.

29. Joshi, V.; Dacruz, C. P.; Cuchiaro, J. D.; Araujo, C. A.; Zuleeg, R. 1997. Analysis of C-V and I-V data of BST thin films. *Int. Ferro.* 14(1–4):133–140.

30. Yi, G.; Wu, Z.; Sayer, M. 1988. Preparation of $Pb(Zr,Ti)O_3$ thin films by sol gel processing: Electrical, optical, and electro-optic properties. *J. Appl. Phys.* 64:2717–2724.

31. Hennings, D.; Rosenstein, G.; Schreinemacher, H. 1991. Hydrothermal preparation of barium titanate from barium-titanium acetate gel precursors. *J. Euro. Ceram. Soc.* 8:107–115.

32. Hoffmann, S.; Waser, R. M. 1997. Dielectric properties, leakage behavior, and resistance degradation of thin films in the solid solution series $Ba(Ti_{1-y}Zr_y)O_3$. *Int. Ferro.* 17(1–4):141–152.

33. Hasenkox, U.; Hoffmann, S.; Waser, R. 1998. Influence of precursor chemistry on the formation of $MTiO_3$ (M = Ba, Sr) ceramic thin films. *J. Sol-Gel Sci. Tech.* 12:67–79.

34. Bradley, D. C.; Mehrotra, R. C.; Gaur, D. P. 1978. *Metal Alkoxides*. Academic Press, New York.

35. Ramamurthi, S. D.; Payne, D. A. 1990. Structural investigations of prehydrolyzed precursors used in the sol-gel processing of lead titanate. *J. Am. Ceram. Soc.* 73:2547–2551.

36. Morrison, R. T.; Boyd, R. N. 1992. *Organic Chemistry*, 6th Edition. Prentice Hall, Englewood Cliffs, NJ.

37. Haertling, G. H. 1991. PLZT thin films prepared from acetate precursors. *Ferro.* 116:51–63.

38. Haertling, G. H. 1992. An acetate process for bulk and thin film PLZT. *Proc. 7th Intl. Symp. Appl. Ferro. (Urbana; June 6–8)*. pp. 292–295.

39. Li, S.; Condrate, Sr., R. A.; Spriggs, R. M. 1988. A FTIR and raman spectral study of the preparation of lead titanate ($PbTiO_3$) by a sol-gel method. *Spectro. Lett.* 21:969–980.

40. Kato, K.; Zheng, C.; Finder, J. M.; Dey, S. K.; Torii, Y. 1998. Sol-Gel route to ferroelectric layer-structured perovskite $SrBi_2Ta_2O_9$ and $SrBi_2Nb_2O_9$ thin films. *J. Am. Ceram. Soc.* 81:1869–1875.

41. Kato, K. 1998. Low-temperature synthesis of $SrBi_2Ta_2O_9$ ferroelectric thin films through the complex alkoxide method: Effects of functional group, hydrolysis and water vapor treatment. *Jpn. J. Appl. Phys.* 37:5178–5184.

42. Kato, K.; Zheng, C.; Dey, S. K.; Torii, Y. 1997. Chemistry of the alkoxy-derived precursor solutions for layer-structured perovskite thin films. *Int. Ferro.* 18(1–4):225–235.

43. Bailey, J. K. 1992. Solvent quality effects in sol-gel processing. In *Better Ceramics Through Chemistry V*, edited by Hampden-Smith, M. J.; Klemperer, W. G.; Brinker, C. J. *Mat. Res. Soc. Symp. Proc.* 271:219–224.

44. Birnie, D. P. 2001. Rational solvent selection strategies to combat striation formation during spin coating of thin films. *J. Mat. Res.* 16:1145–1154.

45. Haas, D. E.; Birnie, D. P.; Zecchino, M. J.; Figueroa, J. T. 2001. The effect of radial position and spin speed on striation spacing in spin on glass coatings. *J. Mat. Sci. Lett.* 20:1763–1766.

46. Assink, R. A.; Schwartz, R. W. 1993. ^{1}H and ^{13}C NMR investigations of $Pb(Zr,Ti)O_3$ thin film precursor solutions. *Chem. Mat.* 5:511–517.

47. Schwartz, R. W.; Assink, R. A.; Headley, T. J. 1992. Solution chemistry effects in PZT thin film processing: spectroscopic and microstructural characterization. In *Ferroelectric Thin Films II*, edited by Kingon, A. I.; Myers, E. R.; Tuttle, B. *Mat. Res. Soc. Symp. Proc.* 243:245–254.

48. Schwartz, R. W.; Assink, R. A.; Dimos, D.; Sinclair, M. B.; Boyle, T. J.; Buchheit, C. D. 1995. Effects of acetylacetone additions on PZT thin film processing. In *Ferroelectric Thin Films IV*, edited by Tuttle, B. A.; Desu, S. B.; Ramesh, R.; Shiosaki, T. *Mat. Res. Soc. Symp. Proc.* 361:377–387.

49. Schwartz, R. W.; Reichert, T. L.; Clem, P. G.; Dimos, D.; Liu, D. 1997. A comparison of diol and methanol-based chemical solution deposition routes for PZT thin film fabrication. *Int. Ferro.* 18(1–4):275–286.

50. Garino, T. J. 1990. The cracking of sol-gel films during drying. In *Better Ceramics Through Chemistry IV*, edited by Zelinski, B. J. J.; Brinker, C. J.; Clark, D. E.; Ulrich, D. R. *Mat. Res. Soc. Symp. Proc.* 180:497–502.

51. Schwartz, R. W.; Boyle, T. J.; Voigt, J. A.; Buchheit, C. D. 1994. Densification and crystallization of zirconia thin films prepared by sol-gel processing. *Ceram. Trans.* 43:145–163.

52. Budd, K. D.; Dey, S. K.; Payne, D. A. 1986. The effect of hydrolysis conditions on the characteristics of $PbTiO_3$ gels and thin films. In *Better Ceramics Through Chemistry II*, edited by Brinker, C. J.; Clark, D. E.; Ulrich, D. R. *Mat. Res. Soc. Symp. Proc.* 73:711–716.

53. Hayashi, T.; Ohji, N.; Hirohara, K.; Fukunaga, T.; Maiwa, H. 1993. Preparation and properties of ferroelectric $BaTiO_3$ thin films by sol-gel process. *Jpn. J. Appl. Phys.* 32:4092–4094.

54. Hoffmann, S.; Waser, R. 1999. Control of the morphology of CSD-prepared $(Ba,Sr)TiO_3$ thin films. *J. Euro. Ceram. Soc.* 19:1339–1343.

55. Kakihana, M. 1996. "Sol-gel" preparation of high temperature superconducting oxides. *J. Sol-Gel Sci. Tech.* 6:7–55.

56. Hirano S. I.; Kato, K. 1988. Formation of $LiNbO_3$ films by hydrolysis of metal alkoxides. *J. Non-Cryst. Sol.* 100:538–541.

57. Liu, D. D. H.; Mevissen, J. P. 1997. Thick layer deposition of lead perovskites using diol-based chemical solution approach. *Int. Ferro.* 18(1–4):263–274.

58. Barrow, D. A.; Petroff, T. E.; Tandon, R. P.; Sayer, M. 1997. Characterization of thick lead zirconate titanate films fabricated using a new sol gel based process. *J. Appl. Phys.* 81:876–881.

59. Lu, Y. T.; Milne, S. J. 1996. Processing and characterization of $Pb(Zr,Ti)O_3$ films, up to $10\,\mu m$ thick, produced from a diol sol-gel route. *J. Mat. Res.* 11:2556–2564.

60. Lu, Y. T.; Calzada, M. L.; Phillips, N. J.; Milne, S. J. 1996. Synthesis and electrical characterization of thin films of PT and PZT made from a diol-based sol-gel route. *J. Am. Ceram. Soc.* 79:441–448.

61. Phillips, N. J.; Calzada, M. L.; Milne, S. J. 1992. Sol-gel-derived lead titanate films. *J. Non-Cryst. Sol.* 147/148:285–290.

62. Merklein, S.; Sporn, D.; Schönecker, A. 1992. Crystallization behavior and electrical properties of wet-chemically deposited lead zirconate titanate. In *Ferroelectric Thin Films III*, edited by Tuttle, B. A.; Myers, E. R.; Desu, S. B.; Larsen, P. K. *Mat. Res. Soc. Symp. Proc.* 310:263–268.

63. See for example, *Better Ceramics Through Chemistry I–VI. Mat. Res. Soc. Symp. Proc.* 32, 73, 121, 180, 271, and 346.

64. Brinker, C. J.; Scherer, G. W. 1984. Relationships between the sol-to-gel and gel-to-glass conversions. In *Ultrastructure Processing of Ceramics, Glasses, and Composites*, edited by Hench, L. L.; Ulrich, D. R. John Wiley & Sons, Inc., New York. pp. 43–59.

65. See, for example, *Ferroelectric Thin Films I–XII, Mat. Res. Soc. Symp. Proc.*, 200, 243, 310, 361, 433, 493, 541, 596, 655, 688, 748, and 784.

66. See, for example, *Int. Ferro., Proc. of the International Symposia on Integrated Ferroelectrics*, 1–20, (1990–2000).

67. Schwartz, R. W.; Lakeman, C. D. E.; Payne, D. A. 1990. The effects of hydrolysis conditions, and acid and base additions, on the gel-to-ceramic conversion in sol-gel derived $PbTiO_3$. In *Better Ceramics Through Chemistry IV*, edited by Zelinski, B. J. J.; Brinker, C. J.; Clark, D. E.; Ulrich, D. R. *Mat. Res. Soc. Symp. Proc.* 180:335–340.

68. Schwartz, R. W.; Payne, D. A.; Holland, A. J. 1989. The effects of hydrolysis and catalysis conditions on the surface area and decomposition behavior of polymeric sol-gel derived $PbTiO_3$ powders. In *Ceramic Powder Processing Science*, edited by Hausner, H.; Messing, G. W.; Hirano, S. *Deutsche Keramische Gesellschaft.* pp. 165–172.

69. Gagliardi, C. D.; Dunuwila, D.; Van Vlierberge-Torgerson, B. A.; Berglund, K. A. 1992. Reaction kinetics for the hydrolysis of titanium isopropoxide carboxylate complexes. In *Better Ceramics Through Chemistry V*, edited by Hampden-Smith, M. J.; Klemperer, W. G.; Brinker, C. J. *Mat. Res. Soc. Symp. Proc.* 271:257–262.

70. Budd, K. D. 1986. Structure evolution in sol-gel derived lead titanate-based materials, and application to the processing of thin dielectric layers. PhD dissertation. University of Illinois at Urbana-Champaign.

71. Schwartz, R. W. 1989. Chemical processing of $PbTiO_3$ by co-precipitation and sol-gel methods: the role of powder and gel characteristics on crystallization behavior. PhD dissertation. University of Illinois at Urbana-Champaign.

72. Klemperer, W. G.; Ramamurthi, S. D. 1988. Molecular pathways in silica sol-gel polymerization. In *Better Ceramics Through Chemistry III*, edited by Brinker, C. J.; Clark, D. E.; Ulrich, D. R. *Mat. Res. Soc. Symp. Proc.* 121:1–14.

73. Armarego, W. L. F.; Perrin, D. D. 1997. *Purification of Laboratory Chemicals*. Butterworth-Heinemann, Oxford, UK.

74. Lakeman, C. D. E.; Guistolise, D. J.; Tani, T.; Payne, D. A. 1994. Effect of thermal processing conditions on the structure and properties of sol-gel derived PZT thin layers. *Br. Ceram. Proc.* 52:70–76.

75. Lakeman, C. D. E.; Payne, D. A. 1992. Processing effects in the sol-gel processing of ferroelectric PZT powders and thin layers. *J. Am. Ceram. Soc.* 75:3091–3096.

76. Sayer, M.; Sedlar, M. 1995. Comparative sol gel processing of PZT thin films. *Int. Ferro.* 7(1–4):247–258.
77. Doeuff, S.; Henry, M.; Sanchez, C.; Livage, J. 1987. Hydrolysis of titanium alkoxides: modification of the molecular precursor by acetic acid. *J. Non-Cryst. Sol.* 89:206–216.
78. Boyle, T. J.; Dimos, D.; Schwartz, R. W.; Alam, T. M.; Sinclair, M. B.; Buchheit, C. D. 1997. Aging characteristics of a hybrid sol-gel $Pb(Zr,Ti)O_3$ precursor solution. *J. Mat. Res.* 12:1022–1030.
79. Alam, T. M.; Boyle, T. J.; Buchheit, C. D.; Schwartz, R. W.; Ziller, J. W. 1994. Formation, structure, and material properties from the reaction product of $M(OCHMe_2)_4$ (M = Ti, Zr) and HOAc. In *Better Ceramics Through Chemistry VI*, edited by Cheetham, A. K.; Brinker, C. J.; MeCartney, M. L.; Sanchez, C. *Mat. Res. Soc. Symp. Proc.* 346:35–40.
80. Boyle, T. J.; Pearson, A. T.; Schwartz, R. W. 1994. Synthesis and characterization of Group IV metal adamantanol alkoxides as potential PZT precursors. *Ceram. Trans.* 43:79–87.
81. Boyle T. J.; Schwartz, R. W. 1994. An Investigation of Group (IV) alkoxides as property controlling reagents in the synthesis of ceramic materials. *Comments Inorg. Chem.* 16:243–278.
82. Boyle, T. J.; Schwartz, R. W.; Doedens, R. J.; Ziller, J. W. 1995. Synthesis and structure of novel Group IV tridentate alkoxide complexes and ceramic thin films derived therefrom. X-ray structures of $H_3CC(CH_2\text{-}\mu_3\text{-}O)(CH_2\text{-}\mu\text{-}O)_2)_2Ti_4(OCH(CH_3)_2)_{10}$, $(H_3CCH_2C(CH_2\text{-}\mu_3\text{-}O)(CH_2\text{-}\mu\text{-}O)_2)_2Ti_4(OCH(CH_3)_2)_{10}$, and $(H_3CC(CH_2\text{-}\mu\text{-}O)_3)_2Zr_4(\mu\text{-}OCH(CH_3)_2)_2(OCH(CH_2)_2)_8$. *Inorg. Chem.* 34:1110–1120.
83. Chae, H. K.; Payne, D. A.; Xu, Z.; Ma, L. 1994. Molecular structure of a new lead titanium bimetallic alkoxide complex $[PbTi_2(\mu_4\text{-}O)(OOCCH_3)(OCH_2CH_3)_7]_2$: Evolution of structure on heat treatment and the formation of thin-layer dielectrics. *Chem. Mat.* 6:1589–1592.
84. Ma, L.; Payne, D. A. 1994. Studies on the nature of a lead zirconate titanate (PZT) precursor solution. Isolation and structural characterization of $[PbZr_2(O)(OOCCH_3)_2(OCH_2CH_3)_6]_2$. *Chem. Mat.* 6:875–877.
85. Lipeles, R. A.; Coleman, D. J. 1988. Effect of drying and annealing on metallo-organic solution deposition of PZT films. In *Ultrastructure Processing of Advanced Ceramics*, edited by Ulrich, D. R.; Mackenzie, J. D. John Wiley & Sons, Inc., New York. pp. 919–933.
86. Baythoun, M. S. G.; Sale, F. R. 1982. Production of strontium-substituted lanthanum manganite perovskite powder by the amorphous citrate process. *J. Mat. Sci.* 17:2757–2769.
87. Van Bael, M. K.; Nelis, D.; Hardy, A.; Mondelaers, D.; Van Werde, K.; D'Haen, J.; Vanhoyland, G.; Van Den Rul, H.; Mullens, J.; Van Poucke, L. C.; Frederix, F.; Wouters, D. J. 2002. Aqueous chemical solution deposition of ferroelectric thin films. *Int. Ferro.* 45(1):113–122.
88. Tummala, R. R. 2001. *Fundamentals of Microsystems Packaging*. McGraw Hill, New York.
89. Brinker, C. J.; Hurd, A. J. 1994. Fundamentals of sol-gel dip-coating. *J. Phys. III France.* 4:1231–1242.

90. Scriven, L. E. 1988. Physics and applications of dip coating and spin coating. In *Better Ceramics Through Chemistry III*, edited by Brinker, C. J.; Clark, D. E.; Ulrich, D. R. *Mat. Res. Soc. Symp. Proc.* 121:717–729.
91. Huffmann, M. 1995. Liquid source misted chemical deposition (LSMCD)—a critical review. *Int. Ferro.* 10(1–4):39–53.
92. Losego, M. D.; Trolier-McKinstry, S. 2004. Mist deposition of micron thick lead zirconate titanate thick films. In *Ferroelectric Thin Films XII*, edited by Hoffmann-Eifert, S.; Funakubo, H.; Kingon, A. I.; Koutsaroff, I.; Joshi, V. *Mat. Res. Soc. Symp. Proc.* 784(C11.28):1–6.
93. McMillan, L. D.; Huffman, M.; Roberts, T. L.; Scott, M. C.; Paz De Araujo, C. A. 1995. Deposition of $Ba_{1-x}Sr_xTiO_3$ and $SrTiO_3$ via liquid source CVD (LSCVD) for ULSI DRAMS. *Int. Ferro.* 4(4):319–324.
94. Lakeman, C. D. E.; Fleig, P. F. 2002. High resolution integrated passives using microcontact printing. *Proc. SPIE Intl. Microelec. Packaging Soc. (IMAPS).* pp. 755–759.
95. Clem, P. G.; Jeon, N.-L.; Nuzzo, R. G.; Payne, D. A. 1997. Monolayer-mediated deposition of tantalum (V) oxide thin film structures from solution precursors. *J. Am. Ceram. Soc.* 80:2821–2827.
96. Nagata, H.; Ko, S. W.; Hong, E.; Randall, C. A.; Trolier-McKinstry, S.; Pinceloup, P.; Skamser, D.; Randall, M.; Tajuddin, A. 2006. Micro-contact printed $BaTiO_3$ and $LaNiO_3$ thin films for capacitors. *J. Am. Ceram. Soc.* 89(9):2816–2821.
97. Kumar, A.; Whitesides, G. M. 1993. Features of gold having micrometer to centimeter dimensions can be formed through a combination of stamping with an elastomeric stamp and an alkanethiol "ink" followed by chemical etching. *Appl. Phys. Lett.* 63:2002–2004.
98. Fu, Y.; Bryan, N. 2002. One-step transfer of diffractive structure from a designed pattern to a replica by use of a hybrid solgel film. *Optics Express* 10:436–442.
99. Brennecka, G.; Tuttle, B. 2007. Deposition of ultrathin film capacitors fabricated by chemical solution deposition. *J. Mat. Res.* 22:2868–2874.
100. Barrow, D. A.; Noteboom, R.; Sayer, M. 1995. Design and fabrication of macroscopic piezoelectric actuators based on thick PZT films. *Int. Ferro.* 8:1–11.
101. Reaney, I. M.; Brooks, K.; Klissurka, R.; Pawlaczyk, C.; Setter, N. 1994. Use of transmission electron microscopy for the characterization of rapid thermally annealed solution-gel, lead zirconate titanate films. *J. Am. Ceram. Soc.* 77:1209–1216.
102. Griswold, E. M.; Weaver, L.; Sayer, M.; Calder, J. D. 1995. Phase transformations in rapid thermal processed PZT. *J. Mat. Res.* 10:3149–3159.
103. Vasant Kumar, C. V. R.; Sayer, M.; Pascual, R.; Amm, D. T.; Wu, Z.; Swanston, D. M. 1991. Lead zirconate titanate films by rapid thermal processing. *Appl. Phys. Lett.* 58:1161–1163.
104. Schwartz, R. W.; Payne, D. A. 1988. Crystallization behavior of chemically prepared and rapidly solidified $PbTiO_3$. In *Better Ceramics Through Chemistry III*, edited by Brinker, C. J.; Clark, D. E.; Ulrich, D. R. *Mat. Res. Soc. Symp. Proc.* 121:199–206.
105. Cooper, A. R. 1986. Differences between gel-derived melts and those prepared by batch melting. In *Better Ceramics Through Chemistry II*, edited by Brinker, C. J.; Clark, D. E.; Ulrich, D. R. *Mat. Res. Soc. Symp. Proc.* 73:421–430.

106. Schwartz, R. W.; Voigt, J. A.; Tuttle, B. A.; DaSalla, R. S.; Payne, D. A. 1997. Comments on the effects of solution precursor characteristics and thermal processing conditions on the crystallization behavior of sol-gel derived PZT thin films. *J. Mat. Res.* 12:444–456.

107. Tuttle, B. A.; Voigt, J. A.; Headley, T. J.; Potter, Jr., B. G.; Dimos, D.; Schwartz, R. W.; Dugger, M. T.; Michael, J.; Nasby, R. D.; Garino, T. J.; Goodnow, D. C. 1994. Ferroelectric thin film microstructure development and related property enhancement. *Ferro.* 151(1–4):11–20.

108. Voigt, J. A.; Tuttle, B. A.; Headley, T. J.; Lamppa, D. L. 1995. The pyrochlore-to-perovskite transformation in solution-derived lead zirconate titanate thin films. In *Ferroelectric Thin Films IV*, edited by Tuttle, B. A.; Desu, S. B.; Ramesh, R.; Shiosaki, T. *Mat. Res. Soc. Symp. Proc.* 361:395–402.

109. Hoffmann, M.; Hofer, C.; Schneller, T.; Böttger, U.; Waser, R. 2002. Preparation and aging behavior of chemical-solution-deposited $(Pb(Mg_{1/3}Nb_{2/3})O_3)_{1-x}$–$(PbTiO_3)_x$ thin films without seeding layers. *J. Am. Ceram. Soc.* 85:1867–1869.

110. Schneller, T.; Waser, R. 2007. Chemical modifications of $Pb(Zr_{0.3},Ti_{0.7})O_3$ precursor solutions and their influence on the morphological and electrical properties of the resulting thin films. *J. Sol-Gel Sci Tech.* 42:337–352.

111. Eichorst, D. J.; Payne, D. A. 1988. Sol-gel processing of lithium niobate thin-layers on silicon. In *Better Ceramics Through Chemistry III*, edited by Brinker, C. J.; Clark, D. E.; Ulrich, D. R. *Mat. Res. Soc. Symp. Proc.* 121:773–778.

112. Coffman, P. R.; Dey, S. K. 1994. Structure evolution in the $PbOZrO_2TiO2$ sol-gel system: Part I—characterization of prehydrolyzed precursors. *J. Sol-Gel Sci. Tech.* 1:251–265.

113. Coffman, P. R.; Barlingay, C. K.; Gupta, A.; Dey, S. K. 1996. Structure evolution in the PbO-ZrO_2-TiO_2 sol-gel system: Part II—pyrolysis of acid and base-catalyzed bulk and thin film gels. *J. Sol-Gel Sci. Tech.* 6:83–106.

114. Lockwood, S. J.; Schwartz, R. W.; Tuttle, B. A.; Thomas, V. A. 1993. Solution chemistry optimization of sol-gel processed PZT thin films. In *Ferroelectric Thin Films III*, edited by Tuttle, B. A.; Myers, E. R.; Desu, S. B.; Larsen, P. K. *Mat. Res. Soc. Symp. Proc.* 310:275–280.

115. Melnick, B. M. 1992. Statistical investigation of sol-gel lead titanate (PT) processing. *Int. Ferro.* 2(1–4):255–268.

116. Lefevre, M. J.; Speck, J. S.; Schwartz, R. W.; Dimos, D.; Lockwood, S. J. 1996. Microstructural development in sol-gel derived PZT thin films: the role of precursor stoichiometry and processing environment. *J. Mat. Res.* 11:2076–2084.

117. Tani, T.; Lakeman, C. D. E.; Li, J.-F.; Xu, Z.; Payne, D. A. 1994. Crystallization behavior and improved properties for sol-gel derived PZT and PLZT thin layers processed with a lead oxide cover coating. *Ceram. Trans.* 43:89–106.

118. Fe, L.; Malic, B.; Norga, G.; Kosec, M.; Wouters, D. J.; Bartic, T. A.; Maes, H. E. 1999 Role of precursor chemistry in the ferroelectric properties of donor doped $Pb(Zr,Ti)O_3$ thin films. In *Ferroelectric Thin Films VII*, edited by Jones, R. E.; Schwartz, R. W.; Summerfelt, S.; Yoo, I. K. *Mat. Res. Soc. Symp. Proc.* 541:369–374.

119. Norga, G. J.; Fe, L.; Vasiliu, F.; Fompeyrine, J.; Locquet, J.-P.; Van der Biest, O. 2004. Orientation selection in functional oxide thin films. *J. Euro. Ceram. Soc.* 24:969–974.

120. Hoffmann, S.; Haxenkox, U.; Waser, R.; Jia, J. L.; Urban, K. 1997. Chemical solution deposition of $BaTiO_3$ and $SrTiO_3$ with columnar microstructures. In *Science and Technology of Semiconductor Surface Preparation*, edited by Hagashi, G. S.; Hirose, M.; Ragahavan, S.; Verhaverbeke, S. *Mat. Res. Soc. Symp. Proc.* 477:9–14.

121. Roy, R. 1969. Gel route to homogeneous glass preparation. *J. Am. Ceram. Soc.* 52:344.

122. Halder, S.; Schneller, T.; Waser, R.; Thomas, F. 2007. Microstructure and electrical properties of $BaTiO_3$ and $(Ba,Sr)TiO_3$ ferroelectric thin films on nickel electrodes. *J. Sol-Gel Sci. Tech.* 42:203–207.

123. Ihlefeld, J. F.; Borland, W.; Maria, J. P. 2005. Synthesis and properties of barium titanate thin film on copper substrates. In *Ferroelectric Thin Films XIII*, edited by Ramesh, R.; Maria, J. P.; Alexe, M.; Joshi, V. *Mat. Res. Soc. Symp. Proc.* 902:7–14.

124. Wu, A.; Vilarinho, P. M.; Srinivasan, S.; Kingon, A. I.; Reaney, I. M.; Woodward, D.; Ramos, A. R.; Alves, E. 2006. Microstructural studies of PZT thick films on Cu foils. *Acta Materialia.* 54:3211–3220.

CHAPTER 3

Solution Processing of Chalcogenide Semiconductors via Dimensional Reduction

DAVID B. MITZI

IBM T. J. Watson Research Center, Yorktown Heights, NY

3.1 INTRODUCTION

Metal chalcogenide films play a pivotal role in a variety of electronic devices, including solar cells,[1–4] rewritable memory,[5–8] thermoelectrics,[9,10] and transistors.[11,12] Two leading thin-film photovoltaic technologies, for example, based on CdTe and Cu(In,Ga)Se_2 layers, employ the electronic properties of metal chalcogenide semiconductors to enable the highest reported efficiencies in single-junction polycrystalline photovoltaic devices.[1] Phase-change materials, particularly those in the Ge-Sb-Te and Ag-In-Sb-Te systems, have found extensive application in optical memory products as a result of the reversible and very fast crystalline–amorphous transition that can be induced in films with laser heating.[5–7] Early thin-film transistors (TFTs) were also based on metal chalcogenides, such as cadmium sulfide, before being supplanted by silicon-based devices.[11]

Each prospective application relies not only on the electronic properties of the metal chalcogenide under consideration but also on the ability to conveniently deposit films with a suitable composition profile, phase purity, and grain structure. Typically, metal chalcogenide films are deposited using vacuum-based techniques such as evaporation or sputtering.[1–3,7] One advantage of these vapor-phase techniques is the versatility afforded by the possible use of multiple evaporation/sputtering sources, thereby providing considerable control over the film composition and phase profile. However, vacuum conditions necessitate a confined space, which is not conducive to the cost-effective handling of large-area substrates. Vacuum-based techniques also suffer from relatively slow throughput, poor materials utilization (much of the deposited material ends up on the chamber walls), and considerable energy expenditure to heat or sputter from the target sources. Chemical vapor deposition (CVD) is an alternative vapor-based approach, which offers substantial flexibility, assuming the identification of suitable volatile source materials.[13–15] Practical constraints on reactor size, however, limit large-area deposition using this approach, and material utilization also presents a problem (much of the material flowing through the system does not end up on the substrate). High substrate temperatures (>350 °C) are also often required to achieve good crystallinity in films produced using the CVD approach.[13,14] Clearly, alternative cost-effective, high-throughput, low-temperature film deposition processes are desired to enable more pervasive application of metal chalcogenide films (especially for "macroelectronics" applications that rely on a large area or flexible form factor).

Particularly desirable among film deposition processes are solution-based techniques, because of the relative simplicity and potential economy of these approaches. However, the covalent character of the metal chalcogenides, which provides the benefit of the desired electronic properties (e.g., high electrical mobility), represents an important barrier for solution processing. Several methods have been developed to overcome the solubility problem, including spray deposition, bath-based techniques, and electrochemical routes, each of which will be discussed in later chapters. In this chapter, a very simple "dimensional reduction" approach will be considered as a means of achieving a convenient solution-based route to film deposition.

3.2 DIMENSIONAL REDUCTION

The term "dimensional reduction" refers to the process of dismantling the extended metal–anion (M—X) framework of a parent compound MX_a upon reaction with an ionic reagent A_bX, to form a new compound $A_{nb}MX_{a+n}$.[16] The "A" cations, which are more electropositive than the metal (M), do not form strong covalent bonds with the "X" anions and, therefore, may be treated as lying outside the M—X framework. The introduced "X" ions (for this chapter, X = S, Se, or Te) terminate connections between metal centers in the structure, yielding a less tightly connected or lower dimensional framework, which may locally retain aspects of the characteristic metal coordination and polyhedron connectivity of the original parent structure. For each additional A_bX unit added to the parent structure (denoted by "n"), additional M—X—M metal bond linkages are broken and the effective dimensionality of the MX_{a+n} framework is progressively reduced (Fig. 3.1). As an example, a series of compounds develop from the interaction between $SnSe_2$ and K_2Se, including $K_4Sn_3Se_8$ (n = 2/3),[17] K_2SnSe_3 (n = 1),[18] $K_6Sn_2Se_7$ (n = 1.5),[19] and K_4SnSe_4 (n = 2),[20] which exhibit varying degrees of condensation among $SnSe_4$ tetrahedra. Although

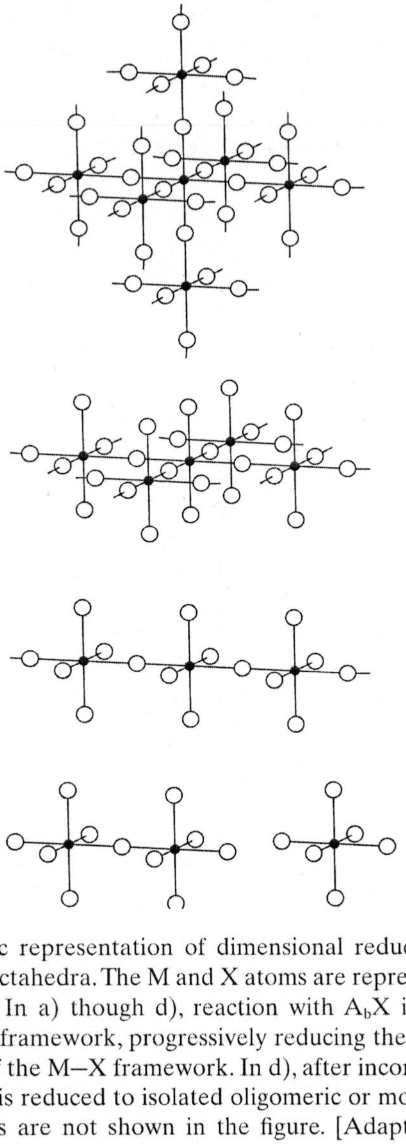

Figure 3.1. Schematic representation of dimensional reduction for a framework of corner-sharing MX_6 octahedra. The M and X atoms are represented by black and white spheres, respectively. In a) though d), reaction with A_bX incorporates additional X atoms into the M–X framework, progressively reducing the connectedness and effective dimensionality of the M–X framework. In d), after incorporating "n" units of A_bX ($n > 2$), the structure is reduced to isolated oligomeric or monomeric components. For clarity, the "A" atoms are not shown in the figure. [Adapted with permission from [Ref. 16]. Copyright 2001 American Chemical Society.]

$K_4Sn_3Se_8$ (n = 2/3) contains trimeric $Sn_3Se_8^{4-}$ ions (edge-sharing tetrahedra), K_2SnSe_3 (n = 1) consists of dimeric $Sn_2Se_6^{4-}$ (edge-sharing tetrahedra), progressively tending toward a collection of discrete $SnSe_4^{4-}$ ions in K_4SnSe_4 (n = 2). Note that $K_2Sn_2Se_5$ (n = 1/2) exhibits an extended anion of edge- and corner-sharing $SnSe_5$ distorted trigonal bipyramids.[21] As a result of reducing the extended three-dimensional (3D) MX_a structure into more isolated MX_{a+n} ionic moieties, the compounds $A_{nb}MX_{a+n}$ often exhibits higher solubility with increasing "n," thereby enabling the preparation of reasonably concentrated $A_{nb}MX_{a+n}$ solutions.

Besides monochalcogenide X^{2-} (X = S, Se, Te) ions, the chalcogenide ions within the A–M–X compound can be incorporated as polychalcogenide X_n^{2-} ($n \geq 2$) species, in which there are direct X–X bonds. There has been significant interest in developing soluble metal polychalcogenide species, often stabilized with bulky organic ions such as R_4N^+ and R_4P^+.[22] Examples include $(Ph_4P)_4[In_2(Se_4)_4(Se_5)]$,[23] $(Et_4N)_3[Tl_3Se_3(Se_4)_3]$,[23] $(Et_4N)_2[Mo(Se_4)]$,[24] $[Ph_4As]_2[WSe(Se_4)_2]$,[24] $(Ph_4P)_4[Cu_2(Se_4)(Se_5)_2]$,[25,26] $(Ph_4P)_2[M(Se_4)_2]$ (M = Zn, Cd), and $(Ph_4P)_2[Sn(Se_4)_3]$.[27] Similar to the monochalcogenide systems, the extra chalcogens in the polychalcogenide structures aid in breaking up M–X–M linkages and introduce more ionic character in the structure, thereby yielding a more soluble material. Note, then, that for the purpose of this chapter, we will use the term "dimensional reduction" in a most general sense, including systems in which the resulting structure after chalcogen incorporation bears little resemblance to the parent metal chalcogenide structure in terms of local coordination and polyhedron connectivity. The key point is the incorporation of extra chalcogen and/or "A" species to break up the extended poorly soluble metal chalcogenide framework into more soluble discrete metal chalcogenide units.

The ability to prepare appropriate precursors for metal chalcogenide film deposition relies on selecting a volatile "A" cation to separate the metal chalcogenide anions, so that a short and low-temperature heat treatment can be used to transform the precursor into the desired product. Upon heating $A_{nb}MX_{a+n}$, the "A" moiety must decompose from the sample, along with a suitable quantity of "X" to maintain charge balance, thereby yielding the desired MX_a product. A second requirement for effective film deposition is that "A" must not be too bulky, in order to avoid excessive disruption of the film during decomposition. In the polychalcogenide systems $(Ph_4P)_4[In_2(Se_4)_4(Se_5)]$ or $(Ph_4P)_4[Cu_2(Se_4)(Se_5)_2]$, for example, when the precursors decompose to their respective parent metal chalcogenide compositions (i.e., In_2Se_3 and Cu_2Se), approximately 86% and 92% of the original material (by weight) is lost, respectively. The large weight and correspondingly volume loss from the precursor leads to films that are generally cracked or discontinuous.[26] The excessive weight loss is a result of both the bulky "A" moiety, as well as the incorporation of polychalcogenide anions, which must be decomposed from the sample before the single-phase parent metal chalcogenide can be achieved. Polychalcogenide species (especially Se and Te) also generally require increased temperatures for dissociation from the sample (typically >400 °C).[26] Likely then, monochalcogenide compounds with small "A" moieties are to be preferred relative to polychalcogenide systems and those with large "A" cations when trying to design a suitable metal chalcogenide precursor for thin-film deposition.

The process for film formation using a dimensional reduction approach requires three conceptually simple steps (Fig. 3.2). The first step involves

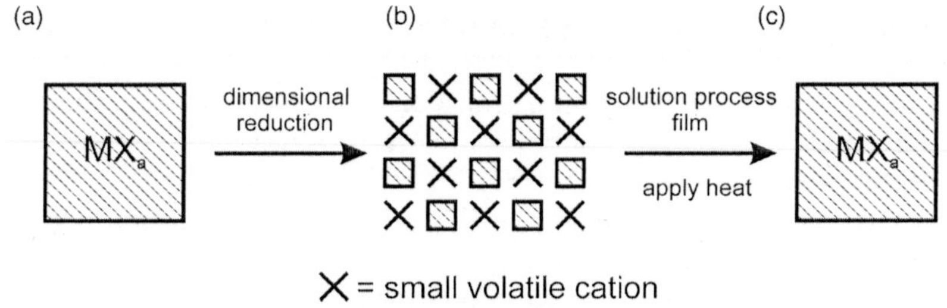

Figure 3.2. Film formation using a dimensional reduction approach involves three steps: 1) breaking up the insoluble extended inorganic framework (a) into more soluble-isolated anionic species, which are separated by some small and volatile cationic species (b). 2) Solution-processing thin films of the precursor (b). 3) Heating the precursor films such that the cationic species and corresponding chalcogen anions are dissociated, leaving behind the targeted inorganic semiconductor (c).

breaking up the extended metal chalcogenide framework into discrete metal chalcogenide anions in solution. This process can either occur by first isolating the metal chalcogenide precursor and then dissolving it in a suitable solvent or by directly dissolving the parent metal chalcogenide in a reactive solvent, which can facilitate the precursor formation. The second step involves depositing the precursor solution on a substrate and forming a solid precursor film. In this chapter, most examples will involve spin coating as the solution-processing technique of choice for film deposition. Having achieved a precursor solution, however, other solution-based techniques could equally be employed, including printing, stamping, doctor blading, drop casting, or dip coating. In each case, as the solvent of the precursor solution evaporates, a film of the precursor remains behind. The third step involves heating the precursor film to decomposition, leaving behind a film of the targeted metal chalcogenide. Ideally, this heating step should be achievable at a temperature below ~400 °C, which approximately represents the highest sustained processing temperature that can be tolerated by high-temperature polymer substrates (e.g., Kapton).[28]

Although, in principal, the steps described above are straightforward, other requirements for successful film deposition include selection and preparation of the substrate surface so that the precursor solution will adequately wet the surface during deposition. Without this condition being met, a uniform film will not be possible. Even assuming a suitably chosen volatile "A" cation, so that the precursor cleanly decomposes to the desired metal chalcogenide, an additional consideration is the pathway for decomposition and how this influences the grain structure of the resulting films. For some applications, amorphous films are sufficient (e.g., rewritable optical memory disks), whereas for others (e.g., transistors and solar cells), crystalline films are generally required. Control over grain structure represents one of the key challenges of solution-based processing techniques (as described for CSD in Chapter 2).

3.3 HYDRAZINE PRECURSOR ROUTE

The solvent properties of hydrazine on metal chalcogenides (especially sulfides) have been previously considered, with little solubility noted for most systems examined.[29] Hydrazine hydrate has been used as a medium to precipitate first- and second-row transition metal sulfides and selenides.[30] The simple approach described in this chapter generally employs additional chalcogen (S, Se, Te) in anhydrous hydrazine to improve directly the solubility and film-forming properties of selected metal chalcogenides, including $SnSe_{2-x}S_x$, In_2Se_3, GeS_2, $GeSe_2$, Cu_2S, Sb_2Se_3, Sb_2Te_3, $CuInSe_2$, and Ga_2Se_3.[31–38] Dissolution generally proceeds by formation of metal chalcogenide anions accompanied by hydrazinium cations. The additional chalcogen added to the hydrazine facilitates the disruption of the metal chalcogenide framework by breaking up M—X—M linkages, in analogy to the dimensional reduction examples described above. For ZnTe (and perhaps selected other metal tellurides), dissolution occurs through the breakdown of the 3D ZnTe structure into one-dimensional (1D) ZnTe chains separated by coordinated neutral molecules.[34] The process of dimensional reduction is different for zinc telluride. However, the result on solubility is similar to the examples that involve salt formation. In many respects, hydrazine is an ideal solvent, since it is both a small and volatile molecule and also tends not to coordinate strongly.[29,39] Hydrazine is, therefore, relatively easy to dissociate from the precursor at low temperature. In addition, the absence of carbon in the precursor and solvent implies that cleaner film deposition may be possible after the heat treatment (i.e., absence of carbon impurities). *Hydrazine is, however, also toxic and explosive and should therefore be used only with adequate protective measures in place to avoid contact with either the vapor or the liquid and/or ignition sources.*[40,41]

3.3.1 $SnSe_{2-x}S_x$ Films

As a first example of the hydrazine-based approach, $SnSe_{2-x}S_x$ can be deposited and employed in TFT devices as the active channel layer.[31] SnS_2 [$SnSe_2$], is an n-type semiconductor with a band gap of 2.1 eV [1.0 eV] and bulk Hall effect mobility measured in the range of 18 cm²/V-s [27 cm²/V-s].[42,43] A soluble $SnSe_{2-x}S_x$ hydrazine-based precursor was isolated by stirring mixtures of tin(IV) sulfide, tin(IV) selenide, and sulfur in hydrazine under an inert atmosphere at room temperature. Dissolution occurs primarily through the overall reaction:

$$5N_2H_4 + 2X + 2SnX_2 \rightarrow N_2(\text{gas}) + 4N_2H_5^+ + Sn_2X_6^{4-} \ (X = S, Se) \qquad (1)$$

Evaporation of the solution under flowing nitogen gas yields crystalline $(N_2H_5)_4Sn_2S_6$ (X = S) or $(N_2H_4)_3(N_2H_5)_4Sn_2Se_6$ (X = Se), two analogous structures comprising dimers of edge-sharing SnX_4 tetrahedra ($Sn_2X_6^{4-}$) alternating with hydrazinium cations and, for X = Se, neutral hydrazine molecules

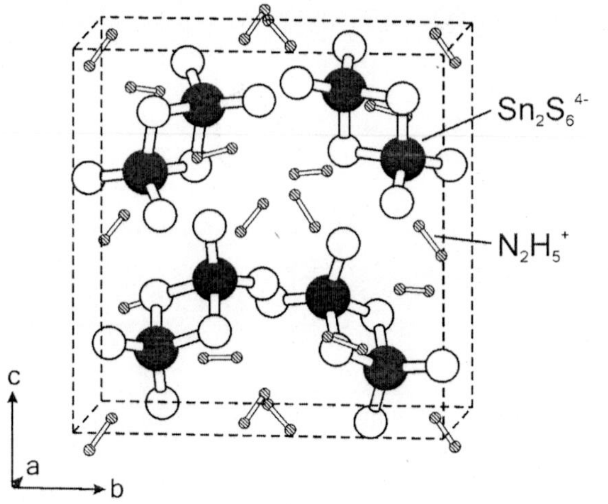

Figure 3.3. The crystal structure of $(N_2H_5)_4Sn_2S_6$ is composed of edge-sharing dimers of SnS_4 tetrahedra ($Sn_2S_6^{4-}$) separated by hydrazinium cations.[31] The analogous tin(IV) selenide structure, $(N_2H_4)_3(N_2H_5)_4Sn_2Se_6$, consists of similar $Sn_2Se_6^{4-}$ anions, separated by both neutral hydrazine and hydrazinium cations.[33] [Adapted from [Ref. 31]. Copyright 2004 Nature Publishing Group.]

(Fig. 3.3).[31,33] The observed crystal structures are similar to that reported[18] for $K_4Sn_2Se_6$ with, however, a small volatile hydrazinium species replacing the relatively nonvolatile alkali metal cation. Note that the sulfur-to-selenium ratio can be tuned in the final product by varying the amount of SnS_2 and $SnSe_2$ employed.

Upon heating, decomposition of the precursor to the crystalline $SnSe_{2-x}S_x$ semiconductor occurs through the pathway (neglecting any neutral hydrazine that may be in the structure, which would evolve first at low temperature if present):

$$(N_2H_5)_4Sn_2X_6 \rightarrow 4N_2H_4 + 2H_2X + 2SnX_2 \ (X = S, Se) \qquad (2)$$

At suitably high temperatures, hydrazine and H_2X decomposition products will leave the sample in a gaseous form, although the H_2X product can further decompose to H_2 and X (increasing probability across the series S, Se, Te). For X = S or Se, the decomposition process is complete by approximately 350 °C (Fig. 3.4), highlighting the low-temperature nature of this process.[33] Note the significantly lower decomposition temperature (<200 °C) for X = S, primarily because of the more volatile nature of sulfur relative to selenium. For low-temperature deposition of metal chalcogenides using a precursor that consists of metal chalcogenide anions and hydrazinium cations, sulfides will generally facilitate lower processing temperatures compared with selenides

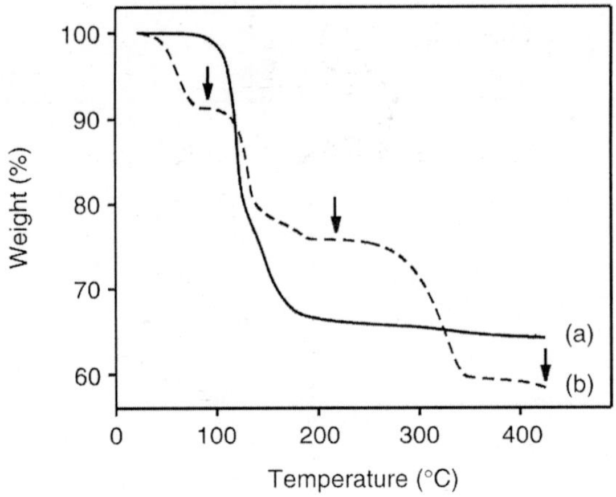

Figure 3.4. Thermogravimetric analysis (TGA) of the precursors (a) $(N_2H_5)_4Sn_2S_6$ (solid line) and (b) $(N_2H_4)_3(N_2H_5)_4Sn_2Se_6$ (dashed line), run at 2 °C/min in a flowing N_2 atmosphere. In (b) the arrows denote the three primary steps in the decomposition of $(N_2H_4)_3(N_2H_5)_4Sn_2Se_6$: loss of neutral hydrazine (100 °C), loss of remaining hydrazine (200 °C), and loss of remaining extra selenium (350 °C). [Reproduced with permission from [Ref. 33]. Copyright 2005 American Chemical Society.]

and especially tellurides, because of the requirement of dissociating the excess chalcogen.

Continuous crystalline tin chalcogenide films, as thin as 40 Å, can be formed by spin coating from the hydrazine-based SnS_2 and $SnSe_2$ solutions.[31] The solutions are spun onto cleaned hydrophilic surfaces. A hydrophilic surface is required to enable adequate wetting of the substrate by the hydrazine solution. For a thermally oxidized silicon substrate, a hydrophilic surface can be achieved by cleaning the substrate using a piranha approach (hydrogen peroxide/sulfuric acid). Film thickness can be controlled by varying spin-speed during the coating process (i.e., see Eq. 8 in Chapter 2), by selecting the concentration of the metal chalcogenide in the solution, as well as by performing multiple deposition cycles to build up thickness. The resulting hydrazinium precursor films are decomposed at temperatures of <350 °C using a short (~20 min) anneal in an inert atmosphere. The relatively low-temperature nature of the process offers the possibility of using a wide range of substrates, including high-temperature-compatible flexible plastic substrates (most notably polyimides). A wider range of plastic substrates might be employed with a rapid laser-based annealing process,[44] since the semiconducting film on top of the substrate can be selectively and rapidly annealed without substantial heating to the substrate.

An important aspect of this process is that the $SnSe_{2-x}S_x$ films are continuous and relatively uniform after decomposition. A cross-sectional transmission

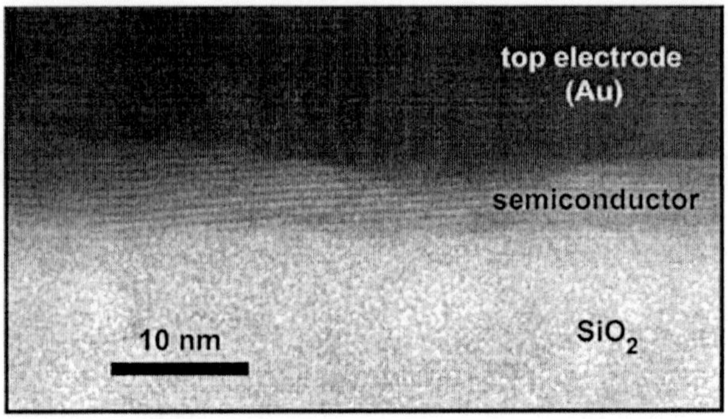

Figure 3.5. TEM cross-sectional image of a $SnS_{1.4}Se_{0.5}$ film prepared by spin coating from a hydrazine-based solution. The final thermal treatment for this film was at 270 °C for 20 minutes. [Reproduced with permission from [Ref. 31]. Copyright 2004 Nature Publishing Group.]

electron microscope (TEM) image (Fig. 3.5) demonstrates the ultrathin, crystalline nature of a typical $SnS_{1.4(1)}Se_{0.5(1)}$ film spun on thermally oxidized silicon, with an average film thickness of ~50 Å—i.e., only 8–10 unit cells thick (film thickness and chemical composition were determined using medium-energy ion scattering–MEIS).[31] Substantial preferred orientation is visible, with the metal chalcogenide layers oriented nominally parallel to the substrate surface (corresponding to the fringes observed in the TEM images). The high degree of solubility, the ability to decompose at relatively low temperature, and the small volume of the dissociating hydrazinium species in the current process are all presumably important factors enabling the formation of high-quality, continuous ultrathin chalcogenide films using the spin-coating process.

Using a standard TFT device configuration (Fig. 3.6a), spin-coated $SnS_{1.4(1)}Se_{0.5(1)}$ channel layers have been electrically evaluated.[31] A representative plot of drain current I_D versus drain voltage V_D is shown in Fig. 3.6b as a function of the applied gate voltage V_G for a TFT with a spin-coated $SnS_{1.4(1)}Se_{0.5(1)}$ channel (same film as for Fig. 3.5). The device operates as an n-channel transistor, working in accumulation mode upon application of a positive gate bias. Application of a negative gate bias depletes the channel of electrons and shuts the device off. At low V_D, the TFT shows typical transistor-like behavior as I_D increases linearly with V_D. Current saturation is observed at high V_D as the accumulation layer is pinched off near the drain electrode. Current modulation (I_{on}/I_{off}) and saturation regime field-effect mobility (μ_{sat}) are calculated from the plot of I_D and $I_D^{1/2}$ versus V_G (Fig. 3.6c),[45] yielding $I_{ON}/I_{OFF} > 10^6$ and $\mu_{sat} = 12.0\,cm^2/V\text{-}s$, respectively, for a –60 to 85 V gate sweep and $V_D = 85\,V$. The linear regime mobility derived from the plots in Fig. 3.6b is $\mu_{lin} = 2.4\,cm^2/V\text{-}s$. Note that use of a thinner or higher dielectric constant gate insulator (relative to the 210-nm SiO_2 layer currently used) enables significant reduction in the device operating voltage over that

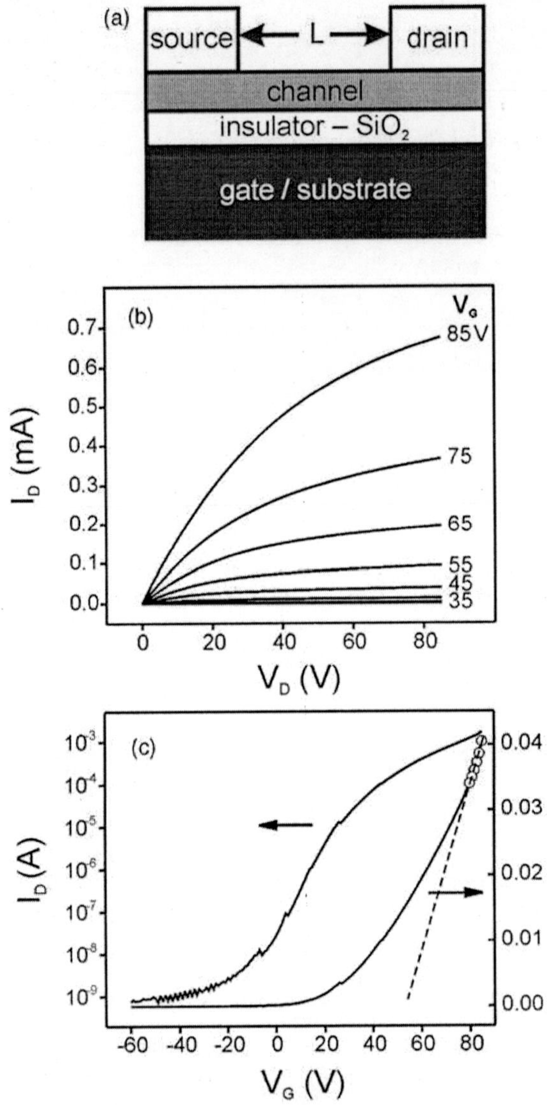

Figure 3.6. (a) The TFT device structure (not shown to scale) consists of a heavily n-doped silicon wafer as the substrate/gate, a thermally grown oxide gate dielectric, the spin-coated chalcogenide channel layer, and patterned Au source/drain electrodes. (b) Drain current I_D versus drain voltage V_D as a function of gate voltage V_G for a spin-coated $SnS_{1.4}Se_{0.5}$ channel layer, employing a 210-nm SiO_2 gate oxide and a channel length/width of L = 14 μm/W = 250 μm, respectively. (c) Plots of I_D and $I_D^{1/2}$ versus V_G at constant V_D = 85 V, from the same device, used to calculate current modulation, I_{ON}/I_{OFF} and saturation regime mobility, μ_{sat}. [Reproduced with permission from [Ref. 31]. Copyright 2004 Nature Publishing Group.]

demonstrated in these early devices (e.g., see Section 3.3.2).[46] Substantially better electrical characteristics were provided by devices employing Se-substituted thin films (i.e., as for the $SnS_{1.4}Se_{0.5}$ films described) versus pure SnS_2 analogs, perhaps because of the improved grain structure in these films or because of modification of the electronic structure brought about by this substitution.[31]

Given the toxic nature of hydrazine, it is clearly desirable to reduce the amount of hydrazine used in the spinning process. To this end, high-quality tin chalcogenide films spun using a mixture of water and hydrazine (20% hydrazine in water by volume) have also been demonstrated.[47] The electrical properties of these films are comparable with those achieved from films deposited from pure hydrazine.

3.3.2 In_2Se_3 Films

In an effort to explore the need for hydrazine during the spin-coating process, In_2Se_3 has been deposited using a route that first isolates the hydrazinium-based metal chalcogenide precursor, followed by dissolving the precursor in a non-hydrazine-based solvent for spin coating.[32] The attractiveness of this approach for large-scale production is that, assuming the identification of a suitable alternative solvent, hydrazine use can be limited to the step of precursor synthesis (presumably to be carried out by a chemical manufacturer that handles hydrazine). During device fabrication, more environmentally benign solvents might be employed.

The In_2Se_3 precursor is isolated by dissolving In_2Se_3 and Se in hydrazine, as for the tin(IV) chalcogenides.[32] Evaporation of the resulting solution yields a nominally amorphous hydrazinium precursor with approximate composition $(N_2H_4)_2(N_2H_5)_2In_2Se_4$. Decomposition of the precursor upon heating in an inert atmosphere follows a multistep process (Fig. 3.7), with completion of the bulk transformation to crystalline γ-In_2Se_3 (Fig. 3.7, inset) by approximately 350 °C. As for the $SnSe_{2-x}S_x$ system, incorporation of selenium results in a decomposition temperature in excess of 300 °C, presumably as a result of the low volatility of Se at temperatures below this point.

Combinations of ethanolamine and dimethylsulfoxide (DMSO) work effectively as an alternative solvent for the hydrazinium precursor (individually these solvents are not as effective) and as a good wetting agent for the thermally oxidized silicon substrates. Although still corrosive and/or moderately toxic, the safety characteristics of ethanolamine and DMSO (often used as a vehicle for drug delivery) are substantially more desirable than those of hydrazine.[48,49] Indium selenide precursor films are formed by dissolving the hydrazinium-based precursor in the mixed-solvent system under nitrogen, spin coating the solution onto thermally oxidized silicon substrates, and performing a short low-temperature decomposition step with a maximum temperature of between 320 °C and 350 °C. The resulting stoichiometry of the deposited films is $In_{2.0(1)}Se_{2.8(1)}$, as determined by MEIS. Continuous films with thicknesses

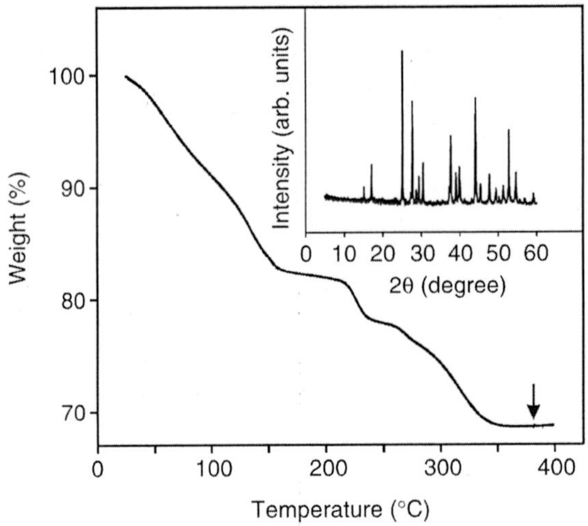

Figure 3.7. TGA scan for the In_2Se_3 precursor, run using a 2°C/min heating rate and a flowing nitrogen atmosphere. Inset: The room-temperature powder XRD data (Cu Kα radiation) for the product, obtained by heating the precursor to 380°C (see arrow in TGA plot) in a nitrogen atmosphere. The observed diffraction peaks match those for the γ-In_2Se_3 structure (Powder Diffraction File [PDF] Card No. 40-1407). [Reproduced with permission from D. B. Mitzi et. al., "Low-voltage transistor employing a high-mobility spin-coated chalcogenide semiconductor," *Adv. Mater.* **17**, 1285 (2005) [Ref. 32]. Copyright Wiley-VCH Verlag GmbH & Co. KGaA.]

ranging from ~30 to 120 Å have been successfully deposited using this process, by changing solution concentration and spin speed. X-ray diffraction from one of the thicker films confirms that the films are γ-In_2Se_3, with a substantial amount of *c*-axis preferred orientation (Fig. 3.8).

TFTs based on the spin-coated In_2Se_3 films have been fabricated employing a relatively thin (250 Å) thermal SiO_x gate insulator (~8× thinner than for the $SnSe_{2-x}S_x$ devices described earlier) and coevaporated gold/indium contacts. Figure 3.9a displays the electrical characteristics for the n-channel indium-selenide-based TFT. Current modulation (I_{ON}/I_{OFF}), saturation regime field-effect mobility (μ_{sat}), linear-regime mobility (μ_{lin}), and subthreshold swing (S) are calculated using standard equations[45] from the data in Figs. 3.9a and 3.9b, yielding $I_{ON}/I_{OFF} = 2 \times 10^6$, $\mu_{sat} = 16.5\,cm^2/V\text{-}s$, $\mu_{lin} = 4.9\,cm^2/V\text{-}s$, and S = 450 mV/decade (Fig. 3.9c). Operating voltages (≤8 V) are an order of magnitude lower than those for the previously discussed devices based on spin-coated tin(IV) chalcogenides,[31] demonstrating the effectiveness of a thinner gate insulator.[46] The observed subthreshold swing can be compared with the theoretical lower limit of ~60 mV/decade at room temperature[45] and is similar in magnitude to that found in high-performance amorphous-silicon-based devices.[50] The calculated mobilities ($V_G = 8\,V$) are consistent with the lower end of the $20 \le \mu \le 60\,cm^2/V\text{-}s$ Hall mobility range previously established in vacuum-evaporated

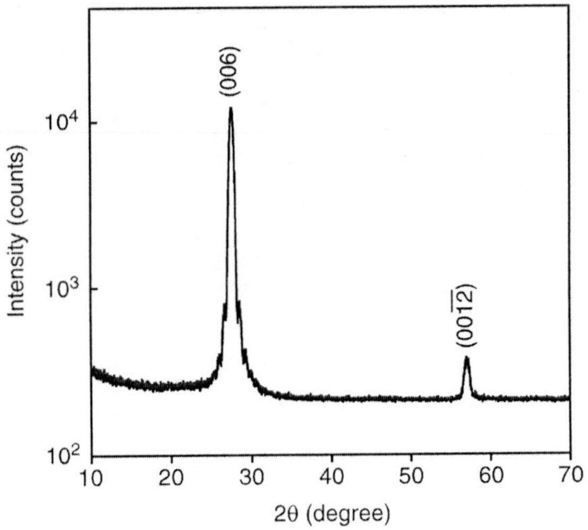

Figure 3.8. Powder XRD data (Cu Kα radiation) for the film achieved by spin-coating the In_2Se_3 precursor, with a final anneal at 350 °C in an inert atmosphere (film thickness ~110 Å).

polycrystalline γ-In_2Se_3 films.[51] These factors suggest that higher mobility values might be possible with application of higher dielectric constant gate insulators and/or by gaining better control over film grain structure.

3.3.3 $CuInTe_2$, $CuInSe_2$, and $Cu(Ga_{1-x}In_x)Se_2$ Films

Although the examples discussed so far fall into the category of n-type semiconductors, p-type films are also desirable for many applications. Solar cells, for example, rely on the formation of a junction between n- and p-type layers (to drive charge separation at the interface), and the availability of both n- and p-type transistors is important for the realization of CMOS (complementary metal-oxide semiconductor) technology. Demonstration of p-type films using the hydrazine-based process was pursued using the ternary chalcopyrite systems, $CuInTe_2$ and $CuInSe_2$.[35,37] $CuInTe_2$ proved to be a particularly interesting example because the zinc-blende analog is not substantially soluble in hydrazine, thereby requiring the development of a modified approach for deposition, and it also represents a first example of deposition of a telluride-based system using the hydrazine-based approach (another example is ZnTe).[34] Crystals of $CuInTe_2$ also offer a band gap of ~1 eV and relatively high Hall mobilities (p-type transport), ranging from 30 to 150 cm^2/V-s, depending on preparation conditions.[52–54]

Since $CuInTe_2$ does not significantly dissolve in hydrazine, a new stepwise process was devised for dissolution, which involves separate indium- and

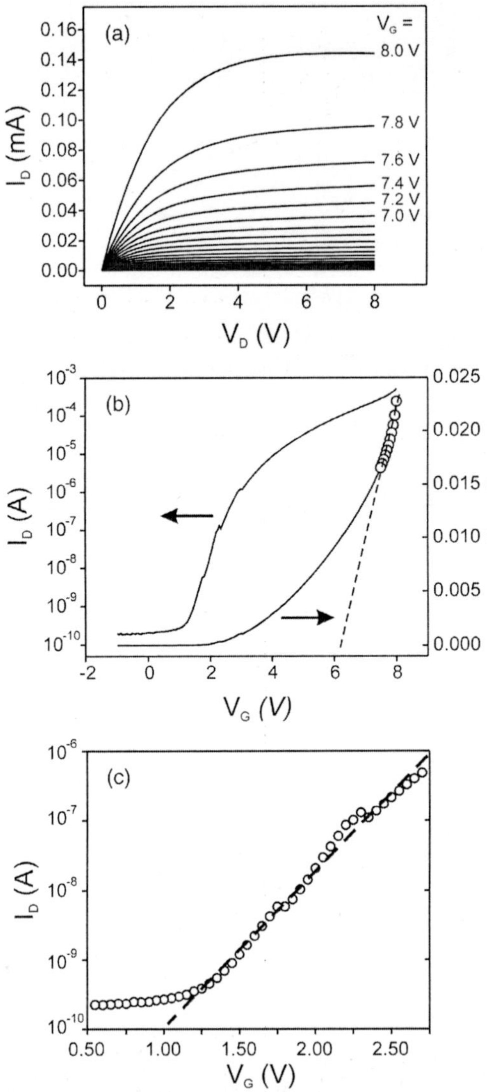

Figure 3.9. (a) Drain current I_D versus drain voltage V_D as a function of gate voltage V_G for a spin-coated In_2Se_3 channel layer, deposited from a non-hydrazine-based solvent and employing a 25-nm-thick SiO_2 gate oxide and a channel length/width of L = 13 μm/W = 1500 μm. (b) Plots of I_D and $I_D^{1/2}$ versus V_G at constant V_D = 8 V, from the same device, used to calculate current modulation, I_{ON}/I_{OFF}, saturation regime mobility, μ_{sat}, and subthreshold swing, S. (c) Enlarged plot of I_D versus V_G at constant V_D = 8 V, in the subthreshold region. The dashed line shows the approximately logarithmic V_G dependence of I_D over several orders of I_D. [Reproduced with permission from D. B. Mitzi et. al., "Low-voltage transistor employing a high-mobility spin-coated chalcogenide semiconductor," *Adv. Mater.* **17**, 1285 (2005) [Ref. 32]. Copyright Wiley-VCH Verlag GmbH & Co. KGaA.]

copper-based precursor solutions.[37] One component, In_2Te_3, readily goes into solution at a concentration of 0.05 M in hydrazine. The other prospective component, Cu_2Te, is not substantially soluble in hydrazine. However, the solubility of the analogous sulfide system enables the preparation of a Cu_2S solution,[38] which provides the soluble source of Cu. At the time of deposition, the Cu_2S solution is combined with the In_2Te_3 solution (~1:1 molar ratio of Cu:In), along with sufficient Te to replace the sulfur from the Cu_2S precursor. It is expected that the sulfur will preferentially dissociate from the material during the heat treatment, given its higher volatility relative to tellurium. The relatively slow precipitation process for Cu_2Te from the mixed precursor solution enables films of the ternary $CuInTe_2$ chalcopyrite to be spin coated from a freshly mixed solution, after a short (~20 min) inert atmosphere thermal decomposition step at 340 °C. It is important to use the mixed precursor solution within a short time frame after mixing of the In_2Te_3/Te and Cu_2S solutions to avoid precipitation of Cu_2Te.

The resulting $CuInTe_2$ films are crystalline and oriented (Fig. 3.10a,b). The fringes in the cross-sectional TEM image of the film, which run parallel to the

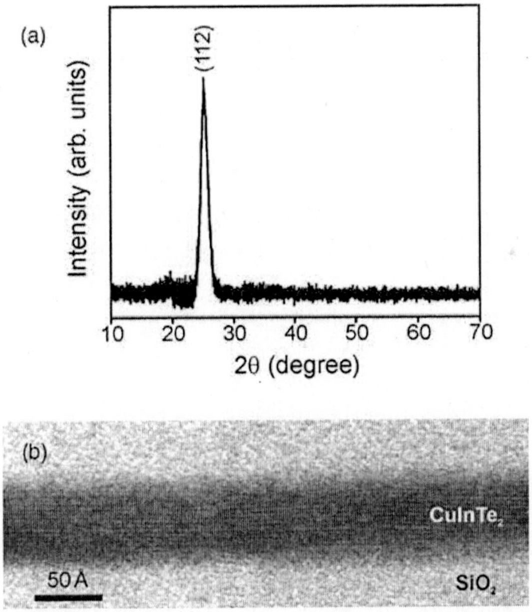

Figure 3.10. (a) Powder X-ray diffraction data (Cu Kα radiation) for the film achieved by spinning the $CuInTe_2$ precursor on quartz, with a final anneal at 340 °C in an inert atmosphere. The reflection indexing is from PDF card 34–1498 for $CuInTe_2$. (b) Cross-sectional TEM image of an analogous $CuInTe_2$ film on thermally oxidized silicon. [Reproduced with permission from D. B. Mitzi et. al., "High-mobility p-type transistor based on a spin-coated metal telluride semiconductor," *Adv. Mater.* **18**, 2448 (2006) [Ref. 37]. Copyright Wiley-VCH Verlag GmbH & Co. KGaA.]

substrate (Fig. 3.10b), correspond to a lattice spacing of 3.54(7) Å, in good agreement with the value for the $CuInTe_2$ (112) planes and with the diffraction peak of Fig. 3.10a. The composition of the spin-coated films is $Cu_{0.9(1)}In_{1.2(1)}Te_{2.0(1)}$, as determined by MEIS and X-ray photoelectron spectroscopy (XPS). The sulfur content in the films is estimated as <2 atomic % from the MEIS analysis, indicating that the sulfur from the initial Cu_2S has largely been removed from the sample. The composition of the films, particularly the Cu-to-In ratio, can be tailored by changing the quantities of the component solutions used during the spin-coating process.

TFTs based on spin-coated $CuInTe_2$ films have been fabricated (the same basic structure as in Fig. 3.6a) employing a 400-Å thermal SiO_x gate insulator and evaporated gold source and drain contacts. Fig. 3.11a displays the device characteristics for the p-channel $CuInTe_2$-based TFT. Current modulation (I_{ON}/I_{OFF}), saturation-regime field-effect mobility (μ_{sat}), and linear-regime mobility (μ_{lin}) were calculated[45] from the data in Figs. 3.11a and 3.11b, yielding $I_{ON}/I_{OFF} = 10^5$, $\mu_{sat} = 11.8\,cm^2/V\text{-}s$, and $\mu_{lin} = 7.0\,cm^2/V\text{-}s$. Electrical measurements performed on vacuum evaporated films of the same material have demonstrated room-temperature field-effect mobilities of as high as $50\,cm^2/V\text{-}s$, indicating the potential for improvement in electrical characteristics for the solution-processed $CuInTe_2$ films with additional process refinement.[55]

In contrast to the relatively stable electrical measurements for $SnS_{2-x}Se_x$ and In_2Se_3 films,[31,32] the $CuInTe_2$ devices exhibited time-dependent electrical character. Electrical measurements performed one week after device fabrication yielded higher drain currents I_D for a given gate (V_G) and drain (V_D) voltage. For example, in the device shown in Fig. 3.11, this led to a reduction in the I_{ON}/I_{OFF} ratio to 10^3. The saturation-regime mobility (μ_{sat}) also decreased by ~2×, whereas μ_{lin} changed less ($6.0\,cm^2/V\text{-}s$). The reason for this variability is still under investigation, but it may be related to Cu^+ ionic mobility in the generally nonstoichiometric chalcopyrite structure.[56,57]

Using the general approach outlined above for $CuInTe_2$, in which multiple stable solutions are mixed immediately prior to spin coating to prevent preciptitation of an insoluble metal chalcogenide component, it is envisioned that other examples of poorly soluble chalcogenides might also be spin coated using hydrazine-based precursors, thereby significantly extending the applicability of this approach. Note that analogous $CuInSe_2$ films have also been deposited using essentially the same process to that described above for the $CuInTe_2$ system.[35] For the selenide analog, however, the mixed precursor solutions are relatively stable, and in fact, the precursor solution can be prepared in a single reaction vessel (with no precipitation of the product from the solution, at least over a period of up to one week). The resulting $CuInSe_2$ films, achieved after spin coating of the precursor solution followed by a 350 °C anneal, are crystalline and (112)-oriented, as for the $CuInTe_2$ system, but they yielded lower mobilities in a TFT configuration.

For solar cell applications, doping of $CuInSe_2$ with Ga (CIGS) yields vacuum-evaporated absorber layers that produce solar cells with the highest

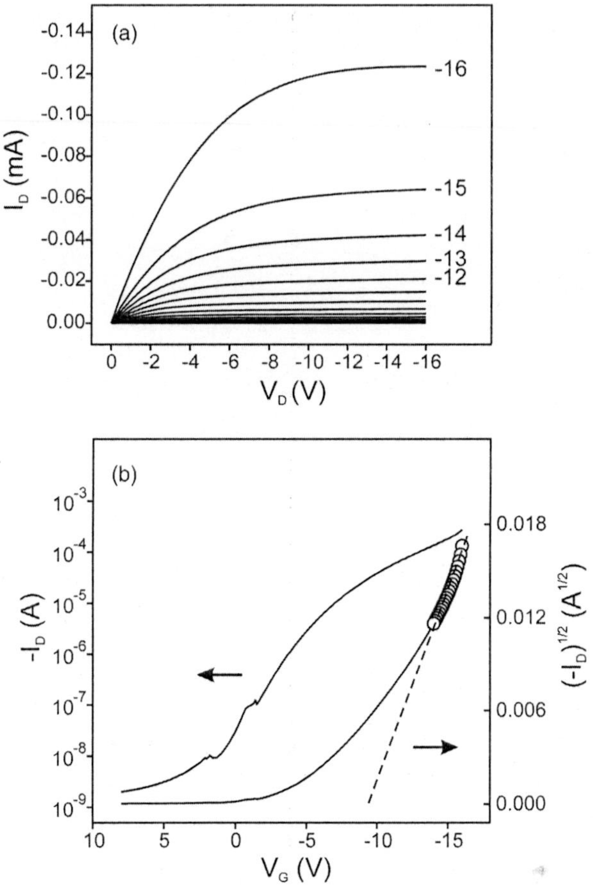

Figure 3.11. (a) Drain current I_D versus drain voltage V_D as a function of gate voltage V_G (listed along the right side of the graph in volts) for a spin-coated $CuInTe_2$ channel layer, employing a 40-nm-thick SiO_2 gate oxide and a channel length/width of L = 95 μm/W = 1000 μm. (b) Plots of I_D and $I_D^{1/2}$ versus V_G at constant $V_D = -16$ V, from the same device, used to calculate current modulation, I_{ON}/I_{OFF}, and saturation regime mobility, μ_{sat}. [Reproduced with permission from D. B. Mitzi et. al., "High-mobility p-type transistor based on a spin-coated metal telluride semiconductor," *Adv. Mater.* **18**, 2448 (2006) [Ref. 37]. Copyright Wiley-VCH Verlag GmbH & Co. KGaA.]

power conversion efficiencies (~19%) among the chalcopyrite family. However, for the TFT application described above, it has been desirable to keep film thicknesses very thin (~10 nm). For solar cell devices, much thicker films are required for the CIGS layer (~1 μm) in order to absorb a large fraction of the incident solar radiation. Deposition of such thick films using the solution-based process requires the transport of a substantial quantity of decomposition products throughout the volume of the deposited film to the film/environment interface, therefore leading to potential issues with film density, homogeneity,

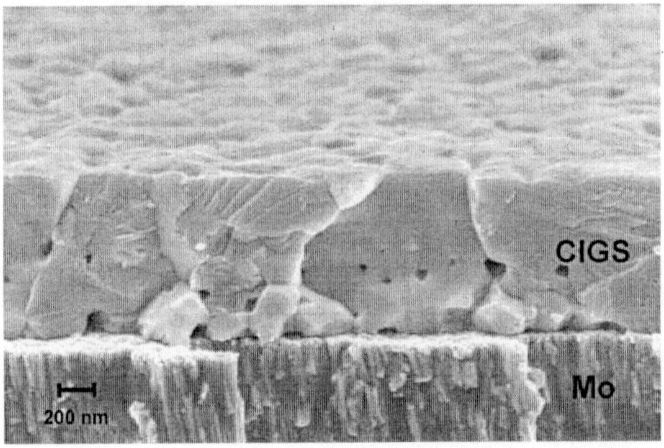

Figure 3.12. SEM image of a $CuIn_{0.7}Ga_{0.3}Se_2$ film deposited using a hydrazine-based spin-coating process. The film is deposited on a ~500-nm-thick layer of Mo on a glass substrate.

and morphology (e.g., cracks and delamination). As seen in Fig. 3.12, micrometer-thick layers, with reasonable grain structure and $CuIn_{0.7}Ga_{0.3}Se_2$ stoichiometry, can be deposited using the hydrazine-precursor approach.[58] These films are expected to be useful in the fabrication of lower cost CIGS-based solar cells.

3.3.4 Cu_2S Precursor

The deposition process for $CuInSe_2$ requires using a hydazine-based precursor of Cu_2S, $N_4H_9Cu_7S_4$, as well as an analogous In_2Se_3 precursor. The same copper(I) precursor was used to deposit films of $CuInTe_2$, with sulfur being replaced by tellurium during the final heat treatment.[37] The structure of $N_4H_9Cu_7S_4$ is found to be composed of extended $\mathbf{Cu_7S_4^-}$ sheets, separated by a mixture of hydrazinium and hydrazine molecules,[38] making this the first example among the reported structurally characterized hydrazinium-based salts with an extended two-dimensional (2D) metal chalcogenide anion (Fig. 3.13), rather than isolated metal chalcogenide anions. The reduced dimensionality (2D rather than 3D) network, nevertheless, apparently aides in the solubility of the Cu_2S precursor. Additionally, $N_4H_9Cu_7S_4$ is found to decompose at a substantially lower temperature than previously studied hydrazinium-based precursors, completing the decomposition process to copper(I) sulfide by ~120 °C (Fig. 3.14) and potentially enabling very low-temperature thin-film deposition for materials containing the copper(I) species.

3.3.5 KSb_5S_8 Films

Phase-change materials (PCMs) represent another important application for metal chalcogenide films, as a result of emerging memory technologies, includ-

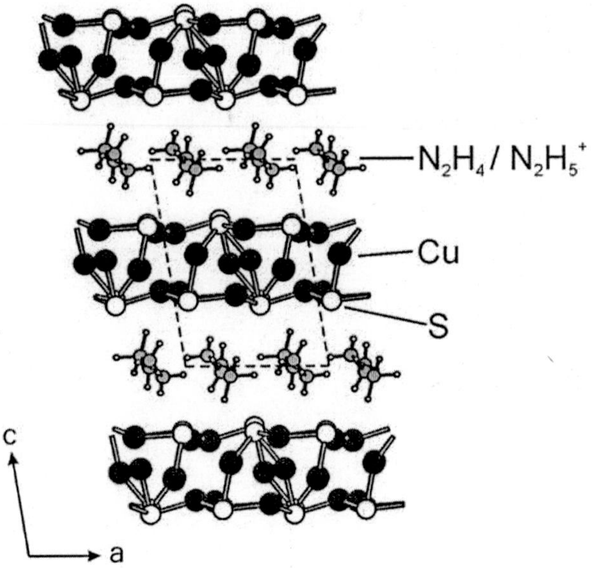

Figure 3.13. Crystal structure of $N_4H_9Cu_7S_4$ viewed down [010], with the unit cell shown using a dashed line. For clarity, atoms are represented as spheres, with uniform sizes selected for each atom type. Within the $Cu_7S_4^-$ slabs, Cu—Cu bonding is not shown to improve clarity.

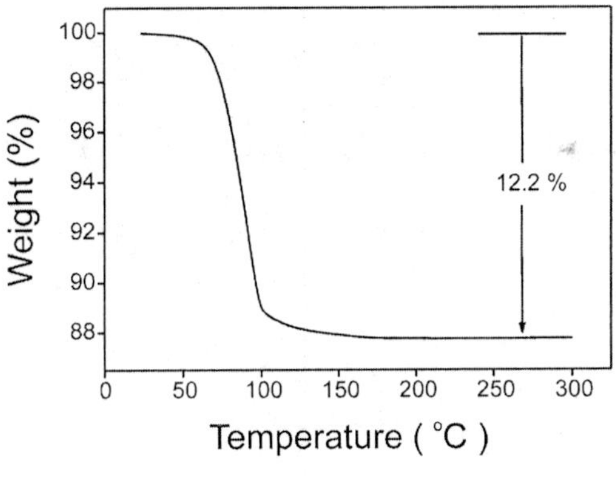

Figure 3.14. TGA of $N_4H_9Cu_7S_4$ (5 °C/min ramp rate; nitrogen atmosphere). Weight loss at 270 °C is shown. [Reproduced with permission from [Ref. 38]. Copyright 2007 American Chemical Society.]

ing commercially available rewritable optical media (e.g., CD-RW, DVD-RW, and DVR)[5,6,36,59,60] and the development of nonvolatile phase-change memory (PRAM).[61,62] A PCM film must be switchable between two physical states (i.e., usually amorphous and crystalline), with the change in state being detectable using a physical measurement, such as optical absorption, reflectivity, or electrical resistivity. Important characterisitics for a PCM include good reversibility (ability to cycle between the two states many times) and very fast (i.e., nanosecond) switching times. Crystallization temperatures also ideally should be above 150 °C to avoid accidental change of device state caused by environmental heating issues (e.g., device placed in a hot car with windows closed). Commercial examples of PCMs include the Ge-Sb-Te and Ag-In-Sb-Te compositions.

In principle, the hydrazine-based film deposition approach is particularly well suited for PCM films because of the reduced importance of grain structure in the resulting films compared with films for applications such as transistors and solar cells. Indeed, whether the resulting film is amorphous or crystalline is of less consequence given that the films will be cycled between these two states during operation. Application of the hydrazine-based film deposition process to PCMs has been demonstrated using KSb_5S_8 as an example.[36] Unlike many chalcogenide glasses, KSb_5S_8 is a congruently melting stoichiometric compound, with a relatively high glass-to-crystallization temperature (287 °C) and large optical band gap (i.e., 1.82 eV for crystalline and 1.67 eV for glass forms).[63]

The KSb_5S_8 precursor solution was prepared by *carefully* dissolving elemental potassium, antimony, and sulfur in hydrazine under an inert atmosphere. The ability to prepare the precursor solution at ambient temperature in a single vessel from the elements, without the need to first isolate the metal chalcogenide, represents a simplification of the hydrazine precursor process and has been demonstrated with several other related binary chalcogenide systems (e.g., SnS_2, In_2Te_3, and Ga_2Se_3). *It should be noted, however, that the addition of alkali metals such as K to hydrazine involves a highly energetic (i.e., explosive) reaction and should only be done with extreme care (i.e., very slow addition and in small quantities). For larger scale production of an alkali-containing precursor, isolation of the alkali chalcogenide species (e.g., K_2S) and use of the metal chalcogenides for starting materials, rather than the elements, would be necessary from a safety standpoint.*

Films are formed by spin coating the precursor solution onto thermally oxidized silicon substrates and then performing a short decomposition step with maximum temperature of ~250 °C. As is typical for sulfides, temperatures above 300 °C are generally not required for the decomposition step because of the facile volatility of hydrogen sulfide and/or sulfur decomposition products. The resulting composition for spin-coated films is $K_{1.2(1)}Sb_{4.9(1)}S_{8.0(1)}$, as determined by MEIS, which is consistent with the targeted KSb_5S_8 stoichiometry. Films with thicknesses ranging from ~10 to 90 nm have been deposited using this process by changing solution concentration and spin speed. However, film surface roughness (see SEM image in Fig. 3.15a) seems to be substantially larger than that observed for earlier discussed metal chalcogenide films, perhaps because of grain growth during the exothermic decomposition or because of less favorable wetting characteristics for the KSb_5S_8 materials relative to those for other examined chalcogenide systems on the oxidized silicon surfaces. As evidence for the poor wetting, an analogous film to that shown

Figure 3.15. (a) Cross-sectional SEM of a spin-coated KSb_5Se_8 film, prepared using a 0.12 M precursor solution (in hydrazine) with 5000 rpm spin speed and deposited on silicon coated with ~100 nm of thermal SiO_x. (b) Similarly prepared KSb_5S_8 film, subsequently heated to 475 °C for 2 min on a hot plate and cooled to room temperature.

in Fig. 3.15a was briefly heated above the melting point for KSb_5S_8 (448 °C) and then cooled, yielding the surface shown in Fig. 3.15b. The chalcogenide material appears to have balled up (i.e., dewetted) the surface. Better film morphology might be expected with a more compatible choice of substrate surface.

X-ray diffraction from the deposited films confirms that they are of the same phase as crystalline bulk KSb_5S_8,[63,64] with little preferred orientation (Fig. 3.16). Quenching films from above the melting temperature (460 °C to below 100 °C in ~1 min) brings them into an amorphous state. Subsequent heating of the amorphous KSb_5S_8 film at 1 °C/s leads to crystallization at ~280(5) °C (Fig. 3.17a), in good agreement with the expected bulk crystallization transition at 287 °C.[63] Further heating to above 400 °C leads to remelting of the film. Three thermal cycles are shown in Figs. 3.17a–c, demonstrating the reversibility of the amorphous-to-crystalline transition.

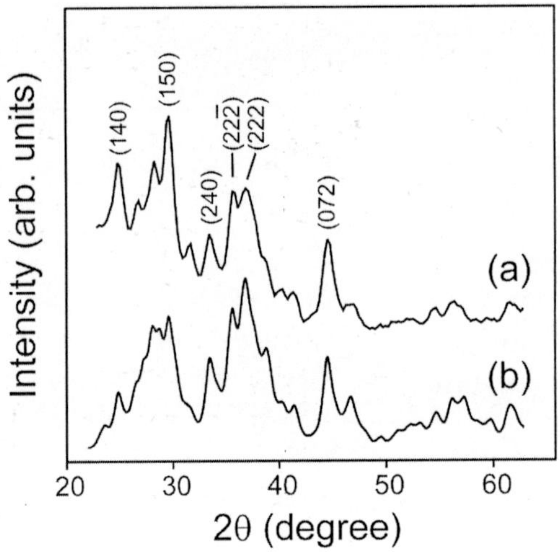

Figure 3.16. (a) X-ray diffraction (λ = 1.797 Å) from a spin-coated KSb_5S_8 film, demonstrating good agreement with (b) a simulated diffraction pattern for bulk KSb_5S_8 (using 1° peak fwhm in the modeling). [Reproduced with permission from [Ref. 36]. Copyright 2006 American Chemical Society.]

KSb_5S_8 films are much easier to melt quench into an amorphous state than most PCMs. Other phase change materials such as $Ge_2Sb_2Te_5$ crystallize in nanosecond timescales[65] and, thus, require cooling rates of as high as 10^9–10^{10} °C/s. The relative ease with which KSb_5S_8 can be quenched into the amorphous state reflects the correspondingly slow crystallization rate of these films. Despite demonstrating a reproducible amorphous-crystallization transition and a reasonably high crystallization temperature, the crystallization kinetics for KSb_5S_8 films seem too slow to render them useful for high-data-rate transfer applications (they could still be useful for write once or write few-time applications). However, although we have demonstrated the spin-coating technique for KSb_5S_8, we expect this process to be more generally applicable to the film deposition and bulk preparation of other chalcogenide-based PCMs. As an example, Ge–Sb–Se films have been recently deposited using an analogous solution-based approach.[66] These films exhibited a faster crystallization time of order 100 ns. In addition, a benefit of solution processing, namely the ability to fill holes and vias effectively, was employed to produce arrays of Ge–Sb–Se nanodots (~30 nm in diameter), which crystallize at a reduced temperature compared with a blanket film.[66]

3.3.6 Other Metal Chalcogenide Systems

Besides $CuInTe_2$, several other metal tellurides have been deposited using the hydrazine-based approach. Among them, ZnTe is somewhat unique in that the precursor was isolated in single crystal form, and therefore, the structure

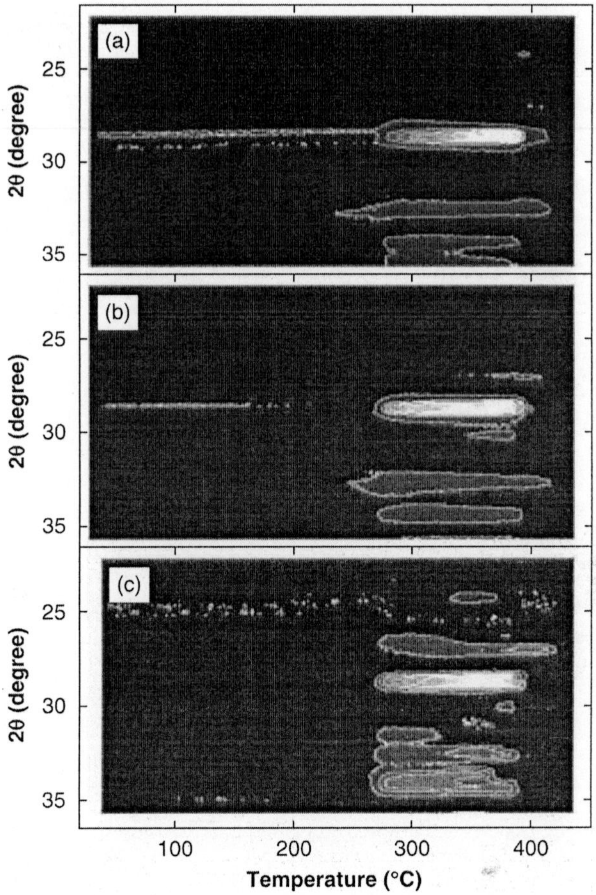

Figure 3.17. Three sequential variable-temperature X-ray scans (λ = 1.797 Å) performed on the same spin-coated KSb_5S_8 film after an initial heat and quench treatment. In each case, the heating rate is 1 °C/s and the sample is quenched to room temperature at the end of the heating portion of the cycle. White coloration in the plot corresponds to higher X-ray intensity.

could be elucidated (Fig. 3.18).[34] In contrast to the crystalline sulfide and selenide precursors discussed earlier, which consisted of metal chalcogenide anions separated by hydrazinium cations, in $(N_2H_4)_2ZnTe$, the structure is composed of extended ZnTe chains with covalently coordinated neutral hydrazine molecules (two per Zn atom). Two distinct structures, α-$(N_2H_4)_2ZnTe$ and β-$(N_2H_4)_2ZnTe$, have been isolated with a slightly different conformation for the extended ZnTe chains. The resulting dimensional reduction relative to bulk ZnTe for each polymorph yields a soluble ZnTe precursor. Perhaps more importantly, the absence of extra tellurium in the compound (which would be present if this precursor were a metal chalcogenide salt) enables very

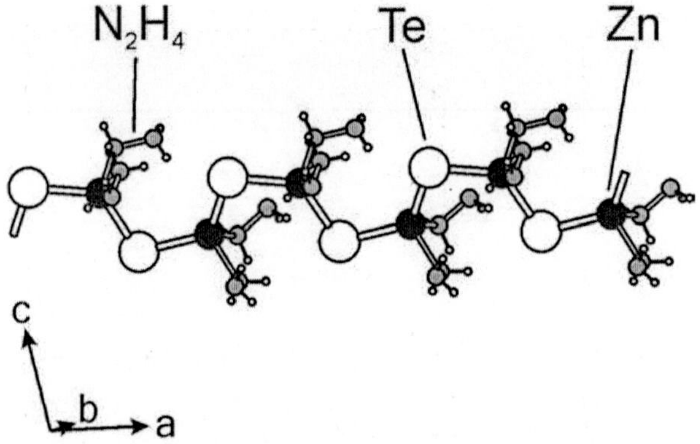

Figure 3.18. Structure of the $(N_2H_4)_2ZnTe$ chains in α-$(N_2H_4)_2ZnTe$, with the atom labeling shown. For clarity, atoms are represented as spheres, with uniform sizes selected for each atom type.

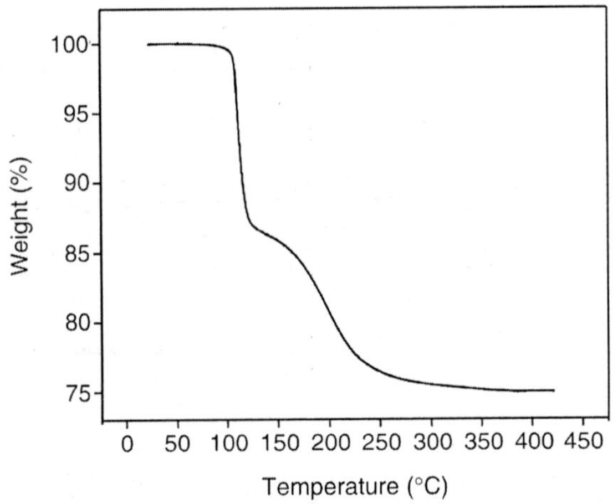

Figure 3.19. TGA of $(N_2H_4)_2ZnTe$, performed using a 5 °C/min heating rate in a nitrogen atmosphere. [Adapted with permission from [Ref. 34]. Copyright 2005 American Chemical Society.]

low temperature decomposition (Fig. 3.19). If extra tellurium were present in the sample, temperatures in excess of 450 °C would likely be required to dissociate the Te from the film.

Despite facile formation of p-type ZnTe from the precursor, films of ZnTe formed from a drop cast precursor solution had a small grain structure (using conditions that we have so far explored), which limited prospective TFT device performance. Note that we expect a similar chemistry to that for ZnTe to also be operative for selected other metal telluride precursors. In contrast to the ZnTe films, well-crystallized films of In_2Te_3 have been formed using

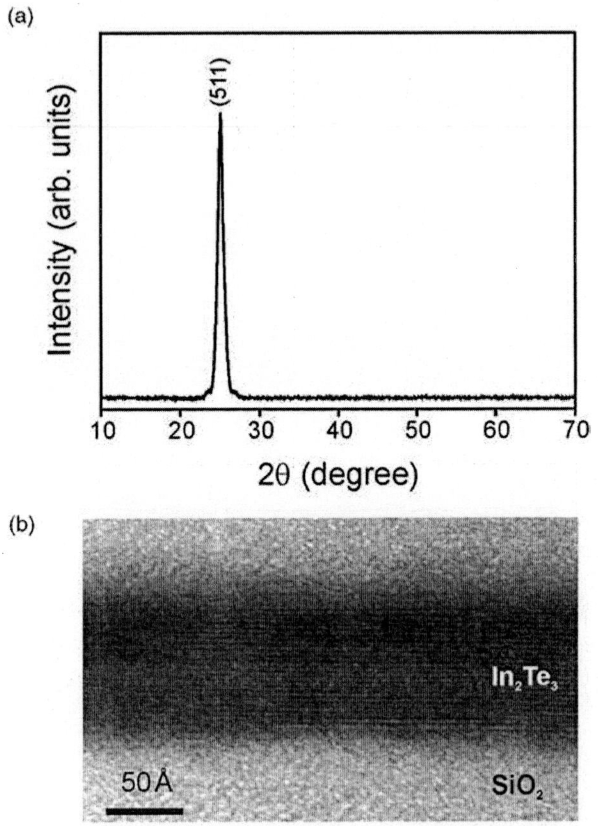

Figure 3.20. (a) Powder X-ray diffraction (Cu Kα radiation) for a spin-coated In_2Te_3 film on quartz. The background from the quartz disk has been separately measured and subtracted. (b) Cross-sectional TEM image of an analogous film deposited on thermally oxidized silicon. The observed lattice fringes, which run parallel to the substrate, correspond to a lattice spacing of 3.53(7) Å, in good agreement with the expected value for the In_2Te_3 (511) planes. [Reproduced with permission from D. B. Mitzi et. al., "High-mobility p-type transistor based on a spin-coated metal telluride semiconductor," *Adv. Mater.* **18**, 2448 (2006) [Ref. 37]. Copyright Wiley-VCH Verlag GmbH & Co. KGaA.]

spin-coated hydrazine-based solutions (Fig. 3.20), which have also been useful in experiments designed to fill vias and holes on a substrate (Fig. 3.21). Attempts to produce thick films by using a much more concentrated solution or by using a slower spin speed (or by drop casting instead of spin coating) generally yielded films that exhibited a substantial amount of porosity, as a result of the need to remove gaseous products during the decomposition process. Therefore, multiple deposition processing is preferable for achieving a thick coating and more complete filling of the vias.

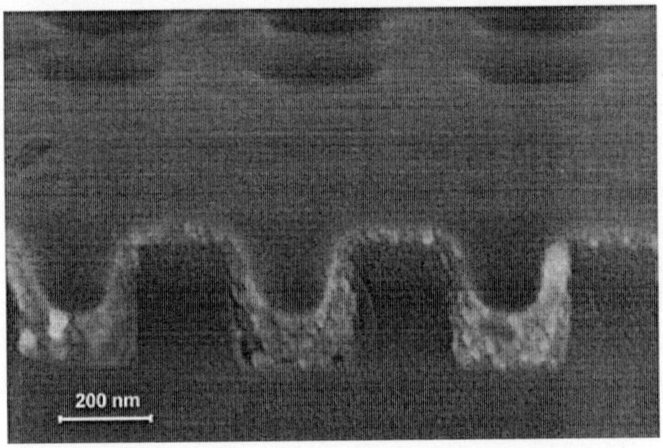

Figure 3.21. Cross-sectional SEM image of an In_2Te_3-coated substrate, with conformal filling of holes. Three iterations of spin-coated In_2Te_3 deposition were employed.

Finally, although the target for most of the systems described above has been for the deposition of crystalline films (with the possible exception of the PCMs), for certain applications, amorphous films are actually desired. As an example, solid-state electrolytic memory devices are based on a chalcogenide-based solid electrolyte glass, such as germanium sulfide, with silver diffused into the material.[67] The devices operate by the electrochemical growth and removal of nanoscale metallic pathways in thin electrolytic films. Using a process analogous to that described for the $SnSe_{2-x}S_x$ films, amorphous GeS_2 film deposition has been accomplished. Even for an annealing temperature of as high as 400 °C, the films remain amorphous, as demonstrated by X-ray diffraction studies performed on the films and decomposed bulk precursor.

3.4 SIMILAR APPROACHES WITHOUT HYDRAZINE

The hydrazine-based approach described above is based on a simple dimensional reduction concept. Namely, the not-very-soluble 3D metal chalcogenide system is transformed into a more soluble lower dimensional array of, for example, metal chalcogenide anions, separated by small, volatile cationic species. The cationic species can readily be removed from the sample (along with corresponding extra chalcogen) using a low-temperature heat treatment, leaving behind the desired metal chalcogenide framework. Although hydrazine has the advantage of being small, volatile, and non-carbon-containing (no carbon impurities left in the film after heating), other solubilizing species are also fully compatible with the idea presented above. For example, films of crystalline MoS_2 have been spin coated from a solution of $(NH_4)_2MoS_4$ in an organic diamine.[68] In $(NH_4)_2MoS_4$, the extended framework of MoS_2 is broken up into discrete MoS_4^{2-} moieties separated by NH_4^+ cations.[69] However, high-temperature post-deposition anneals are generally required to achieve crystalline films (600–800 °C), rendering the process incompatible with organic-based flexible substrate materials. A similar approach with $(CH_3NH_3)_2MS_4$ (M = W, Mo) dissolved in organic solvents and an inert atmosphere anneal led to highly conducting and textured n-type MoS_2 (~50 $\Omega^{-1}cm^{-1}$) and WS_2 (~7 $\Omega^{-1}cm^{-1}$) films.[70] Amorphous As_2S_3 and As_2Se_3 films have also been deposited using an analogous process,[71] but attempts to deposit other main-group metal chalcogenides, such as Sb_2S_3 and GeS_x, have not been successful, because of the low solubility of the precursors in the diamine solvents.[72] The use of hydrazine as a solvent, therefore, offers some benefits in terms of providing for a wider range of materials that may be deposited, and perhaps with respect to a cleaner (no carbon) and lower temperature decomposition pathway.

3.5 FUTURE PROSPECTS

The basic idea presented in this chapter is that of taking an extended metal chalcogenide framework and breaking it up into smaller (lower dimensional) moieties, which are separated by small and volatile solublizing species. Several metal chalcogenides have been solution-processed using this approach, including $SnSe_{2-x}S_x$, In_2Se_3, GeS_2, $GeSe_2$, Cu_2S, Sb_2Se_3, Sb_2Te_3, $CuInSe_2$, $Cu(In_{1-x}Ga_x)Se_2$, $CuInTe_2$, Ga_2Se_3, and ZnTe. The application of this idea to metal chalcogenides has the advantage of providing a simple, relatively low-temperature approach to the film deposition, as well as offering the additional benefit of flexibility to fill holes and vias during the deposition process (important for the implementation of several advanced device designs). Although the approach is described here for chalcogenides, the basic idea might be expected to work equally well for oxides and other inorganic systems.

Despite being a viable and attractive approach for large-area and low-cost deposition on a variety of substrate materials (including potentially those that are flexible and plastic), the application of this solution-based process is still at a very early stage of development. One issue that needs to be addressed for certain applications is the deposition of thick films versus thin films. So far, the process has been primarily used to deposit films with thickness less than 100 nm. For some applications (e.g., solar cells), deposition of metal chalcogenide layers with thickness of order 1 μm is required (note the early results along these lines in Section 3.3.3). In addition, it is often desirable to deposit multilayers composed of different material systems (e.g., n-type and p-type materials with selected band gap). An important issue is therefore the ability to deposit multiple layers using this approach, without disturbing (e.g., partially dissolving) the underlying layers. Although, in many respects hydrazine represents an ideal solvent for the deposition process (i.e., it is a small and volatile molecule with no carbon or oxygen incorporated), the toxicity and reactivity of this solvent suggest the need to identify more environmentally benign alternatives (see Sections 3.3.2 and 3.4 for some possibilities). Finally, many of the chalcogenide systems discussed are inherently non-stoichiometric compounds. For example, the system SnS_2 can accommodate a moderate range of sulfur stoichiometry and $CuInSe_2$ similarly provides for a range of metal cation and chalcogenide vacancies and substitutions. The potential for variations in stoichiometry, while providing the opportunity for tuning the properties of the resulting films, also leads to challenges in producing films with easily reproducible properties. More work on characterizing the effects of solution aging and preparation conditions on film properties would also be useful in the effort to advance this approach toward commercial utility.

REFERENCES

1. Contreras, M. A.; Ramanathan, K.; AbuShama, J.; Hasoon, F.; Young, D. L.; Egaas, B.; Noufi, R. 2005. Diode characteristics in state-of-the-art ZnO/CdS/$Cu(In_{1-x}Ga_x)Se_2$ solar cells. *Prog. Photovolt.: Res. Appl.* 13:209–216.
2. Herrmann, D.; Kessler, F.; Klemm, U.; Kniese, R.; Friedlmeier, T. M.; Spiering, S.; Witte, W.; Powalla, M. 2005. Flexible, monolithically integrated $Cu(In,Ga)Se_2$ thin-film solar modules. *Mater. Res. Soc. Symp. Proc.* 865:F15.1.1–F15.1.8.
3. Dhere, N. G.; Dhere, R. G. 2005. Thin-film photovoltaics. *J. Vac. Sci. Technol. A.* 23:1208–1214.
4. Siebentritt, S. 2002. Wide gap chalcopyrites: material properties and solar cells. *Thin Solid Films* 403–404:1–8.
5. Feinleib, J.; deNeufville, J.; Moss, S. C.; Ovshinsky, S. R. 1971. Rapid reversible light-induced crystallization of amorphous semiconductors. *Appl. Phys. Lett.* 18:254–257.
6. Ohta, T. 2001. Phase-change optical memory promotes the DVD optical disk. *J. Optoelectron. Adv. Mater.* 3:609–626.
7. Lee, B.-S.; Abelson, J. R.; Bishop, S. G.; Kang, D.-H.; Cheong, B.-k.; Kim, K.-B. 2005. Investigation of the optical and electric properties of $Ge_2Sb_2Te_5$ phase change material in its amorphous, cubic, and hexagonal phases. *J. Appl. Phys.* 97: 093509-1–093509-8.
8. Oh, H.-r.; Cho, B.-h.; Cho, W. Y.; Kang, S.; Choi, B.-g.; Kim, H.-j.; Kim, K.-s., Kim, D.-e.; Kwak, C.-k.; Byun, H.-g.; Jeong, G.-t.; Jeong, H.-s.; Kim, K. 2006. Enhanced write performance of a 64-Mb phase-change random access memory. *IEEE J. Solid-State Circ.* 41:122–126.
9. Bottner, H.; Nurnus, J.; Gavrikov, A.; Kuhner, G.; Jagle, M.; Kunzel, C.; Eberhard, D.; Plescher, G.; Schubert, A.; Schlereth, K.-H. 2004. New thermoelectric components using microsystem technologies. *J. Microelectromech. Systems* 13: 414–420.
10. Venkatasubramanian, R.; Colpitts, T.; O'Quinn, B.; Liu, S.; El-Masry, N.; Lamvik, M. 1999. Low-temperature organometallic epitaxy and its application to supperlattice structures in thermoelectrics. *Appl. Phys. Lett.* 75:1104–1106.
11. Weimer, P. K. 1962. The TFT—a new thin-film transistor. *Proc. IRE.* 50: 1462–1469.

12. Meth, J. S.; Zane, S. G.; Sharp, K. G.; Agrawal, S. 2003. Patterned thin film transistors incorporating chemical bath deposited cadmium sulfide as the active layer. *Thin Solid Films*. 444:227–234.

13. Fischer, D.; Dylla, T.; Meyer, N.; Beck, M. E.; Jäger-Waldau, A.; Lux-Steiner, M. Ch. 2001. CVD of $CuGaSe_2$ for thin film solar cells employing two binary sources. *Thin Solid Films*. 387:63–66.

14. Chang, K. J.; Lahn, S. M.; Chang, J. Y. 2006. Growth of single-phase In_2Se_3 by using metal organic chemical vapor deposition with dual-source precursors. *Appl. Phys. Lett.* 89:182118-1–182118-3.

15. Shinn, G. B.; Gillespie, P. M.; Wilson, W. L., Jr.; Duncan, W. M. 1989. Laser-assisted metalorganic chemical vapor deposition of zinc selenide epitaxial films. *Appl. Phys. Lett.* 54:2440–2442.

16. Tulsky, E. G.; Long, J. R. 2001. Dimensional reduction: A practical formalism for manipulating solid structures. *Chem. Mater.* 13:1149–1166.

17. Sheldrick, W. S. 1988. The edge-bridged tritetrahedral selenostannate(IV) anion $[Sn_3Se_8]^{4-}$. Preparation and structure of $K_4Sn_3Se_8$. *Z. Naturforsch. B* 43:249–252.

18. Eisenmann, B.; Hansa, J. 1993. Crystal structure of tetrapotassium hexaselenodistannate, $K_4Sn_2Se_6$. *Z. Kristallogr.* 203:299–300.

19. Eisenmann, B.; Hansa, J. 1993. Crystal structure of hexapotassium heptaselenodistannate, $K_6Sn_2Se_7$. *Z. Kristallogr.* 203:303–304.

20. Klepp, K. O. 1992. Na_4SnSe_4 and K_4SnSe_4, two new selenostannates with discrete anions. *Z. Naturforsch. B* 47:411–417.

21. Klepp, K. O. 1992. Preparation and crystal structure of $K_2Sn_2S_5$ and $K_2Sn_2Se_5$. *Z. Naturforsch. B* 47:197–200.

22. Kanatzidis, M. G. 1990. Soluble polychalcogenides of the late transition and main group elements. *Comments Inorg. Chem.* 10:161–195.

23. Dhingra, S. S.; Kanatzidis, M. G. 1993. Polyselenide chemistry of indium and thallium in dimethylformamide, acetonitrile, and water. Syntheses, structures and properties of the new complexes $[In_2(Se_4)_4(Se_5)]^{4-}$, $[In_2Se_2(Se_4)_2]^{2-}$, $[In_3Se_3(Se_4)_3]^{3-}$, $[Tl_3Se3(Se_4)_3]^{3-}$. *Inorg. Chem.* 32:1350–1362.

24. Wardle, R. W. M.; Mahler, C. H.; Chau, C.-N.; Ibers, J. A. 1988. New soluble monomeric polyselenide anions, $[MQ(Se_4)_2]^{2-}$ (M = Mo, W; Q = O, S, Se). *Inorg. Chem.* 27:2790–2795.

25. Müller, U.; Ha-Eierdanz, M.-L.; Kräuter, G.; Dehnicke, K. 1990. Synthesis and crystal structure of $(PPh_4)_4[Cu_2Se_{14}]$. *Z. Naturforsch. B* 45:1128–1132.

26. Dhingra, S.; Kanatzidis, M. G. 1990. The use of soluble metal-polyselenide complexes as precursors to binary and ternary solid metal selenides. *Mater. Res. Soc. Symp. Proc.* 180:825–830.

27. Kanatzidis, M. G.; Huang, S.-P. 1994. Coordination chemistry of heavy polychalcogenide ligands. *Coord. Chem. Rev.* 130:509–621.

28. MacDonald, W. A. 2004. Engineered films for display technologies. *J. Mater. Chem.* 14:4–10.

29. Welsh, T. W. B.; Broderson, H. J. 1915. Anhydrous hydrazine. III. Anhydrous hydrazine as a solvent. *J. Am. Chem. Soc.* 37:816–824.

30. Vecht, A.; Davies, D. A.; Smith, D. 1998. A novel method for the preparation of sulfides and selenides and its application for phosphor synthesis. *Mat. Res. Innovat.* 2:176–180.

31. Mitzi, D. B.; Kosbar, L. L.; Murray, C. E.; Copel, M.; Afzali, A. 2004. High-mobility ultrathin semiconducting films prepared by spin coating. *Nature* 428:299–303.

32. Mitzi, D. B.; Copel, M.; Chey, S. J. 2005. Low-voltage transistor employing a high-mobility spin-coated chalcogenide semiconductor. *Adv. Mater.* 17:1285–1289.

33. Mitzi, D. B. 2005. Synthesis, structure, and thermal properties of soluble hydrazinium germanium(IV) and tin(IV) selenide salts. *Inorg. Chem.* 44: 3755–3761.

34. Mitzi, D. B. 2005. Polymorphic one-dimensional $(N_2H_4)_2ZnTe$: Soluble precursors for the formation of hexagonal or cubic zinc telluride. *Inorg. Chem.* 44: 7078–7086.

35. Milliron, D. J.; Mitzi, D. B.; Copel, M.; Murray, C. E. 2006. Solution-processed metal chalcogenide films for p-type transistors. *Chem. Mater.* 18:587–590.

36. Mitzi, D. B.; Raoux, S.; Schrott, A. G.; Copel, M.; Kellock, A.; Jordan-Sweet, J. 2006. Solution-based processing of the phase-change material KSb_5S_8. *Chem. Mater.* 18:6278–6282.

37. Mitzi, D. B.; Copel, M.; Murray, C. E. 2006. High-mobility p-type transistor based on a spin-coated metal telluride semiconductor. *Adv. Mater.* 18:2448–2452.

38. Mitzi, D. B. 2007. $N_4H_9Cu_7S_4$: A hydrazinium-based salt with a layered $Cu_7S_4^-$ framework. *Inorg. Chem.* 46:926–931.

39. Audrieth, L. F.; Ogg, B. A. 1951. The Chemistry of Hydrazine. John Wiley & Sons, New York.

40. Sotaniemi, E.; Hirvonen, J.; Isomäki, H.; Takkunen, J.; Kaila, J. 1971. Hydrazine toxicity in the human. Report of a fatal case. *Ann. Clin. Res.* 3:30–33.

41. Toxicological Profile for Hydrazines. 1997. U.S. Dept. of Health and Human Services (Agency for Toxic Substances and Disease Registry), Atlanta, GA. Accessed: http://www.atsdr.cdc.gov/toxprofiles/tp100.pdf.

42. Domingo, G.; Itoga, R. S.; Kannewurf, C. R. 1966. Fundamental optical absorption in SnS_2 and $SnSe_2$. *Phys. Rev.* 143:536–541.

43. Shibata, T.; Muranushi, Y.; Miura, T.; Kishi, T. 1991. Electrical characterization of 2H-SnS_2 single crystals synthesized by the low temperature chemical vapor transport method. *J. Phys. Chem. Solids* 52:551–553.

44. Carey, P. G.; Smith, P. M.; Theiss, S. D.; Wickboldt, P. 1999. Polysilicon thin film transistors fabricated on low temperature plastic substrates. *J. Vac. Sci. Technol. A* 17:1946–1949.

45. Sze, S. M. 1981. Physics of Semiconductor Devices. Wiley, New York.

46. Dimitrakopoulos, C. D.; Purushothaman, S.; Kymissis, J.; Callegari, A.; Shaw, J. M. 1999. Low-voltage organic transistors on plastic comprising high-dielectric constant gate insulators. *Science* 283:822–824.

47. Mitzi, D. B. 2004. Solution-processed inorganic semiconductors. *J. Mater. Chem.* 14:2355–2365.

48. Santos, N. C.; Figueira-Coelho, J.; Martins-Silva, J.; Saldanha, C. 2003. Multidisciplinary utilization of dimethyl sulfoxide: pharmcological, cellular, and molecular aspects. *Biochem. Pharmacol.* 65:1035–1041.

49. Knaak, J. B.; Leung, H.-W.; Stott, W. T.; Busch, J.; Bilsky, J. 1997. Toxicology of mono-, di-, and triethanolamine. *Rev. Environ. Contam. Toxicol.* 149:1–86.

50. Chen, C.-Y.; Kanicki, J. 1996. High field-effect-mobility a-Si:H TFT based on high deposition rate PECVD materials. *IEEE Electron Device Lett.* 17:437–439.

51. Chaiken, A.; Nauka, K.; Gibson, G. A.; Lee, H.; Yang, C. C.; Wu, J.; Ager, J. W.; Yu, K. M.; Walukiewicz, W. 2003. Structural and electronic properties of amorphous and polycrystalline In_2Se_3 films. *J. Appl. Phys.* 94:2390–2397.

52. Shay, J. L. 1975. *Ternary Chalcopyrite Semiconductors: Growth, Electronic Properties, and Applications. International Series of Monographs in the Science of the Solid State*, Vol. 7. Pergamon, New York.

53. Wasim, S. M.; Sánchez Porras, G.; Tomlinson, R. D. 1982. Some electrical characteristics of copper- and indium-doped $CuInTe_2$. *Phys. Status Solidi A* 71:523–530.

54. Davis, J. G.; Bridenbaugh, P. M.; Wagner, S. 1978. Electrical and optical properties of copper indium ditelluride crystal grown from near-stoichiometric compositions. *J. Electron. Mater.* 7:39–45.

55. Dawar, A. L.; Kumar, A.; Kumar, P.; Mathur, P. C. 1984. Field-effect studies on *p*-type $CuInTe_2$ metal-insulator-semiconductor structures. *J. Appl. Phys.* 55:3695–3698.

56. Dagan, G.; Ciszek, T. F.; Cahen, D. 1992. Ion migration in chalcopyrite semiconductors. *J. Phys. Chem.* 96:11009–11017.

57. Kleinfeld, M.; Wiemhöfer, H.-D. 1988. Chemical diffusion coefficients and stability of $CuInS_2$ and $CuInSe_2$ from polarization measurements with point electrodes. *Solid State Ionics.* 28–30:1111–1115.

58. Mitzi, D. B.; Yuan, M.; Liu, W.; Kellock, A.; Chey, S. J.; Deline, V.; Schrott, A. G. 2008. A high-efficiency solution-deposited thin-film photovoltaic device. *Adv. Mater.* 20:3657–3662.

59. Kolobov, A. V.; Fons, P.; Frenkel, A. I.; Ankudinov, A. L.; Tominaga, J.; Uruga, T. 2004. Understanding the phase-change mechanism of rewritable optical media. *Nature Mater.* 3:703–708.

60. Zhou, G.-F. 2001. Materials aspects in phase change optical recording. *Mater. Sci. Eng. A* 304–306:73–80.

61. Ovshinsky, S. R. 1970. An introduction to ovonic research. *J. Non-Cryst. Solids.* 2:99–106.

62. Lee, H.; Kim, Y. K.; Kim, D.; Kang, D.-H. 2005. Switching behavior of indium selenide-based phase-change memory cell. *IEEE Trans. Magnetics.* 41:1034–1036.

63. Kyratsi, T.; Chrissafis, K.; Wachter, J.; Paraskevopoulos, K. M.; Kanatzidis, M. G. 2003. KSb_5S_8: A wide bandgap phase-change material for ultra high density rewritable information storage. *Adv. Mater.* 15:1428–1431.

64. Berlepsch, P.; Miletich, R.; Armbruster, Th. 1999. The crystal structures of synthetic KSb_5S_8 and $(Tl_{0.598}K_{0.402})Sb_5S_8$ and their relation to parapierrotite ($TlSb_5S_8$). *Z. Kristallogr.* 214:57–63.

65. Khulbe, P. K.; Wright, E. M.; Mansuripur, M. 2000. Crystallization behavior of as-deposited, melt quenched, and primed amorphous states of $Ge_2Sb_{2.3}Te_5$ films. *J. Appl. Phys.* 88:3926–3933.

66. Milliron, D. J.; Raoux, S.; Shelby, R. M.; Jordan-Sweet, J. 2007. Solution-phase deposition and nanopatterning of GeSbSe phase-change materials. *Nature Mater.* 6:352–356.

67. Kozicki, M. N.; Balakrishnan, M.; Gopalan, C.; Ratnakumar, C.; Mitkova, M. 2005. Programmable metallization cell memory based on Ag-Ge-S and Cu-Ge-S solid electrolytes. *Proceedings of the Non-Volatile Memory Technology Symposium (Dallas, TX; Nov. 7–10, 2005)*, pp. 83–89.

68. Pütz, J.; Aegerter, M. A. 1998. Spin deposition of MoS_x thin films. *Thin Solid Films* 351:119–124.

69. Schwarz, D. E.; Rauchfuss, T. B.; Wilson, S. R. 2003. Aggregation of PMe_3-stabilized molybdenum sulfides and the catalytic dehydrogenation of H_2S. *Inorg. Chem.* 42:2410–2417.

70. Ki, W.; Huang, X.; Li, J.; Young, D. L.; Zhang, Y. 2007. Highly conductive group VI transition metal dichalcogenide films by solution-processed deposition. *J. Mater. Res.* 22:1390–1395.

71. Chern, G. C.; Lauks, I. 1982. Spin-coated amorphous chalcogenide films. *J. Applied Phys.* 53:6979–6982.

72. Pütz, J.; Aegerter, M. A. 1998. Spin-coating of MoS2 thin films. *Proc. of International Congress on Glass (San Francisco, CA; July 5–10, 1998)*, vol. 18, pp. 1675–1680.

CHAPTER 4

Oxide Dielectric Films for Active Electronics

DOUGLAS A. KESZLER, JEREMY T. ANDERSON, and STEPHEN T. MEYERS
Department of Chemistry, Oregon State University, Corvallis, OR

4.1 INTRODUCTION

In this chapter, we consider the solution deposition of high-quality oxide dielectric thin films for the fabrication of active electronic devices. The importance of such insulating films is readily appreciated by considering the pervasive use of integrated circuits built from Si-based CMOS devices, where utility is largely predicated on the superior insulating properties of SiO_2 as opposed to the modest semiconductor performance of Si.[1] Although the advances in solution-processed semiconductors detailed in this volume are impressive, it is important to note that their device implementation through additive processing will generally require suitable oxide dielectric materials. But solution-processed oxide films, because of their deposition, morphological characteristics, and performance limitations, are often found to be poor mimics of vapor-deposited counterparts. Such has certainly been the case for integrated high-performance insulators, especially in the context of enabling direct methods of high-speed printing and patterning. Although establishing capabilities to print and integrate high-quality oxide dielectrics poses a very significant challenge for solution processing, success is very likely to open many opportunities in fabricating active electronics, as well as in providing new approaches to the production of a variety of unique optical and optoelectronic devices.

4.2 GATE DIELECTRIC MATERIALS SELECTION

The function, utility, and design parameters for an insulator in a field-effect thin-film transistor (TFT) are conveniently exemplified by considering the

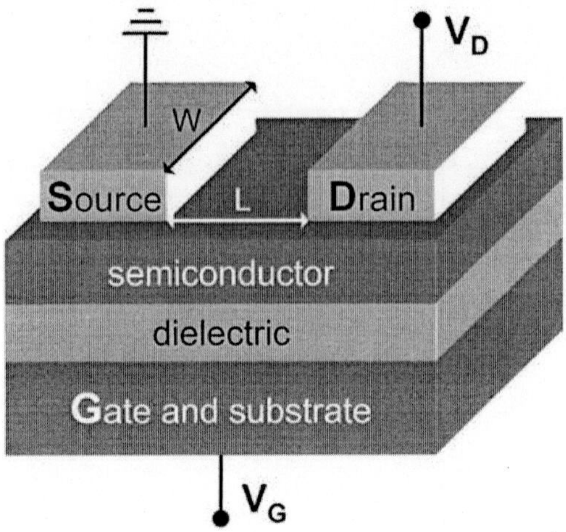

Figure 4.1. Schematic representation of a staggered, bottom-gate, thin-film transistor.

bottom-gate staggered device shown in Fig. 4.1. In this transistor, source-to-drain current through the semiconductor (channel) is modulated by a field imposed on the semiconductor via the metal-insulator gate.

Here, the semiconductor is assumed to be *n*-type, and the source voltage is set to ground. Under ideal conditions, the carrier concentration in the channel is sufficiently low such that no measurable current flows on application of a small positive voltage between the source and drain (V_{DS}) while maintaining a zero gate-to-source voltage (V_{GS}). For V_{GS} and $V_{DS} > 0$, electrons in the semiconductor accumulate near the semiconductor–insulator interface, trap states fill, and drain-to-source current (I_{DS}) flows. At a fixed value of V_{GS} and a small V_{DS}, I_{DS} increases in a linear fashion according to Ohms law (Fig. 4.2). At higher values of V_{DS}, a charge depletion zone forms near the drain, effectively pinching off the accumulation layer and producing the observed saturation in $I_{DS} \equiv I_{DSAT}$ when $V_{DS} \geq V_{GS} - V_{ON}$. Also, as observed in Fig. 4.2, the magnitude of I_{DSAT} is determined by V_{GS} and the associated charge density in the accumulation layer. (Note: Additional details on transistor operation and characterization can be found in Chapter 1.)

In a high-performance transistor, I_{DSAT} is large even for small V_{GS}, whereas only a small leakage current flows through the gate insulator. I_{DSAT} is given by

$$I_{DSAT} = \frac{W}{2L}\mu C_G (V_{GS} - V_{ON})^2$$

where, (cf., Fig. 4.1) W is the width of the transistor, L is the source-to-drain separation, μ is the channel device mobility, C_G is the gate-capacitance density,

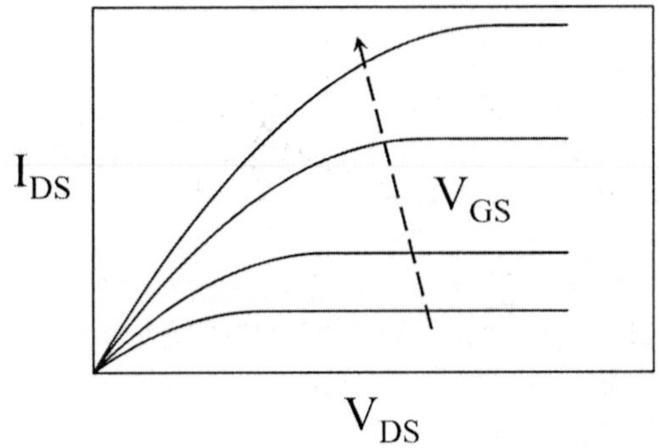

Figure 4.2. Idealized I–V curves for a TFT at increasing gate voltages V_{GS}.

and V_{ON} is the turn-on voltage (represented as the value of V_{GS}, where I_{DS} rises above the gate leakage current). I_{DSAT} can be maximized by minimizing L, using a semiconductor with a large μ or maximizing C_G.

For our purposes, we are interested in maximizing C_G, which is described by

$$C_G = \frac{\varepsilon_1 \varepsilon_0}{d_I}$$

where ε_I is the relative dielectric constant of the insulator and d_I is the thickness of the insulator. Higher values of C_G, which lead to enhanced transistor performance, will thus be observed with thin insulators having high dielectric constants. But these characteristics must be tempered by the need to maintain small leakage currents, which increase dramatically via tunneling for very thin films. Such leakage currents can represent a significant source of power loss in driving devices, which in the case of solution-processed and printed electronics is of considerable concern in the context of the dimensionally large transistors envisioned for use in macroelectronic applications.

At this point, it is useful to consider the important film features and performance metrics that will produce an optimal gate dielectric. The surface of the dielectric should be atomically smooth. A dielectric with a rough surface will lead to an irregular insulator–semiconductor interface, impeding the flow of charge through the semiconductor and reducing channel mobility. As noted, thin films with large relative dielectric constants (>~10) and small leakage current densities (<~10 nA/cm^2 @ 1 MV/cm) are desirable. At the same time, large breakdown fields (>~4 MV/cm) are required to preserve the device function. To achieve the necessary leakage currents and breakdown fields, films must be as dense as possible and exhibit no pores or cracks. Both from the perspective of surface smoothness and the need for high breakdown fields and low leakage currents, amorphous films are generally preferred for the fabrication of gate dielectric layers.

Because of the challenges in producing such insulators via solution methods, it should not be surprising that most oxide semiconductor[2–5] and solution-processed inorganic TFTs[6–8] have been fabricated by using binary oxide gate insulators, e.g., SiO_2, Al_2O_3, Y_2O_3, or HfO_2, formed via thermal oxidation or vapor deposition; TFTs containing inorganic semiconductors in conjunction with solution-deposited thin organic/inorganic hybrid dielectrics have also been described.[9–11] Although binary oxides will continue to be used for TFT gate dielectric applications, they do not represent an optimal approach to realizing high-performance devices. Binary oxides have a tendency to crystallize,[1,12] many at low process temperatures, producing grain boundaries that contribute to enhanced impurity interdiffusion and high leakage currents. An important gate insulator figure-of-merit is provided by the product of the dielectric constant and breakdown field.[13] It is important to note that binary oxides with high dielectric constants have small band gaps, and binary oxides with small dielectric constants have wide band gaps. Wide band-gap oxides are desirable for gate dielectrics, since breakdown fields scale with the magnitude of the band gap. But, these are generally the materials with small dielectric constants. Hence, selectron of a binary oxide as an insulator generally involves a compromise between dielectric constant and breakdown field.

One approach to the production of high-performance dielectrics relies on the use of mixed-metal, multiple-component oxides. These oxides provide convenient means for controlling the dielectric-constant breakdown-field product through incorporation of components that specifically contribute to performance via dielectric constant or breakdown. At the same time, the mixed materials can inhibit crystallization, resulting in deposition of amorphous films with extremely flat surfaces. Common candidates, base oxides for tuning these properties, are listed in Table 4.1.

The oxides SiO_2, Al_2O_3, and related wide band-gap binary oxides are attractive for use at the channel–semiconductor interface, because their large gaps and attendant conduction-band discontinuities contribute to high breakdown fields and suppression of charge injection. Oxides such as HfO_2, Ta_2O_5, and TiO_2 exhibit relatively small band gaps, but they are highly polarizable, which translates into high and desirable dielectric constants. But, their modest band gaps and high electron affinities contribute to small conduction-band offsets

TABLE 4.1. Properties of Selected Binary Oxide Gate Dielectrics[1,9,10]

Oxide	Dielectric Constant	Band gap (eV)
SiO_2	4	8.9
Al_2O_3	9	8.7
HfO_2	25	5.7
Ta_2O_5	26	4.5
TiO_2	80	3.0

at the insulator–semiconductor interface, providing small carrier injection barriers, low breakdown-field capabilities, and high leakage current densities.

Multiple-component oxides can be used in three ways to meet the performance requirements of gate dielectrics in TFTs. A single homogeneous dielectric can be produced by combining selected wide band-gap materials with those exhibiting smaller gaps and higher dielectric constants. For example, the mixtures HfO_2–SiO_2[14] and HfO_2–Al_2O_3[15] have been extensively studied as gate dielectrics in Si CMOS devices. Alternatively, wide and small-gap materials can be interleaved to form multilayered structures, as demonstrated by stacked layers of TiO_2 and Al_2O_3 produced via atomic layer deposition. The presence of sharp dielectric interfaces in such structured materials provides a means to improve dielectric-breakdown fields. Finally, a compositionally graded material dominated by a high dielectric-constant material at the metal–insulator interface and a high band-gap material at the dielectric–semiconductor interface provides an additional alternative.

4.3 PRODUCING HIGH-QUALITY FILMS FROM SOLUTION

The production of high-quality insulators is predicated on the deposition of pore-free, crack-free oxide films. The fundamental challenges in depositing oxide thin films from solution are associated with the processes of conversion of soluble precursors into dense solids. This statement should be intuitively obvious, but without appropriate reaction pathways between the liquid and solid states, high-quality films cannot be produced. The prevalence of morphologically coarse oxide films in the literature reveals that suitable chemistries have not been applied to many oxide systems. The failure of many precursor systems results from the addition of a variety of condensation inhibitors and surface-capping groups that effectively stabilize reactant species under a variety of processing conditions. In the majority of such cases, the modifications produce high energy barriers to reaction, precluding low-temperature conversion to solid oxide and promoting the production of high-surface-area, porous, and rough films. By placing greater emphasis on conversion pathways from precursor to oxide, low-energy reactions should be devised that allow condensation to proceed uniformly. Especially for electronic applications, thin oxide films must retain density, homogeneity, and uniformity during condensation. Identification of appropriate conversion pathways, therefore, represents the primary prerequisite for success.

Typically, in solution-phase film deposition, a precursor coating is applied to a surface and heated, whereupon the elimination of solvent leads to localized crystallization and separation of solid particles. To prevent these processes, metal-organic precursors can be commonly employed for "sol-gel" deposition (see Chapter 2), wherein the organic ligands essentially act as vehicles to promote glass formation. Ultimately, embedded organic ligands must be removed from thin-film layers, either by rapid combustion or slow diffusion. Rapid combustion generates a high density of reactive surfaces, which leads to pinholes and voids that must be eliminated via high-temperature densification. Alternatively, the organic ligands may be removed by slow diffusion. Diffusion occurs most readily near the surface of the film, creating a condensed layer of oxide that tends to encapsulate ligands at depths of greater than a few nm. Additional removal of organic groups can only occur through cracks and voids in the film. So the required mass transport limits the quality of deposited films, their deposition rates, or thickness per deposition cycle.

In principle, high-quality films could be deposited at a high rate if appropriate precursors were available. We have designed a class of aqueous inorganic precursors that exhibit chemical reactivities that are especially suitable for high-speed processing. Namely, the precursors described in the following sections undergo rapid condensation and they resist crystallization without the addition of organic ligands.

4.4 HafSOx THIN-FILM DIELECTRICS[16]

Hafnium oxide sulfate (HafSOx) was chosen as a candidate oxide dielectric because the system exhibits diverse chemistries in solution, and it resists crystallization in the solid phase. In addition, as noted in Section 4.2, oxides containing HfO_2 are well known to have relatively high dielectric constants. The strong interactions between hafnium-hydroxo groups (and analogous zirconium-hydroxo groups) and sulfato ligands have been known for a century,[17] but solution species have only been characterized under special conditions or inferred from structural characterization of precipitated crystals. Through variations in concentration, temperature, and time, the nature of solution-phase condensation can be controlled to be molecular, colloidal, gelatinous, or nanocrystalline.[18–21] All of these forms of hafnium oxide sulfate become amorphous as they dehydrate, and in fact, crystallized phases have not been synthesized by solid-state reaction. Therefore, the material can readily be prepared in the amorphous state and remain so over a great range of conditions, providing a convenient path to the desired morphology of a dielectric film.

A simplified series of reactions between a hafnium salt and sulfuric acid is given in Fig. 4.3. The reactions showcase important facets of thin-film synthesis (but do not address the precise identities of intermediates or complexities of aqueous hafnium chemistry.) In the first step, a hafnium oxide chloride crystal hydrate is dissolved in water to disperse small hafnium-hydroxo molecular clusters. Sulfato ligands are subsequently added in the form of sulfuric acid. Since sulfato binds more strongly than chloro, hafnium-hydroxo-sulfato aqueous species are created. Under mild heating, these species readily poly-

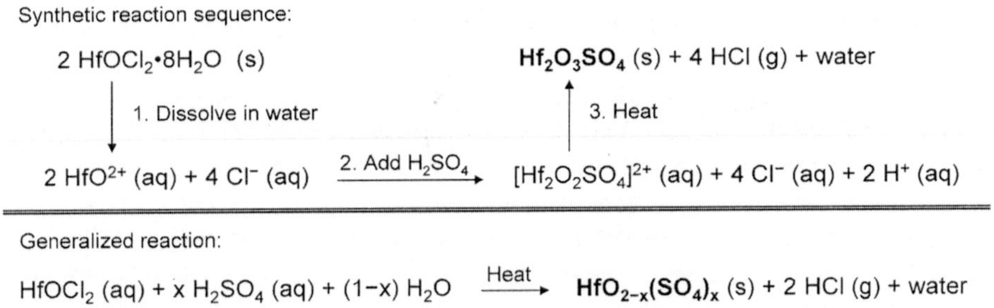

Figure 4.3. Chemical sequence representing deposition of hafnium oxide sulfates and generalized formulation for "HafSOx."

Figure 4.4. SEM of HafSOx thin film on Ta metal layer. Top HafSOx layer deposited by spin coating and annealing at 325 °C for 10 min in air. [Reproduced with permission. Anderson, J. T.; Munsee, C. L.; Hung, C. M.; Phung, T. M.; Herman, G. S.; Johnson, D. C.; Wager, J. F.; Keszler, D. A. 2007. Solution-processed HafSOx and ZircSOx inorganic thin-film dielectrics and nanolaminates. *Adv. Funct. Mater.* 17:2117–2124. Copyright 2007 Wiley–VCH Verlag GmbH & Co. KGaA.]

merize through dehydration to form amorphous thin films. During drying, changes in pH accentuate condensation of already reactive clusters. Hence, long-range covalent bonding proceeds before localized rearrangements and "mud cracking" can occur.

The condensation chemistry allows films of various compositions, as the addition of sulfate renders the materials amorphous over a range of concentrations as implied by the acronym HafSOx, where "x" typically assumes values of 0.3–1 (refer to Fig. 4.3, where the top reaction sequence represents $x = 0.5$.) The amorphous character and structural integrity are retained until the material decomposes with stoichiometric loss of $SO_3(g)$ at approximately 700 °C. The smoothness and uniformity of deposited films are illustrated by the Scanning electron microscope (SEM) images in Fig. 4.4. Rapid kinetics, absence of organics, and facile condensation all play important roles in the deposition of these dense HafSOx films.

The rapid condensation processes also allow for a variety of metal-atom substitutions within the amorphous matrix, including but not limited to Al, Ca, Ce, and La. HafSOx thin films and several atomically mixed compositions at thicknesses of 150–250 nm were studied as dielectrics in metal–insulator–metal (MIM) capacitor structures. Such MIM structures were constructed by depositing the insulator on Ta metal, (cf., Fig. 4.4) and by adding an array of Al contacts of 1.2 mm diameter via thermal evaporation. At a frequency of 1 kHz, dielectric permittivities of 9–12, and loss tangents <1% have been established for all variations, exhibiting little dependence on composition. Breakdown fields of 4–6 MV cm^{-1} have been demonstrated with leakage currents < 10 nA cm^{-2} at 1 MV cm^{-1}. Selected materials have been evaluated as gate dielectrics within TFT structures, where the other component films of the TFTs have been deposited by conventional vapor methods. For this application, the dielectric must provide effective modulation of current in the adjacent semiconductor layer; any significant mobile charge in the dielectric will hamper operation. In the example provided here, La atoms were included in the matrix (HafSOx:La) in an attempt to nominally raise solution pH without premature particle agglomeration, thereby lowering levels of residual chloride and minimizing mobile charge in the dielectric. Operational characteristics of a resulting TFT are illustrated in Fig. 4.5a. These results represent the first successful incorporation of a solution-deposited oxide gate dielectric in a functioning inorganic TFT. The device behaves in a qualitatively ideal manner. Current levels are effectively modulated with applied gate voltage; and at each gate voltage, drain current increases predictably until saturation. More importantly, gate leakage currents are so low that a large ON/OFF ratio (~10^7) in the transfer curve (Fig. 4.5b) can be attributed to effective switching rather than to undesirable current across the dielectric. Transfer and measured gate leakage (I_G) curves, such as those given in Fig. 4.5b, are only occasionally reported in the literature; yet they represent a fundamental contribution to establishing the veracity of a reported TFT function. The ordinate I_D, here and in any I–V curve for a transistor, simply represents the total drain current, which originates from both the desired charge flow through the semiconductor channel and any unwanted leakage through the dielectric. As a result, a device with a poor dielectric and a high gate leakage current can produce apparent ideal transistor behavior similar to that illustrated in Fig. 4.5a. In such devices, derived parameters such as channel mobility may be erroneously derived.

The combined characteristics of HafSOx films make them effective dielectrics. Smooth interfaces and homogeneous morphology, permittivity of approximately 10, and low leakage currents collectively allow moderate voltage operation, while alleviating energy losses from gate leakage. The preparation temperature of 325 °C is relatively low, considering that organic sol-gel precursors usually require heating to ≥500 °C and/or oxygen plasma ashing for conversion to oxides.

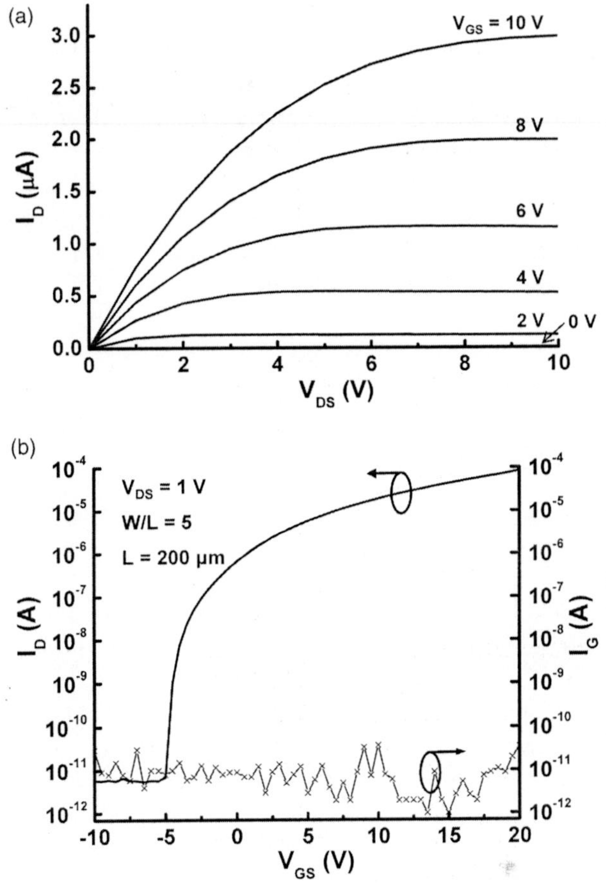

Figure 4.5. Device characteristics for bottom-gate TFT with HafSOx:La gate dielectric. The gate is evaporated Ta metal on a Si substrate. The gate dielectric is spin-coated HafSOx:La, deposited by spin-coating and annealing at 325 °C. Sputtered zinc indium oxide serves as the semiconductor layer, and source/drain contacts are evaporated Al. [Reproduced with permission. Anderson, J. T.; Munsee, C. L.; Hung, C. M.; Phung, T. M.; Herman, G. S.; Johnson, D. C.; Wager, J. F.; Keszler, D. A. 2007. Solution-processed HafSOx and ZircSOx inorganic thin-film dielectrics and nanolaminates. *Adv. Funct. Mater.* 17:2117–2124. Copyright 2007 Wiley–VCH Verlag GmbH & Co. KGaA.]

4.5 ALPO THIN-FILM DIELECTRIC

The system $Al_2O_{3-3x}(PO_4)_{2x}$ (AlPO) offers unique opportunities for realizing high-performance dielectrics, both in terms of precursor chemistries and solid phases.[22] Hydrolysis of Al^{+3} solutions is known to generate molecular hydroxo clusters through condensation,[23,24] whereas the addition of phosphate oxoanions produces heteropolymeric species, although these have been studied in much less detail.[25] Precursors from these reactions may be readily synthesized over a wide composition range, relying on the dissolution of aluminum hydroxide in a mineral acid, as facilitated by polymeric interactions of Al^{+3} with phosphate ligands supplied in the form of phosphoric acid. Thin liquid films deposited from these solutions undergo prompt condensation and partial dehydration with mild heating, forming a dense oxide framework as solvent molecules are eliminated. Kinetically rapid condensation and nondisruptive hydroxo diffusion facilitate a flexible, additive deposition strategy that is not accessible with kinetically hindered sol-gel processes. The nominal reaction occurring during the coating and curing of a fully dehydrated film with an electrically useful composition may be expressed as follows:

$$\mathrm{Al(OH)Cl_2(aq)} + \frac{1}{2}\mathrm{H_3PO_4(aq)} \rightarrow \mathrm{AlO_{3/4}(PO_4)_{1/2}(s)} + 2\mathrm{HCl(g)} + \frac{1}{4}\mathrm{H_2O(g)}$$

In cured films, as in the solid state, stoichiometric $AlPO_4$ is physically and structurally similar to various phases of SiO_2, whereas covalent Al-rich oxide compositions maintain the morphology of a glass under aggressive thermal processing (>1000 °C). In the SEM images shown in Fig. 4.6a, a dense amorphous AlPO film cured at 275 °C is distinguishable from the high-quality thermally grown SiO_2 substrate only by the small difference in electron density. Because a long-range, strong covalent framework results from condensation at low temperatures, the films are structurally coherent with only mild heating. Thus, although short-range relaxations are possible, long-range diffusion, segregation, and roughening do not accompany subsequent solvent loss, which is apparent from the retention of film quality after nondisruptive dehydration (Fig. 4.6b). Here, the film is ramped nearly instantaneously from 275 °C to 600 °C without cracking or pore formation, all while undergoing a 15% volume reduction. Importantly, these glasses may also accommodate a large fraction

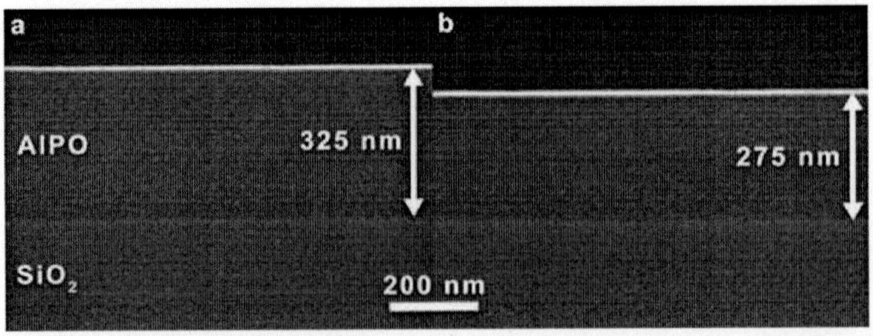

Figure 4.6. Cross-sectional SEM images of an AlPO film deposited on SiO_2 and cured at (a) 275 °C, and (b) flash annealed to 600 °C. [Reproduced with permission. Meyers, S. T.; Anderson, J. T.; Hong, D.; Hung, C. M.; Wager, J. F.; Keszler, D. A. 2007. Solution processed aluminum oxide phosphate thin-film dielectrics. Chem. Mater. 19:4023–4029. Copyright 2007 American Chemical Society.]

(ca. 33 at%) of a less acidic oxide, such as La_2O_3, which provides enhanced polarizability and dielectric constant, albeit at the expense of phase segregation above 800 °C.

The insulator–semiconductor interface is critical for transistor performance, as the number of charge trapping defects scales with interfacial area. The surfaces of AlPO dielectrics produced through inorganic condensation are essentially featureless, making them suitable for bottom gate TFT configurations. Rapid high-temperature processing does not alter film morphology across the entire temperature range 275–1000 °C; contact-mode atomic force microscopy (AFM) surface and line-profile scans of an AlPO film annealed at 1000 °C are shown in Fig. 4.7. Root-mean-squared (rms) roughness remains <0.1 nm, and a line-scan over 7 μm indicates a maximum height variation <1 nm. These surfaces compare favorably with thermally grown SiO_2 dielectrics processed near 1000 °C,[26] all the more impressive given the low-temperature deposition (<275 °C) and subsequent stability with respect to rapid high-temperature processing.

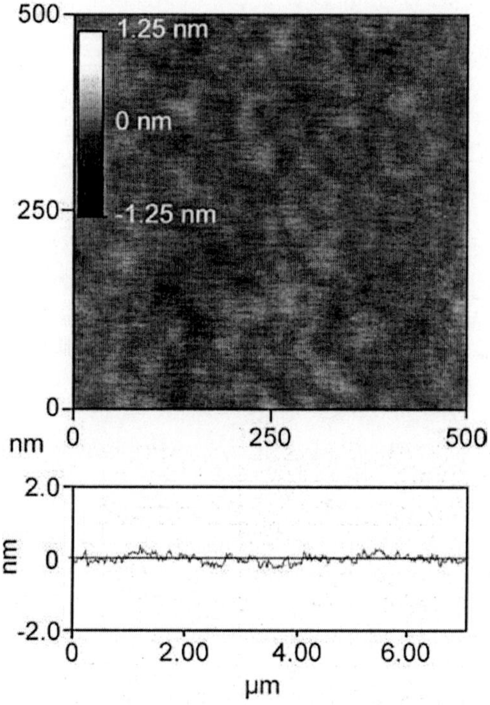

Figure 4.7. Contact-mode AFM surface and line-profile scans of a 148-nm AlPO film on Si after a 1000 °C anneal for 5 min. [Reproduced with permission. Meyers, S. T.; Anderson, J. T.; Hong, D.; Hung, C. M.; Wager, J. F.; Keszler, D. A. 2007. Solution processed aluminum oxide phosphate thin-film dielectrics. *Chem. Mater*. 19:4023–4029. Copyright 2007 American Chemical Society.]

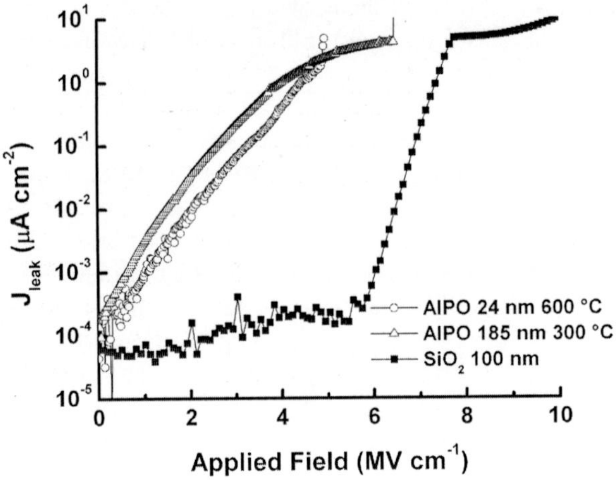

Figure 4.8. Current voltage curves for selected AlPO capacitor structures. A high-quality thermally oxidized SiO_2 dielectric in an identical structure is included for reference. Top contacts are 0.011-cm^2 Al dots thermally evaporated via shadow mask. Bottom contact is made via conductive substrate: p++ Si in the case of 600 °C AlPO and SiO_2 capacitors, and sputtered Ta metal for 300 °C AlPO devices.

Morphology and structure are relevant only so far as they influence dielectric performance, which can be definitively analyzed only through device integration. Capacitance-voltage, capacitance-frequency, and current-voltage measurements are typically performed using MIM or metal–insulator–semiconductor (MIS) capacitor structures.[27] Such characterization, although useful for extracting basic leakage current density, permittivity, and breakdown data are not, by themselves, sufficient analysis of functionality. Furthermore, because of widely varying requirements with respect to device area, test structures must be of an appropriate size to reflect potential applications. Selected I–V curves extracted from MIM and MIS devices with spin-coated AlPO insulators are depicted in Fig. 4.8. Leakage current density (J_{leak}) at a field strength of $1\,MV\,cm^{-1}$ is $<10\,nA\,cm^{-2}$ for 185-nm-thick AlPO films processed at 300 °C, with accompanying breakdown fields $\geq 6\,MV\,cm^{-1}$. The thermal flexibility of the multicomponent system is evident in that rapid annealing to 1000 °C results in *reduced* leakage with improved dehydration. Although MIS devices are not ideal test structures because of the potential growth of an interfacial SiO_2 layer, current profiles of MIM capacitors with M = Ta and Al are essentially identical to those on Si for temperatures where the metals are chemically and morphologically stable (T ≤ 350 °C).

This thermal stability comes with a price, however, as low polarizability of the covalent matrix imparts a modest dielectric constant of ~5 for $Al_2PO_{5.5}$ compositions (although substitution of 33 at% La can boost this value to ~8.5). This suboptimal value can be somewhat mitigated by high breakdown fields and low leakage currents. As demonstrated by use in Si CMOS devices for more than 30 years, low-leakage insulators with (relatively) low permittivity, e.g., SiO_2, can offer capacitance densities suitable for high-performance applications, provided they can be deposited sufficiently thin without compromising breakdown and leakage characteristics.

There are very few methods, aside from expensive and slow atomic layer deposition (ALD), that are suitable for depositing high-quality, ultrathin films over large areas and hence the interest in solution processing as a potential path to high-performance oxide dielectrics. In fact, thin (<100 nm) insulators build on a major strength of film deposition through aqueous inorganic condensation, i.e., nanodimensional control in the z-direction. As discussed in Section 4.6, this methodology is eminently suited for facile nanolaminate fabrication requiring repetitive deposition of layers as thin as ~3 nm with subnanometer precision. Obviously the device sizes (and attendant leakage) compatible with manufacturable solution patterning methods for macroelectronics may preclude extremely thin dielectrics on the scale of CMOS processing. Nevertheless, AlPO films as thin as 25 nm offer insulator characteristics superior to those of many thicker (>100 nm), high-permittivity oxide films deposited via conventional vapor methods. We have successfully characterized such films in MIS devices having capacitor areas of 1.1 mm^2, i.e., device areas that are easily consistent with the dimensional capabilities of ink-jet, microcontact, and other low-cost printing methods (see Chapter 12). Low-field leakage in these thin insulators can actually be lower than that observed in thicker films (cf. Fig. 4.8), which seems to be aided by enhanced dehydration of ultrathin layers at intermediate temperatures. Dielectric breakdown, although diminished, remains acceptable at 3–5 $MV\,cm^{-1}$. In comparison, J_{Leak} and breakdown values (although not dielectric constant) are equivalent or superior to ALD-deposited dielectrics of similar thickness that are being considered for use in large-area oxide electronics fabricated via conventional physical vapor deposition (PVD).[4]

Integration in active-channel TFTs is the decisive test of gate dielectric performance, as both current modulation and minimal leakage are required. By combining a single solution-processed layer with other components deposited through conventional vacuum methods, a comparative analysis of substituted layers is possible. Even limited by the difficult dehydration common to alumina-based systems, AlPO gate insulators ~185 nm thick offer qualitative TFT performance in rf-sputtered ZnO-channel TFTs for processing temperatures as low as 300 °C. For the device curve shown in Fig. 4.9, an ON-to-OFF ratio of 10^5 is achieved in concert with hard saturation in the I_D–V_{DS} curves. Importantly, leakage current at 40 V_{GS} is ~0.1 nA. A small amount of counterclockwise hysteresis is observed in the reverse V_{GS} sweep of the transfer curve, which may be attributed to residual hydration as acidic protons are repelled from the dielectric–semiconductor interface. Similar residual solvent or counterion contamination remains a major barrier to realizing metal oxide dielectrics with processing temperatures <300 °C. Gate-bias-induced carrier trapping

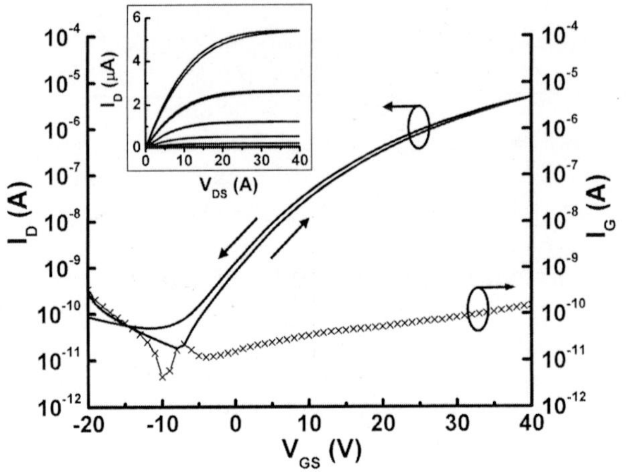

Figure 4.9. Log(I_D)-V_{GS} (V_{DS} = 20V) and (inset) I_D-V_{DS} characteristics for an RF-sputtered ZnO TFT on a spin-coated AlPO dielectric annealed at 300 °C. V_{GS} is stepped from 0 to 40V in 5-V increments for the I_D-V_{DS} curves; device W/L = 5; L = 100 μm. [Reproduced with permission. Meyers, S. T.; Anderson, J. T.; Hong, D.; Hung, C. M.; Wager, J. F.; Keszler, D. A. 2007. Solution processed aluminum oxide phosphate thin-film dielectrics. *Chem. Mater.* 19:4023–4029. Copyright 2007 American Chemical Society.]

can impart a similar, opposite hysteresis, raising the possibility that this effect, although nominally undesirable, could be useful to stabilize or counteract the I–V effects of such trapping.[22]

Higher processing temperatures permit more complete dehydration, especially when curing thinner films, and mute the aforementioned hysteresis. I–V characteristics of a TFT integrating a 40-nm AlPO gate dielectric cured at 600 °C and an rf-sputtered zinc tin oxide[3] (ZTO) channel are shown in Fig. 4.10a. An identical device on 100 nm of thermally grown SiO_2 is shown in Fig. 4.10b. Hysteresis for the AlPO device is small, and I_{DSAT} is > 600 μA at comparatively low operating voltages (V_{GS} = V_{DS} = 10V). Importantly, I_G values remain more than five orders of magnitude lower (~1 nA). As expected, current drive at equivalent V_{GS} is higher for the AlPO device, which has a capacitance density ~3× that of the SiO_2. Current increases are amplified in disordered semiconductors such as ZTO, where field-effect mobilities are strongly dependent on gate bias (accumulated charge density).[28,29] In this particular case, when biased at $V_{GS} - V_{on}$ = 10V, ZTO field-effect mobility on the SiO_2 dielectric is ~15 $cm^2V^{-1}s^{-1}$, whereas on the AlPO dielectric, it is ~25 $cm^2V^{-1}s^{-1}$.

Potential chemical interactions across dissimilar solvent systems and the sensitivity of oxide semiconductors to redox conditions and surface adsorption combine to make integration of solution-processed dielectrics and semiconductors a nontrivial undertaking. By maintaining appropriate aqueous

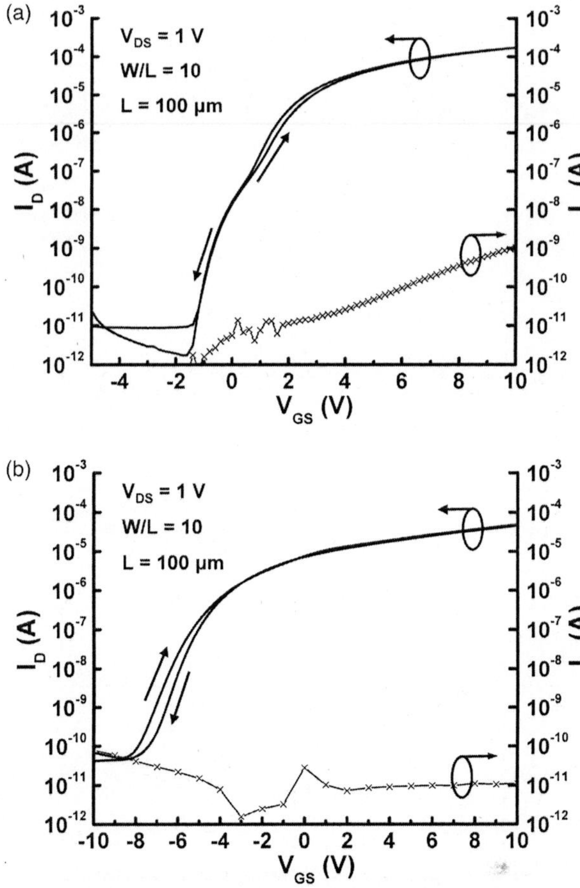

Figure 4.10. Representative $\log(I_D)$-V_{GS} behavior of an RF-sputtered ZTO-channel TFT on (a) 40-nm AlPO and (b) 100-nm SiO_2 gate dielectrics on unpatterned p++ Si gate electrodes. Channel and Al S/D contacts were defined with shadow masks. Dielectric films were annealed at 600 °C prior to channel deposition, whereas the completed stack was annealed at 300 °C.

deposition conditions for each component, however, such difficulties can be minimized. Application of related solvation and condensation principles to multicomponent oxide semiconductors has allowed fabrication of high-performance oxide TFTs with solution-processed channel and dielectric. The field-effect behavior of a bottom-gate TFT composed of a 60-nm AlPO dielectric layer and an indium gallium zinc oxide[30] (IGZO) channel is illustrated in Fig. 4.11, where both layers were deposited by spin coating, and all processing and testing was performed in air at a maximum temperature of 450 °C. The TFT operation remains exceptional, as the relatively high mobility (>4 $cm^2 V^{-1} s^{-1}$)

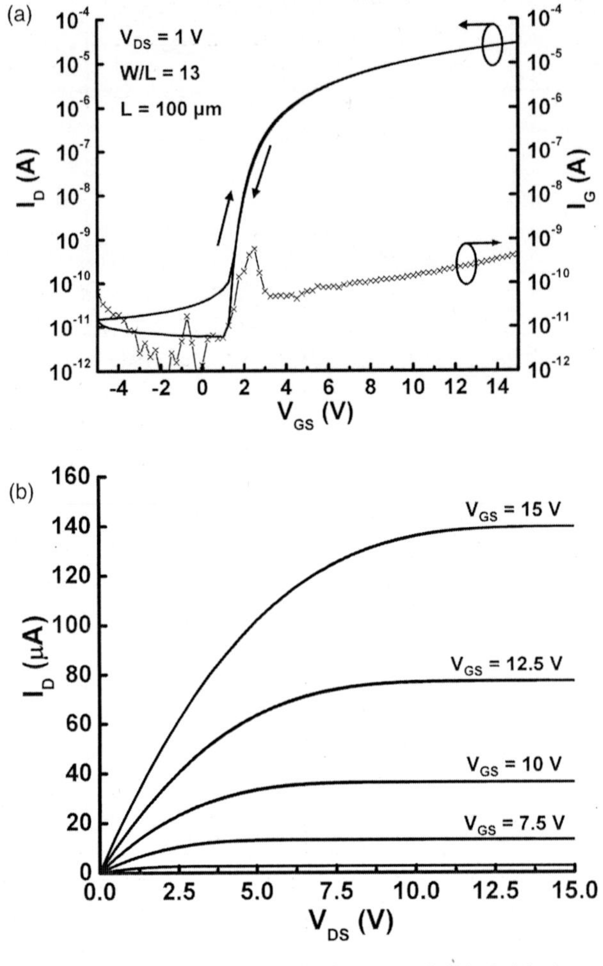

Figure 4.11. (a) Transfer curve and gate leakage and (b) field characteristics for a bottom-gate TFT with solution-processed IGZO-channel and 60-nm AlPO gate dielectric on an unpatterned p++ Si gate electrode. Maximum processing temperature for the device is 450 °C.

and current drive (140 μA at $V_{GS} = V_{DS} = 15$ V) are coupled with gate leakage < 1 nA, negligible hysteresis, and moderate operating voltages. These and related devices offer the highest reported performance for oxide TFTs with integrated solution-processed channels and dielectrics, and they clearly demonstrate the potential for solution-deposited oxide electronics.

Although current processing temperatures for AlPO dielectrics remain above the threshold for plastic substrates, they are well within the thermal budget of several other substrates of interest for large-area applications.

Furthermore, by emphasizing film quality and the chemical prerequisites for multicomponent oxide film deposition, the potential for qualitatively ideal solution-deposited dielectrics has been established. Additionally, new processing conditions and chemical modifications that will facilitate lower dehydration and precursor decomposition temperatures are readily envisioned.

4.6 COMPOSITIONALLY GRADED AND LAMINATED STRUCTURES

As noted in Section 4.2, additional methods to control dielectric behavior include compositional grading and highly structured laminates. The laminates should contain multiple, discrete layers of different materials that are selected to maximize the product of dielectric constant and breakdown field. On the basis of high-temperature approaches and compromised morphologies of traditionally processed solution-derived films, the multilayer, laminated approach to structured dielectrics would be deemed impractical. The processing conditions and film qualities described in this chapter, however, have provided a means to deposit for the first time via solution such laminates, exhibiting qualities that are unsurpassed by any other known deposition technique. As shown by the transmission electron microscopy (TEM) image of Fig. 4.12, films of HafSOx and its zirconium analog ZircSOx can be interleaved to form very high-resolution, high-order multilayer structures. An accurate bilayer thickness within the laminate can be readily determined from X-ray reflectivity (XRR) measurements, which also provide confirmation of the abrupt interfaces ($<0.2\,nm$ by XRR) that are visually evident in Fig. 4.12. Although we have not fully characterized the physical properties of the laminates, they do offer unique opportunities for development of highly structured dielectrics.

For tuning purposes, individual film thickness can already be varied in a controllable manner from 3 to several hundred nanometers. In addition, the film stack can be deposited with variable levels of hydration, providing explicit control over film interdiffusion. For example, high-order dehydrated stacks exhibiting stability to high temperatures can be produced by dehydrating individual films during each deposition step. Alternatively, stacks can be initially deposited at low temperatures with retention of water and then subsequently heated at higher temperatures to promote intermixing, providing pathways to the formation of selected compositions or compositionally graded films; in addition, differential crystallization of individual films can be harnessed to fabricate crystalline high-dielectric films interleaved with amorphous low-leakage films. Here, the unique precursor chemistries and condensation processes are providing an entirely new, cost-effective bench-top approach to the deposition, study, and development of oxide films with capabilities currently unknown with conventional vapor-based methodologies.

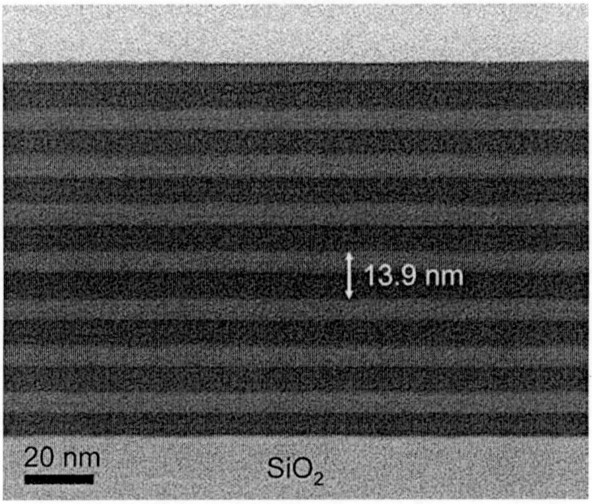

Figure 4.12. TEM cross-sectional image of a HafSOx/ZircSOx nanolaminate, consisting of 16 alternating layers (8 bilayers) with HafSOx on the SiO_2/Si substrate. HafSOx layers of 8.6 nm and ZircSOx layers of 5.3 nm together compose bilayers of 13.9-nm total thickness. [Reproduced with permission. Anderson, J. T.; Munsee, C. L.; Hung, C. M.; Phung, T. M.; Herman, G. S.; Johnson, D. C.; Wager, J. F.; Keszler, D. A. 2007. Solution-processed HafSOx and ZircSOx inorganic thin-film dielectrics and nanolaminates. *Adv. Funct. Mater.* 17:2117–2124. Copyright 2007. Wiley–VCH Verlag GmbH & Co. KGaA.]

4.7 SUMMARY AND PERSPECTIVE

We have provided a brief overview of activities directed toward solution processing of high-quality oxide dielectrics. By addressing issues in precursor formation and condensation processes, oxide films of high morphological quality and performance have been successfully incorporated into functioning TFTs. Under appropriate conditions, performance characteristics of solution-processed films meet or exceed those achievable with the most advanced vapor-deposition technologies. We have demonstrated an approach for depositing, on the bench top, ALD-quality structures by using beaker chemistries, pipettes, and inexpensive deposition tools. Many opportunities await examination in the continuing development and study of these materials, through both the generation of new precursors and film compositions and the continued evolution of processing and printing conditions. The new ability to deposit high-quality laminates having controlled levels of hydration provides an unprecedented method for fabricating uniquely structured dielectrics. It is expected that rearrangements at interfaces of chemically distinct hydrates will create regions of unique ordering. Such self-assembly would augment the nanostructure of deposited layers or lead to complete blending as multinary phases. Of course, this film deposition and development vehicle has many implications beyond structured dielectrics.

Future, widespread use of such solution-derived oxide films will likely be dictated by continued progress in lowering process temperatures and in demonstrating effective printing and patterning techniques for the fabrication of stable devices and circuits. At present, we have identified no fundamental issues restricting the application of aqueous-based inorganic inks via common patterning techniques that cover length scales from nano to macro. For all-oxide electronic devices and circuits, where the performance characteristics of the component oxide films are likely to be considerably affected by the use of conventional photolithography with polymer resists, direct additive processing via printing could provide an important opportunity to demonstrate the power of solution processing and patterning. An all-oxide approach also eliminates vacuum processing, while producing films and devices that are stable, chemically robust and environmentally benign. The demonstrated ability to readily deposit ultrathin films in air at modest temperatures also provides a unique platform for fabrication of high-performance transistor devices, where, for example, exceptionally short source-to-drain lengths (<10 nm) could be realized through vertical integration of robust and simple materials sets.

REFERENCES

1. Robertson, J. 2006. High dielectric constant gate oxides for metal oxide Si transistors. *Rep. Prog. Phys.* 69:327–396.
2. Hoffman, R. L.; Norris, B. J.; Wager, J. F. 2003. ZnO-based transparent thin-film transistors. *Appl. Phys. Lett.* 82:733–735.
3. Chiang, H. Q.; Wager, J. F.; Hoffman, R. L.; Jeong, J.; Keszler, D. A. 2005. High mobility transparent thin-film transistors with amorphous zinc tin oxide channel layer. *Appl. Phys. Lett.* 86:013503/1–013503/3.
4. Carcia, P. F.; McLean, R. S.; Reilly, M. H. 2006. High performance ZnO thin-film transistors on gate dielectrics grown by atomic layer deposition. *Appl. Phys. Lett.* 88:123509/1–123509/3.
5. Yabuta, H.; Sano, M.; Abe, K.; Aiba, T.; Den, T.; Kumomi, H.; Nomura, K.; Kamiya, T.; Hosono, H. 2006. High-mobility thin-film transistor with amorphous $InGaZnO_4$ channel fabricated by room temperature rf-magnetron sputtering. *Appl. Phys. Lett.* 89:112123/1–112123/3.
6. Norris, B. J.; Anderson, J.; Wager, J. F.; Keszler, D. A. 2003. Spin-coated zinc oxide transparent transistors. *J. Phys. D: Appl. Phys.* 36:L105–L107.
7. Mitzi, D. B.; Copel, M.; Murray, C. E.; Kosbar, L. L.; Afzali, A. 2005. Thin-film transistors based on spin-coated chalcogenide semiconductor channels. *Proceedings—Electrochemical Society, 2004–15 (Thin Film Transistor Technologies (TFTT VII)).* pp. 189–199.
8. Lee, D-H.; Chang, Y-J.; Herman, G. S.; Chang, C-H. 2007. A general route to printable high-mobility transparent amorphous oxide semiconductors. *Adv. Mater.* 19:843–847.

9. Wang, L.; Yoon, M-H.; Lu, G.; Yang, Yu; Facchetti, A.; Marks, T. J. 2006. High-performance transparent inorganic-organic hybrid thin-film n-type transistors. *Nat. Mater.* 5:893–900.

10. Sanghyun, J.; Lee, K.; Janes, D. B.; Yoon, M-H.; Facchetti, A.; Marks, T. J. 2005. Low operating voltage ZnO nanowire field-effect transistors enabled by self-assembled organic gate nanodielectrics. *Nano Lett.* 5:2281–2286.

11. Lin, H. C.; Ye, P. D.; Xuan, Y.; Lu, G.; Facchetti, A.; Marks T. J. 2006. High-performance GaAs metal-insulator-semiconductor field-effect transistors enabled by self-assembled nanodielectrics. *Appl. Phys. Lett.* 89:142101/1–142101/3.

12. Ono, Y. A. 1995. *Electroluminescent Displays*. World Scientific, Singapore.

13. Wilk, G. D.; Wallace, R. M.; Anthony, J. M. 2001. High-k gate dielectrics: current status and materials properties considerations. *J. Appl. Phys.* 89:5243–5275.

14. Rittersma, Z. M.; Roozeboom, F.; Verheijen, M. A.; van Berkum, J. G. M.; Dao, T.; Snijders, J. H. M.; Vainonen-Ahlgren, E.; Tois, E.; Tuominen, M.; Haukka, S. 2004. $HfSiO_4$ dielectric layers deposited by ALD using $HfCl_4$ and $NH_2(CH_2)_3Si(OC_2H_5)_3$ precursors. *J. Electrochem. Soc.* 151:C716–C722.

15. Londergan, A. R.; Ramanathan, S.; Vu, K.; Rassiga, S.; Hiznay, R.; Winkler, J.; Velasco, H.; Matthysse, L.; Seidel, T. E.; Ang, C. H.; Yu, H. Y.; Li, M. F. 2002. Process optimization in atomic layer deposition of high-k oxides for advanced gate stack engineering. *Proceedings—Electrochemical Society, 2002–11 (Rapid Thermal and Other Short-Time Processing Technologies III).* pp. 163–175.

16. Anderson, J. T.; Munsee, C. L.; Hung, C. M.; Phung, T. M.; Herman, G. S.; Johnson, D. C.; Wager, J. F.; Keszler, D. A. 2007. Solution-processed HafSOx and ZircSOx inorganic thin-film dielectrics and nanolaminates. *Adv. Funct. Mater.* 17:2117–2124.

17. Ruer, R. 1904. Zirconium salts. Constitution of normal zirconium salts. *Z. Anorg. Chem.* 42:87–99.

18. Clearfield, A. 1964. Structural aspects of zirconium chemistry. *Rev. Pure Appl. Chem.* 14:91–108.

19. Mark, W.; Hansson, M. 1975. The crystal structure of $Hf_{18}O_{10}(OH)_{26}(SO_4)_{13}(H_2O)_{33}$. *Acta Crystallogr., Section B: Struct. Crystallogr. Cryst. Chem.* B31:1101–1108.

20. Chiavacci, L. A.; Santilli, C. V.; Pulcinelli, S. H.; Craievich, A. F. 1997. Thermo-stimulated sol-gel transition in suspensions of sulfate-zirconium oxychloride. *J. Appl. Cryst.* 30:750–754.

21. Cölfen, H.; Schnablegger, H.; Fischer, A.; Jentoft, F. C.; Weinberg, G.; Schlögl, R. 2002. Particle growth kinetics in zirconium sulfate aqueous solutions followed by dynamic light scattering and analytical ultracentrifugation: implications for thin film deposition. *Langmuir.* 18:3500–3509.

22. Meyers, S. T.; Anderson, J. T.; Hong, D.; Hung, C. M.; Wager, J. F.; Keszler, D. A. 2007. Solution processed aluminum oxide phosphate thin-film dielectrics. *Chem. Mater.* 19:4023–4029.

23. Fu, G.; Nazer, L. F.; Bain, A. D. 1991. Aging processes of aluminum sol-gels: Characterization of new polyoxocations by aluminum-27 NMR spectroscopy. *Chem. Mater.* 3:602–610.

24. Pophristic, V.; Klein, M. L.; Holerca, M. N. 2004. Modeling small aluminum chlorohydrate polymers. *J. Phys. Chem. A.* 108:113–120.

25. Lima, E. C.; Neto, J. M.; Fujiwara, F. Y.; Galembeck, F. 1995. Aluminum polyphosphate thermoreversible gels: A study of ^{31}P and ^{27}Al NMR spectroscopy. *J. Coll. Interface Sci.* 176:388–396.

26. Iacona, F.; Raineri, V.; La Via, F.; Rimini, E. 1998. Roughness of thermal oxide layers grown on ion implanted silicon wafers. *J. Vac. Sci. Tech.* B 16:699–624.

27. Schroder, D. K. 1998. *Semiconductor Materials and Device Characterization*. John Wiley and Sons, New York.

28. Greve, D. W. 1998. *Field Effect Devices and Applications* Prentice Hall, Englewood Cliffs, NJ.

29. Hoffman, R. L. 2004. ZnO-channel thin-film transistors: channel mobility. *J. Appl. Phys.* 95:5813–5819.

30. Nomura, K.; Ohta, H.; Takagi, A.; Kamiya, T.; Hirano, M.; Hosono, H. 2004. Room-temperature fabrication of transparent flexible thin-film transistors using amorphous oxide semiconductors. *Nature* 432:488–492.

CHAPTER 5

Liquid Silicon Materials

MASAHIRO FURUSAWA and HIDEKI TANAKA
SEIKO EPSON Corporation, Frontier Device Research Center, Nagano-ken, Japan

5.1 INTRODUCTION

The solution processing of electronic devices has been receiving considerable attention recently, in an effort to reduce processing costs compared with conventional vacuum-based processes and vapor deposition. The liquid materials used in solution processing are also suitable for device printing without conventional photolithography, depositing the materials only where needed. Compared with the conventional processes [e.g., sputtering, evaporation, and chemical vapor deposition (CVD)], solution processing improves material utilization efficiency, simplifies the manufacturing processes, and reduces the size/cost of the manufacturing apparatus. The solution-based approach also has the potential to be a more ecological or "green" process, conserving both resources and energy.

The preparation of a functional liquid, a stable solution containing materials for a target film, is the critical first step in the solution process. The difficulty of this step mostly determines the difficulty of the solution process. Organic materials, such as organic semiconductors and light emitting polymers, are compatible with solution processing, because these materials are soluble in common solvents yet retain their original function. Liquid organic materials have thus been used to develop electroluminescent displays,[1,2] transistors,[3–5] solar cells,[6] and other devices, all of which have been fabricated using inkjet printing or other coating methods. Solution processing has also been applied to inorganic materials, as described in the other chapters of this book. For example, metal nanoparticle dispersions have recently come under intensive research as printable precursors of metal thin films.[7]

The development of solution processing for semiconductor materials has mostly focused on organic semiconductors.[3–5] Organic transistors can be fabricated at a fairly low cost using a solution-based process and printing techniques. Although the performance of the organic thin-film transistors (TFTs) recently has become comparable with that of amorphous silicon (a-Si) TFTs,[5] the stability and the reliability of the organic TFTs are insufficient for use in many practical applications. On the other hand, inorganic semiconductors are stable and have higher mobilities, but they are difficult to introduce into appropriate precursor solutions. Several experimental results have been reported on metal chalcogenide semiconductors,[8,9] which exhibit field-effect mobilities greater than $10\,cm^2/Vs$, an order of magnitude higher than for a-Si TFTs (e.g., see Chapter 3). However, silicon has received very little attention so far as a target for solution processing.

Recently, we have developed a silane-based liquid precursor for silicon films.[10,11] Spin coating this precursor, we have formed high-quality silicon films and transistors that operated with a mobility of as high as $108\,cm^2/Vs$. The fluidic parameters of the precursor, such as viscosity and surface tension, are adjustable to also be compatible with ink-jetting. A TFT formed using an ink-jetted silicon island operated with a mobility of $6.5\,cm^2/Vs$. Furthermore, the liquid silicon precursor is also suitable for the deposition of SiO_2 insulator films and doped (both n- and p-type) silicon films. In this chapter, we describe the synthesis and properties of the liquid silicon material, the formation process and characteristics of amorphous and poly-silicon films, as well as the performance of solution-processed, silicon-based transistors.

5.2 LIQUID SILICON MATERIAL

We have been pursuing the development of a novel liquid precursor (herein referred to as "liquid silicon material") that can be used in a solution process to form a silicon film.[10] Since the films produced from this precursor must consist of high-purity silicon, potential candidates are limited to carbon- and oxygen-free hydrogenated silicon compounds. Typical hydrogenated silicon compounds are either of the straight-chain (Si_nH_{2n+2}) or cyclic (Si_nH_{2n}) forms. For $n \geq 3$, these compounds are liquid at room temperature and decompose to form a-Si when heated to 300 °C or higher. However, for $n < 10$, boiling points are less than 300 °C, and such compounds evaporate before thermal decomposition, which makes solution processing difficult. Oligomeric and polymeric hydrogenated polysilanes, $-(SiH_2)_n-$, have received little attention since first being synthesized by the Kipping method,[12,13] because of their limited solubility in organic solvents. Nevertheless, hydrogenated polysilanes are potentially ideal liquid silicon materials, provided a suitable solvent can be found. Since cyclic silanes are known to undergo ring-opening polymerization,[14,15] this reaction can be used for the development of liquid silicon materials. We have applied photo-induced, ring-opening polymerization to obtain pure hydrogenated polysilanes from purified hydrogenated cyclic silanes. Among the hydrogenated cyclic silanes, we chose to focus on cyclopentasilane

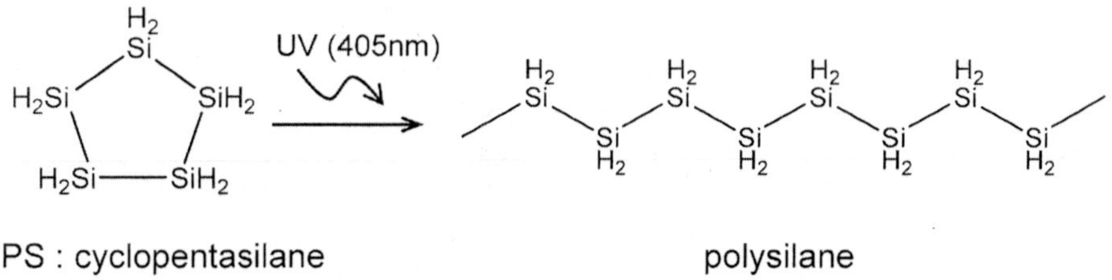

Figure 5.1. Synthetic process for polysilane.

TABLE 5.1. Solubility of CPS and Polysilane into Organic Solvents

Solute	CPS	Polysilane	Polysilane	Polysilane
Solvent	Organic solvent	Organic solvent	CPS	CPS + organic solvent
Solubility	Soluble at any rate	Insoluble	Soluble	Soluble under certain conditions

(CPS), Si_5H_{10}, since it is relatively stable and exhibits a high photoreactivity upon irradiation with ultraviolet (UV) light.

Using the method developed by Hengge et al.,[16,17] a CPS monomer (a clear and colorless liquid under ambient conditions) was synthesized. It has a boiling point of 194 °C and is soluble in most organic solvents. After purification, the CPS was exposed to 405-nm UV light to induce photo-polymerization as shown in Fig. 5.1. During exposure, the CPS gradually became cloudy and viscous. After sufficient exposure to the UV source, the liquid was transformed into a white solid, presumably made up of a mixture of hydrogenated polysilanes. Although this solid is known to be insoluble in all common organic solvents, it proved to be soluble in the CPS monomer precursor. The hydrogenated polysilane was also found to dissolve in a mixture of CPS and an organic solvent, as indicated in Table 5.1. This solvent mixture plays an important role in controlling the wettability, coating properties, and thickness of the resulting silicon films, all of which are difficult to control when CPS is used alone. In the actual process, UV irradiation is halted before the CPS completely polymerizes such that polysilanes of various molecular weights are dissolved in unreacted CPS. By diluting the solution with an organic solvent and then filtering out those insoluble polysilanes that precipitated as a result of dilution, we obtained the solution that we refer to as liquid silicon material.

The polymerization process of CPS was investigated using gel permeation chromatography (GPC).[10] Figure 5.2 shows the results of the GPC measurements of the CPS (Fig. 5.2a) and the UV-irradiated CPS (Fig. 5.2b). Peaks corresponding to CPS and toluene were observed in both nonirradiated and irradiated samples. In addition, in Fig. 5.2b, a broad peak was also observed

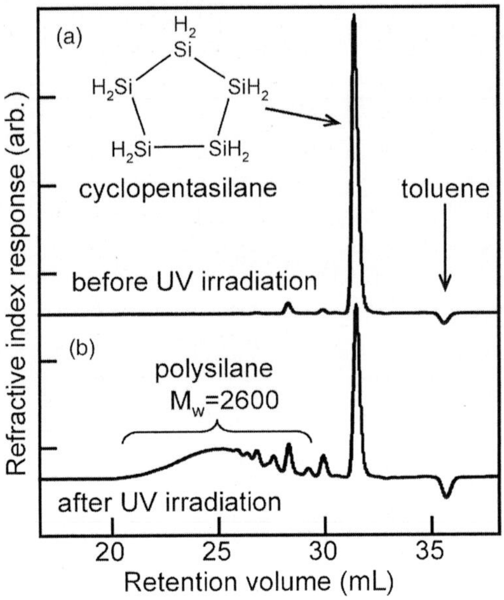

Figure 5.2. Gel permeation chromatogram (GPC) of the liquid silicon precursor for Si film formation for (a) cyclopentasilane (CPS) and (b) UV-irradiated CPS, both of which were diluted with toluene (1 vol.%) before GPC measurements. The UV-irradiation conditions were 405 nm, 100 mW/cm^2, and a 10-min irradiation for 1 cm^3 of CPS. The broad peak around $M_w = 2600$ corresponds to polysilanes of various molecular weights, as a result of the photo-induced polymerization of CPS. [Reproduced with permission from Ref. 10. Copyright 2006 Nature Publishing Group.]

around $M_w = 2600$. This peak corresponds to polysilanes of various molecular weights, indicating CPS polymerization. The GPC measurements were also used to control and optimize the molecular weight distribution of the polysilanes, which significantly affects the wettability of the precursor solution to a glass substrate.

5.3 FORMING SILICON FILMS FROM THE LIQUID SILICON MATERIALS

The next step after forming the precursor solution and spin coating a film (or forming it using some other solution-based approach) is forming an a-Si film by heating the polysilane film to induce thermal decomposition. We investigated the thermal decomposition process through the following experiments.[10] A 30-vol.% toluene solution of UV-irradiated CPS was spin-coated onto quartz substrates so that approximately 100-nm-thick a-Si films are obtained after heat treatment. Then, three samples a, b, and c were prepared with the following heating conditions on a hot plate: 300 °C for 10 min, 300 °C for

120 min, and 540 °C for 120 min, respectively. All experiments were carried out in a nitrogen-filled dry box with a residual oxygen concentration of less than 0.5 ppm, which was monitored using a galvanic fuel cell sensor. Raman scattering spectra of these samples were recorded to confirm the a-Si nature of the films using the typical a-Si peak around 480 cm^{-1}. Impurity concentrations in the films were investigated using secondary ion mass spectrometry (SIMS). The SIMS results confirmed that the film was composed almost entirely of silicon, with only a few trace impurities. Alkali and alkali-earth metal impurities, which negatively affect TFT characteristics, were present at a level of less than 1 ppm. The carbon content in the resulting film was surprisingly low—only 200 ppm—considering the organic solvent included in the starting liquid silicon material. By strictly controlling the oxygen content in the dry box to less than 0.5 ppm, we were able to limit the oxygen content of the film to less than 2000 ppm.

The above three samples were analyzed using thermal desorption spectroscopy (TDS) to investigate the process by which polysilane is converted to a-Si.[10] In TDS, a sample is heated in a vacuum and the gases that are desorbed from the sample are analyzed using mass spectroscopy. The relationship between the preheating conditions (as described above) of polysilane and the amount of relevant gases (H_2, SiH_2, and SiH_3) desorbed during the postannealing phase of TDS are shown in Fig. 5.3. Sample a, prebaked at 300 °C for

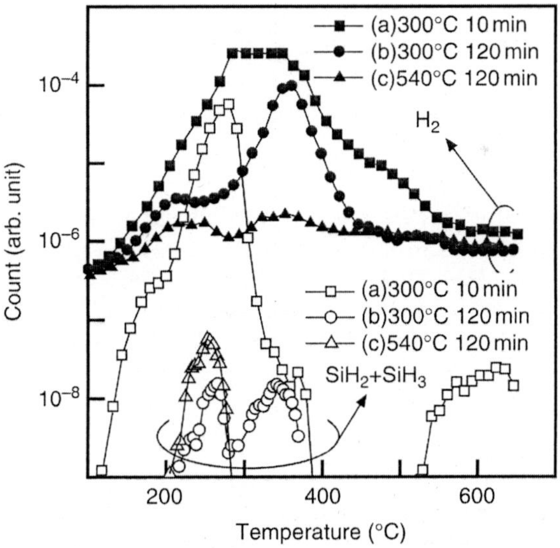

Figure 5.3. TDS of solution-processed a-Si films. Three samples were prepared by the thermal decomposition of polysilane under the following conditions: sample a, 300 °C for 10 min; sample b, 300 °C for 120 min; and sample c, 540 °C for 120 min. Desorbed gases from the samples were analyzed using mass spectroscopy while the samples were heated in a vacuum. [Reproduced with permission from Ref. 10. Copyright 2006 Nature Publishing Group.]

10min, reveals a SiH_2 and SiH_3 desorption peak at around 280 °C, followed by intensive hydrogen desorption between 300 °C and 400 °C, indicating that the polysilane is not completely converted to a-Si when baked at 300 °C for 10min. Given the binding energies of Si–Si (224kJ/mol) and Si–H (318kJ/mol),[18] the Si–Si bonds in the polysilane break first at a temperature lower than 280 °C, followed next by the breaking of the Si–H bonds, where a three-dimensional silicon network starts to form. In sample b, prebaked at 300 °C for 2h, SiH_2 and SiH_3 desorption is three or four orders of magnitude lower than that of sample a. Sample c released much less desorption gas than samples a and b, since it had almost completely converted to a-Si during preheating at 540 °C for 2h. Using TDS analysis, we estimate the total amount of hydrogen atoms (the atomic ratio of H/Si) in samples a, b, and c, before TDS measurement, to be more than 22%, 3%, and 0.3%, respectively. Thus, the formation process of a-Si is inferred to be as follows. As the spin-coated polysilane film is heated, volatile components such as toluene and CPS evaporate first. Next, the Si–Si bonds in the polysilane begin to break at a temperature below 280 °C, and a fraction of the polysilane is released as SiH_2 and SiH_3. After this, at around 300 °C, the Si–H bonds break, resulting in a three-dimensional a-Si network.

To test the electrical properties of a solution-processed a-Si film, we fabricated simplified bottom-gate TFTs using the a-Si film.[10] The typical mobility of these TFTs was 10^{-3}–10^{-4} cm^2/Vs, which is three or four orders of magnitude smaller than that of TFTs using conventional a-Si films formed by plasma-enhanced chemical vapor deposition (PE-CVD). Such poor mobility is attributed to the low concentration of hydrogen atoms that terminate the dangling bonds in the film. Unlike a-Si films formed by conventional PE-CVD, which contain 5–20% hydrogen, the spin-coated a-Si film baked at 540 °C for 2h contains only 0.3% hydrogen and appreciable quantities of dangling bonds that hinder the mobility. In contrast, the film spin-coated and baked at the temperature of 300 °C or less contains more than 20% hydrogen and could function somewhat as a semiconductor. However, such a low-temperature processed film might no longer be described as amorphous silicon, as it is easily oxidized in air. Some kind of technical solution must be found to strike a balance between a lower processing temperature and preventing oxidation.

In addition to considering a-Si as a target film material, we have also studied the crystallization of the film in order to demonstrate the fundamental potential of this solution processing approach for low-temperature poly-Si (LTPS) TFT applications.[10] Low-hydrogen a-Si films are suitable for excimer laser crystallization, which is the standard crystallization method adopted during the commercial production of LTPS TFTs.[19] The a-Si films, formed by coating the liquid silicon material and baking it at 540 °C, were irradiated by a 308-nm XeCl excimer laser (28-ns pulse, single shot) at various laser energies. As the laser energy increases, the color of the films changes from light auburn to light yellow, suggesting a conversion from amorphous to polycrystalline. The

Figure 5.4. A TEM image of a solution-processed poly-Si film. The film was formed by spin coating and baking of the liquid silicon materials, followed by laser crystallization. The high-resolution TEM image inserted in the figure clearly highlights the atomic image of the silicon crystal. The micrograph also illustrates that the grain size in the film is about 300 nm, which is comparable with that of a conventional CVD-formed poly-Si film. [Reproduced with permission from Ref. 10. Copyright 2006 Nature Publishing Group.]

transmission electron microscope (TEM) image in Fig. 5.4 clearly shows the crystalline growth as a result of laser irradiation. Using Raman spectroscopy,[20] this crystallization process was confirmed to be almost identical to that of the a-Si film formed by conventional CVD. As shown in Fig. 5.5, the full-width at half-maximum of the crystalline peak in the Raman spectrum decreases sharply as the laser energy increases, reaching a minimum value of 6.3 cm^{-1} at ~300 mJ/cm^2; thereafter the peak slightly broadens, reflecting microcrystallization.

5.4 FABRICATION OF A TFT USING A SOLUTION-PROCESSED SILICON FILM

Next, we fabricated n-channel TFTs, whose structure is schematically illustrated in Fig. 5.6a, using the coating-formed poly-Si films followed by standard fabrication steps for conventional LTPS TFTs.[10] First, a SiO_2 underlayer was formed by PE-CVD on a quartz substrate. After cleaning the substrate surface using 172-nm UV-irradiation at 10 mW/cm^2 for 10 min, the liquid silicon material (12-vol.% toluene solution of UV-irradiated CPS) was then spin-coated at 2000 rpm in a nitrogen-filled dry box. The spin-coated substrate was immediately placed on a hot plate heated at 200 °C, and the temperature was raised to 400 °C within 10 min. After maintaining the temperature at 400 °C for

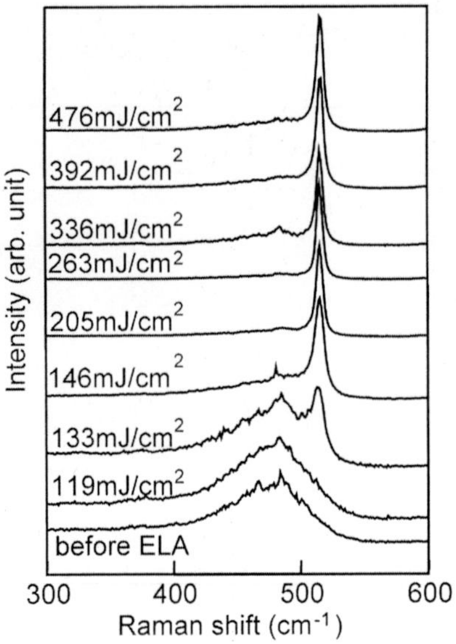

Figure 5.5. Raman spectroscopy of laser crystallized solution-processed silicon films. The plot shows laser intensity versus Raman shift of the film. The crystallinity is estimated from the intensity and width of the crystalline peak at 520 cm^{-1}. [Reproduced with permission from Ref. 11. Copyright 2006 The Japan Society of Applied Physics.]

30 min, it was further increased to 540 °C over another 10-min period and held for 2 h to form a 50-nm-thick a-Si film. Next, the amorphous film was converted to a polycrystalline one by irradiating it with a 308-nm excimer laser at 345 mJ/cm^2 (23-ns pulse, 20 shots with 250 Hz). After the poly-Si had been etched to create islands, a 120-nm-thick SiO_2 gate insulator (Gox in Fig. 5.6a) was formed by PE-CVD, followed by tantalum sputtering and etching to form gate electrodes. The source and drain regions were formed by self-aligned ion implantation of phosphorus ions using the gate electrodes as a mask. Finally, we made the TFTs accessible for measurement by forming an interlayer insulator, opening contact holes in the insulator to reveal the source and drain regions, and then sputtering aluminum to form electrodes. The channel width and length of the TFTs were both 10 μm. The coating-formed silicon film did not present any notable problems or difficulties during the above fabrication steps, as the silicon film was of semiconductor-grade purity and the film processing conditions—etching rate, laser conditions, and so on—were nearly the same as those used for the CVD-produced silicon film. For comparison, conventional TFTs were fabricated by the same process, except that the silicon film was formed by CVD.

As shown in Fig. 5.6b, these TFTs exhibit good electrical characteristics, with field-effect mobility (calculated from the transconductance in the

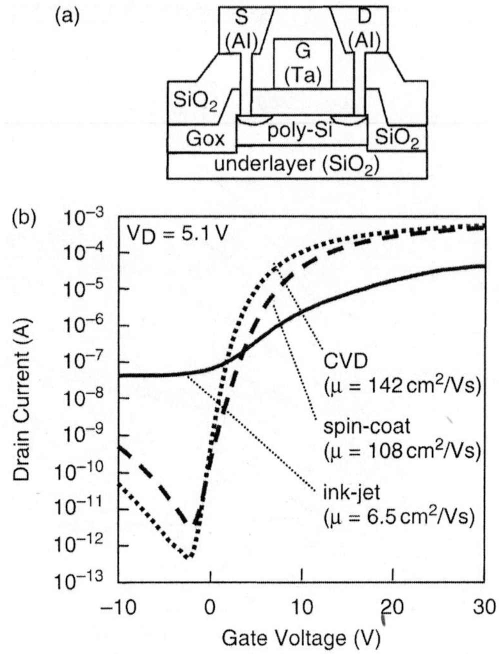

Figure 5.6. (a) The structure of a fabricated solution-processed LTPS TFT. (b) The transfer characteristics of LTPS TFTs, whose silicon film was formed by CVD (dotted line), spin coating (dashed line), and inkjet printing (solid line), respectively. The drain current of the inkjetted TFT is normalized to have the same channel width and length as the CVD-formed and spin-coated TFTs, for comparison. [Reproduced with permission from Ref. 10. Copyright 2006 Nature Publishing Group.]

saturation region) ranging from 74 to 108 cm^2/Vs in 15 transistors randomly selected among a 4-inch substrate. The transistor of mobility 108 cm^2/Vs, whose output characteristics are shown in Fig. 5.7, also possesses an ON/OFF ratio of seven digits (seven orders of magnitude), a threshold voltage V_{th} of 5.0 V and 0.83 V per decade for the s-factor (subthreshold swing).

The mobility was strongly affected not only by the conditions of laser crystallization but also by the amount of oxygen in the silicon film.[10] In the poly-Si film that exhibited the mobility of 108 cm^2/Vs, the oxygen concentration was 1100 ppm. Since the CPS and polysilanes are fairly oxygen-sensitive,[21] several precautions had to be taken during the preparation and treatment of the liquid precursor. The coarsely synthesized CPS was distilled repeatedly to remove impurities, including oxides before photo-polymerization. Furthermore, the organic diluent such as toluene was carefully deoxidized by leaving it in a dry box for several days before use. The oxygen concentration in the box was maintained strictly below 0.5 ppm during all processes, from spin coating to baking. Such precautions enabled control of the oxygen content of the film to

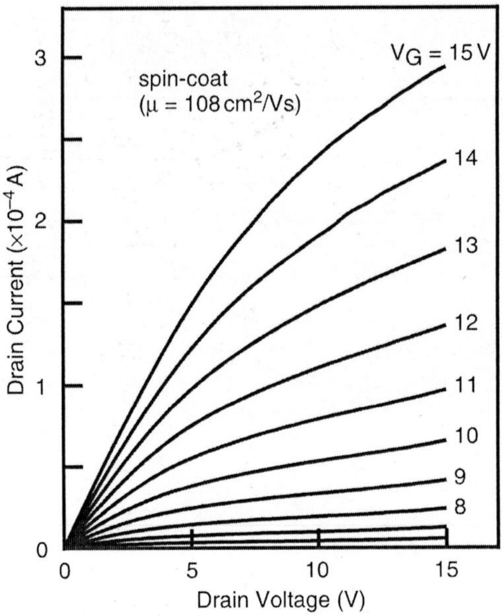

Figure 5.7. The output characteristics of the TFT formed using a spin-coated silicon film and whose transfer properties are shown in Figure 5.6. [Reproduced with permission from Ref. 10. Copyright 2006 Nature Publishing Group.]

less than 2000 ppm. The TFTs resulting from such low-oxygen films exhibited mobility as high as 50–100 cm^2/Vs, whereas no significant correlations were observed between the oxygen concentration and mobility (at these low oxygen levels), since manufacturing variations such as the laser processing conditions are the dominant factors. However, a slight failure in controlling the oxygen level in the dry box yielded a silicon film with an oxygen concentration of 8000 ppm and TFTs with a maximum mobility of 20 cm^2/Vs. By intentionally maintaining the oxygen level in the dry box at 10 ppm during all processing, we obtained an insulating a-Si film that contained 7% oxygen.

5.5 FABRICATION OF TFT USING INKJET-PRINTED SILICON FILM

The compatibility of the liquid silicon material with printing technology was also examined by fabricating TFTs with the same structure as above using ink-jet printing instead of photolithography to form the channel silicon island.[10] A 10% toluene solution of UV-irradiated CPS was inkjet-printed onto a glass substrate in a nitrogen-filled dry box. This solution was suitable for ejection from a piezo-driven print head.[1,7] Since its viscosity is almost the same as that of toluene and its stability is such that it can be kept for months in a dark place at room temperature, the ejection of the solution from the print head was

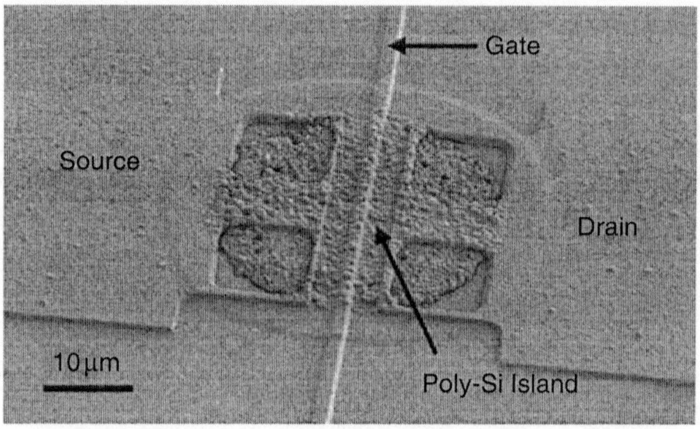

Figure 5.8. A SEM photograph of a fabricated LTPS TFT made using an inkjetted silicon film. [Reproduced with permission from Ref. 10. Copyright 2006 Nature Publishing Group.]

stable and reproducible. Three droplets, each 10 ng in weight, were deposited at the location where the channel island was to be formed. The droplets were converted into a poly-Si island of diameter 30–40 μm by baking at 540 °C (using the same steps as for the spin-coated film) followed by 308-nm excimer laser crystallization at 450 mJ/cm^2 (28-ns pulse, single shot). The island was 300 nm thick at the center and became thinner toward the periphery of the sample. After the gate insulator had been formed, the same steps were applied as for the spin-coated TFTs. The channel width and length of the resulting TFTs were 36 μm and 2 μm, respectively.

In the scanning electron microscope (SEM) image of an ink-jet-fabricated TFT in Fig. 5.8, the silicon island can be seen as the disk-shaped element with a rough surface. The nature of the external disk-shaped phase with a smooth surface is still unknown, but it is presumably the residual remains of the droplet, formed during the drying and shrinking processes. Since the wettability of the liquid silicon material was not sufficiently well understood to enable control of the shrinking behavior of the ink droplet, the resulting a-Si island became too thick for optimal laser crystallization. A thicker film generally requires a more intense laser for crystallization. The actual laser intensities used for the crystallization of 60-nm–thick spin-coated and 300-nm-thick ink-jet-treated Si films were 345 mJ/cm^2 and 450 mJ/cm^2, respectively. The latter intensity is not optimized, but it is determined roughly from the sample thickness. With such an intense laser source, surface roughness is easily induced, as shown in Fig. 5.8.

The inkjet-printed (or "inkjetted") TFT operated with a mobility of 6.5 cm^2/Vs and an ON/OFF ratio of three digits as shown in Fig. 5.6b. This low mobility is attributed to the poor crystallinity and rough surface, whereas the large off current, which was confirmed to be the current between source and drain rather than a leakage current through the gate insulator, is also attributed to the thick silicon film. Such a thick film contains many dangling bonds and defects near the substrate where crystallization is incomplete. The formation of a thin and uniform a-Si film using an ink-jet process would improve the surface morphology and electrical properties of the poly-Si films. However, the wettability and behavior of microdroplets are known to differ significantly from those of macroscopic droplets.[22,23] A thorough understanding of microdroplets will be needed to realize the ink-jet printing of liquid silicon material for practical applications.

5.6 FORMING SiO_2 FILMS FROM THE LIQUID SILICON MATERIALS

Just as silane gas is convertible to both silicon and silicon dioxide films using CVD, the liquid silicon material is also convertible to both of these compounds.[24] The conversion of polysilane into an oxide film requires a bake in an atmosphere that contains oxygen. However, baking the polysilane in air too soon after the coating process makes a porous SiO_2 film, since the CPS and the polysilane are fairly oxygen-sensitive.[21] On the other hand, after an a-Si film is formed completely, high-temperature thermal oxidation conditions exceeding 800 °C are needed to oxidize the film. If the oxidation of polysilane starts at an early stage of a-Si formation, as for the route indicated in Fig. 5.9 (arrowed broken line), a high-quality SiO_2 film may be formed, at a lower temperature that is compatible with selected plastic substrates.

To test this idea, we tried a two-step baking approach[24] as follows. In the first step, the polysilane film is baked at a temperature below 300 °C in a nitrogen atmosphere so that the film is not completely converted to a-Si. In the second step, the film is baked in air to convert it into SiO_2. Three samples

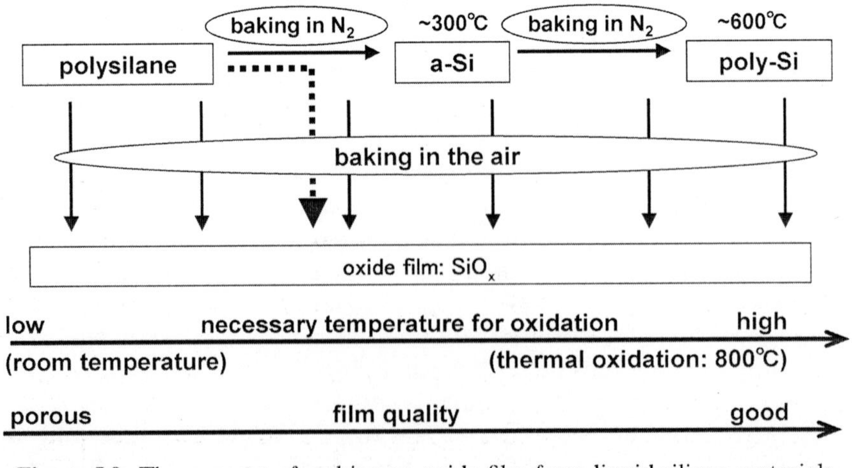

Figure 5.9. The concept of making an oxide film from liquid silicon materials.

TABLE 5.2. Baking Conditions and Colors of Resulting SiOx Films for Samples 1–3

	Sample 1	Sample 2	Sample 3
1st step (in N_2)	220 °C 60 min.	270 °C 60 min.	315 °C 60 min.
2nd step (in air)	410 °C 30 min.	410 °C 30 min.	410 °C 30 min.
Film color	Colorless	Light auburn	Auburn

[Reproduced with permission from Ref. 24. Copyright 2007 Society for Information Display.]

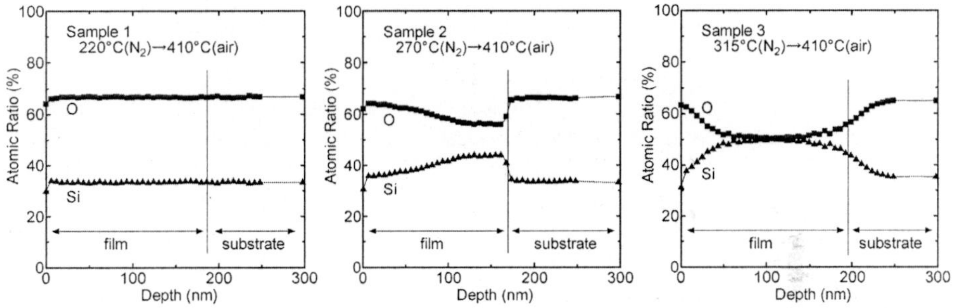

Figure 5.10. Depth profiles of atomic composition in two-step baked polysilane films prepared under various conditions. [Reproduced with permission from Ref. 24. Copyright 2007 Society for Information Display.]

were prepared by spin-coating a 20-vol.% toluene solution of UV-irradiated CPS onto a quartz substrate in a nitrogen atmosphere, to make polysilane films. Then, these films were baked in nitrogen (first step) under the conditions 220 °C for 60 min, 270 °C for 60 min, and 315 °C for 60 min for samples 1, 2, and 3, respectively. In the second bake, all samples were baked at 410 °C for 30 min in air. All heat treatments were carried out on a hot plate. Table 5.2 summarizes the baking conditions and the colors of the resulting films; sample 1 is colorless, whereas samples 2 and 3 are light auburn in color. The color differences suggest that samples 2 and 3 are partially converted to a-Si or Si-rich SiO_x films.

The Si:O composition ratios for these samples were investigated using X-ray photoelectron spectroscopy (XPS). As shown in Fig. 5.10, sample 1 is completely converted into a SiO_2 film, whereas the deep areas of samples 2 and 3 are not oxidized sufficiently. In samples 2 and 3, the three-dimensional connections of the Si atoms (i.e., a-Si) would be sufficiently developed during the first baking in nitrogen to withstand oxidation during the second baking step. The colors of the samples were not changed by the additional baking of samples 2 and 3 at 450 °C for 60 min in the air.

Figure 5.11 shows cross-sectional TEM images of the annealed polysilane films on poly-Si islands. After spin coating polysilane under the same condi-

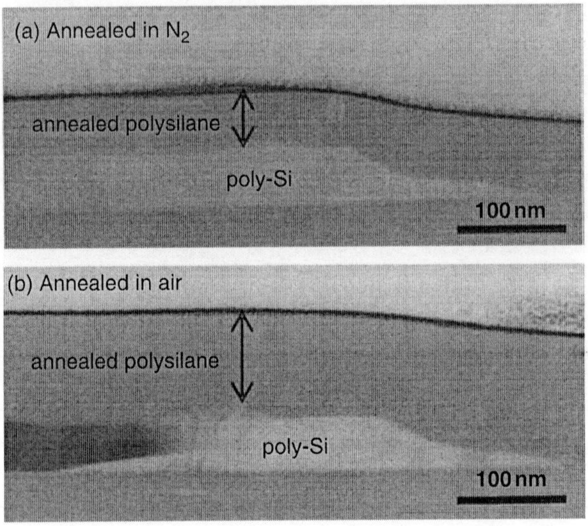

Figure 5.11. Cross-sectional TEM images of annealed polysilane films. (a) Amorphous Si film baked in nitrogen. (b) Two-step baked SiO_2 film. [Reproduced with permission from Ref. 24. Copyright 2007 Society for Information Display.]

tion, a film was converted to an a-Si film by baking it in nitrogen. On the other hand, another film was two-step baked in the same way as the sample 1 (see Table 5.2) to convert it to a SiO_2 film. The film annealed in air [Fig. 5.11b, SiO_2 film] is about 1.8–2.0 times thicker than the film annealed in nitrogen [Fig. 5.11a, a-Si film]. This ratio is an appropriate value, since the thermal oxide film becomes about 1.8 times as thick as the original silicon film.

5.7 LTPS FABRICATION USING SOLUTION-PROCESSED SiO_2 FILMS

To evaluate the solution-processed SiO_2 (SP-SiO_2) film, we fabricated three kinds of LTPS TFTs, with the structure shown in Fig. 5.6a. In the first TFT (TFT-1), the SP-SiO_2 was used only for the gate dielectric. In the second TFT (TFT-2), the SP-SiO_2 was used only for the underlayer dielectric (ULD). The third TFT (TFT-3) was a conventional TFT as a reference, with all dielectrics formed by CVD. The gate dielectric (120 nm) of TFT-1 and the ULD (200 nm) of TFT-2 were fabricated by the solution process as follows. The liquid silicon material (a toluene solution containing 12-vol.% UV-irradiated CPS for the gate dielectric and 20 vol.% for the ULD) was spin-coated at 2000 rpm in a nitrogen-filled dry box. Then, the spin-coated substrate was two-step baked with the conditions of sample 1, as shown in Table 5.2. All other steps are conventional ones (CVD, sputter, and photolithography) described in the previous sections.

Figure 5.12 and Table 5.3 show the electrical characteristics of the three TFTs (W/L = 10 μm/10 μm). The mobilities of TFT-1 (gate dielectric: SP-SiO_2) and TFT-2 (ULD: SP-SiO_2) are 108.9 cm^2/Vs and 110.6 cm^2/Vs, respectively. They are comparable with that of TFT-3 (conventional). These TFTs also possess an ON/OFF ratio of nine digits. The leakage current through the 120-nm-thick gate dielectric (TFT-1) of a 50-μm × 50-μm area was measured as

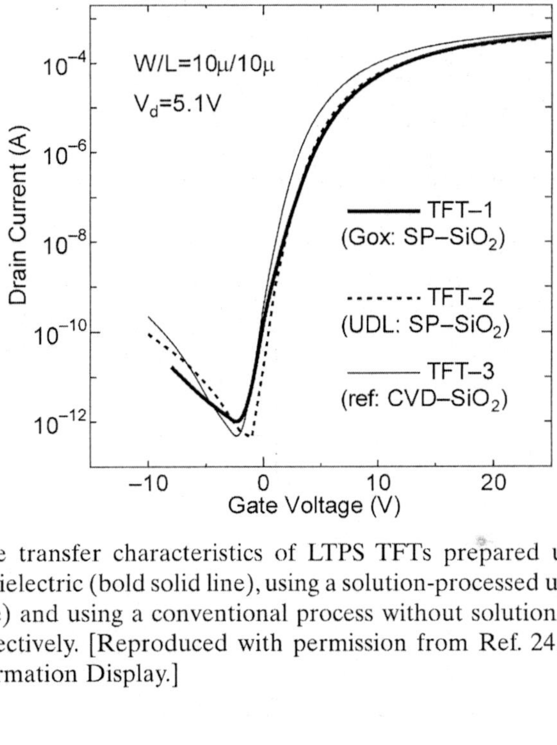

Figure 5.12. The transfer characteristics of LTPS TFTs prepared using a solution-processed gate dielectric (bold solid line), using a solution-processed underlayer dielectric (broken line) and using a conventional process without solution processing (thin solid line), respectively. [Reproduced with permission from Ref. 24. Copyright 2007 Society for Information Display.]

TABLE 5.3. Characteristics of TFTs with Solution-Processed Gate Dielectric and Underlayer Dielectric

	TFT-1	TFT-2	TFT-3
ULD	CVD	SP	CVD
Channel Si	CVD	CVD	CVD
Gate dielectric	SP	CVD	CVD
Mobility (cm^2/Vs)	108.9	110.6	147.8
V_{th} (V)	4.3	4.0	3.1
S-factor (V/dec)	0.75	0.57	0.59

[Reproduced with permission from Ref. 24. Copyright 2007 Society for Information Display.]

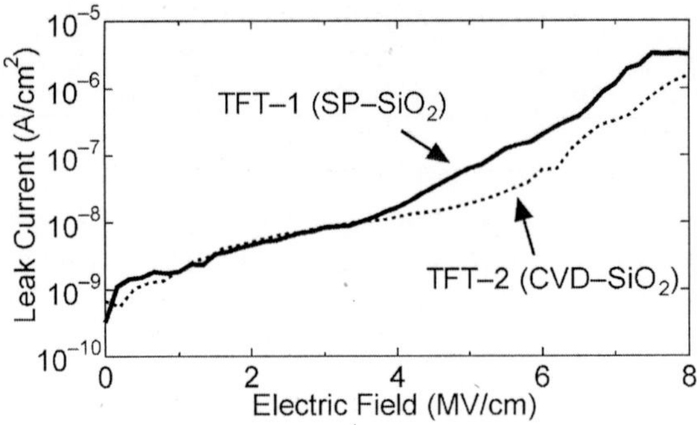

Figure 5.13. Leakage current through a solution-processed SiO_2 film (solid line) and a conventional CVD formed SiO_2 film (broken line). [Reproduced with permission from Ref. 24. Copyright 2007 Society for Information Display.]

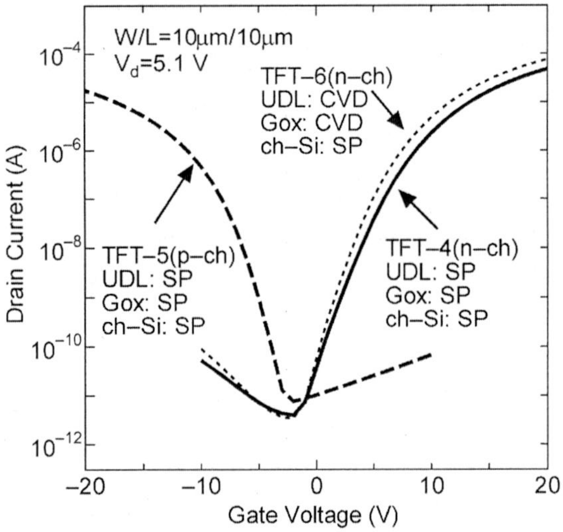

Figure 5.14. The transfer characteristics of LTPS TFTs prepared using a solution-processed gate dielectric, underlayer dielectric, and channel silicon (TFT-4, n-ch, solid line: TFT-5, p-ch, dashed line), respectively. TFT-6 (n-ch, broken line) is a reference TFT prepared using solution processing only for the channel silicon. [Reproduced with permission from Ref. 24. Copyright 2007 Society for Information Display.]

shown in Fig. 5.13. The leakage current of the SP-SiO_2 and that of the CVD formed-SiO_2 are the same in lower electric fields.

Next, we fabricated TFTs whose ULD, channel Si, and gate dielectric were all solution-processed. The fabricated TFTs (TFT-4, 5, and 6) have similar solution-processed 50-nm-thick silicon films,[10,11] the details of which are described in Section 5.4. In addition, TFT-4 (n-channel) and TFT-5 (p-channel) have the SP-SiO_2 as both ULD and gate dielectric, which are fabricated using

TABLE 5.4. Characteristics of TFTs with Solution-Processed Gate Dielectric, Underlayer Dielectric, and Channel Silicon

	TFT-4 (n-ch)	TFT-5 (p-ch)	TFT-6 (n-ch)
ULD	SP	SP	CVD
Channel Si	SP	SP	SP
Gate dielectric	SP	SP	CVD
Mobility (cm^2/Vs)	23.0	9.9	32.0
V_{th} (V)	7.9	11.0	7.1
S-factor (V/dec)	1.4	1.1	1.1

[Reproduced with permission from Ref. 24. Copyright 2007 Society for Information Display.]

the same conditions as those for TFT-1 and TFT-2. The p-channel TFT was fabricated using ion implantation of boron ions instead of phosphorus ions, as described in Section 5.4 for the n-channel device.

Figure 5.14 and Table 5.4 show the electrical characteristics of the fabricated TFTs (W/L = 10 μm/10 μm). TFT-4 and 5 (Gox, UDL and channel Si are solution-processed) have the mobility values, 23.0 cm^2/Vs and 9.9 cm^2/Vs, respectively. They are lower than that of TFT-6 (only the channel silicon was solution-processed). In this experiment, however, the mobility of the reference TFT (TFT-6) is also relatively poor, as expected, because the laser power and other conditions under which the channel silicon was solution-processed were not optimized. Thus, the mobilities of TFT-4 and TFT-5 were also affected by the channel silicon and were much lower than the mobilities of TFT-1 and TFT-2. With optimization of the conditions under which the channel silicon is deposited, we believe that higher mobility values can be achieved in the devices with solution-processed Gox, UDL, and channel Si.

5.8 FORMING DOPED SILICON FILMS

To apply solution processing for producing various silicon devices, doped silicon films (both n- and p-type) as well as non-doped films should be solution-processed using the liquid silicon material or its derivative. In this section, we describe the solution processing of n-type doped silicon films,[25] using a liquid precursor based on doped polysilane. White phosphorus (P_4) is a good dopant source to be added to the liquid silicon material, because white phosphorus dissolves well into CPS and organic solvent. We have tried following two methods to interfuse phosphorus atoms into the liquid silicon material.[25] (1) In the post-polymerization addition method, white phosphorus is simply added and dissolved into the liquid silicon material solution after the photo-polymerization process. (2) In the copolymerization method, after white phosphorus is dissolved into the CPS monomer, the solution is UV-irradiated to polymerize the CPS, so that the phosphorus atoms are included in the polymer structure. These doped precursors, prepared by either method, could be treated similarly (coating and baking) as the non-doped precursor to form doped silicon films.

The details of the copolymerization method are as follows. First, white phosphorus was dissolved into CPS at 1 wt%. This solution was exposed to 405-nm UV light of 3500 mW/cm^2 with various irradiation times to induce photo-polymerization. All these treatments were carried out in an inert nitrogen atmosphere with an oxygen concentration of less than 1.0 ppm. Among the various irradiation times, five conditions were chosen for analysis using a gas chromatograph, namely, an initial (nonirradiated) solution, 20-min-, 25-min-, and 30-min-irradiated solutions, as well as a white phosphorus alone solution for comparison, all of which were diluted with toluene before the analysis. Figure 5.15 shows the gas chromatograms of these solutions. The intensities of the peaks originating from both white phosphorus and CPS decrease with UV irradiation time. Although the white phosphorus peak vanishes in the first 25 min, the CPS peak decreases only 20% in the same

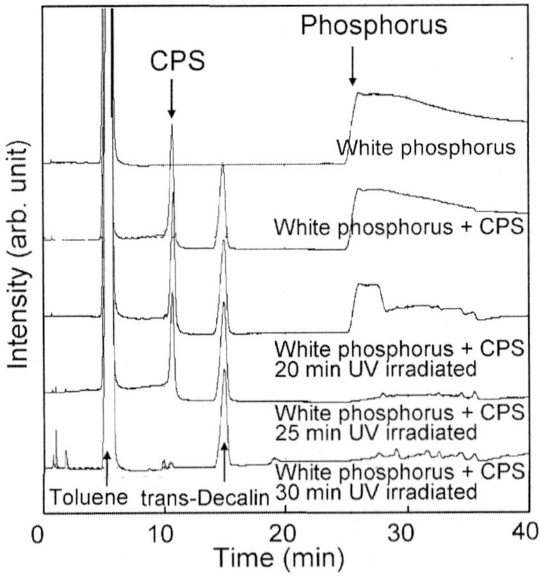

Figure 5.15. Gas chromatograms of liquid precursors for producing n-type Si films, namely, solution of CPS with white phosphorus dissolved at 1 wt%; initial (nonirradiated) solution; 20-min-, 25-min-, and 30-min-irradiated solutions; and white phosphorus alone solution, for comparison, all of which were diluted with toluene before analysis. Trans-decalin was used as a standard for a quantitative estimation of an unreacted CPS monomer. [Reproduced with permission from Ref. 25. Copyright 2007 The Japan Society of Applied Physics.]

period. The presence of the white phosphorus seems to block CPS polymerization, since the UV dose for this process is 30 times larger than that for the polymerization of CPS without phosphorus. Moreover, decomposed white phosphorus and CPS would react in some manner, such as through copolymerization, since they are decomposed at the same time.

The concentration of phosphorus atoms in the resulting doped silicon film was estimated using SIMS. To compare the post-polymerization addition and copolymerization methods, two types of doped silicon films were formed using the following two solutions. The first solution was prepared using the copolymerization method described above with 1 wt% phosphorus and 30-min irradiation. The second solution was prepared using the post-polymerization addition method, whereby white phosphorus was simply dissolved in the liquid silicon material at the same concentration as the first solution (1 wt%). The concentration (wt%) is the weight ratio of initial P_4 to CPS and 1 wt% corresponds to 0.83 atm% of P to Si. Moreover, a series of solutions with different concentrations were prepared by diluting the first solution with non-doped liquid silicon material (UV-irradiated non-doped CPS) to obtain phosphorus concentrations of 0.1, 0.01, and 0.001 wt%. These solutions were diluted with toluene, spin-coated onto a quartz substrate, and annealed at 500 °C for 2 hours on a hot plate to form doped silicon films.

Figure 5.16 shows the results of SIMS analysis of these films. In the case of copolymerization, the phosphorus concentration in the Si film remains at the

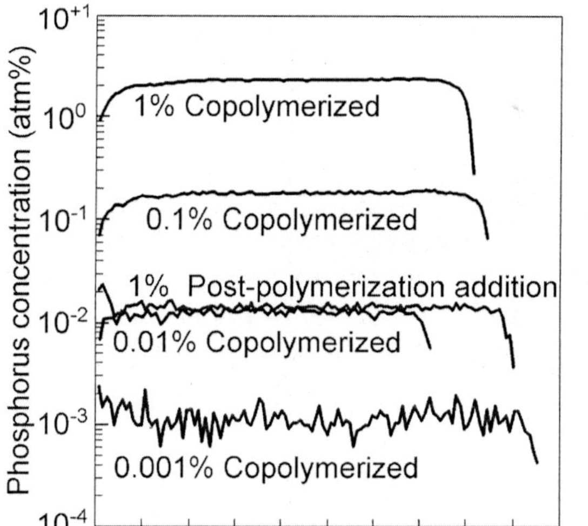

Figure 5.16. SIMS of a solution-processed phosphorus-doped a-Si film. The phosphorus concentration in the films is almost proportional to that of the initial solutions in the case of the copolymerized method. [Reproduced with permission from Ref. 25. Copyright 2007 The Japan Society of Applied Physics.]

same level as that in the original solution. However, in the case of post-polymerization addition, the phosphorus concentration in the film is only 1% of that in the film made by copolymerization using a solution with the same phosphorus concentration, which means that nearly all the phosphorus was depleted from the film during the coating and annealing steps. This difference supports the assumption that phosphorus atoms in a copolymerized solution are coupled to the polysilane molecules, unlike the white phosphorus in the post-polymerization addition solution, which is merely dissolved in the solution.

The solution-processed doped silicon films described above (baked at 500 °C for 2 hr) exhibited high electrical resistivity (greater than 300 Ωcm), which is the measurement limit of the instrument we used. To lower the resistivity, we tried an additional rapid thermal annealing (RTA) of the film prepared from the copolymerized solution with 1 wt% phosphorus concentration. In this RTA, the SiC plate on which the sample was placed was irradiated with infrared (IR) light from a 1-kW IR lamp. The RTA conditions were 600 °C for 2 hr, 650 °C for 20 min, 700 °C for 5 min, and 750 °C for 5 min; these temperatures were that of the SiC plate, and the temperature of the Si film is estimated to be several dozens of degrees lower than that.

We measured the electric resistivity and Raman scattering of the films after RTA to investigate the relationship between resistivity and crystallinity. The Raman spectra in Fig. 5.17 indicate that only the 750 °C 5-min-processed

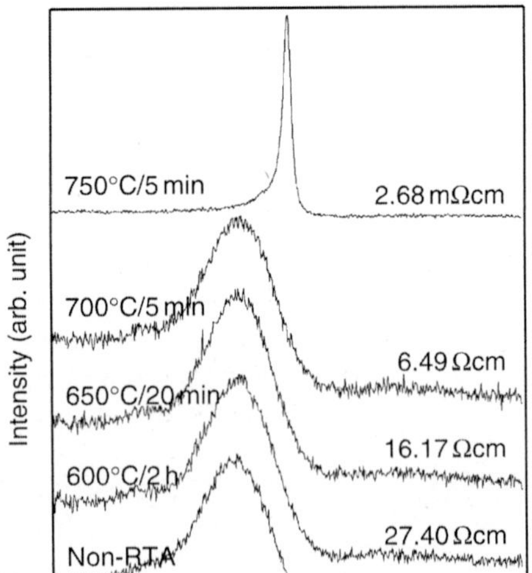

Figure 5.17. Electrical resistivity and Raman scattering spectroscopy of doped-Si films formed from the copolymerized solution. [Reproduced with permission from Ref. 25. Copyright 2007 The Japan Society of Applied Physics.]

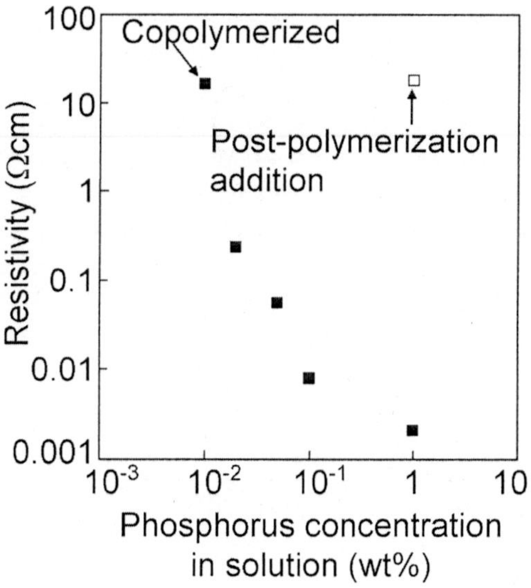

Figure 5.18. Relationship between white phosphorus concentration of the starting solutions and electrical resistivity of resulting doped-Si films. [Reproduced with permission from Ref. 25. Copyright 2007 The Japan Society of Applied Physics.]

sample was converted into a poly-Si film, with the other samples remaining amorphous. The resistivity decreased rapidly just as the film became polycrystalline. However, the dopant atoms are activated and the resistivity decreases somewhat even in the amorphous films formed by RTA.

Figure 5.18 shows the relationship between the resistivity and phosphorus concentration of the initial solution for the film formed from various solutions and heated under the same polycrystallizing RTA conditions (750 °C for 5 min). As the initial phosphorus concentration increases, the resistivity decreases down to 2.1 mΩcm. The film formed from a 1-wt% post-polymerization addition solution and the film formed from a 0.01-wt% copolymerized solution exhibit almost the same resistivity, which is reasonable since the two films have almost the same amount of phosphorus atoms, as shown in Fig. 5.16. To apply these doped-silicon films to the source and drain regions of poly-Si TFTs, the initial concentration of 0.1–1 wt% will be sufficient in the case of the copolymerized solution for this heating condition.

Finally, we tried to activate dopant atoms using pulsed laser irradiation, which is effective in lowering the process temperature. The light source was a 308-nm XeCl excimer laser, which is a standard source for crystallizing a-Si films used in the LTPS process.[19] A test sample of 76 nm thickness, prepared from the copolymerized solution (1-wt% phosphorus, 30-min UV irradiation, 500 °C 2 hr annealing), was irradiated using a XeCl laser at various intensities to activate the dopant atoms. Figure 5.19 shows the relationship between the

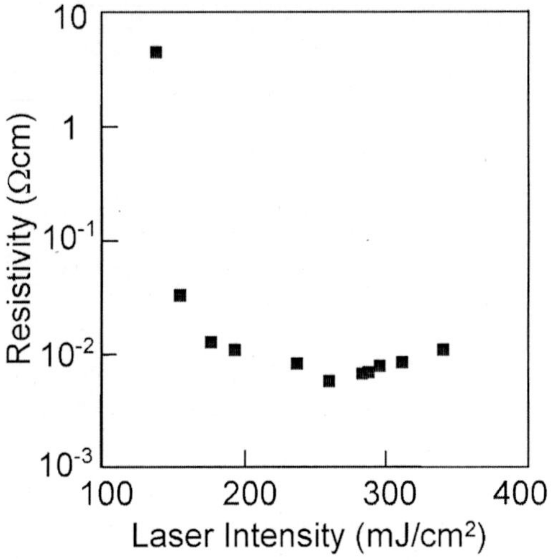

Figure 5.19. Relationship between activation laser intensity and electrical resistivity of the resulting film. [Reproduced with permission from Ref. 25. Copyright 2007 The Japan Society of Applied Physics.]

laser intensity and the electric resistivity of the resulting film. The resistivity decreases rapidly with increasing laser intensity around 150 mJ/cm^2 and exhibits about the same resistivity (lower than 10 mΩcm) over the wide range of laser intensities between 200 and 350 mJ/cm^2. The laser intensity required to crystallize a 76-nm-thick a-Si film completely is generally around 300 mJ/cm^2, which is much higher than the intensity that induces the rapid resistivity decrease. This is because the surface of the film is partially crystallized even at a lower irradiating intensity in the case of laser crystallization. Laser irradiation may be considered a suitable method for activating phosphorus dopants, because of the wide process window for source and drain fabrication in LTPS.

In summary, phosphorus-doped liquid silicon material can be used to solution-process n-type silicon films. The liquid silicon material is phosphorus-doped by dissolving white phosphorus into CPS and copolymerizing them using UV irradiation. The phosphorus concentration in the film, and the resistivity of the film, can be controlled by altering the initial phosphorus concentration in the copolymerized solution. The resulting doped-silicon film showed a resistivity in the range of 6.5–27 Ωcm (a-Si film) and 2.0–10 mΩcm (poly-Si film). These values are sufficiently low for the source and drain regions of TFTs. The detailed process and the structure of the copolymerized polymer (polysilane) as well as the activation at lower temperature are issues for future inquiry. P-type doping would be similarly likely possible, using the same method as described above for the n-type system, but with an appropriate dopant source, such as a boron or indium compound instead of the phosphorus.

5.9 CONCLUSIONS

The constituent elements of a silicon device—i.e., intrinsic silicon, doped silicon and silicon oxide films—have become possible targets of solution processing. Using the described liquid silicon materials, we have demonstrated high-performance intrinsic silicon and silicon dioxide films, which are applicable to the active layer and gate dielectric for LTPS TFTs, respectively. Doped n-type silicon films have also been formed, with low enough resistivity for the source and drain region of TFTs. Solution processing using these materials provides an opportunity for low-cost production of silicon-based devices, such as large-size flat panel displays and solar cells.

However, direct patterning using ink-jet printing currently has a size limitation of about 20 μm, and this approach is not yet generally suitable for microfabrication. Another major issue for solution processing using liquid silicon materials is the processing temperature, which is currently higher than 300 °C to form even amorphous films. The extremely low oxygen atmosphere required to handle the liquid silicon material, less than 0.5 ppm, might also be an obstacle for manufacturing equipment. These issues should be addressed in the future in order to use this approach more generally for realistic applications.

Finally, silicon-based polymers, especially with hydrogen lateral groups, are very interesting, but they are not yet explored sufficiently. There are many unknown properties in these materials, including the details of the photopolymerization process and a-Si formation from polysilane. Additional academic work in this field is expected and necessary to make the solution processing of silicon devices more convenient and reliable.

ACKNOWLEDGMENTS

This work is partially supported by a grant from the New Energy and Industrial Technology Development Organization (NEDO) to a collaboration project between Seiko Epson Corporation and JSR Corporation.

REFERENCES

1. Shimoda, T.; et al. 1999. Multicolor pixel patterning of light-emitting polymers by ink-jet printing. *Tech. Digest of SID '99*. pp. 376–379.
2. Miyashita, S.; et al. 2001. Full color displays fabricated by ink-jet printing. *Proc. of Asia Display/IDW '01*. pp. 1399–1402.

3. Sirringhaus, H.; et al. 2000. High-resolution inkjet printing of all-polymer transistor circuits. *Science* 290:2123–2126.
4. Kawase, T.; Sirringhaus, H.; Friend, R. H.; Shimoda, T. 2000. All-polymer thin film transistors fabricated by high-resolution ink-jet printing. *Tech. Digest of IEDM*. pp. 623–626.
5. Gelinck, G. H.; et al. 2004. Flexible active-matrix displays and shift registers based on solution-processed organic transistors. *Nature Mater*. 3:106–110.
6. Peumans, P.; Uchida, S.; Forrest, S. R. 2003. Efficient bulk heterojunction photovoltaic cells using small-molecular-weight organic thin films. *Nature* 425: 158–162.
7. Furusawa, M.; et al. 2002. Inkjet-printed bus and address electrodes for plasma display. *Tech. Digest of SID '02*. pp. 753–755.
8. Ridley, B. A.; Nivi, B.; Jacobson, J. M. 1999. All-inorganic field effect transistors fabricated by printing. *Science* 286:746–749.
9. Mitzi, D. B.; Kosbar, L. L.; Murray, C. E.; Copel, M.; Afzali, A. 2004. High-mobility ultrathin semiconducting films prepared by spin coating. *Nature* 428:299–303.
10. Shimoda, T.; et al. 2006. Solution-processed silicon films and transistors. *Nature* 440:783–786.
11. Tanaka, H.; et al. 2006. Solution-processed silicon films and transistors using novel liquid silicon materials. *Digest of Technical Papers AM-FPD 06*. pp. 27–30.
12. Kipping, F. S. 1924. Organic derivatives of silicon. Complex silicohydrocarbons $[SiPh_2]_n$. *J. Chem. Soc.* 125:2291–2297.
13. John, P.; Oher, I. M.; Wood, J. 1983. The electrical conductivity of polysilane, $(SiH_2)_x$. *J. Chem. Soc. Chem. Commun.* 1496–1497.
14. Suzuki, M.; Kotani, J.; Gyobu, S.; Kaneko, T.; Saegusa, T. 1994. Synthesis of sequence-ordered polysilane by anionic ring-opening polymerization of phenylnonamethycyclopentasilane. *Macromolecules* 27:2360–2363.
15. Cypryk, M.; Gupta, Y.; Matyjaszewski, K. 1991. Anionic ring-opening polymerization of 1,2,3,4-tetramethy-1,2,3,4-tetraphenylcyclotetrasilane. *J. Am. Chem. Soc.* 113:1046–1047.
16. Hengge, E.; Bauer, G. 1973. Cyclopentasilan, das erste unsubstituierte cyclische Siliciumhydrid. *Angew. Chem.* 85:304–305.
17. Hengge, E.; Bauer, G. 1975. Darstellung und eigenschaften von cyclopentasilan. *Monatsh. Chem.* 106:503–512.
18. Raabe, G.; Michl, J. 1985. Multiple bonding to silicon. *Chem. Rev.* 85:419–509.
19. Sameshima, T.; Usui, S.; Sekiya, M. 1986. XeCl excimer laser annealing used in the fabrication of poly-Si TFTs. *IEEE Electron Device Lett.* 7:276–278.
20. Kitahara, K.; Yamazaki, R.; Kurosawa, T.; Nakajima, K.; Moritani, A. 2002. Analysis of stress in laser-crystallized polysilicon thin films by Raman scattering spectroscopy. *Jpn. J. Appl. Phys.* 41:5055–5059.
21. Chatgilialoglu, C.; et al. 1998. Autoxidation of poly(hydrosilane)s. *Organometallics* 17:2169–2176.
22. Morii, K.; et al. 2000. Characterization of light-emitting polymer devices prepared by ink-jet printing. *Proc. 10th Int. Workshop on Inorganic and Organic Electroluminescence.* pp. 357–360.

23. Morii, K.; Masuda, T.; Ishida, M.; Hotta, S.; Shimoda, T. 2004. The direct patterning of crystalline organic-semiconductor films on a substrate by ink-jet printing. In *Proc. Int. Conference on Synthetic Metals*. pp. 126–127.

24. Tanaka, H.; et al. 2007. Solution-processed SiO_2 films using hydrogenated polysilane based liquid materials. *Tech. Digest of SID '07*. pp. 188–191.

25. Tanaka, H.; et al. 2007. Spin-on n-type silicon films using phosphorous-doped polysilane. *Jpn. J. Appl. Phys.* 46:L886–L888.

CHAPTER 6

Spray CVD of Single-Source Precursors for Chalcopyrite I–III–VI_2 Thin-Film Materials

ALOYSIUS F. HEPP,* KULBINDER K. BANGER,† MICHAEL H.-C. JIN,** JERRY D. HARRIS,†† JEREMIAH S. MCNATT,* and JOHN E. DICKMAN*

*Photovoltaic & Power Technologies Branch, NASA Glenn Research Center, Cleveland, OH; †Ohio Aerospace Institute, Cleveland, OH; **Department of Materials Science and Engineering, University of Texas at Arlington, Arlington, TX; ††Department of Chemistry, Northwest Nazarene University, Nampa, ID

6.1 INTRODUCTION

The development of new technologies that will enable the exploration of the universe has been one of the major goals of the National Aeronautics and Space Administration (NASA) since its founding in 1958. New power technologies will enhance the performance of rovers, satellites, and other payloads as the agency advances its broad-based charter in aeronautics and space exploration. With the increasing cost of launching payloads into space, currently estimated to be approximately $20k per kg, lightweight materials for space power systems are now being considered. Use of flexible solar cells, which allows for efficient storage in launch vehicles for later controlled deployment in space, will be mission enabling for several proposed NASA space programs (Fig. 6.1).

As part of its efforts to fulfill these goals, NASA Glenn Research Center (GRC) has been working on the development of novel fabrication technologies to enable deployment of lightweight, thin-film solar cells for space power generation.[1–9] The ability to deposit thin films on lightweight, flexible substrates will lead to more design flexibility, lower launch costs, and provide extra space and mass for payloads.

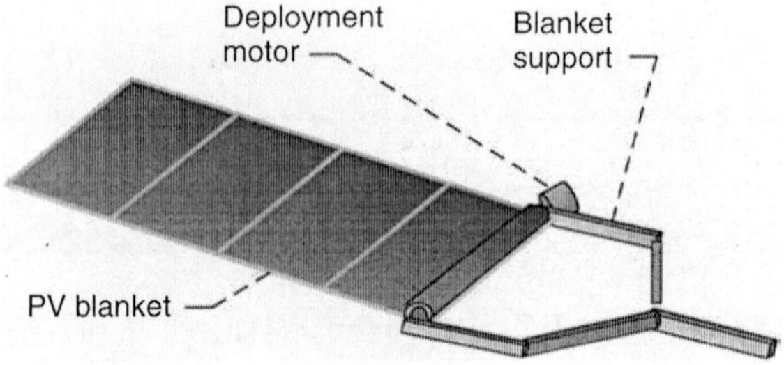

Figure 6.1. Lightweight flexible PV array.

In fact, the development of thin-film, flexible lightweight solar cells is important for both terrestrial and space applications. Thin-film solar cells use 30–100 times less semiconducting material and are less expensive to manufacture than conventional crystalline silicon cells. For many terrestrial applications, thin-film photovoltaic technologies are advantageous because of their flexible lightweight construction, permitting them to be "molded" onto non-rigid, or uniform, structures for innovative power systems. Importantly for space applications, their lightweight structure enables them to achieve much higher mass specific power (Wkg^{-1}) than conventional single crystalline device technologies.[10]

Current thin-film photovoltaic (PV) research encompasses development of CdTe, $Cu(In:Ga)(S:Se)_2$ (CIGSSe), and thin-film silicon-based solar cells.[11] Photovoltaic devices based on the ternary chalcopyrite absorber layers, CIGSSe or more generally $I–III–VI_2$, have been the particular focus of intense investigation for over two decades. The use of chalcopyrite absorbers is advantageous since the band gaps of many of these alloys are a good match for the maximum photon power density in the solar spectrum for both terrestrial (AM1.5) and space applications (AM0) (Fig. 6.2), while displaying long-term stability and excellent radiation tolerance.[12–14] Additionally, by adjusting the percent atomic composition of either Ga for In and/or S for Se, the band gap can be tuned from 1.0 eV to 2.4 eV, thus permitting fabrication of large or graded band gaps.[15]

Polycrystalline $CuInS_2$ (CIS) was chosen as the target absorber layer for the effort at NASA GRC because of its high absorption coefficient and its direct band gap of 1.5 eV, which is near optimum for AM0 conditions. So far, the highest total area efficiency achieved by a $CuInS_2$ cell using vacuum-based co-evaporation techniques has been 12.5%.[16] Other alloys of CIGSSe, such as $Cu(In,Ga)Se_2$ (CIGS), have shown conversion efficiencies of as high as 19.2% under AM1.5 illumination.[17] However, the 1.5-eV direct band gap of $CuInS_2$ makes it ideal for solar radiation conversion in space. $CuInS_2$ thin films are also more resistant to radiation in space than other CIGS alloys,[12,13] they can be used as a top cell in a tandem structure with CIGS,[18] and they are less toxic to process than selenium-containing alloys.

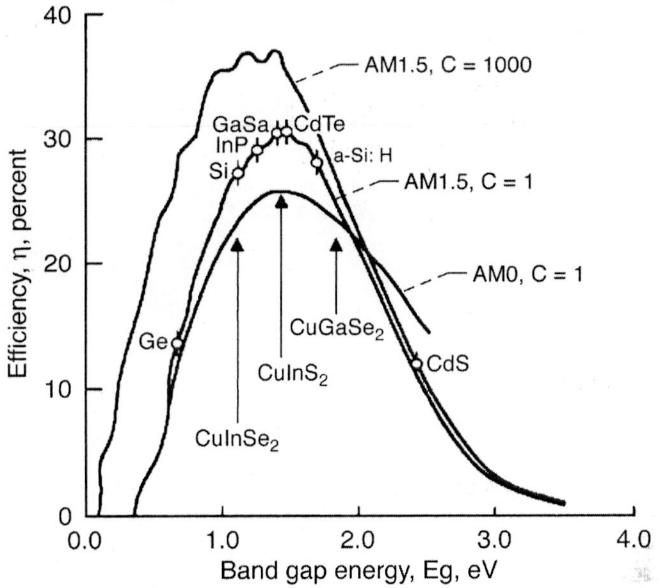

Figure 6.2. Predicted efficiency versus band gap for thin-film photovoltaic materials for solar spectra in space (AM0) and on the surface of the Earth (AM1.5) at 300K, with unconcentrated (C = 1) and high concentration (C = 1000) sunlight.

A key technical issue outlined in the 2001 U.S. photovoltaic road map is the need to develop low-cost, high-throughput manufacturing for high-efficiency thin-film solar cells. Thus, an important process to target for device fabrication for thin-film solar cells is deposition onto flexible, lightweight substrates such as polyimides. Current methods for depositing ternary polycrystalline compounds include coevaporation of elements[19–21] or alloys,[22] electrodeposition,[23] reactive-sintering,[24] and flash evaporation,[25] which are often followed by toxic sulphurization/selenization steps at increased temperatures. Furthermore, under these conditions, loss of volatile In/Ga chalcogenides is common.[19,26] The high-temperature requirements make this protocol incompatible with all currently known flexible polyimides or other polymer substrates. In addition, the use of toxic reagents is a limiting factor. The use of multisource inorganic/organometallic precursors in a chemical vapor deposition (CVD) type process is more appealing because of milder process parameters. However, stoichiometric control of deposited films can be difficult to achieve, and film contamination has been reported.[27,28]

A novel alternative approach is the use of ternary single-source precursors (SSPs), which have the I–III–VI_2 stoichiometry "built in" and are suitable for low-temperature deposition (Fig. 6.3). Although a rich and diverse array of binary SSPs[29–33] are known, characterized, reviewed, and tested, the number of known ternary SSPs is limited, as is their use in deposition processes.[29] SSPs

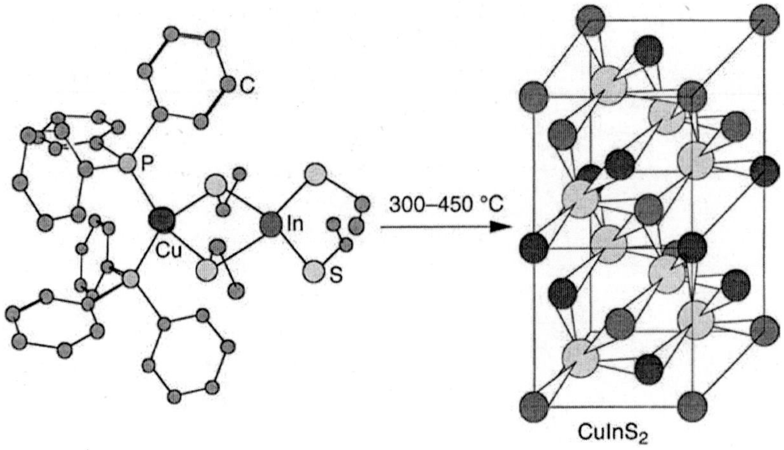

Figure 6.3. An example of the pyrolysis of a SSP, [{PPh_3}$_2$Cu(SEt)$_2$In(SEt)$_2$], to produce the semiconductor $CuInS_2$.

for chalcopyrite (I–III–VI_2) materials possibly offer a cleaner or "greener" approach for depositing thin films. One advantage is that they offer several tunable sites within the complex, allowing various combinations for depositing chalcopyrite films of different compositions. Early work on SSPs was performed by Nomura and colleagues[34–37] and by Kanatzidis and colleagues.[38] Buhro and Hepp along with their colleagues later demonstrated that the thiolate-bridged bimetallic complex, [{PPh_3}$_2$Cu(μ-SEt)$_2$In(SEt)$_2$], could be used in a spray CVD process for depositing $CuInS_2$ thin films, with processing temperatures under 400 °C.[39–41] A review of work on chemical precursors was recently reported.[3]

Films at NASA GRC were deposited using homemade spray or aerosol-assisted chemical vapor deposition (AACVD) reactors to exploit the lower deposition temperature enabled by the simpler decomposition chemistry for the SSPs.[6–9] AACVD is a simple and inexpensive process that offers the advantage of a uniform, large-area deposition, just like metal organic CVD (MOCVD), while also offering the low-temperature solution reservoir typical of spray pyrolysis methods.

Following growth of films and completion of devices, characterization techniques were employed to determine the properties and performances of the films and devices. Film thicknesses were measured using a profiler (Sloan Dektak IIA or KLA-Tencor HRP 75). Film morphologies were studied using a scanning electron microscope (SEM) (Hitachi S-3000N and S-800). The Hitachi S-3000N includes built-in energy dispersive spectroscopy (EDS) (EDAX, Falcon), used to examine film compositions. Compositional measurements were performed in different areas across each cell to determine the compositional uniformity. The Cu/In ratios were obtained by quantifying the Cu K and In L emission lines using the ZAF standardless method. X-ray diffraction (XRD) (PANalytical X'pert Pro) was used to identify the phases, and glancing angle X-ray diffraction (GAXRD) was performed using the same instrument to identify the phases present at the film surfaces. The optical band gaps of the films were determined by analyzing transmittance measurements taken in an ultraviolet (UV)/visible (VIS) spectrophotometer (Perkin-Elmer Lamda-19). Electrical measurements were performed using a four-point probe system (Bio-Rad HL5500PC) operated in the van der Pauw configuration, and a custom hot-point probe system—a commercial soldering iron (Weller TC201)—was used as the hot probe.

Throughout the NASA GRC studies, various processing and postprocessing parameters were modified to determine how these affected morphology, stoichiometry, crystallography, electrical properties, and optical properties of deposited thin films. Some of the modifications made included varying the reactor type and configuration, deposition temperatures, location of the substrate within the reactor, annealing times, and annealing atmospheres. We will discuss several deposition designs, the details of processing experiments, and their impact on film properties. Complete working devices were also fabricated and tested. Hence, it is the goal of this review to highlight recent advances from NASA GRC and other groups, summarizing the current state-of-the-art of this highly promising technique for thin-film growth, exploiting molecular design of single-source precursors and monitoring film quality as it relates to and is predictive of device performance.

6.2 SINGLE-SOURCE PRECURSOR STUDIES

6.2.1 Background

In early studies, Nomura et al. reported that an equimolar mixture of $[Bu^i_2InSPr^n]$ and $[Cu(S_2CNBu^n_2)_2]$ decomposed to afford $CuInS_2$ powders.[34] On this basis, solution pyrolysis of this mixture dissolved in *p*-xylene was used to deposit thin-film $CuInS_2$ at 350 °C onto glass substrates. Film composition, analyzed by EDS, revealed that the ratios of In/Cu and S/Cu decreased with substrate temperature, and XRD showed that a second phase was present for films deposited at 350 °C. Grain size was estimated to be in the range of 50–100 nm as determined by SEM. It was later realized that the equimolar reaction mixture of $[Bu^i_2InSPr^n]$ and $[Cu(S_2CNBu^n_2)_2]$ (as used in solution pyrolysis) afforded the single-source precursor $[Bu^i_2In(SPr^n)Cu(S_2CNBu^n_2)]$ before decomposing to the chalcopyrite matrix (see equation 6.1).[35] Analytical and spectral data confirmed that the mixture of $[Bu^i_2InSPr]$ and $[Cu(S_2CNBu_2)_2]$ yielded an SSP, although the nature of the chemical bridge was not well defined. Other analogous CIS precursors were also synthesized by the reaction of alkyl indium thiolates with copper dithiocarbamates.[36] However, only $[Bu_2In(SPr^i)Cu(S_2CNPr^i_2)]$ was successfully implemented for depositing pure $CuInS_2$ by low-pressure MOCVD. In the case of

$[BuIn(SPr^i)_2Cu(S_2CNPr^i_2)]$, tetragonal $CuIn_5S_8$ was deposited (i.e., equation 6.2).[37]

$$2Bu_2In(SR) + 2Cu(S_2CNR'_2)_2 \rightarrow 2[Bu_2In(SR)Cu(S_2CNR'_2)] + (R'_2NCS_2)_2$$
$$R = Pr^n;\ R' = Bu^n \text{ or } R = R' = Pr^i \qquad (6.1)$$

$$[BuIn(SPr^i)_2\,Cu(S_2CNPr^i_2)] \xrightarrow{400\,°C/0.6\,torr} CuIn_5S_8 \qquad (6.2)$$

Several years later, Kanatzidis et al. reported the preparation of heterobinuclear complexes of tetrahedrally arranged Cu and In centers, with two bridging thiolato or selenolato groups, $[\{PPh_3\}_2Cu(\mu\text{-}QEt)_2In(SEt)_2]$, (Q = S or Se) (i.e., equation 6.3).[38] Pyrolysis studies undertaken revealed that the Se derivative could be converted into $CuInSe_2$ at 400–450 °C at 0.01 mm Hg (see equation 6.4). However, none of the precursors had been evaluated in a thin-film deposition study.

$$[\{Ph_3P\}_2Cu(MeCN)_2]^+ + [In(QR)_4]^- \xrightarrow{MeOH}$$
$$[\{Ph_3P\}_2Cu(\mu - QR)_2In(QR)_2] + 2MeCN \qquad (6.3)$$

$$[\{Ph_3P\}_2Cu(\mu - SeEt)_2In(SeEt)_2] \xrightarrow{400-450\,°C/0.01\,mmHg}$$
$$CuInSe_2 + 2PPh_3 + 2EtSeEt \qquad (6.4)$$

In continuing the NASA-funded work, Buhro and Hepp, along with their colleagues, were able to demonstrate that In and Cu binuclear thiolato or selenolato complexes could be used in a spray CVD process for depositing thin-film $CuInS_2$ below 400 °C.[39–41] Thin films were deposited using a dual-solvent system of toluene and dichloromethane (CH_2Cl_2) as the carrier solvent. Single-phase (112)-oriented $CuInS_2$ films were successfully deposited at a range of temperatures from 300 to 400 °C, whereas at increased temperatures (>500 °C), $CuIn_5S_8$ phase films could be deposited. Rutherford backscattering spectroscopy (RBS), EDS, and XPS analysis showed that the films were free from any detectable impurities and were highly crystalline, thus concluding that the precursor decomposes cleanly. During the course of the study, the morphology of the deposited films was found to be temperature and carrier solvent dependent. Films deposited at 300 °C and 350 °C yielded a grain size of 400–800 nm, with smaller finer particles of 50–200 nm residing on top. At a higher deposition temperature of 400 °C, the films consisted of more angular grains with a more uniform size of approximately 200 to 400 nm. Photoluminescence (PL) data and optical transmission measurements confirmed that the deposited $CuInS_2$ thin films were direct band-gap semiconductors.

In the course of our investigations for improved single-source precursors for the spray CVD of CIGSSe chalcopyrite thin films, we have continued to expand the molecular design of SSPs based on the $[\{ER_3\}_2Cu(\mu\text{-}QR')_2M(QR')_2]$ (Q = S, Se; E = P, As, Sb) architecture (hereafter the symbol μ, which indicates a bridging geometry, will be eliminated for the sake of simplicity and clarity).[38]

Furthermore, the number of "tunable" sites within the complex allows for their utility in preparing several ternary chalcopyrites of varying composition, in addition to engineering the SSP to match a given spray CVD process (Fig. 6.4).

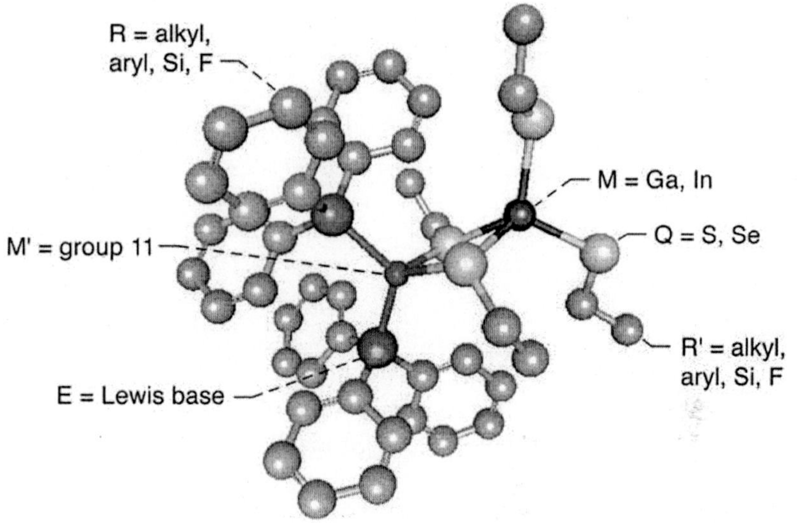

Figure 6.4. Molecular structure diagram of $[\{PPh_3\}_2Cu(SEt)_2In(SEt)_2]$, indicating "*tunable*" sites.

6.2.2 Chemical Synthesis of SSPs

The SSPs are prepared by the reaction of a stabilized Cu(I) cation, with an indium(III) or gallium(III) chalcogenide anion, prepared *in situ* by reaction of the conjugate acid of the thiol or selenol with NaOMe in methanol (Scheme 6.1 below).[1,3]

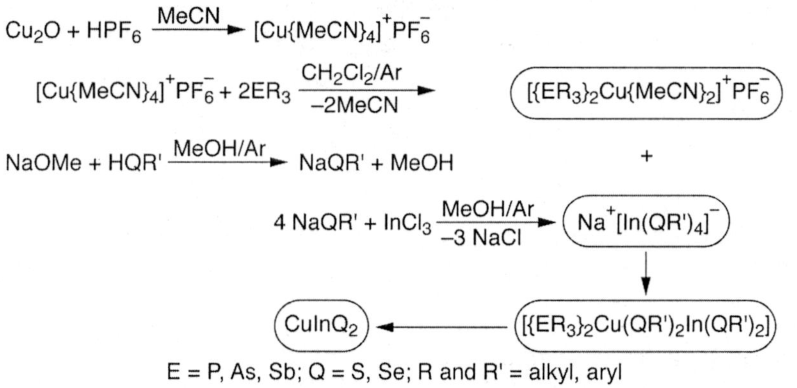

Scheme 6.1. Synthesis of ternary single-source precursors and decomposition to solid state materials.

The versatility of this synthetic pathway can be illustrated by the ability to modulate the physical properties of the precursor and composition at any of the intermediate synthetic steps by either

- Adjusting the Lewis acid–base interaction (E→M, M = Group 11 metal)
- Adjusting the accessibility of the lone pair of electrons on the neutral donor ligand by variation of R
- Adjusting the bond strength between the chalcogenide with either In/Ga and Cu metal centers
- Selecting either S, Se, or Te group 16 analogs
- Preparing either indium or gallium derivatives

The Lewis acid–base interaction is a valuable component to the overall stability of the molecule, given that the ability of the Lewis base to dissociate from the cation at lower energies is pertinent to the degradation of the precursor at reduced temperatures. Hence, the Drago–Wayland approximation[42] can be used for ternary single-source precursor design, to quantitatively estimate the strength of the Lewis acid–base interaction between the copper center and the neutral donor. The cleavage of chalcogen–R′ bonds also plays an important role, since this allows the chalcogen to be released for incorporation into the ternary chalcopyrite matrix. Hence, the use of a sterically demanding R′ group of good "leaving ability" would promote the facile release of the chalcogen.

6.2.3 Thermal Analysis and Characterization of SSPs

Initial studies focused on basic modification of the SSPs, and their influence on precursor stability. Multinuclear nuclear magnetic resonance (NMR) data demonstrated that the precursors were free from any starting reagents. Thermogravimetric analyses (TGAs) were performed at ambient pressure in platinum pans on samples of the precursors, heated at a rate of 10 °C/min under a nitrogen atmosphere. Weight loss was associated with decomposition of the complexes. Calculation of the derivative maximum rate of weight loss (%/°C), listed as MRW in Table 6.1, shows a range from a low of 225%/°C for **8** to a high of 325%/°C for **5**. The measured weight loss, based on the residual (pyrolysate) material from TGA experiments, was within 5% (see Fig. 6.5a) of expected total weight loss, assuming complete conversion to $CuInS_2$ or $CuInSe_2$.

An additional example of the flexibility of the $[\{ER_3\}_2Cu(QR')_2M(QR')_2]$ architecture to direct adjustment of these precursors are the SSPs **7** and **8**, which represent the first liquid single-source precursors for the deposition of $CuInS_2$.[1] An added benefit of a charge neutral liquid precursor is the potential for greater solubility in nonpolar organic solvents and the possibility of direct

TABLE 6.1. Thermal Data for Ternary Single-Source Precursors

	TGA			DSC	
Single source precursors	Extrap. onset °C	MRW %/°C	Residue %	M.P. °C	Decomp. °C
1 $[\{PPh_3\}_2Cu(SEt)_2In(SEt)_2]$	236	269	25	122	266
2 $[\{AsPh_3\}_2Cu(SEt)_2In(SEt)_2]$	205	233	18	47	276
3 $[\{SbPh_3\}_2Cu(SEt)_2In(SEt)_2]$	212	239	26	45	271
4 $[\{PPh_3\}_2Cu(SPr^i)_2In(SPr^i)_2]$	215	254	29	163	260
5 $[\{PPh_3\}_2Cu(SPh)_2In(SPh)_2]$	261	325	22	117	280
6 $[\{PPh_3\}_2Cu(SePh)_2In(SePh)_2]$	223	253	22	53	219
7 $[\{PBu^n_3\}_2Cu(SEt)_2In(SEt)_2]$	189	238	31	–	264
8 $[\{PBu^n_3\}_2Cu(SPr^n)_2In(SPr^n)_2]$	171	225	29	–	239
9 $[\{PPh_3\}_2CuGa(SEt)_x]$*	#	310	25	#	#

*Structure not eludicated (x = 3, or 4).
#Not recorded.

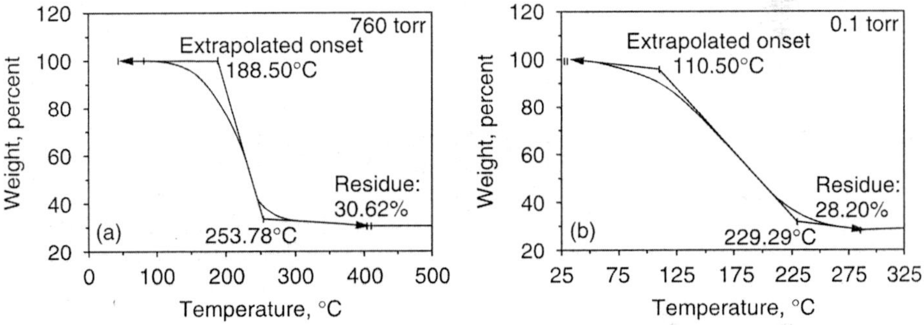

Figure 6.5. TGA profiles at (a) 760 torr and (b) 0.1 torr for $[\{PBu^n_3\}_2Cu(SEt)_2In(SEt)_2]$, **7**.

delivery without the need for a carrier solvent. Both these enhancements permit the use of higher molarity solutions, which can help achieve thicker CIS films without an increase in deposition run times. Because of their semi-viscous nature, the physical properties of the clear neat liquid precursors were studied.

Low-temperature differential scanning calorimetry (DSC) was used to investigate the liquid phase for **7** and **8**. In separate studies, samples **7** and **8** were subjected to both quench cooling and slow-controlled cooling before being heated at 10 °C and 5 °C/min.[42] In low-temperature DSC experiments using controlled and quench cooling, both samples **7** and **8** were found not to show an endotherm assignable to a melting-phase transition before decomposition, thus supporting their liquid state at ambient temperatures (see Fig. 6.6).

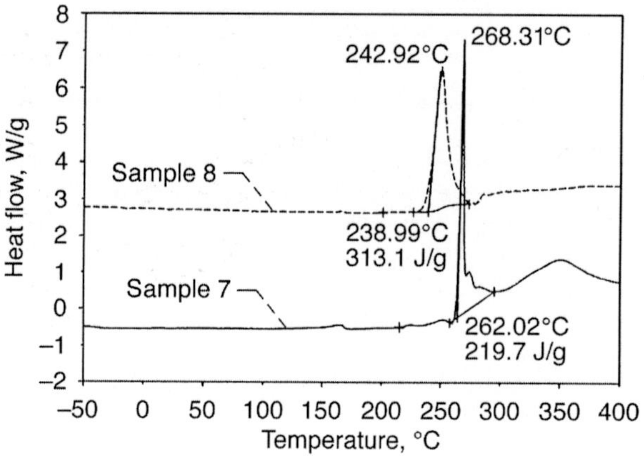

Figure 6.6. Low-temperature DSC of $[\{PBu^n_3\}_2Cu(SEt)_2In(SEt)_2]$, **7** and $[\{PBu^n_3\}_2Cu(SPr^n)_2In(SPr^n)_2]$, **8**.

Examination of the other phase transitions reveals that the main exothermic events for **7** and **8** begin with extrapolated onset temperatures of 262 °C and 239 °C, which represent the decomposition of the samples. The lower decomposition temperature of **8** can be explained, since an increase in chain length and/or steric "bulk" of the alkyl groups is known to decrease the stability of a complex.[29] In addition, preliminary vacuum-TGA studies for the thermal profile for the SSP **7** show that the extrapolated onset can be lowered by ~80 °C, thus lowering the degradation temperature window and making these precursors highly suitable for use in low-temperature/pressure-spray CVD on space-qualified substrates such as Kapton™ (Fig. 6.5b). Remarkably both liquid precursors show excellent solubility in both polar and nonpolar solvents, which can be attributed to their ionic structure and to the nonpolar alkyl groups resident on the tertiary phosphine.

The ability of the new precursors to decompose thermally to yield single-phase CIS was investigated by powder XRD analysis and EDS on the non-volatile solids from the TGA experiments of selected compounds. Furthermore, using TGA-evolved gas analysis (EGA), the volatile components from the degradation of the SSPs could be analyzed via real-time fourier transform infrared (FTIR) and mass spectrometry (MS), thus providing information for the decomposition mechanism.[3] The real-time FTIR spectrum for **7** and **8** shows absorptions at approximately 3000, 1460, 1390, 1300, and 1250 cm^{-1} (see Fig. 6.7).

Correlation with the EGA-mass spectra allows for the assignment to the initial loss of diethyl sulfide, as supported by the library fit and from the assignment of the fragment and parent ions (m/z = 90) (see Fig. 6.8a). After approximately 15 min, mass-spec EGA shows the absence of peaks assignable to Et_2S

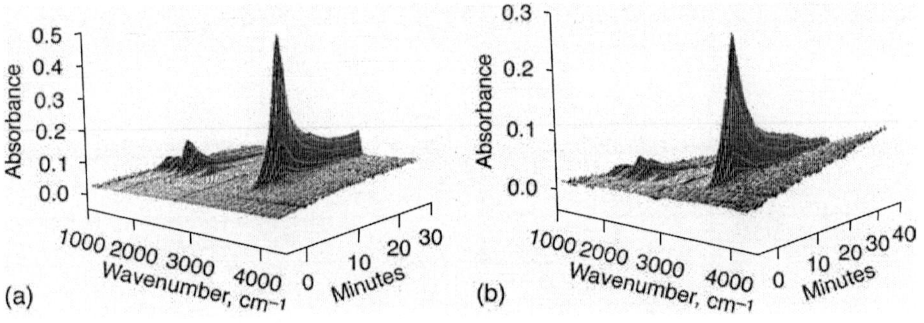

Figure 6.7. EGA–FTIR spectra for (a) $[\{PBu^n_3\}_2Cu(SEt)_2In(SEt)_2]$, **7** and (b) $[\{PBu^n_3\}_2Cu(SPr^n)_2In(SPr^n)_2]$, **8**.

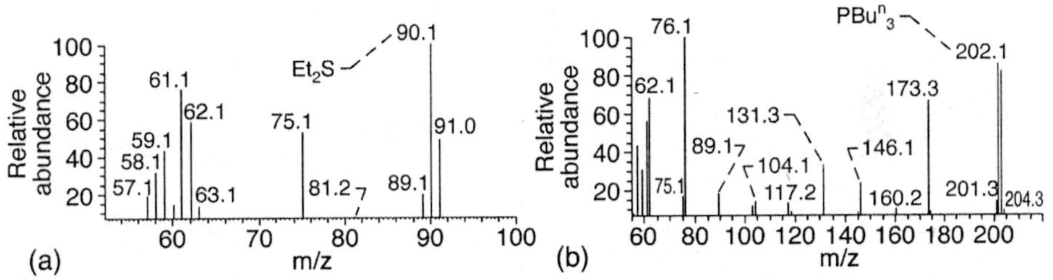

Figure 6.8. EGA–MS (EI) data, after (a) ~5 min and (b) ~15 min, for $[\{PBu^n_3\}_2Cu(SEt)_2In(SEt)_2]$, **7**.

and the occurrence of fragment ions with a mass-to-charge ratio (m/z) greater than 90, with an intense peak at m/z = 202. These can be assigned to the successive loss of PBu_3 on the basis of its library fit of 92% and assignment of the fragment ions (see Fig. 6.8b). Thus, the use of TGA-EGA provides conclusive evidence for the mechanism of decomposition for the single-source precursors to occur via the initial loss of a chalcogenide moiety, followed by loss of the neutral donor ligand.

6.2.4 Preparation of I–III–VI$_2$ Powders from SSPs

The XRD spectra for the nonvolatile material produced from the pyrolysis of **7**, with the Joint Committee on Powder Diffraction Standards (JCPDS) reference patterns for $CuInS_2$ (27–0159), confirmed it to be single-phase $CuInS_2$ (see Fig. 6.9). Examination of the EDS spectra for the same samples shows predominant emissions from Cu, In, and S edges, with the approximate percentage atomic composition of 27, 23, and 50 for **7** and 28, 23, and 49 for **8**, respectively, thus supporting the formation of $CuInS_2$.

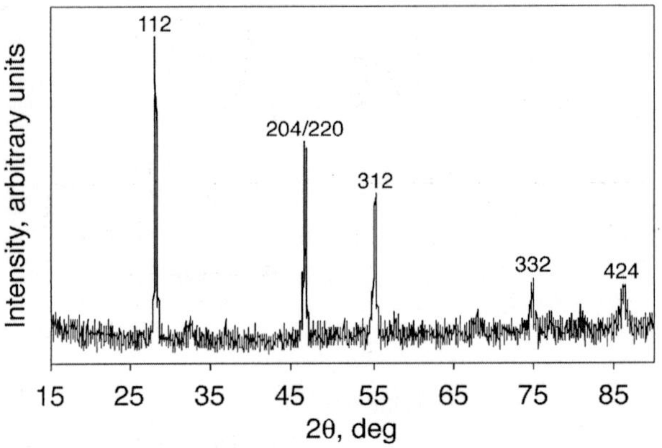

Figure 6.9. Powder XRD pattern for the nonvolatile residue formed from pyrolysis of $[\{PBu^n_3\}_2Cu(SEt)_2In(SEt)_2]$, **7** (Cu K_α, 1.541 Å).

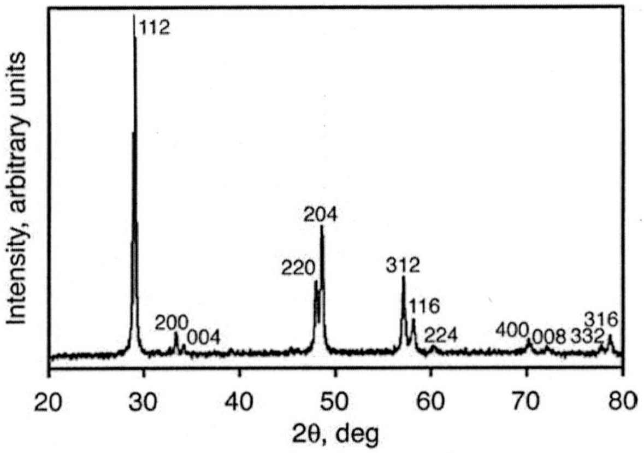

Figure 6.10. XRD pattern of the nonvolatile solid obtained from bulk pyrolysis of $[\{PPh_3\}_2CuGa(SEt)_x]$ **9** (600 °C, 300 min). All reflections correspond to gallite—$CuGaS_2$ (JCPDS #25-0279).

The preparation of the first SSP to the semiconductor $CuGaS_2$ was also investigated.[3] Although the molecular structure for SSP **9** was not fully characterized, its suitability to afford bulk and thin-film $CuGaS_2$ was studied by TGA (see Table 6.1) and XRD. In addition, the optical properties of the deposited thin films were also investigated. Powder XRD analysis on the nonvolatile product from bulk pyrolysis, (N_2, 600 °C), confirmed the collected material to be single-phase (112)-oriented $CuGaS_2$ (see Fig. 6.10), having an average coherence length of 28 nm.

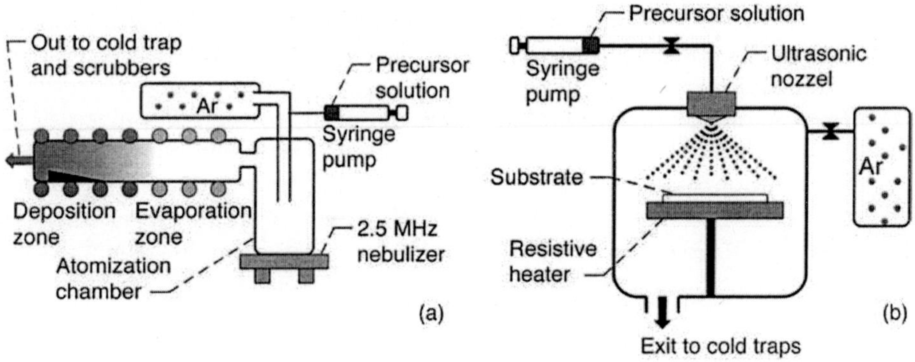

Figure 6.11. Schematic of two of the reactors used: (a) atmospheric pressure horizontal hot-wall reactor (Reactor A) and (b) vertical cold-wall reactor (Reactor B).

6.3 SPRAY OR ATMOSPHERE-ASSISTED CVD PROCESSING

Spray CVD has become an increasingly used processing method; a room-temperature precursor solution is ultrasonically nebulized and is swept into a two-zone, hot-wall reactor (see Fig. 6.11a). The carrier solvent is evaporated in the warm zone, and the gaseous precursors are decomposed in the hot zone, where film growth occurs, as in conventional CVD. Spray or AACVD maintains the most desirable features of MOCVD and spray pyrolysis, such as film growth in inert atmospheres, large-area deposition, laminar flow over the substrate, and a low-temperature solution reservoir, while avoiding the major difficulties of each.[43,44] It minimizes the high volatility and temperature requirements for the precursor, which are essential in MOCVD, by delivering the precursor to the furnace as an aerosol propelled by a fast-flowing carrier gas from a low-temperature precursor reservoir analogous to that employed in spray pyrolysis. The latter feature is an important benefit that can prevent premature precursor decomposition when using thermally labile precursors. Most importantly for electronic device growth, the solid/vapor phase interface and precursor decomposition supports the growth of thin films directly instead of the intermediate powders or gel films that are more typical of spray pyrolysis processing.[45]

6.3.1 AACVD Reactor Design

Three different reactors were used to deposit $CuInS_2$ films via AACVD. Reactor A, shown schematically in Fig. 6.11a, was primarily used in the parametric studies described below. This is a horizontal, atmospheric pressure, hot-wall reactor with a plate-type 2.5-MHz ultrasonic nebulizer from Sonaer Ultrasonics. The precursor (1.5–3.5 g) was dissolved into distilled toluene (50–400 ml) and fed into the nebulizer using a syringe pump. The nebulizer created an aerosol that was swept into the reactor by an Ar carrier gas at a flow rate of 4 l/min (or SLPM). The reactor is a two-zone furnace consisting of a warm precursor evaporation zone and a hot deposition zone. The aerosol is converted to precursor vapor and mixes uniformly through the reactor gas in the evaporation zone. A solid graphite susceptor coated with SiC was placed in the deposition zone. The second reactor is a vertical atmospheric pressure, cold-wall reactor (reactor B, Fig. 6.11b) with a commercial ultrasonic nozzle (Sono-Tek 120 kHz). The third reactor (not shown) is a horizontal, low-pressure, hot-wall tool, which is similar in design to reactor A but with a pulsed aerosol injection system using a commercial automotive fuel-injector (Ford 2M2EA7B).

Even though all three reactors share the same precursor delivery system, each tool offers specific advantages. For example, a cold-wall reactor (reactor B) helps prevent decomposition of the precursor before it reaches the substrate. A pulsed aerosol injection system at low pressure (reactor C) allows the film to grow under better-defined conditions than in a continuous process (reactor A) because of the minimization of undesirable transient effects caused by the high volatility of the solvents used.[46] A more detailed description of each of the conditions for film growth, including reactor type, precursor type, delivery method, deposition temperature, growth time, and other parameters are summarized in Table 6.2. Depositions were done on bare and Mo-coated

TABLE 6.2. Growth Parameters for AACVD Deposited $CuInS_2$ Films

Film	I	II	III	IV	V
Reactor	A	A	B	C	C
Precursor*	Solid	Solid	Liquid	Liquid	Liquid
Concentration (M)	0.01	0.01	0.01	0.01	0.01
Precursor delivery method	Carrier gas	Carrier gas	Carrier gas	Vacuum-driven	Vacuum-driven
Delivery rate (ml/min)	~1.7	~1.7	1.5	0.4	0.4
Temperatures (°C)					
Evaporation zone:	128	120	no zones	250	150
Deposition zone:	390	360	400	250	250
Substrate:	390#	360#		420	400
Pressure (torr)	atm.	atm.	atm.	~12	~10
Ar flow rate (l/min)	4	4	4	0.08‡	0.15‡
Substrate	Mo/glass (Corning 2947)	Mo foil	Glass (Corning 7059)	Mo/glass (Corning 2947)	Mo/glass (Corning 2947)
Substrate (cm × cm × 0.1 cm)	0.87 × 7.6	0.87 × 7.6	2.5 × 2.5	2.5 × 7.6	2.5 × 7.6
Growth time (min)	80	60	60	250	250
Thickness (μm)	~1	~1	~4	~1	~1

*Solid precursor: $[\{PPh_3\}_2Cu(SEt)_2In(SEt)_2]$, **1**. Liquid precursor: $[\{PBu^n_3\}_2Cu(SEt)_2In(SEt)_2]$, **7**.
#Substrate is located within the deposition zone.
‡Direct reading from MKS flow controller (not calibrated); Ar flow is required for the aerosol creation mechanism.

glass slides from Corning (Corning 2947 and 7059). Commercial Mo foil substrates were also used. Postdeposition annealing of film III was performed in a tube furnace (Lindberg 54477) at 600 °C for 10 min under N_2 flow (4 SLPM).

6.3.2 Preliminary Thin-Film Deposition Studies

6.3.2.1 $CuInS_2$ Growth on Molybdenum. Well-adhering dark blue/black $CuInS_2$ (CIS) films were deposited on a molybdenum substrate by spray CVD at 390 °C using $[\{PBu_3\}_2Cu(SEt)_2In(SEt)_2]$ (SSP **7** in Table 6.1). As deposited, the $CuInS_2$ films were preferentially (112) oriented (Fig. 6.12); this is significant for producing solar cells, since (112)-oriented films tend to have a low series resistance.[47]

Measurements of the EDS emission were limited to Cu and In edges since the EDS emissions for sulfur and molybdenum overlapped. SEM-EDS data on several regions on the film gave atomic percents representative of $CuInS_2$ (Table 6.3), and no evidence of phosphorous could be detected under the typical detection limit of EDS (0.02 wt%), verifying that the precursor decomposes cleanly, as evidenced in EGA–TGA studies.

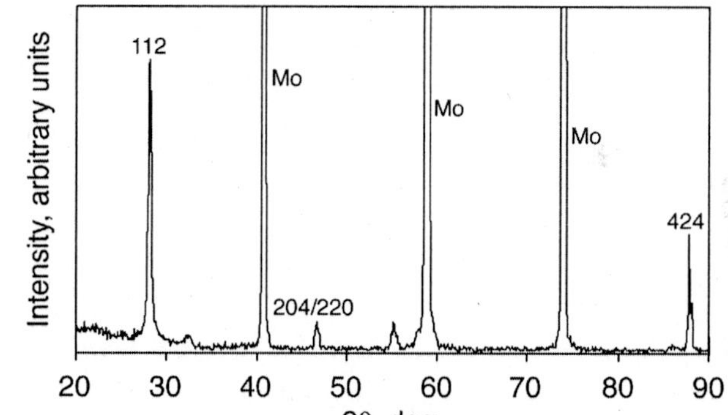

Figure 6.12. X-ray powder diffraction spectra of a $CuInS_2$ thin film on Mo, prepared using SSP **7**.

TABLE 6.3. Atomic Composition of $CuInS_2$ Thin-Film Deposited Using SSP 7

	At.%, (±3%)		
	Front 1	Front 2	Back
Cu	50	51	50
In	50	49	50

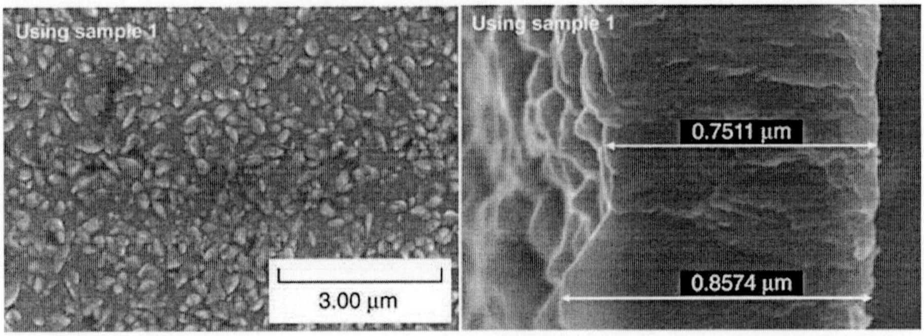

Figure 6.13. SEM images of plane-view (left) and cross-sectional (right) views of deposited CIS film using [{P(PPh)$_3$}$_2$Cu(SEt)$_2$In(SEt)$_2$], **1**.

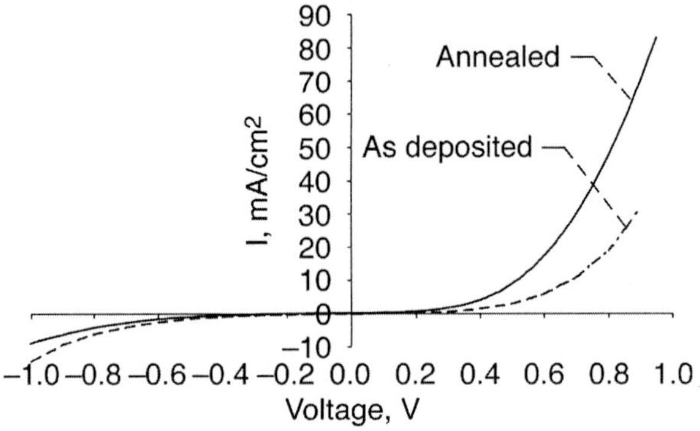

Figure 6.14. Current density versus voltage plot of a Schottky barrier made from evaporated Al on $CuInS_2$ on Mo foil using SSP **1**.

6.3.2.2 Schottky Diode Growth. The electrical properties of the films deposited using SSP **1** (Fig. 6.13) were evaluated by current versus voltage (I–V) measurements recorded for the thin films using thermally evaporated aluminum contacts (10 mm^2), to make Schottky barrier diodes (see Fig. 6.14).

The Schottky barriers were excellent diodes for films annealed at 600 °C, with "turn on" voltages of 0.6–0.8 V and minimal reverse bias leakage.[48] However, many of the contacts on the as-deposited films gave large reverse bias currents and nearly ohmic responses. This behavior is indicative of degeneracy of the semiconductor because of a high carrier density resulting from native defects. The improvement in the diode behavior of the annealed films is attributed to enhanced crystallinity and reduction of defects.

6.3.2.3 Growth and Characterization of $CuGaS_2$ Films. Spray CVD deposition studies using the new Ga SSP **9**, in a toluene/CH_2Br_2 solution (0.01 M:86 vol% toluene; 1.3 mmol; SSP **9**), afforded well-adhered dense films (450

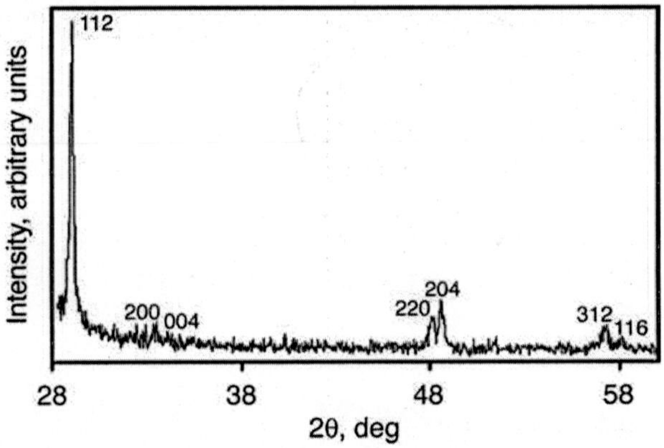

Figure 6.15. XRD pattern of a spray CVD grown $CuGaS_2$ film on Si (111), using SSP 9. Reflections correspond to those reported for gallite in JCPDS reference #25-0279.

TABLE 6.4. Comparison of Thin-Film and Single-Crystal $CuGaS_2$ Lattice Parameters (a and c), c/a, and the Distortion Parameter x

a (Å)	c (Å)	c/a	x	Thin film fabrication method
5.353	10.495	1.9606	0.0394	Spray-CVD deposited film
5.35	10.48	1.959	0.0410	Evaporated film[49]
5.351	10.484	1.9593	0.0407	JCPDS card 25–0297: Single crystal prepared from the elements

± 5 °C, 4.0 L/min). The films were visually smooth and optically transparent, exhibiting a pink and green surface tint. XRD analysis confirmed the film to be highly (112) oriented, tetragonal, single-phase $CuGaS_2$ (see Fig. 6.15).

The 220/204 reflections and the 312/116 reflections were split, which is consistent with the tetragonal distortion of the crystal lattice[49] (Fig. 6.15). Lattice parameters a and c were calculated from X-ray d spacings according to Eq. 6.5,

$$\frac{1}{d^2} = \frac{h^2 + k^2}{a^2} + \frac{l^2}{c^2} \tag{6.5}$$

where h, k, and l refer to the Miller indices of individual reflections.[50] The d spacing of the 220 reflection was used to calculate a, and the d spacing of the 112 reflection was used, with the calculated lattice parameter a, to determine c. The lattice parameters, c/a ratio, and the distortion parameter x (where $x = 2 - c/a$, given that a hypothetical c/a ratio of 2 would result in the *absence* of any tetragonal distortion) are presented in Table 6.4. Comparison of the data collected from the $CuGaS_2$ film shows they are in good agreement with the JCPDS reference values for single-crystal $CuGaS_2$ and with those reported in the literature.[51]

SEM images reveal that the films are dense, with an average grain size of 410 nm and predominantly columnar grains on the surface of the substrate (see Fig. 6.16c and d). The surface microstructure consisted of faceted grains, many of which exhibited a trigonal shape (Fig. 6.16a and b), which occurs as a result of close-packed intersecting (112) faces of the chalcopyrite lattice. These are the lowest surface-energy faces, and they typically control chalcopyrite morphology.[33,52] The grain sizes observed by SEM are considerably

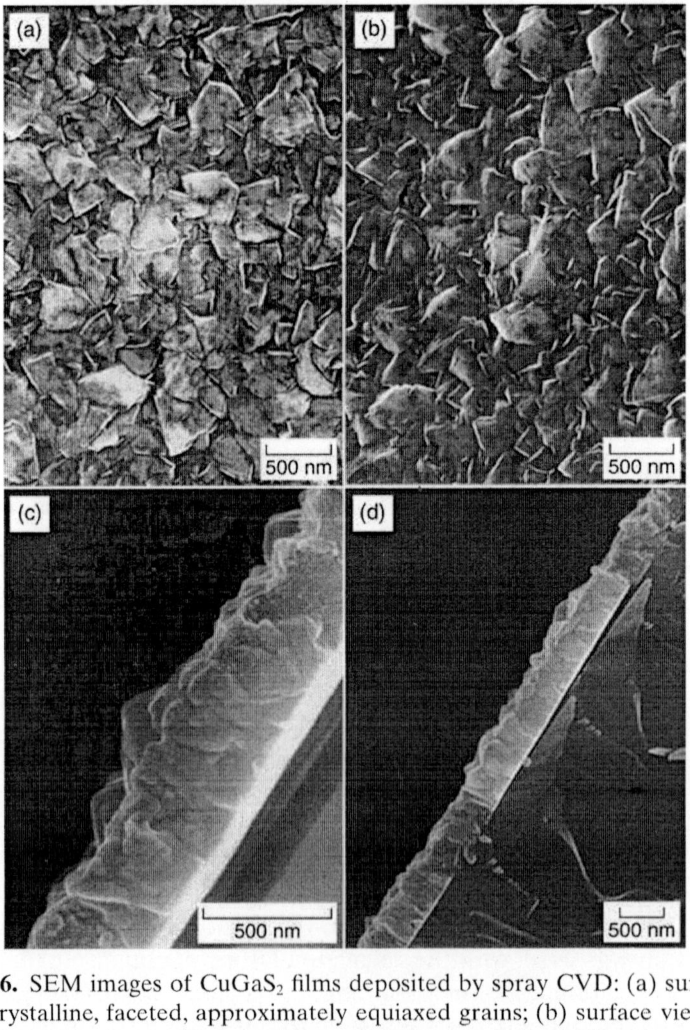

Figure 6.16. SEM images of $CuGaS_2$ films deposited by spray CVD: (a) surface view showing crystalline, faceted, approximately equiaxed grains; (b) surface view; 30° tilt showing triangular shape of many of the grains characteristic of 112-oriented crystals; (c) and (d) edge views showing roughly columnar crystal-growth pattern and the dense nature of the film. Triangular grains protruding from the film surface are visible in (c).

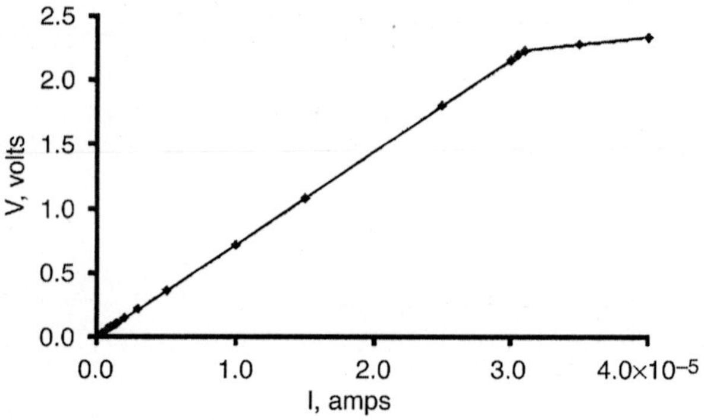

Figure 6.17. I–V curve for $CuGaS_2$ thin film. The linear region below 3.0×10^{-5} amps was used to calculate sheet resistance.

larger than the average coherence lengths calculated from X-ray line broadening, which could have been reduced by a high-defect density in the grains or by strain in the film.

Sheet resistance (R_s) is defined by $\rho = R_s t$, where ρ is resistivity and t is film thickness. For $CuGaS_2$ thin-film samples, R_s was determined using the four-point probe method[53] and calculated from current-voltage data (Fig. 6.17) using

$$R_s = K_p \cdot V/I \tag{6.6}$$

where K_p is a constant. For a probe of the type used here and for a sample with planar dimensions that can be considered infinite relative to the probe spacing, $K_p = \pi/\ln 2$. The resistivity determined for the film deposited at 450 °C on fused silica was found to be 15.6(4) Ω·cm, which is comparable with values reported in the literature.[53]

The optical band gaps of the films were determined from optical transmittance data, which showed the films to absorb incident light below 480 nm (Fig. 6.18a). Transmittance was observed to change most dramatically in the region from 480 to 540 nm, which is the region of the direct absorption edge. Calculation of the derivative for the transmission curve provided a more precise method to determine the band gap within this region,[51,54,55] yielding a maxima at 2.42 eV (Fig. 6.18b). This value is in very good accord with the known direct band gap of $CuGaS_2$, $E_g = 2.43$ eV, as reported in the literature.[55]

6.3.2.4 Dual-Source AACVD Study. Initial studies using the two analogous Ga and In SSPs for fabrication of an alloy film, $Cu(In:Ga)S_2$, in a dual-source spray CVD were also investigated. A film was deposited from a mixed toluene solution of $[\{Ph_3P\}_2Cu(\mu\text{-SEt})_2In(SEt)_2]$ **1** (0.0075 M) and the Ga analog **9** (0.0025 M).[3] Although a thin film was deposited, composition and microstructure varied along the length of the film, which may be attributed to

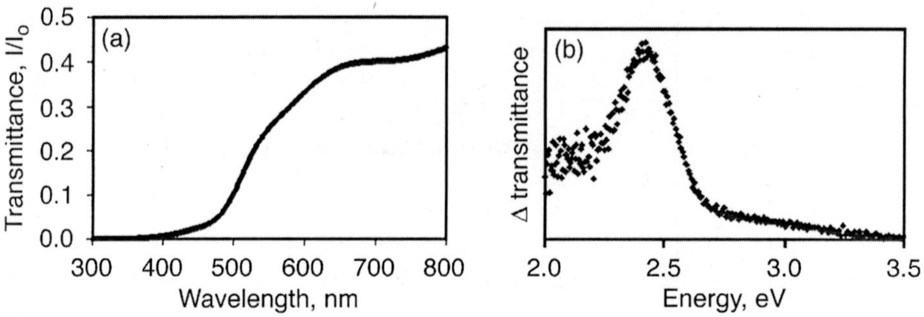

Figure 6.18. (a) Transmittance versus wavelength for a $CuGaS_2$ thin film (I = transmitted power, I_o = incident power); (b) plot of the derivative of the transmission data versus energy.

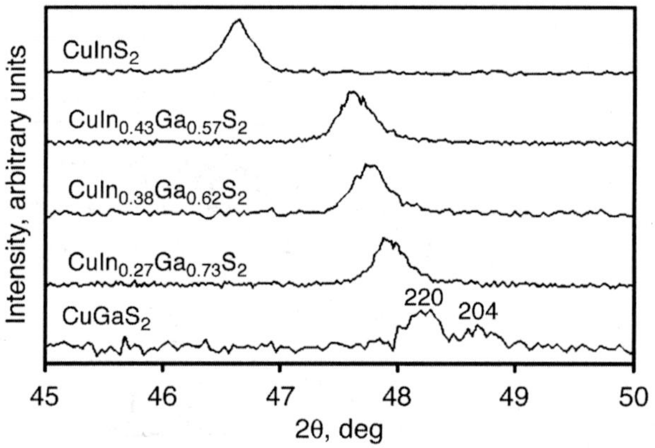

Figure 6.19. XRD spectra highlighting the 220/204 reflections of a $CuGaS_2$ film (bottom; T_s = 450 °C), a $CuInS_2$ film (top; T_s = 400 °C) and alloy films having In_xGa_y contents in the range: $In_{0.27}Ga_{0.73}$—$In_{0.43}Ga_{0.57}$. All films were deposited on fused silica.

the "boundary layer" model for velocity, temperature, and reactant concentration for horizontal CVD reactors.[33] XRD reflections representing the 112 planes were broad and complicated by the presence of an unidentified reflection in that region. The 220/204 planes were represented by a single unresolved reflection that yielded an average grain size of ~40 nm. The relative contribution of Ga and In to the multinary structure was determined by comparing the 2-theta values for this reflection in the multinary pattern with those in the patterns of the ternary end-members, $CuInS_2$ and $CuGaS_2$ (Fig. 6.19). The composition of each metal was assumed to vary linearly with 2-theta from 100% In to 100% Ga based on Vegard's Law. The tetragonal splitting was neglected in the $CuGaS_2$ pattern by averaging the 2-theta values for the 220 and the 204 reflections.

The atomic percent of Ga in the film was found to increase along the substrate from the front to rear of the film. The film deposited in the first centimeter had a composition of $CuIn_{0.43}Ga_{0.57}S_2$. In the middle portion of the film, the composition was $CuIn_{0.38}Ga_{0.62}S_2$ (statistically indistinguishable from the first centimeter), whereas the last centimeter of the film had the highest contribution of Ga (i.e., $CuIn_{0.27}Ga_{0.73}S_2$). A uniform composition over large areas was therefore not achieved, and interestingly, none of the compositions were close to the expected $In_{0.75}Ga_{0.25}$ ratio in accord with the precursor molar ratio. SEM images of the films (Fig. 6.20) revealed that the microstructure was

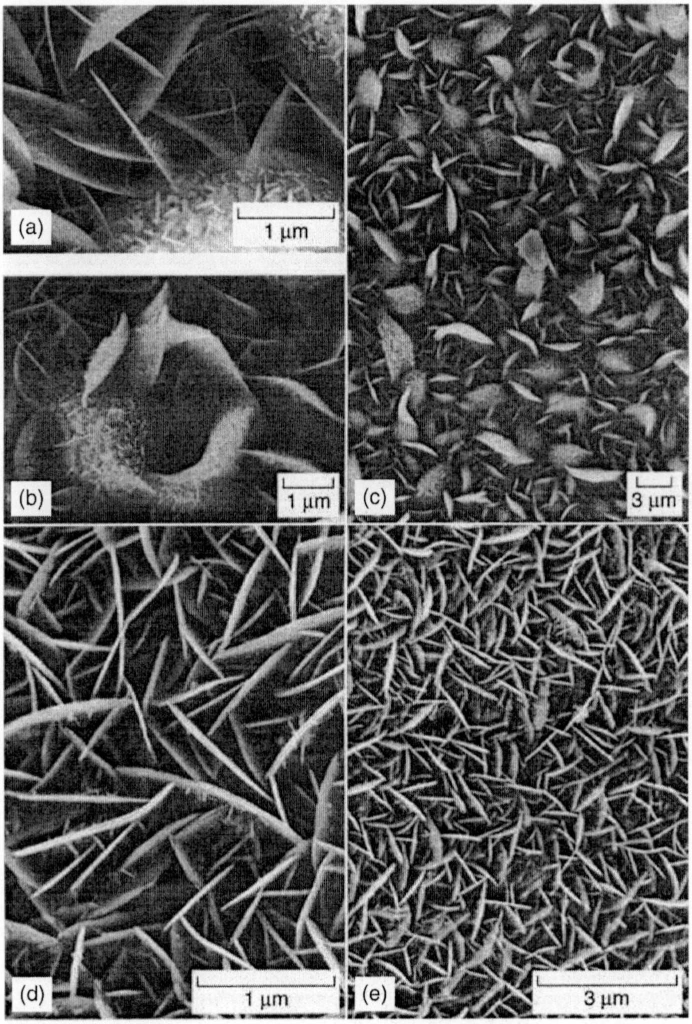

Figure 6.20. SEM images of the alloy film showing the variation in microstructures with composition. (a–c) Film deposited in the first centimeter: $CuIn_{0.43}Ga_{0.57}S_2$. (d and e) Film deposited in the last centimeter: $CuIn_{0.27}Ga_{0.73}S_2$.

similar to that observed for $CuInS_2$ deposited at 450 °C.[40] The variation in film composition is understandable since the two precursor thermal decomposition profiles are dissimilar. Therefore, using two SSPs with matching thermal profiles might provide a more effective means for depositing quaternary films.

6.3.3 Impact of Reactor Design on $CuInS_2$ Film Growth

In another precursor decomposition study, three different reactors were used to grow $CuInS_2$ films (See Table 6.2). Hot-wall reactors A (film I) and C (films IV and V) produced dense, columnar grain growth, whereas the cold-wall reactor B (film III) yielded porous nanostructures (i.e., see SEM micrographs in Fig. 6.21 and Table 6.2 for deposition conditions). The thickness of the films produced from each reactor where similar in magnitude, thus supporting the notion that film thickness is directly proportional to solution molarity. Film roughness was found to vary between reactors, but more studies are needed

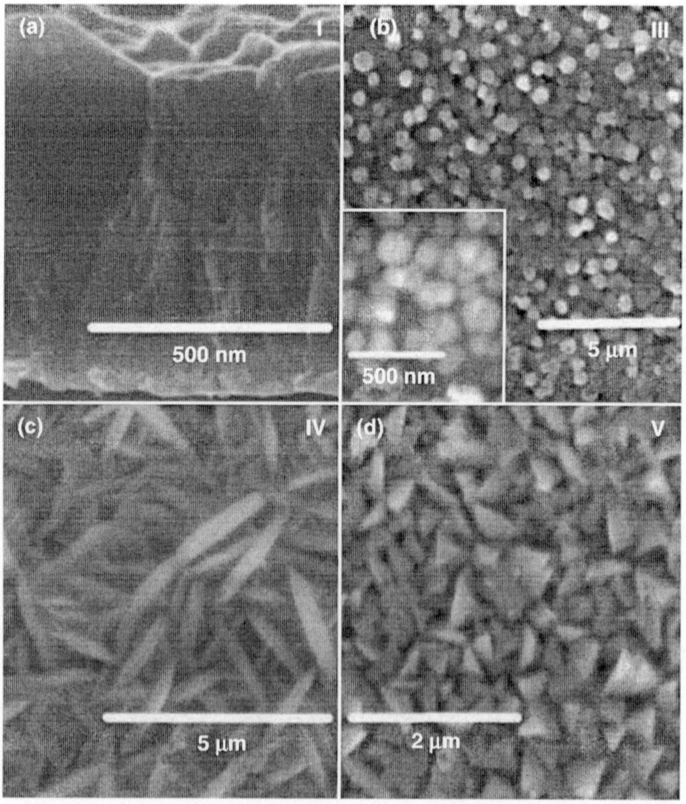

Figure 6.21. SEM micrographs of AACVD grown $CuInS_2$ thin films: (a) cross-sectional image of film I, (b) plane-view image of film III (inset at higher magnification), (c) plane-view image of film IV, and (d) plane-view image of film V.

to determine whether reactor design was responsible or whether thin-film fabrication process parameters were not optimized. Figure 6.21(a), the cross-sectional image of film I, shows a columnar grain structure. The columnar grain structure lowers the cross section for photo-excited carrier recombination at grain boundaries, thus improving the performance of the solar cells.

Figure 6.21(b) shows a plane-view image of film III together with an inset of the same region at a higher magnification. Under low magnification, only the round structures appeared. However, when the magnification was increased, smaller nanoparticles (~100 nm) were observed. The nanoparticles may have been created above the hot substrate by pyrolysis of nanodroplets generated from the aerosol.[56] Under this model, the ultrasonically excited aerosol continues to flow through until it breaks into numerous nanodroplets. Other proposed models include nanoparticles coagulating before landing on the growth surface[57] and liquid deposition taking place instead of vapor deposition from the partial evaporation of the liquid droplet.[58]

Film IV is shown in Fig. 6.21(c). It exhibits a dendritic microstructure with nonfaceted, elongated grains. It was previously proven that this was a consequence of diffusion-limited growth.[39] By increasing the flow rate, dense and trigonal-shaped faceted grains were obtained for film V [Fig. 6.21(d)]. This shape is from the intersection of {211} faces in the chalcopyrite structure. The largest grain sizes obtained using SSPs were of the order of 0.5 μm for film V.

XRD patterns revealed that the films were either (112) or (204/220) oriented. Differentiation between the chalcopyrite and the sphalerite phases was made by differences in XRD patterns.[59] XRD patterns for all five films are shown in Fig. 6.22(a). It was found that films often contained a secondary phase ($2\theta = 26.5°$), believed to be indium-rich (In-rich).[7] The more In-rich the film, the more likely it was that the secondary phase was present, with the film being more (204/220) oriented. EDS measurements (not shown) indicate that the films containing the secondary phase were In-rich and that the Cu/In ratio increased when the secondary phase was reduced upon annealing [Fig. 6.22(b)]. To confirm the In-rich nature of the secondary phase, GAXRD was performed on the films. The GAXRD pattern [Fig. 6.22(c)] revealed that the secondary phase was concentrated at the surface. To find out the chemical nature of the phase, films containing the phase were etched in a 10% aqueous KCN solution for 2 min. It is common to etch Cu-rich $CuInS_2$ films in this solution prior to CdS deposition to remove undesired CuS_x compounds segregated on the surface during $CuInS_2$ deposition. In addition, it is also known that the etch rate of Cu-rich compounds in KCN solutions is much higher than for In-rich compounds.[60] After etching, GAXRD still showed the presence of the secondary phase on the surface without any change in its diffraction intensity, which means that the phase is not CuS_x. The nature of the secondary In-rich phase was also confirmed by Raman spectroscopy.[7]

Processing conditions were sought to prevent the formation of the secondary phase, to favor growth conditions of (112)-oriented films, and to avoid the

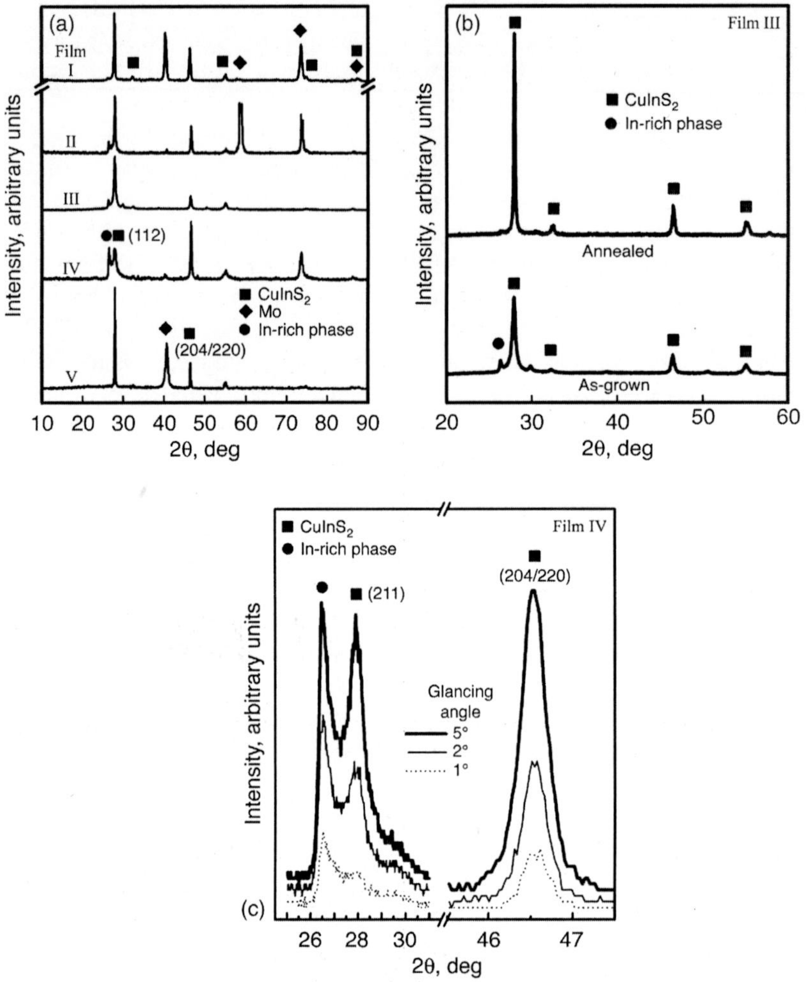

Figure 6.22. XRD spectra of $CuInS_2$ films grown by AACVD: (a) untreated films, (b) film III with postdeposition annealing, and (c) GAXRD spectrum of film IV.

surface kinetic regime where a (204/220)-oriented film is favored. In our setup, for any reactor, the substrate temperature had to stay around 400 °C in order to promote the growth of (112)-oriented films without a high concentration of the In-rich secondary phase. Attempts to grow (204/220)-oriented films without the secondary phase were not successful. Another way to reduce this secondary phase is by postdeposition heat treatment.

Band-gap energies between 1.45 eV and 1.47 eV were obtained for the films. The band-gap energies were estimated using plots of $(\alpha h\nu)^2$ versus E (Fig. 6.23), where α is an absorption coefficient estimated from optical transmittance data and $h\nu$ is the photon energy. Figure 6.23 shows that the band edge

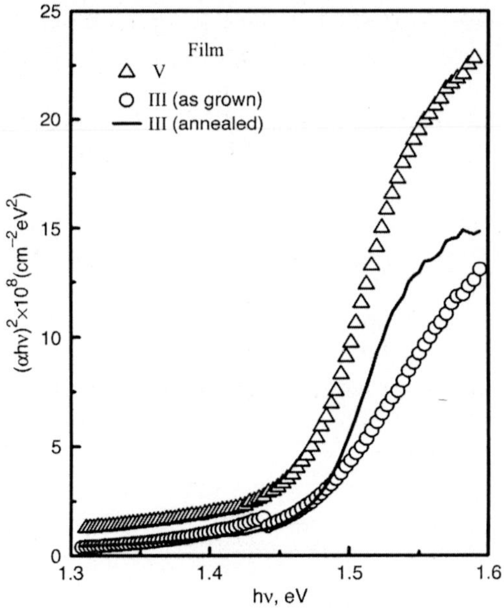

Figure 6.23. Plot of $(\alpha h\nu)^2$ versus E for films III and IV.

sharpens upon postgrowth annealing, which is ascribed to improved crystallinity and densification (film III). The overall absorption of film V was higher than that of film III in the figure, and this could be attributed to the difference in the density between the two films; film V had a denser grain structure than film III, as shown in Fig. 6.21. This finding could also be attributed in part to differences in reflectivity.

All of the films grown showed p-type conduction regardless of the reactor type, Cu/In ratio, or morphology. It was reported that S-rich material shows p-type conduction[61,62] and that the Cu-on-In antisite (Cu_{In}) in Cu-rich films is expected to be the major acceptor for p-type conduction.[47] The bulk resistivities ranged from 0.1 Ω-cm to 30 Ω-cm; Cu-rich films generally had lower resistivity than In-rich films.

6.4 ATMOSPHERIC PRESSURE HOT-WALL REACTOR PARAMETRIC STUDY

6.4.1 Parametric Study Approach

The reactor used for this study was the horizontal atmospheric-pressure hot-wall reactor (reactor A in Fig. 6.11). The susceptor accommodated three substrates side by side at an angle of 15.5° above horizontal. The leading edge of the film experiences a more reactant-rich gas stream and is closer to the bottom of the reactor. The environment of the trailing edge, deeper in the deposition zone, is more product-rich and is closer to the axial center of the reactor tube.

To optimize film and device quality, a four-phase study was conducted. The temperature of the deposition zone, location of the susceptor along the length of the reactor tube, concentration of the precursor solution, and postdeposition annealing conditions were varied individually. Initially, the substrate temperature was varied between 350 °C and 425 °C. In the second phase, the measured distance from the end of the reactor tube up to the trailing edge of the susceptor was varied from 57.15 mm (2.25 in) to 127 mm (5 in). The larger the distance, the closer the susceptor was to the evaporation warm zone. For phase three, the precursor concentration was varied between 0.005 and 0.04 mol/L. The deposition temperatures for the two middle parts of the experiment were maintained at a fixed value of 395 °C. Finally, postdeposition annealing of films was performed in the reactor. The evaporation zone temperature was set to 120 °C, and run-to-run deposition zone temperatures were varied between 450 °C and 580 °C. During annealing, Ar was flowed through the reactor tube at a rate of 10 mL/min. Annealing times varied from 15 min to 17 h. The annealing times began when the annealing temperature was reached and ended when the cooling process started. It took about 30 min to reach the annealing temperature when starting at room temperature and a little over 1 h to cool down to room temperature. For some runs, sulfur vapor was added during annealing by placing sulfur (Strem Chemicals, 99+%) in a powder-filled crucible in the evaporation zone.

As discussed, in this experiment, various reactor parameters were manipulated to study their effects on film growth. In general, the leading edge of the films tended to be smooth, shiny, and dense, whereas the trailing edge tended to be rough, dull, and highly porous. The rough surface morphology can be attributed to the stronger diffusion-limited film growth at the trailing edge.[63] By the time the carrier gas (Ar) reaches the trailing edge of the substrate, it may be less precursor-rich than the gas flowing over the leading edge—reaction occurs on the hot wall of the reactor along the deposition zone. This would make the concentration boundary layer thicker at the trailing edge of the substrate, requiring the precursor to diffuse a greater distance to reach the surface of the trailing edge, and hence, the film growth would be more diffusion-limited. SEM images of a rough, dull film can be seen below in Fig. 6.24(a); the images of a smooth, shiny film can be seen in Fig. 6.24(b).

6.4.2 Variation of Deposition Temperature

The run-to-run deposition zone temperature (350–425 °C) was the first parameter varied. Temperature variation affected film stoichiometry [Fig. 6.25(a)] and crystalline orientation [Fig. 6.25(b)], while not significantly affecting the deposition rate [Fig. 6.25(c)]. From Fig. 6.25(a), we can see that the films were closest to stoichiometry when deposited at 395 °C. Cu-to-In ratios ranged from

Figure 6.24. SEM images of $CuInS_2$ films grown by AACVD: (a) porous, rough, dull films, and (b) smooth, shiny, dense films.

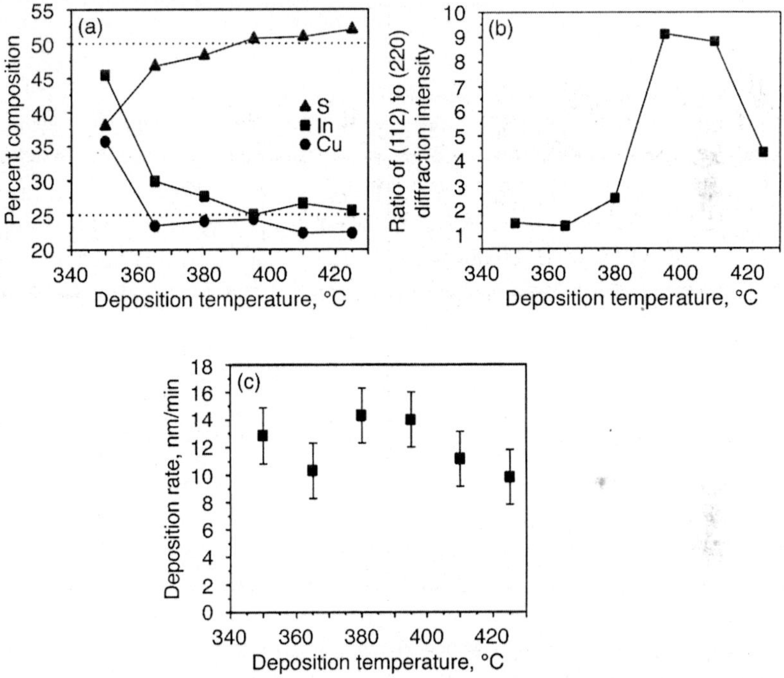

Figure 6.25. Deposition temperature versus (a) elemental composition, (b) crystalline orientation, and (c) deposition rate for $CuInS_2$ films with precursor feed rate of 1.6 g/h. In plot (a), the accuracy limit of 5% per element resulted in an anomalous overall percent composition for the In-rich, S-poor deposited material.

0.79 to 0.98, with the highest ratio occurring also at 395 °C. The higher deposition zone temperature correlated to increased sulfur content in the films. The increased surface mobility at higher temperatures would favor increased inclusion of sulfur.

$CuInS_2$ films deposited in this study were observed to be crystallographically (220/204) oriented or (112) oriented by XRD. Chalcopyrite solar cells had a higher series resistance for the (220/204)-oriented films.[64] From XRD measurements in Fig. 6.25(b), we observed that deposition temperatures of around 395 °C yielded higher ratios of the preferred (112)-oriented films. Finally, deposition rates were not limited by supply of thermal energy (temperature variation). Therefore, film growth was not reaction-limited, as shown in Fig. 6.25(c).

6.4.3 Variation of Susceptor Location and Precursor Concentration

The next parameter varied was the location of the susceptor within the deposition zone. Variations of the susceptor location did not affect film stoichiometry, but they did influence the morphologies of the films. When the susceptor was moved toward the evaporation zone, denser, smoother, and shinier looking films were obtained from a reduced concentration boundary layer (less diffusion-limited) toward the evaporation zone.

The single-source precursor concentration in the carrier solvent was another of the parameters varied. The concentrations of the SSP in toluene were varied between 0.005 and 0.04 mol/L. There were no significant changes in elemental composition of the films as a function of concentration, but increasing the precursor concentrations gave smoother, shinier films than with lower concentrations. This is because, at higher precursor concentrations, more precursor compound was available to the film reaction surface, reducing the effects of a diffusion-limited growth process.

6.4.4 Postdeposition Annealing

The final parameter manipulated was postdeposition annealing. It was determined that postdeposition annealing improves the elemental composition and crystalline structure of films. For example, after a 5-h anneal at 580 °C in a sulfur-rich atmosphere, typical film elemental compositions changed from 51.5% S, 24.5% In, and 24.0% Cu to 52.0% S, 23.5% In, and 24.5% Cu measured by EDS. Although the compositional change in bulk is minimal, annealing might improve the electronic properties of films by removing the volatile residual indium-rich secondary phase and surface states that are detrimental to the performance of solar cells. For example, it is reported that increasing the Cu-to-In ratio improved electrical properties of $CuInS_2$ films[65] and annealing in a S-rich atmosphere passivates S vacancies, an undesirable n-type dopant.[66] Ordering of the Cu and In atoms within the crystalline structure also increased, producing a better ordered chalcopyrite film. $CuInS_2$ films typically occur in either a sphalerite (more disordered) or a chalcopyrite structure.

Evidence of a more ordered chalcopyrite structure can be shown in Fig. 6.26, where the (101) and (211) diffractions at 17.9° and 37.3°, respectively, can be clearly shown in the annealed film. Another characteristic of the chalcopyrite structure not shown in the sphalerite structure is the peak splitting in

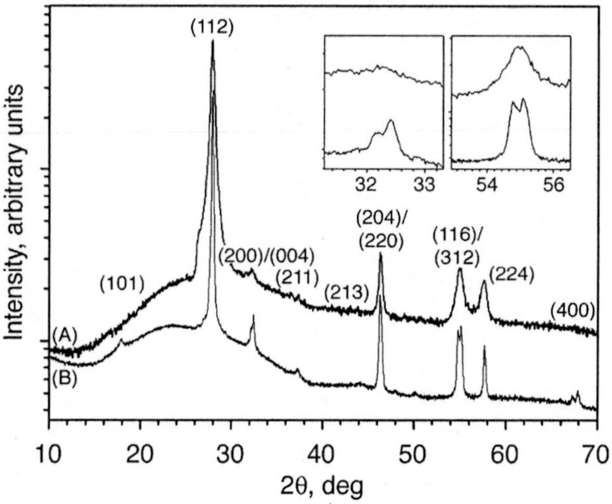

Figure 6.26. XRD pattern from (A) as-deposited films and (B) annealed film. Inset shows expanded scale emphasizing peak splitting.

the (200)/(004) and (116)/(312) diffractions at 32.2/32.4° and 54.8/55.1°, respectively (Fig. 6.26, inset). The highest power conversion efficiency for a cell made with a AACVD-deposited film using SSPs was 1.0%, and the film was annealed under S/Ar at 450°C for about 7h.[8] The highest efficiency for a cell made with a nonannealed film was 0.9%.[8] Both cells had smooth, shiny, and dense $CuInS_2$ absorber layers.

6.4.5 Photoluminescence Studies

Films annealed at 450°C for 7h were characterized using photoluminescence (PL). It has been previously shown that annealing can remove structural defects associated with metal ions in films, thereby improving the electronic properties of the absorber layer.[67,68] $CuInS_2$ films were excited using an Ar ion laser system (Coherent Innova 70) with an excitation wavelength of 514nm. To control the laser power density, neutral density filters were used. The PL emission detection system consisted of a liquid nitrogen-cooled charge-coupled device (CCD) (Horiba Jobin Yvon, Symphony), a germanium detector (EG&G Judson, J16D), and a lock-in system (Stanford Research System, SR810DSP). The detectors were connected to a Spex 1269 monochromator, whereas the samples were mounted on a helium cryostat (Janis, STVP-100) in order to lower the temperature to 4K. Finally, devices were tested using a solar simulator at GRC (Spectrolab X-25 Mark II) to obtain current versus voltage curves under AM0 illumination. Calibration was performed using a standard GaAs single-junction solar cell.

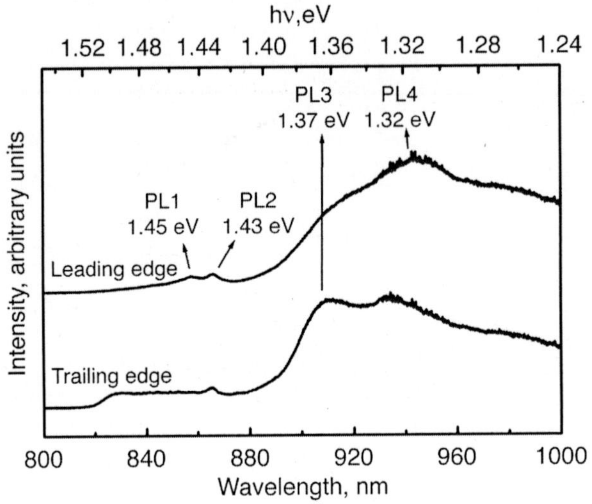

Figure 6.27. Photoluminescence spectra of as-deposited $CuInS_2$ thin films made from an SSP.

PL measurements were performed on 76 mm (3 in) long annealed films grown by the process used in the previous experiment. PL spectra of the leading and trailing edges were recorded for as-deposited, S-annealed, and Ar-annealed films. As mentioned in previous sections, the leading edges were always more In-rich than the trailing edges and therefore had lower Cu-to-In ratios (0.84 compared with the 1.03). Four different emission bands—PL1 at 1.45 eV, PL2 at 1.43 eV, PL3 at 1.37 eV, and PL4 at 1.32 eV—in Fig. 6.27 were identified in both the leading and trailing edges of almost all samples (PL1 was not clearly seen in some samples). For the as-deposited films, a near-band-edge emission can be seen for the trailing edge sample in Fig. 6.27. This is probably from the combination of excitonic recombination and other transitions associated with unidentified shallow defect levels.[68] Clear differences between the leading and trailing edges were observed from emission bands associated with deeper defect levels. For the trailing edges, the intensities of PL3 at 1.37 eV were always higher than those of PL4 at 1.32 eV (Fig. 6.27). Finally, the broad, deep-level emission bands spanning from 1.32 eV to 1.24 eV actually tailed down to 0.9 eV, as measured by a Ge detector.

After postdeposition Ar and Ar/S-annealing, the films were studied again using PL. S-anneals reduced the relative intensities of the PL1, PL4, and broad emission bands, whereas Ar-anneals increased the relative intensity of the PL1 band. This can be seen in Fig. 6.28. We can also see in this figure that S-anneals suppressed the broad near-band-edge emission from the trailing edge samples. When EDS measurements were performed on the films after S-anneals, an

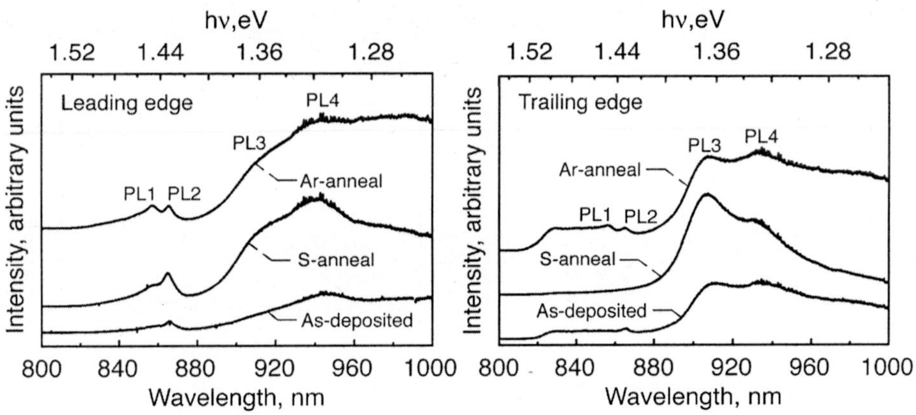

Figure 6.28. Photoluminescence spectra of $CuInS_2$ films prior to annealing and after annealing in either S/Ar or Ar flow at 450 °C for 7 h for leading edge (left) and trailing edge (right) of the deposited film.

extra 1–2 at% S was found incorporated into the films. On the other hand, when Ar-anneals were performed, a 5–6 at% S loss occurred. XRD measurements yielded similar results as above, where more ordered chalcopyrite structures were obtained after S-anneals.[8] S-anneals yielded higher resistivity films, whereas Ar-anneals yielded lower resistivity films. Hot-probe measurements demonstrated that all annealed films showed a p-type conduction.

The excitation intensity was also varied in order to determine the effect this had on the PL spectrum. PL1 and PL3 bands had a blueshift per decade of 3.7 meV and 5.5 meV, respectively, with an increase in excitation intensity. The blueshifts were attributed to donor–acceptor pair (DAP) recombination.[68,69] PL2 did not show any excitation power dependency, whereas the analysis of the PL4 band was not attempted because of the uncertainty in its precise location. The effect of increasing excitation intensity can be clearly observed in Fig. 6.29.

The blueshift of the bands with increasing excitation intensity is explained as follows. The emission energy for the recombination can be given by

$$h\nu(r) = E_g - (E_A + E_D) + \frac{e^2}{\varepsilon r_{DA}} \tag{6.7}$$

where E_g is the band gap, E_A (E_D) is the activation energy of the acceptor (donor), and $e^2/\varepsilon r_{DA}$ is the Coulomb energy for the pair at a distance r_{DA}.[68] When the excitation energy increases, the film is flooded with photogenerated carriers. As a result, the average separation between the donor and the acceptor decreases, causing the blueshift.

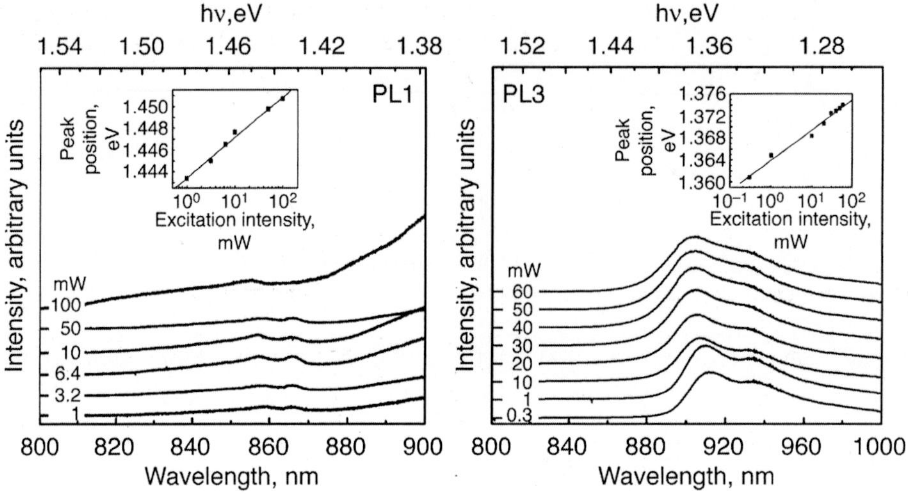

Figure 6.29. Spectra of $CuInS_2$ films with various excitation intensities; the insert shows the blueshift of PL1 (left) and PL3 (right) emission bands.

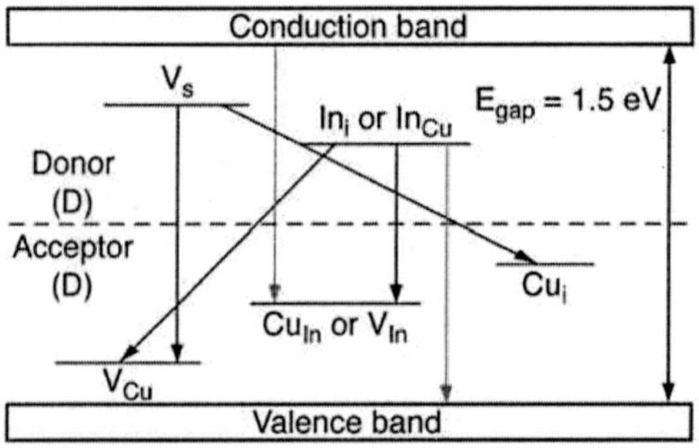

PL band	Energy, eV	Transition, D-A
PL1	1.45	$V_S - V_{Cu}$
PL2	1.43	In_i/In_{Cu} – VB or CB – Cu_{In}/V_{In}
PL3	1.37	$In_i/In_{Cu} - V_{Cu}$ or $In_i/In_{Cu} - Cu_{In}/V_{In}$
PL4	1.32	$V_S - Cu_i$

Scheme 6.2. Diagram and table of tentative assignment of transitions for the different PL bands. Diagram is not drawn to scale.

Shallow donors and acceptors for $CuInS_2$ have been previously studied and identified.[66,70–74] Scheme 6.2 includes some of the most energetically favorable intrinsic defects used in our analysis to assign transitions to the four different emission bands. Shallow donors include sulfur vacancy (V_S), In-on-Cu antisite (In_{Cu}), and indium interstitial (In_i). Acceptors include copper vacancy (V_{Cu}), Cu-on-In antisite (Cu_{In}), and copper interstitial (Cu_i). The PL1 emission band at 1.45 eV was assigned to a donor–acceptor transition, because of its blueshift upon increasing excitation intensity. V_S was attributed to the donor state because of the band's sensitivity to the change of V_S concentration upon annealing, and V_{Cu} was attributed to the acceptor state because the PL1 intensity was relatively stronger in leading edge (In-rich) samples. The PL2 band at 1.43 eV was assigned to a defect-band transition because there was no noticeable blueshift with increasing excitation intensity. The PL3 band at 1.37 eV was assigned to transitions between metal ions.[66,75] Finally, the PL4 band at 1.32 eV was assigned to a V_S–Cu_i transition. When the film was S-annealed, this band was suppressed; S incorporates into the structure reducing the V_S concentration.[75]

6.5 FABRICATION AND TESTING OF CIS SOLAR CELLS

6.5.1 Cell Fabrication at GRC

Complete $CuInS_2$ solar cells were fabricated at GRC using the cell architecture shown in Fig. 6.30. After coating the glass substrates with Mo, $CuInS_2$ films were deposited via AACVD, following procedures optimized for the atmospheric-pressure hot-wall reactor (reactor A). Postdeposition, sulfur/Ar-anneals were performed from 450 °C to 580 °C for up to 17 h. The films were then etched in a 1.5 M KCN solution for 1 min at room temperature to remove residual Cu_2S. Cadmium sulfide (CdS) was then deposited on the films using chemical bath deposition (CBD) to form the collecting heterojunction. The CBD used a solution of 0.001 M $CdSO_4$, 1.5 M NH_4OH, and 0.0075 M thiourea ($H_2NC(S)NH_2$). The bath was heated to 70 °C. Samples were immersed into the bath for 7 to 10 min until the solution started turning yellow. Once the CdS had been deposited, films were placed in an ultrasonic bath to remove particulates. A layer of ZnO:F was deposited from a 16.5 cm (6.5 in) diameter F-doped ZnO target in a RF sputtering system using pure Ar as the sputter gas. Aluminum (Al) was used as the top contact metal. A 0.2-µm-thick layer of Al was deposited by thermal evaporation through a shadow mask. The devices were isolated by mechanical scribing. The final cell area of the GRC-produced devices was typically ~0.4 cm².

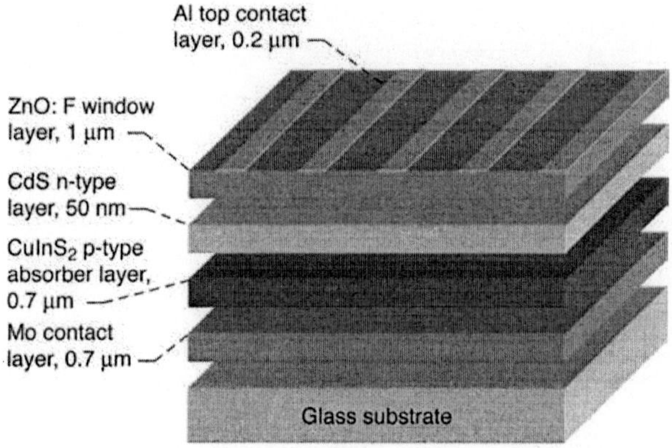

Figure 6.30. Schematic of standard NASA GRC thin-film $CuInS_2$ solar cell architecture.

6.5.2 Cross-Fabrication of Solar Cells

Solar cells were also fabricated at the Institute for Energy Conversion (IEC) following their well-established fabrication process.[76] This alternative fabrication process was done to isolate issues related to AACVD deposition of CIS films from those associated with the rest of the solar cell fabrication at GRC. For the cells prepared at IEC, only CIS layers were deposited at GRC. The rest of the device fabrication, including the coating of the glass substrates with Mo and the mechanical scribing, were performed at the IEC. The final cell area for IEC-produced devices was typically ~0.45 cm^2. Finally, postfabrication anneals were carried out in air on a hot plate (Cimarec 2) at 150 °C for up to 24 h.

6.5.3 Solar Cell Characterization

The best cell fabricated at GRC was made from a 0.7-μm-thick CIS film AACVD-deposited using a 0.01 M solution. After deposition, an S-anneal at 450 °C for 17 h was performed. This cell was not annealed after fabrication. The Mo bottom contact layer had a sheet resistance of 0.5 Ω/sq and a thickness of 0.7 μm. The CdS layer had an optical transmittance of 65% at a photon energy of 1.5 eV. GRC-fabricated cells used a 1 μm thick, n-type ZnO:F window layer with a sheet resistance of 200 Ω/sq and a transmittance of 85% at a photon energy of 1.5 eV. The current-voltage output characteristic (or performance) curve under light exposure for this cell together with cell parameters can be found in Fig. 6.31 and Table 6.5, respectively. For solar cells, it is common to plot the mirror image of the third quadrant of the conventional current-voltage plot to output characteristics (Fig. 6.31).

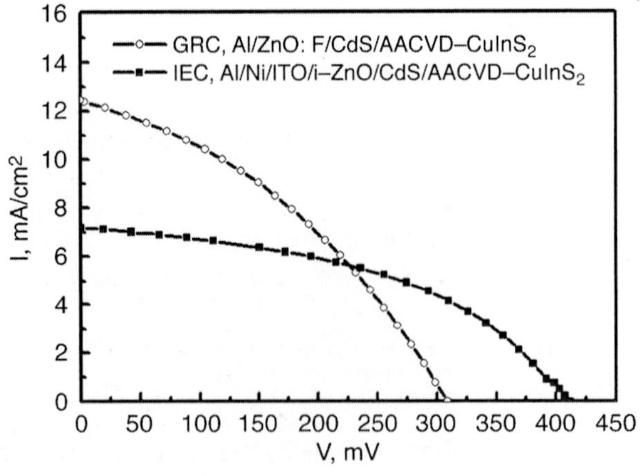

Figure 6.31. Light I–V characteristics of solar cells fabricated with AACVD-deposited $CuInS_2$ films.

TABLE 6.5. AM0 Output Characteristics of Solar Cells Fabricated With AACVD-Deposited $CuInS_2$ Films

Fabricator	V_{oc} (mV)	I_{sc} (mA/cm^2)	Fill factor	η (%)
GRC	309	12.5	0.37	1.0
IEC	412	7.2	0.45	1.0

The cell fabricated at the IEC is also included in Fig. 6.31 and Table 6.5. This cell was fabricated from a 0.7-μm-thick film deposited using a 0.04 M solution and S-annealed at 450 °C for 6 h and 40 min. The completed cell was then annealed under air at 150 °C for about 24 h. This cell's window layer, a 70-nm-thick i-ZnO (intrinsic ZnO) layer, had a transmittance of over 90% at a photon energy of 1.5 eV. The sheet resistance of the double top contact layer (2-μm Al/50-nm Ni) was about 1.8 mΩ/sq. Both cells had nearly the same efficiency (~1%), but the IEC cell had a higher shunt resistance, which improved the fill factor (see Table 6.5). The use of the intrinsic ZnO and the Ni/Al top contact improved the quality of the IEC cells by lowering the series resistance and improving the fill factor, as previously reported in the literature.[18] It should be also noted that most solar cells prepared with the AACVD method in the literature had a superstrate structure.[6] There has been no report of a working device with a typical chalcopyrite solar cell structure (top contact/window layer/CdS/$CuInS_2$/Mo/substrate) except for the authors' previous results.[48]

6.6 CONCLUDING REMARKS

6.6.1 Summary

$CuInS_2$ thin films were grown at low temperatures (<400 °C) via AACVD using SSPs (e.g., $(PPh_3)_2Cu(SEt)_2In(SEt)_2$). During the growth process, it was noticed that (112)-oriented films were close to stoichiometric with no secondary phases present, whereas (204/220)-oriented films were In-rich and always included an In-rich secondary phase. This indium-rich secondary phase was removed by postprocessing annealing at 600 °C, thereby increasing the Cu/In ratio. The resulting $CuInS_2$ films always exhibited p-type conduction and band-gap energies of around 1.46 eV ± 0.02. Films with a grain size up to 0.5 μm were obtained.

Variation of the deposition zone temperature affected the film stoichiometry and crystalline structure while not significantly affecting the deposition rate. A deposition zone temperature was optimized at 395 °C to produce (112)-oriented films without any detectable secondary phases. The susceptor location within the furnace did not affect the stoichiometry of deposited films, but it did alter morphology. Moving the susceptor toward the evaporation zone improved the morphology of the films. Increasing the precursor concentration in the carrier solvent also improved the morphology of the films while not altering the stoichiometry. EDS and XRD analyses indicated that after S-annealing, the stoichiometry and crystalline structure of the films was enhanced. PL studies revealed four major emission bands (1.45 eV, 1.43 eV, 1.37 eV, and 1.32 eV) together with a broad band associated with deep defects. The blue-shift of the 1.45 eV and 1.37 eV emission bands with increasing excitation power revealed DAP transitions. The broad band together with the 1.45-eV and 1.32-eV bands were reduced upon S-annealing because of passivation of V_S.

These films were then further processed to fabricate working solar cells. The best parameters obtained from the cells were the following: The IEC cell had the highest V_{oc} and FF, 412 mV and 0.45, respectively, whereas the GRC cell had the highest I_{sc} with 12.5 mA/cm^2. Devices fabricated at both GRC and IEC showed efficiencies of ~1.0% (AM0). The major challenge has been achieving higher V_{oc}, which seems to be a problem for cells prepared with AACVD deposited films in general. It is not yet clear what exactly causes the low V_{oc} and low fill factor of the cells prepared with AACVD deposited films. Additional study on remaining issues like defects, junction alignment, recombination losses, and undetected process issues remains to be performed to find which material component and/or procedural step is limiting the performance of the solar cells.

6.6.2 Outlook and Future Work

The versatility of the $[\{ER_3\}_2Cu(QR')_2M(QR')_2]$ architecture is clearly demonstrated by the preparation of SSPs for multinary semiconductors. Thermal analysis data substantiates that steric and electronic molecular modification on either the neutral donor or the chalcogenide permits directed adjustment of the solid-state phase and stability of the precursors. Evolved gas analysis confirms that the decomposition mechanism for the SSPs proceeds "cleanly" by the loss of the chalcogenide moiety, followed by loss of the neutral donor. Spray CVD using SSPs is a mild, simple, clean, and scalable technique for depositing $CuInQ_2$ or $CuGaQ_2$ (Q = S, Se) thin films on flexible polymer substrates at reduced temperatures. Although tests for the deposition of the wide band-gap alloy $Cu(Ga{:}In)S_2$ led to an inhomogeneous film composition, it is evident that with the use of two SSPs with similar thermal profiles, consistent film stoichiometry might be achieved.

Clearly the full potential of CIS PV devices has not been fully exploited, since the combination of group I–III–VI_2 elements can result in a variety of end products. Therefore, standards need to be defined that can associate device processing, fabrication, and film composition to cell band gap and efficiency. Spray CVD in conjunction with SSP design provides a proof-of-concept for a reproducible highly manufacturable process. Items that need more investigation include (1) **precursor design:** development of more volatile/thermally labile systems. This can be achieved by the incorporation of fluorinated or silylated functional groups. Importantly, because of the stability of fluorosilyl moieties, incorporation of both elements in the molecule can serve not only to increase volatility, but also as a "self-cleaning" mechanism should the precursor not decompose in a desirable pathway. (2) **Processing parameters:** spray CVD has several tunable variables, such as droplet size, flow-rate, concentration, solvent polarity, which is advantageous to achieve the desired film characteristics. Thus, an in-depth study needs to address these parameters with respect to film composition, morphology, and grain structure. (3) **Device fabrication:** working devices from deposited films need to be tested to aide SSP design and spray CVD process parameters. The work reported here on the molecular design of SSPs for their use in a spray CVD process, although still in its infancy, undoubtedly shows it as a mass producible, cost-effective method for fabricating commercial thin-film PV devices.

ACKNOWLEDGMENTS

The authors gratefully acknowledge NASA for financial support under cooperative agreement NCC3-947 and NCC04AA71A. We also wish to thank Mr. David Scheiman from the Ohio Aerospace Institute for his assistance with the solar simulator at the NASA Glenn Research Center and Mr. Philip Jenkins also from the Ohio Aerospace Institute for his assistance with the PL measurements. We are grateful to Dr. Robert Birkmire, Dr. William Shafarman, and the staff at the Institute of Energy Conversion of the University of Delaware for their help with solar cell fabrication. We also acknowledge the contributions of Drs. William Buhro and Jennifer Hollingsworth from Washington University, Dr. John Scofield from Oberlin College, and Mr. Jonathan Cowen from Case Western Reserve University. Finally, we acknowledge Dr. Kannan Ramanathan of NREL for providing a shadow mask.

REFERENCES

1. Banger, K. K.; Cowen, J. E.; Hepp, A. F. 2001. Synthesis and characterization of the first liquid single-source precursors for the deposition of ternary chalcopyrite ($CuInS_2$) thin film materials. *Chem. Mater.* 13:3827–3829.
2. Banger, K. K.; Harris, J. D.; Cowen, J. E.; Hepp, A. F. 2002. Facile modulation of single source precursors: The synthesis and characterization of single source precursors for deposition of ternary chalcopyrite materials. *Thin Solid Films* 403–404:390–395.
3. Banger, K. K.; Hollingsworth, J. A.; Harris, J. D.; Cowen, J.; Buhro, W. E.; Hepp, A. F. 2002. Ternary single-source precursors for polycrystalline thin-film solar cells. *Appl. Organomet. Chem.* 16:617–627.

4. Banger, K. K.; Jin, M. H. -C.; Harris, J. D.; Fanwick, P. E.; Hepp, A. F. 2003. A new facile route for the preparation of single-source precursors for bulk, thin-film, and nanocrystallite I-III-VI semiconductors. *Inorg. Chem.* 42:7713–7715.

5. Banger, K. K.; Hollingsworth, J. A.; Jin, M. H. -C.; Harris, J. D.; Bohannan, E. W.; Switzer, J. A.; Buhro, W. E.; Hepp, A. F. 2003. Ternary precursors for depositing I-III-VI_2 thin films for solar cells via spray CVD. *Thin Solid Films* 431–432: 63–67.

6. Jin, M. H. -C.; Banger, K. K.; Kelly, C. V.; Scofield, J. H.; McNatt, J. S.; Dickman, J. E.; Hepp, A. F. 2004. Solar cells fabricated with $CuInS_2$ films deposited using single-source precursors. *Proceedings of 19th European Photovoltaic Solar Energy Conference* (WIP, Munich, Germany and ETA, Florence, Italy), pp. 1943–1946.

7. Jin, M. H. -C.; Banger, K. K.; Harris, J. D.; Hepp, A. F. 2005. $CuInS_2$ films deposited by aerosol-assisted chemical vapor deposition using ternary single-source precursors. *Mater. Sci. Eng. B* 116:395–401.

8. Kelly, C. V.; Jin, M. H. -C.; Banger, K. K.; McNatt, J. S.; Dickman, J. E.; Hepp, A. F. 2005. Parametric study on non-vacuum chemical vapor deposition of $CuInS_2$ from a single-source precursor. *Mater. Sci. Eng. B* 116:403–408.

9. Jin, M. H. -C.; Banger, K. K.; McNatt, J. S.; Kelly, C. V.; Dickman, J. E.; Hepp, A. F. 2005. Post-deposition annealing of thin film $CuInS_2$ made from a single-source precursor. *Proceedings of the 31st IEEE Photovoltaic Specialists Conference.* IEEE, Piscataway, NJ. pp. 382–385.

10. Hoffman, D. J.; Kerslake, T. W.; Hepp, A. F.; Jacobs, M. K.; Ponnusamy, D. 2000. Thin-film photovoltaic solar array parametric assessment. *Proceedings of the 35th Intersociety Energy Conversion Engineering Conference* (IECEC 35). IEEE, Piscataway, NJ. pp. 670–680.

11. Contreas, M.; Egaas, B.; Ramanathan, K. 1999. Progress toward 20% efficiency in Cu(In,Ga)Se_2 polycrystalline thin-film solar cells. *Prog. Photovolt. Res. App.* 7:311–316.

12. Bailey, S. G.; Flood, D. J. 1998. Space photovoltaics. *Prog. Photovolt. Res. App.* 6:1–14.

13. Schock, H. W.; Noufi, R. 2000. CIGS-based solar cells for the next Millennium. *Prog. Photovolt. Res. Appl.* 8:151–160.

14. Schock, H. W.; Bogus, K. 1998. Development of CIS solar cells for space applications, edited by Schmid, J.; Ossenbrink, H. A.; Helm, P.; Ehmann H.; Dunlop, E. D. *Proc. 2nd World Conf. on Photovolt. Energy Conf.* (E. C. Joint Res. Center, Luxembourg). pp. 3586.

15. Tarrant, D.; Ermer, J. 1993. I-III-VI_2 multinary solar cells based on $CuInSe_2$. *Proc. 23rd IEEE Photovoltaic Specialist Conference* (Louisville, KY). IEEE, New York. pp. 372–378.

16. Klaer, J.; Bruns, J.; Henninger, R.; Töpper, K.; Klenk, R.; Ellmer, K.; Bräunig, D. 1998. A tolerant two step process for efficient $CuInS_2$ solar cells. *Proceedings of the 2nd World Conference on Photovoltaic Energy Conversion*, Report EUR 18656, Vol. I (*European Commission, Luxemborg*). pp. 537–540.

17. Ramanathan, K.; Contreras, M. A.; Perkins, C. L.; Asher, S.; Hasoon, F. S.; Keane, J.; Young, D.; Romero, M.; Metzger, W.; Noufi, R.; Ward, J.; Duba, A. 2003. Properties of 19.2% efficiency ZnO/CdS/$CuInGaSe_2$ thin-film solar cells. *Prog. Photovolt. Res. Appl.* 11:225–230.

18. Archer, M. D.; Hill, R. 2001. *Clean Electricity from Photovoltaics*. Imperial College Press, Singapore. Chapter 7.
19. Basol, B. M.; Kapur, V. K.; Halani, A.; Leidholm, C.; Sharp, J.; Sites, J. R.; Swartzlander, A.; Matson, R.; Ullal, H. 1996. $Cu(In,Ga)Se_2$ thin films and solar cells prepared by selenization of metallic precursors. *J. Vacuum Sci. Technol. A* 14:2251–2256.
20. Probst, V.; Stetter, W.; Riedl, W.; Vogt, H.; Wendl, M.; Calwer, H.; Zweigart, S.; Ufert, K.; Freienstein, B.; Cerva, H.; Karg, F. H. 2001. Rapid CIS-process for high efficiency PV-modules: development towards large area processing. *Thin Solid Films* 387:262–267.
21. Dimmler, B.; Schock, H. W. 1998. Scalability and pilot operation in solar cells of $CuInSe_2$ and their alloys. *Prog. Photovolt. Res. Appl.* 6:193–199.
22. Park, S. C.; Lee, D. Y.; Ahn, B. T.; Yoon, K. H.; Song, J. 2001. Fabrication of $CuInSe_2$ films and solar cells by sequential evaporation of In_2Se_3 and Cu_2Se binary compounds. *Sol. Energy Mater. Sol. Cells* 69:99–105.
23. Guillen, C.; Herrero, J. 2001. Recrystallization and components redistribution processes in electrodeposited $CuInSe_2$ thin films. *Thin Solid Films* 387:57–59.
24. Eberspacher, C.; Fredric, K.; Pauls, K.; Serra, J. 2001. Thin-film CIS alloy PV materials fabricated using non-vacuum, particles-based techniques. *Thin Solid Films* 387:18–22.
25. Klenk, M.; Schenker, O.; Alberts, V.; Bucher, E. 2001. Properties of flash evaporated chalcopyrite absorber films and solar cells. *Thin Solid Films* 387:47–49.
26. Dzionk, C.; Metzner, H.; Hessler, S.; Mahnke, H. E. 1997. Phase formation during the reactive annealing of Cu-In films in H_2S atmosphere. *Thin Solid Films* 299:38–44.
27. Krunks, M.; Mikli, V.; Bijakina, O.; Hebane, H.; Mere, A.; Varema, T.; Mellikov, E. 2000. Composition and structure of $CuInS_2$ films prepared by spray pyrolysis. *Thin Solid Films* 361–362:61–64.
28. Artaud, M. C.; Ouchen, F.; Martin, L.; Duchemin, S. 1998. $CuInSe_2$ thin films grown by MOCVD: Characterization, first devices. *Thin Solid Films* 324:115–123.
29. Jones, A. C.; O'Brien, P. 1997. *CVD of Compound Semiconductors: Precursors Synthesis, Development & Application*. VCH Press, Weinheim, Germany.
30. Rees, W. S. 1996. Introduction. In *CVD of Non-metals*, edited by Rees, W. S. VCH Press, Weinheim, Germany. pp. 1–35.
31. Keys, A.; Barbarich, T.; Bott, S. G.; Barron, A. R. 2000. Tert-butyl compounds of gallium. *(J. Chem. Soc.) Dalton Trans.* 2000:577–588.
32. Gleizes, A. N. 2000. MOCVD of chalcogenides, pnictides, & heterometallic compounds from single-source molecule precursors. *Chem. Vap. Deposition* 6:155–173.
33. Kodas, T. T.; Hampden-Smith, M. J. 1998. *The Chemistry of Metal CVD*. VCH Press, Weinheim, Germany. pp. 475.
34. Nomura, R.; Kanaya, K.; Matsuda, H. 1988. Preparation of copper-indium-sulfide thin films by solution pyrolysis of organometallic sources. *Chem. Lett.* 17: 1849–1850.

35. Nomura, R.; Fujii, S.; Kanaya, K.; Matsuda, H. 1990. Oxygen- or sulfur-containing organoindium compounds for precursors of indium oxide and sulfide thin films. *Polyhedron* 9:361–366.

36. Nomura, R.; Sekl, Y.; Matsuda, H. 1992. Preparation of copper indium sulfide ($CuInS_2$) thin films by single-source MOCVD process using $Bu_2In(SPr)Cu(S_2CNPr^i_2)$. *J. Mater. Chem.* 2:765–766.

37. Nomura, R.; Sekl, Y.; Matsuda, H. 1992. Preparation of copper indium sulfide ($CuIn_5S_8$) thin films by single-source organometallic chemical vapor deposition. *Thin Solid Films* 209:145–147.

38. Hirpo, W.; Dhingra, S.; Sutorik, A. C.; Kanatzidis, M. G. 1993. Synthesis of mixed copper-indium chalcogenolates. Single-source precursors for the photovoltaic materials $CuInQ_2$ (Q = S, Se). *J. Am. Chem. Soc.* 115:1597–1599.

39. Hollingsworth, J. A.; Hepp, A. F.; Buhro, W. E. 1999. Spray CVD of copper indium disulfide films: Control of microstructure and crystallographic orientation. *Chem. Vap. Deposition* 5:105–108.

40. Hollingsworth, J. A.; Buhro, W. E.; Hepp, A. F.; Jenkins, P. P.; Stan, M. A. 1998. Spray chemical vapor deposition of $CuInS_2$ thin films for application in solar cell devices. *Chemical Aspects of Electronic Ceramics Processing*, edited by Kumta, P. N.; et al., MRS Symp. Proc., Vol. 495, Materials Research Society, Pittsburgh, PA, pp. 171–176.

41. Harris, J. D.; Hehemann, D. G.; Cowen, J. E.; Hepp, A. F.; Raffaelle, R. P.; Hollingsworth, J. A. 2000. Using single source precursors and spray chemical vapor deposition to grow thin-film $CuInS_2$. *Proceedings of the 28th IEEE Photovoltaic Specialists Conference*. IEEE, Piscataway, NJ. pp. 563–566.

42. Cowen, J. E.; Riga, A. T.; Hepp, A. F.; Duraj, S. A.; Banger, K. K.; McClarnon, R. 2004. Synthesis and characterization of $CuInS_2$ single source precursors for chemical vapor deposition. *J. Therm. Analysis and Calor.* 75:929–936.

43. Miyake, H.; Hayashi, T.; Sugiyama, K. 1993. Preparation of $CuGa_xIn_{1-x}S_2$ alloys from In solutions. *J. Cryst. Growth* 134:174–180.

44. Shaukat, A. 1990. Composition-dependent band gap variation of mixed chalcopyrites. *J. Phys. Chem. Solids* 51:1413–1418.

45. Odier, P.; Supardi, Z.; De-Barros, D.; Vergniéres, L.; Ramirez-Castellanos, J.; Gonzales-Calbet, J. M.; Vallet-Regi, M.; Villard, C.; Peroz, Ch.; Weiss, F. 2004. Spray pyrolysis for high T_c superconductor films. *Supercond. Sci. Technol.* 17: 1303–1310.

46. Krumdieck, S.; Sbaizero, O.; Raj, R. 2001. Unique precursor delivery and control afforded by low-pressure pulsed-CVD process with ultrasonic atomization. *J. Phys. IV (Proceedings)* 11:1161–1168.

47. Zhang, S. B.; Wei, S. H.; Zunger, A.; Katayama-Yoshiba, H. 1998. Defect physics of the $CuInSe_2$ chalcopyrite semiconductor. *Phys. Rev. B* 57:9642–9656.

48. Harris, J. D.; Banger, K. K.; Scheiman, D. A.; Smith, M. A.; Jin, M. H. -C.; Hepp, A. F. 2003. Characterization of $CuInS_2$ films prepared by atmospheric pressure spray chemical vapor deposition. *Mater. Sci. Eng. B* 98:150–155.

49. Albin, D.; Noufi, R.; Tuttle, J.; Goral, J.; Risbud, S. H. 1988. Composition-structure relationships for multisource evaporated copper gallium selenide ($CuGaSe_2$) thin films. *J. Appl. Phys.* 64:4903–4908.

50. Edington, J. W. 1976. *Practical Electron Microscopy in Materials Science*. Van Nostrand Reinhold, New York.

51. Soliman, H. S. 1998. Structural and optical properties of thin films of $CuGaS_2$. *J. Phys. D: Appl. Phys.* 31:1516–1521.

52. Scheer, R.; Diesner, K.; Lewerenz, H. J. 1995. Experiments on the microstructure of evaporated $CuInS_2$ thin films. *Thin Solid Films* 268:130–136.

53. Tell, B.; Shay, J. L.; Kasper, H. M. 1972. Room-temperature electrical properties of ten I-III-VI_2 semiconductors. *J. Appl. Phys.* 43:2469–2470.

54. Tiller, W. A. 1964. Dendrites. Understanding of this familiar phenomenon has led to the development of useful man-made materials. *Science*. 146:871–879.

55. Jaffe, J. E.; Zunger, A. 1983. Anion displacements and the band-gap anomaly in ternary ABC_2 chalcopyrite semiconductors. *Phys. Rev. B: Condens. Matter.* 28:5176–5179.

56. Jokanovic, V.; Janackovic, D. J.; Spasic, P.; Uskokovic, D. 1999. Modeling of nanostructural design of ultrafine mullite powder particles obtained by ultrasonic spray pyrolysis. *Nanostruct. Mater.* 12:349–352.

57. Kodas, T. T.; Hampden-Smith, M. J. 1999. *Aerosol Processing of Materials*. Wiley-VCH, New York. Chapter 5.

58. Jin, M. H.-C.; Banger, K. K.; Harris, J. D.; Cowen, J. E.; Hepp, A. F. 2002. Thin film $CuInS_2$ prepared by spray pyrolysis with single-source precursors. *Proceedings of the 29th IEEE Photovoltaic Specialists Conference*. IEEE, Piscataway, NJ. pp. 672–675.

59. Álvarez-García, J.; Pérez-Rodríguez, A.; Romano-Rodríguez, A.; Morante, J. R.; Calvo-Barrio, L.; Scheer, R.; Klenk, R. 2001. Microstructure and secondary phases in coevaporated $CuInS_2$ films: Dependence on growth temperature and chemical composition. *J. Vac. Sci. Technol. A* 19:232–239.

60. Weber, M.; Scheer, R.; Lewerenz, H. J.; Jungblut, H.; Störkel, U. 2002. Micro-roughness and composition of cyanide-treated $CuInS_2$. *J. Electrochem. Soc.* 149:G77–G84.

61. Look, D. C.; Manthuruthil, J. C. 1976. Electron and hole conductivity in copper indium sulfide ($CuInS_2$). *J. Phys. Chem. Solids* 37:173–180.

62. Kazmerski, L. L.; Ayyagari, M. S.; Sandborn, G. A. 1975. Copper indium sulfide ($CuInS_2$) thin films. Preparation and properties. *J. Appl. Phys.* 46:4865–4869.

63. Van den Brekel, C. H. J. 1977. Characterization of chemical vapor deposition processes. I. *Philips Res. Rep.* 32:118–133.

64. Siemer, K.; Klaer, J.; Luck, I.; Braunig, D. 2000. Influence of crystal orientation on device performance of $CuInS_2$ solar cells. *Proceedings of the 28th IEEE Photovoltaic Specialists Conference*. IEEE, Piscataway, NJ. pp. 630–633.

65. Siebentritt, S. 2002. Wide gap chalcopyrites: Material properties and solar cells. *Thin Solid Films* 403–404:1–8.

66. Ueng, H. Y.; Hwang, H. L. 1989. The defect structure of copper indium sulfide ($CuInS_2$). Part I. Intrinsic defects. *J. Phys. Chem. Solids* 50:1297–1305.

67. Gossla, M.; Metzner, H.; Mahnke, H.-E. 2001. $CuInS_2$ thin-films from co-evaporated precursors. *Thin Solid Films.* 387:77–79.

68. Töpper, K.; Bruns, J.; Scheer, R.; Weber, M.; Weidinger, A.; Braunig, D. 1997. Photoluminescence of $CuInS_2$ thin films and solar cells modified by postdeposition treatments. *Appl. Phys. Lett.* 71:482–484.

69. Leite, R. C. C.; DiGiovanni, A. E. 1967. Frequency shift with temperature as evidence for donor-acceptor pair recombination in relatively pure n-type gallium arsenide. *Phys. Rev.* 153:841–843.

70. Binsma, J. J. M.; Giling, L. J.; Bloem, J. 1982. Luminescence of copper (I) indium sulfide. I. The broad band emission and its dependence on the defect chemistry. *J. Luminescence* 27:35–53.

71. Binsma, J. J. M.; Giling, L. J.; Bloem, J. 1982. Luminescence of copper (I) indium sulfide. II. Exciton and near edge emission. *J. Luminescence* 27:55–72.

72. Ueng, H. Y.; Hwang, H. L. 1990. The defect structure of copper indium sulfide ($CuInS_2$). Part II. Thermal annealing defects. *J. Phys. Chem. Solids* 51:1–10.

73. Schön, J. H.; Bucher, E. 1999. Characterization of intrinsic defect levels in $CuInS_2$. *Phys. Stat. Sol. A* 171:511–519.

74. Krustok, J.; Raudoja, J.; Schön, J.-H.; Yakushev, M.; Collan, H. 2000. The role of deep donor-deep acceptor complexes in CIS-related compounds. *Thin Solid Films* 361–362:406–410.

75. Nanu, M.; Schoonman, J.; Goossens, A. 2004. Raman and PL study of defect-ordering in $CuInS_2$ thin films. *Thin Solid Films* 451–452:193–197.

76. Hanket, G.; Paulson, P. D.; Singh, U.; Junker, S. T.; Birkmire, R. W.; Doyle III, F. J.; Eser, E.; Shafarman, W. N. 2000. Fabrication of graded $Cu(InGa)Se_2$ films by inline evaporation. *Proceedings of the 28th IEEE Photovoltaic Specialists Conference*. IEEE, Piscataway, NJ. pp. 499–503.

CHAPTER 7

Chemical Bath Deposition, Electrodeposition, and Electroless Deposition of Semiconductors, Superconductors, and Oxide Materials

RAGHU BHATTACHARYA
National Renewable Energy Laboratory, Golden, CO

7.1 INTRODUCTION

This chapter presents chemical-bath deposition, electrodeposition, and electroless deposition of materials that currently are at the forefront of technology development and industrial applications. Only a select few materials are discussed in this chapter. The emphasis will be on chemical-bath-deposited CdS, ZnS, and CdZnS, which are important systems for photovoltaic applications, and on electrodeposited $CuIn_{1-x}Ga_xSe_2$ (CIGS), high critical-temperature (T_c) oxide superconductors, and related oxide layers. All these materials can be prepared as thin films by low-cost non-vacuum solution growth methods or by conventional vacuum technology (e.g., physical vapor deposition (PVD), sputtering, and pulsed laser deposition).

Some of the many disadvantages of vacuum technology are complicated instrumentation, materials waste, high cost per surface area of deposition, and instability of some compounds at the deposition temperature. On the other hand, distinct advantages of solution growth technology include low cost per surface area of deposition and ease of doping. Solution growth technology also provides more uniform deposits on a variety of substrates, including tubes (interior and exterior) and various nonplanar devices.

7.2 CHEMICAL BATH DEPOSITION

In 1933, Bruckman[1] deposited a PbS thin film by the solution growth method—also known as the chemical bath deposition (CBD) method—from a solution containing lead acetate, thiourea, and sodium hydroxide. Thiourea furnished sulfur by hydrolysis in an alkaline solution, which resulted in sulfide films. The resulting CBD technique does not require expensive capital equipment and is a scalable technique that can be employed for large-area batch processing or continuous reel-to-reel deposition. The only apparatus required for the CBD process are appropriate solution containers and substrate mounting devices. The major drawback for this technology is the significant amount of solution waste generated after each deposition run.

7.2.1 CdS Deposition

In 1965, Kitaev et al.[2] prepared CBD CdS using cadmium salt and thiourea, which suggests that hydroxide particles in solution were necessary to grow good-quality films, with the decomposition of thiourea being stimulated by a solid phase such as cadmium hydroxide. Later, improved CBD chalcogenide thin films were obtained by using only NH_4OH, instead of both NH_4OH and NaOH.[3–12]

The deposition of CBD CdS as a junction layer for solar cell devices has proven to be a very successful industrially acceptable technique. Kessler et al.[13] reported on copper indium gallium diselenide (CIGS) mini-modules (area = $16\,cm^2$) with a conversion efficiency of 16.6%, wherein CBD CdS was used as a junction layer. Basol et al.[14] fabricated 9.3% active-area efficient thin-film flexible $CuInSe_2$ (CIS) solar cells (specific power >1 kW/kg) on light-weight, flexible metallic, and polymeric (polymide-based) substrates using CBD CdS.

The CBD CdS on CIGS or CIS devices produce superior solar cells relative to solar cells made by dry chemical vapor deposition (CVD) or physical vapor deposition (PVD) of CdS. Several groups[15–20] put forward plausible reasons for the superior performance, including selective etching or removal of native oxides by ammonia and also that the CBD process does not cause any physical damage, which could occur during CVD or PVD processes.

The current state-of-the-art CBD process for CdS consists of cadmium ions, thiourea, and ammonia. The deposition of a uniform CBD CdS thin film occurs because of a chemical reaction of controlled sulfur and cadmium ions in the alkaline solution. The slow release of sulfur from thiourea via decomposition and slow release of Cd^{2+} ions from $Cd(NH_3)_4^{2+}$ allow the following chemical reactions:

$$Cd(NH_3)_4^{2+} \rightarrow 4NH_3 + Cd^{2+} \tag{1}$$

$$SC(NH_2)_2 + OH^- \rightarrow SH^- + CH_2N_2 + H_2O \tag{2}$$

$$SH^- + OH^- \rightarrow S^{2-} + H_2O \tag{3}$$

$$Cd^{2+} + S^{2-} \rightarrow CdS \tag{4}$$

The overall chemical reaction is as follows:

$$Cd(NH_3)_4^{2+} + SC(NH_2)_2 + 2OH^- \rightarrow CdS + CN_2H_2 + 4NH_3 + 2H_2O \tag{5}$$

The stability constants K_a of $Cd(NH_3)_4^{2+}$ and K_b of NH_3 can be represented by the following equations:

$$Cd^{2+} + 4NH_3 \leftrightarrow Cd(NH_3)_4^{2+} \tag{6}$$

$$K_a = 7.56 \times 10^{-8} = [Cd^{2+}][NH_3]^4 / [Cd(NH_3)_4^{2+}] \tag{7}$$

$$NH_3 + H_2O \leftrightarrow NH_4^+ + OH^- \tag{8}$$

$$K_b = 1.8 \times 10^{-5} \tag{9}$$

Several research groups[21–23] suggested the following mechanism for CBD CdS film deposition, where $Cd(OH)_2$ species assist the film growth:

$$Cd^{2+} + 2OH^- \rightarrow Cd(OH)_2\text{, adsorbed on substrate surface} \tag{10}$$

$$SC(NH_2)_2 + Cd(OH)_2\text{, adsorbed} \rightarrow C^* \tag{11}$$

$$C^* \rightarrow CdS \text{ (deposited film)} + CdS \text{ (precipitate} \downarrow\text{)} \tag{12}$$

where C* is a reaction intermediate. Note that $Cd(OH)_2$ is readily converted to CdS in the presence of thiourea; this reaction does not occur as easily with $Zn(OH)_2$ during the ZnS CBD process.

In general, the CBD process starts by a quick mixing of stable solutions and proceeds under the influence of temperature control and stirring. The reaction advances as a function of time. The degree of reaction α (the ratio between the amount of substance that has reacted and the total amount able to react) can be expressed by the following equation[24]:

$$\alpha = (C_0 - C)/(C_0 - S) \tag{13}$$

where C_0 = initial concentration (t = 0), C = concentration at reaction time t, and S = solubility (molecules/cm^3).

Most CBD processes start slowly at a specific bath temperature, then accelerate, and eventually slow down again. Nucleation sites appear instantly on the substrates, the moment the solution contacts the substrate. In most cases, it is better to initiate the nucleation sites by inserting the substrate in a metal-ion solution, instead of a solution mixture containing both cations and anions. When cations and anions are mixed together, the resultant compounds begin to precipitate as soon as the ionic product, also called the solubility product, reaches a critical value, independent of the initial concentrations, but only depending on temperature. For example, the solubility product of CdS is 800 $\times 10^{-28}$ (K_{sp} at 25 °C).[25] If the ionic product of $[Cd^{2+}] \times [S^{2-}]$ exceeds the solubility product of CdS, deposition is initiated. The low solubility product of CdS indicates that the CdS precipitation can take place even at the lowest S^{2-} ion concentration because of the high value of sulfide ion hydrolysis equilibrium ($\approx 10^{-17}$). Nevertheless, the supersaturation with respect to an individual phase is important, as well as the kinetics of the growth processes. Under most deposition conditions for CdS films, the ionic product of $[Cd^{2+}] \times [OH^-]$ also exceeds the solubility product of $Cd(OH)_2$, and chemical reactions also occur among the reagents and solvents, which leads to the inclusion of numerous impurities in the CBD thin films.

The impurity content in CBD cadmium sulfide films has been examined by several groups. Kylner et al.[26] identified the impurities by using secondary-ion mass spectroscopy (SIMS), Fourier transform infrared (FTIR) spectroscopy, Rutherford backscattering spectrometry (RBS), and X-ray photoelectron spectroscopy (XPS). They concluded that CBD-grown CdS films contain ~11 atom% O, ~5 atom% N, as well as C and H. They also reported the presence of cadmium carbonate ($CdCO_3$) and water (H_2O) in the CBD CdS films. No clear evidence of cadmium hydroxide [$Cd(OH)_2$] or cadmium oxide (CdO) was found. Carbon-nitrogen bonds originating from any pseudohalogenic ions [e.g., cyanamide(NCN^{2-}), thiocyanate (SCN^-), cyanate (OCN^-), and cyanide (CN^-)] are identified as the main nitrogen impurity. Later, Webb et al.[27] also detected most of these impurities.

It is also reported that anions can be preferentially adsorbed in the CBD process, which leads to different levels of impurities. Hashimoto et al.[28] compared the iodide and chloride impurities in the CBD CdS thin films when the films were deposited from an aqueous solution of iodides and chlorides, respectively. The CdS films deposited using iodide contained 3% iodine, whereas the CdS grown from chlorides contained less than 1% chlorine.

The separation between substrates in batch-produced CBD CdS is also a likely important factor for reproducibility. Arias-Carbajal Readigos et al.[29] studied thin-film yield in the CBD technique as a function of separation between substrates in batch production. Based on a mathematical model, scientists proposed and experimentally verified that, in the case of CdS thin films, the film thickness reaches an asymptotic maximum with an increase in substrate separation. This behavior is explained on the basis of a critical layer of solution that exists near the substrate, within which the relevant ionic species have a higher probability of interacting with the thin-film layer than of contributing to precipitate formation. The critical layer depends on the solution composition and the temperature of the bath, as well as on the duration of deposition.

Surfactants may also influence the quality and thickness of the deposited films. Popescu et al.[30] prepared CBD CdS films in the presence of surfactants such as polyethylene glycol or ether sulfate, at temperatures of 40–60 °C. They studied the influence of these surfactants on the growth kinetics of CdS thin films and observed that surfactants decreased the reaction rate of CdS formation, thus leading to thinner but smoother thin films.

In most cases, CBD thin films deposited on amorphous or polycrystalline substrates/materials are amorphous or polycrystalline. But it has been demonstrated that epitaxial thin films by the CBD process can be deposited on single-crystal-type substrates. Furlong et al.[20] deposited epitaxial CdS thin films from an aqueous ammonia solution containing cadmium ions and thiourea on single-crystal $CuInSe_2$ films prepared by molecular-beam epitaxy (MBE) on Si (111) and GaAs (100) substrates. The structure and quality of the films were investigated by reflection high-energy electron diffraction (RHEED), glancing-angle X-ray diffraction (GAXRD), and high-resolution transmission electron microscopy (HRTEM) in cross section. The films are cubic on (100) substrates and mixed cubic and hexagonal on (111) substrates because of the presence of stacking faults parallel to the substrate. Froment et al.[31] prepared epitaxial cadmium sulfide on InP single crystals by CBD in ammonia solutions.

The CBD CdS was also applied in other areas of device fabrication and application. Mereu et al.[32] fabricated thin-film field-effect transistors using CBD CdS on SiO_2/Si (n-type) substrates. Meth et al.[33] fabricated thin-film transistors (TFTs) that incorporated patterned CBD cadmium sulfide as the active layer. Mobility in the $\approx$1-cm^2/Vs range with ON/OFF ratios of $\approx 10^5$ are needed for practical TFTs (e.g., for displays). Mobility values of 0.1 – 1 cm^2/Vs with ON/OFF ratios exceeding 10^7 have been reported for these types of TFTs using CBD CdS.

7.2.2 ZnS(O,OH) Deposition

Instead of CBD CdS, CBD ZnS(O,OH) has also been used to fabricate high-efficiency CIGS-based solar cells. The higher band gap of ZnS (~3.8 eV) compared with CdS (~2.4 eV) is an added advantage that improves the device efficiency by eliminating absorption loss. Structurally, CdS and ZnS are similar materials, but at equivalent levels of supersaturation, the pH of ZnS is 2.5 lower compared with that of CdS.[34] The degree of supersaturation for zinc with respect to the hydroxide is also greater than for cadmium. If the reaction is carried out at higher pH, about 4 orders of magnitude more thiourea is required for the zinc system compared with the cadmium system. At lower pH, more thiourea is required because the rate of hydrolysis is lower at lower pH.

There are several possible soluble and insoluble zinc species in a chemical bath containing OH^-, S^{2-}, NH_3, and NH_2–NH_2. The probable species and relevant equilibrium of the zinc system in an ammonium-based solution is as follows[35]:

$$Zn^{2+} + S^{2-} \leftrightarrow ZnS; \quad K = 10^{18.9} \qquad (14)$$

$$ZnS\downarrow \leftrightarrow Zn^{2+} + S^{2-}; \quad K = 10^{-24.7} \tag{15}$$

$$NH_3 + H_2O \leftrightarrow NH_4^+ + OH^-; \quad K = 10^{-4.8} \tag{16}$$

$$Zn(OH)_2\downarrow \leftrightarrow Zn^{2+} + 2OH^-; \quad K = 10^{-16.9} \tag{17}$$

$$Zn^{2+} + 4NH_3 \leftrightarrow Zn(NH_3)_4^{2+}; \quad K = 10^{8.9} \tag{18}$$

In an ammonia-based solution, the solution pH can vary between 7 (for an extremely diluted NH_3 solution) and 11.5 (for an NH_3 concentration of 1.2 M).[36] The equilibrium reactions suggest that, in an ammonia solution, a Zn^{2+} ion in the solution will stay in the form of $Zn(OH)_2$ along with $Zn(NH_3)_4^{2+}$, whereas a Cd^{2+} ion in an ammonia solution is predominantly $Cd(NH_3)_4^{2+}$.

Nakada et al.[37] effectively used CBD ZnS(O,OH) on a CIGS absorber to produce high-efficiency solar cells. Three consecutive depositions were required in an ultrasonic environment to obtain device-quality 100-nm layers of CBD ZnS (O,OH). The chemical bath consisted of $ZnSO_4$, ammonia, and thiourea. X-ray photoemission spectroscopy (XPS) analysis showed that a large amount of oxygen was included in the CBD ZnS films in the form of $Zn(OH)_2$ and ZnO. Single-layer CBD ZnS(O,OH) was later developed by Bhattacharya et al.,[38] producing an 18.5%-efficient CIGS-based device (Fig. 7.1). The deposition bath was based on $ZnSO_4$, ammonia, and thiourea dissolved in a mixture of water and methanol solvent. The CBD ZnS(O,OH) was

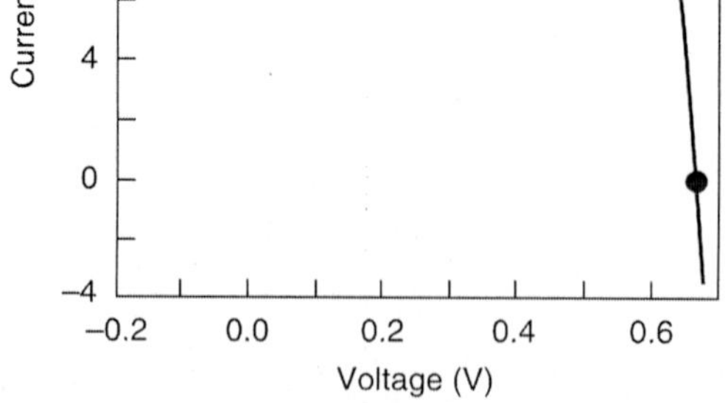

Figure 7.1. The current-voltage plot for an 18.5%-efficient CBD ZnS(O,OH)/CIGS-based photovoltaic device. Device temperature: 25.0 ± 1 °C; Device area = 0.402 cm^2; Irradiance: 1000.0 W/m^2; V_{oc} = 0.6698 V; I_{sc} = 14.122 mA; J_{sc} = 35.112 mA/cm^2; and Fill Factor = 78.78%. [Reproduced with permission from Ref. 38(a). Copyright 2004 Institute of Pure and Applied Physics.]

deposited at about 80°C. The ZnS(O,OH)/CIGS film was cleaned ultrasonically for 2 min in deionized water after CBD of ZnS(O,OH). The CBD ZnS(O,OH) layer was annealed in air for 10 min at 200°C.

A CBD ZnS(O,OH) film was also deposited on a Ni substrate and analyzed by XPS. Figure 7.2 shows the XPS data for single-layer CBD ZnS(O,OH) on a Ni substrate. The estimated thickness of these layers calculated from sputter etching time is about 300 Å. XPS analysis identified two types of oxygen [ZnO and $Zn(OH)_2$], as determined from binding energies. The experimentally measured binding energies for the XPS O1s peaks are 529.8 eV for the ZnO peak and 531.2 eV for the $Zn(OH)_2$ peak.

Figure 7.3 shows the FTIR spectra[38] of CBD ZnS(O,OH) (from the National Renewable Energy Laboratory [NREL]), which had been annealed at 200°C in air. NREL CBD ZnS(O,OH) films were prepared by an NREL-developed chemical bath approach using a solvent mixture of water and methanol, whereas standard ZnS(O,OH) was prepared using only water as a solvent, as described by Nakada et al.[37] For comparison, see the absorbance spectra of ZnS (99.99%), $ZnSO_4$ (99.999%), and ZnO (99.999%), where all commercial reagents were purchased from Alfa Aesar. The $C{=}N^{n-}$ impurities in CBD ZnS(O,OH) were identified by diffuse-reflectance FTIR spectroscopy, a powerful technique for analyzing fine particles and powders. The impurities containing carbon–nitrogen bonds are most likely cyanamide (NCN^{2-}) or thiocyanate (SCN^-), which result from the chemical reaction of thiourea and

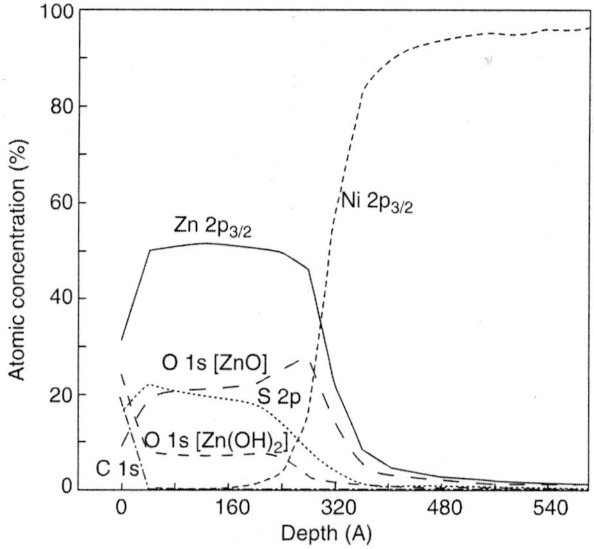

Figure 7.2. XPS data of an as-deposited single-layer CBD ZnS(O,OH) film on Ni substrate. [Reproduced with permission from Ref. 38(a). Copyright 2004 Institute of Pure and Applied Physics.]

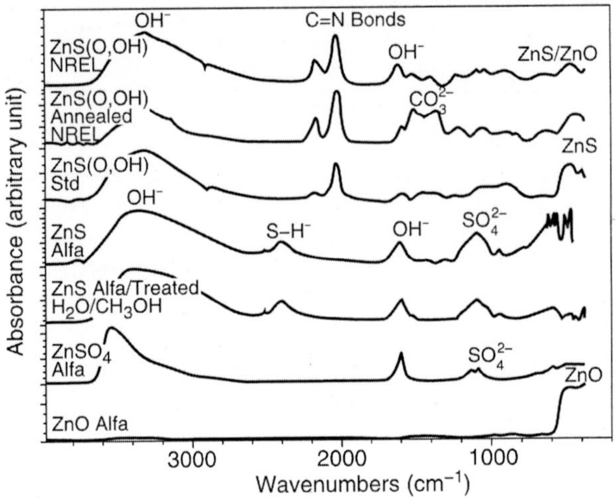

Figure 7.3. FTIR spectra of NREL CBD ZnS(O,OH); annealed NREL CBD ZnS(O,OH); standard ZnS(O,OH); Alfa-ZnS, Alfa-$ZnSO_4$, and Alfa-ZnO (all Alfa chemicals analyzed were as-received from the chemical company). [Reproduced with permission from Ref. 38(b). Copyright 2005 Elsevier Limited.]

ammonia. The OH^- bond indicates the presence of $Zn(OH)_2$ in the CBD ZnS(O,OH). The CBD ZnS(O,OH) annealed at 200 °C in air for 24 h began to pick up additional carbonate (CO_3^{2-}) impurities.

Marcotrigiano et al.[39] suggested that the desulfuration of thiourea in sodium hydroxide results in the formation of cyanamide, amidinourea, and guanidine at a pH below 12; they also suggested that the formed urea converts to ammonium cyanate and finally to ammonium carbonate (by hydrolysis). The proposed chemical reactions are as follows:

$$(NH_2)_2CS + OH^- \rightarrow NCNH_2 + SH^- + H_2O \tag{19}$$

$$NCNH_2 + H_2O \rightarrow O{=}C(NH_2)_2 \tag{20}$$

$$O{=}C(NH_2)_2 \rightarrow NH_4CNO \tag{21}$$

$$OCN^- + 2H_2O \rightarrow HCO_3^- + NH_3 \tag{22}$$

7.2.3 $Cd_{1-x}Zn_xS$ Deposition

Perhaps the best compromise to eliminate the high concentration of hydroxide and oxide impurities could be to use a CBD $Cd_{1-x}Zn_xS$ (denoted as CdZnS) buffer layer instead of a CBD ZnS(O,OH) buffer layer, if Cd does not pose a significant environmental issue. The use of CdZnS has multiple advantages. In high Cd-containing CdZnS, the oxide and hydroxide impurities are mini-

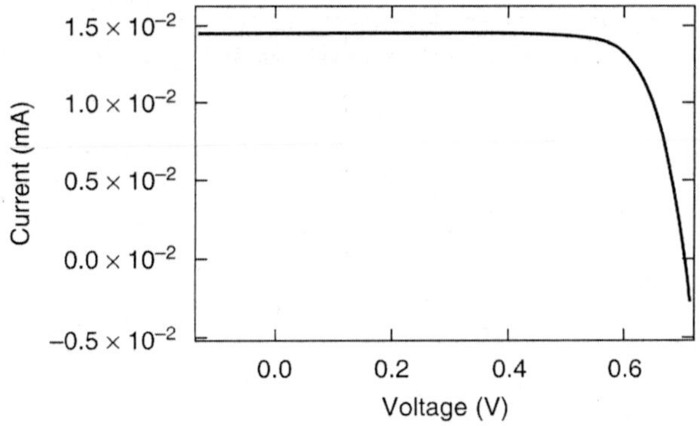

Figure 7.4. The current-voltage (I–V) plot of the 19.52%-efficient CIGS/CdZnS device. Device temperature: 25.0 ± 1°C; Device area = 0.408 cm^2; Irradiance: 1000.0 W/m^2; V_{oc} = 0.7052 V; I_{sc} = 14.49 mA; J_{sc} = 35.515 mA/cm^2; and Fill Factor = 77.90%. [Reproduced with permission from Ref. 40. Copyright 2006 American Institute of Physics.]

mized and the deposited film could be as good as CBD CdS, while maintaining the high band gaps that eliminate any loss in the range of 2.4 to ~3.6 eV.

Bhattacharya et al.[40] have already reported a 19.52%-efficient CIGS-based solar cell that uses a CBD CdZnS buffer layer. The CBD CdZnS thin films were obtained from a solution containing $CdSO_4$, $ZnSO_4$, thiourea, ammonia, water, and iso-propanol. The compositional ratio of Cd:Zn was 80:20, as determined by inductively coupled plasma (ICP) analysis. Figure 7.4 shows the current-voltage (I–V) curve of the 19.52%-efficient CIGS/CdZnS device. It also appears that the commonly used i-ZnO layer is not necessary with the CdZnS buffer, which makes this technique commercially favorable. The CBD CdZnS buffer layer also allows greater numbers of high-energy photons to reach the junction as was observed with the CBD ZnS layer, which increases the short-circuit current above that obtained for the CdS layer. This difference is clearly shown in Fig. 7.5, which compares the absolute external quantum efficiencies of CIGS/CdS and CIGS/CdZnS devices made with identical absorbers. It is evident from these measurements that significant current density (2 mA/cm^2) is gained between 300 and 500 nm when the CdZnS buffer layer is used. Also shown in Fig. 7.5 is a small quantum efficiency difference in the opposite direction for the longer wavelength photons, which suggests that the collection of deeper penetrating photons is slightly less. This phenomenon is probably from the relatively greater number of electronic defects in the depletion region compared with the CBD CdS/CIGS device. The identical long-wavelength cutoff in the two cases helps confirm that the absorbers are in fact identical.

The deposition mechanism of CBD CdZnS thin films under the current stirring conditions are dominated by convection mode (stirring or hydrodynamic transport). In a solution, fluid flow occurs by a natural convection mode

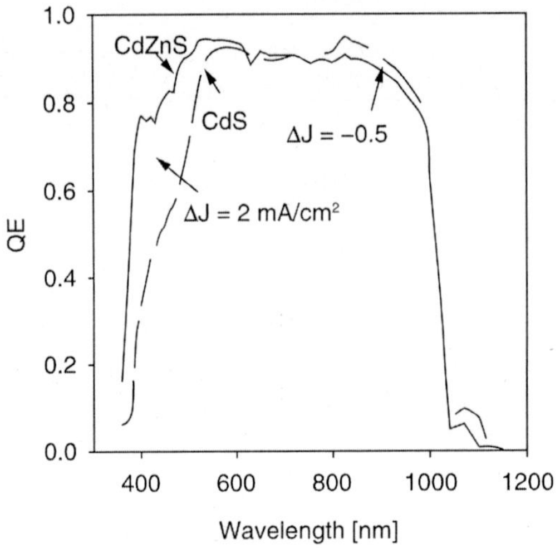

Figure 7.5. The absolute external quantum efficiencies of CIGS/CdS and CIGS/CdZnS devices made with identical absorbers. [Reproduced with permission from Ref. 40. Copyright 2006 American Institute of Physics.]

resulting from density gradients. Stirring conditions impose forced convection in the solution, which could result into laminar flow and turbulent flow. The formation of nucleation sites[41] on the substrates is instantaneous at the moment the cations and anions are mixed, catalyzed by the inhomogeneities in the solution mixture, where the nucleation process is of higher order than film growth. Nucleation takes place at a much higher frequency at high, rather than low, concentrations. Since growth cannot be maintained at the same relative velocity, the nucleation of CdZnS predominates at the current high concentrations. The particles of Cd^{2+},Zn^{2+} in aqueous/nonaqueous suspension will, in general, carry electrostatic charges and therefore repel each other. This phenomenon is responsible for the stability of many colloid suspensions. The principle of electroneutrality demands that the particles be surrounded by clouds of ions of the opposite sign (OH^-,S^{2-}); thus, the particles are, in fact, surrounded by an electric double layer. The addition of electrolytes, especially some organic solvents, may change the size, and even the sign, of the electric charge. When the electrolytes are electroneutral, the particles do not repel but may come into contact and adhere. When this happens, so-called rapid coagulation results in the solution.

Diffusion and convection are both important when transport through the solution controls the growth velocity. At very low stirring rates, a steady-state diffusion layer around the substrates is set up faster than the liquid in the vicinity of the substrate is renewed; thus, convection can be neglected. The solution at time $t = 0$ consists of a homogeneous, supersaturated solution. Soon after, it contains numerous small crystals that are being kept dispersed by stirring, if necessary. In general, the final particles are much larger than the nuclei. The initial deposition process exhibits a significant incubation period, especially in the case of a mononuclear-layer growth mechanism. In diffusion-controlled precipitations, the incubation period is very short and may easily be overlooked. The length of the incubation period also depends greatly on how it is defined or measured.

7.2.4 Other Systems

In general, CBD processes have been investigated by a sizable group of scientists worldwide. The CBD technique has been used to prepare a wide range of sulfides, selenides, and oxides. The following paragraphs only cite selected papers of interest to the author, primarily relating to the field of photovoltaics.

Yamada et al.[42] fabricated a 14.3%-efficient CIGS-based solar cell by using CBD ZnO as a buffer layer. CBD zinc oxide was grown from a solution mixture of zinc acetate and ammonium hydroxide. Mikami et al.[43] prepared a 14.3%-efficient CIGS-based solar cell by using CBD ZnO films as a buffer layer. Hariskos et al.[44] fabricated 15.7%-efficient (active area) CIGS solar cells using CBD $In_x(OH,S)_y$ thin films. Kim et al.[45] fabricated a 12.55%-efficient device using a CBD $In_x(O,OH,S)_y$ buffer layer on a CIGS absorber. Larina et al.[46] fabricated a 10.0%-efficient CIGS solar cell with a 30-nm-thick $In_x(OOH)_yS_z$ buffer layer, whereas a CIGS cell with a 60-nm-thick CdS buffer layer produced a 13.4%-efficient device. Tokita et al.[47] prepared a 14%-efficient CIGS-based device using a CBD $In(OH)_3{:}Zn^{2+}$ buffer layer. The buffer layer was deposited by CBD using $ZnCl_2$, $InCl_3{:}4H_2O$, and thiourea.

Gangopadhyay et al.[48] used a CBD ZnS antireflection coating on large-area commercial monocrystalline silicon solar cells, resulting in a 13.8% efficiency. Ennaoui et al.[49] deposited CBD $Zn(Se,OH)_x$ films on $Cu(In,Ga)(S,Se)_2$. They reported an active-area efficiency of up to 15.7% and a total-area efficiency of up to 13.26% for such devices.

Yamaguchi et al.[50] prepared CBD (Cd,Zn)S with different [Zn]/([Cd] + [Zn]) ratios. The band gaps in the thin films were controlled by varying the mixture ratio of CdI_2 and ZnI_2 solutions used as reactants. The variation of band gaps yielded a bowing parameter of 0.724 eV. Rajebnhonsale and Pawar[51] fabricated solid-solution alloy films of $Cd_{1-x}Zn_xS$ by CBD and reported a decrease of film thickness with an increase in zinc composition. Cheng et al.[52] prepared CBD ZnS thin films with an optical band gap of 3.53 eV using aqueous solutions of zinc acetate, thiourea, and trisodium citrate (where trisodium citrate was used as the complexing agent). Yamaguchi et al.[53] prepared CBD ZnS thin films from zinc acetate–thioacetamide (TAA) aqueous solutions. Film growths proceed by the accumulation of ZnS nanocrystallites (4 nm) formed in the solution, according to a cluster-by-cluster mechanism.

Aguilar et al.[54] prepared CBD ZnS and CuS to develop architectural glazing capable of withstanding hot tropical climates and windstorms. Lipowsky et al.[55] used organic additives to control the growth of CBD zinc oxide particles in solution and prevent the formation of the typical rod-like morphology. Hankare et al.[56] prepared CBD $MoSe_2$ semiconductor thin films using ammonium molybdate, tartaric acid, hydrazine hydrate, and sodium selenosulfate in an aqueous alkaline medium. Mane et al.[57] prepared CBD TiO_2 on indium-doped, tin-oxide-coated glass for dye-sensitized solar cells. Cheng et al.[58] fabricated TFTs using CBD ZnO layers. The ZnO films were fabricated on the defined areas of bottom-gate-type TFT plates by immersing in a chemical bath containing zinc nitrate and dimethylamineborane aqueous solution at 60 °C. Kale and Lokhande[59] reported CBD ZnSe with an optical bandgap of 2.85 eV. Lokhande et al.[60] and Gumus et al.[61] prepared CBD MnS thin films with a band gap of 3.88 eV. Fan et al.[62] prepared CBD MnS with an optical band gap of 3.23.

Bhattacharya[63] prepared CBD CIS films using sodium selenosulfate. Triethanolamine was found to be a suitable complexing agent that forms complexes with both Cu^{2+}/Cu^{+} and In^{3+}, thus preventing the precipitation of the metal hydroxides over a wide range of pH. Later, several research groups prepared CBD CIS using a wide range of source and complexing agents.[64–69] Pillai et al.[64] prepared CBD CIS films from a deposition mixture containing aqueous solutions of copper sulfate, trisodium citrate, sodium selenosulphate, and indium chloride. Pathan and Lokhande[68] prepared CBD CISe thin films from separate solutions of $CuSO_4$ + In_2SO_4 and Na_2SeSO_3 as cationic and anionic sources, respectively. This modified CBD approach, also known as SILAR (successive ionic layer adsorption and reaction), is based on the immersion of the substrate into separately placed cationic and anionic precursors and will be discussed in detail in Chapter 8.

Cao et al.[70] prepared ultraviolet-light-emitting ZnO nanosheets by a CBD method. Ganchev et al.[71] prepared CBD ZnSe from a solution containing zinc chloride, hydrazine hydrate, ammonia, and sodium selenosulfate. Bindu et al.[72] prepared amorphous selenium films at room temperature using CBD, which was achieved with an acidified solution of Na_2SeSO_3 at pH 4.5. Bini et al.[73] converted CBD Cu_xS films into $CuInS_2$ by evaporating indium on Cu_xS with subsequent annealing. Nair et al.[74] reported the CBD techniques of CdS, CdSe, ZnS, ZnSe, PbS, SnS, Bi_2S_3, Bi_2Se_3, Sb_2S_3, CuS, and CuSe. Ito and Nakamura[75] prepared a ZnO film by the flowing-liquid film method.

7.3 DEPOSITION OF CIGS BY ELECTRODEPOSITION AND ELECTROLESS DEPOSITION

Thin-film solar cell devices based on CIGS have already demonstrated an efficiency of 19.52%.[40] The direct energy gap of CIGS results in a large optical absorption coefficient, which, in turn, permits the use of thin (~1 μm) layers of active material. CIGS solar cells are also known for their long-term stability.[76]

The CIGS absorbers are being investigated by several techniques, including vacuum deposition technology, such as PVD and sputtering, and nonvacuum technologies such as electrodeposition and electroless deposition. The PVD technique is expensive and challenging to scale up because of film nonuniformity and low material utilization. Sputtering techniques are suitable for large-area deposition; however, they require expensive vacuum equipment and sputtering targets.

Nonvacuum electrodeposition and electroless deposition techniques have the potential to prepare large-area uniform precursor films using low-cost source materials and low-cost capital equipment. Therefore, these techniques are very attractive for growing CIGS layers for photovoltaic applications.

7.3.1 Electrodeposition of CIGS

Compared with the CBD technique, electrodeposition requires some additional capital equipment (i.e., suitable power supplies and electrodes). Major advantages of the electrodeposition process include the insignificant amount of waste generation. The electrodeposition bath can be reused for an unlimited number of cycle times when salts are replenished in the bath. The major drawback for electrodeposition is that it requires conductive substrates, which limits the application of this process in several key technologies.

Electrodeposition uses electrolysis to deposit a coating of desired form on conducting substrates from a solution ("bath") containing the ions of interest (e.g., Cu^{2+}, In^{3+}, and Ga^{3+}). Almost without exception, one of the two phases contributing to an interface of interest in electrodeposition will be an electrolyte, which is a phase through which charge is carried by the movement of ions. The second phase at the boundary is the substrate, through which charge is carried by electronic movement.

In cathodic electrodeposition, when the potential of the substrate (electrode) is moved from its equilibrium value toward negative potentials, the cation that will be reduced first is the one with the least negative (or most positive) redox potential E°. In a solution containing Cu^{2+}, In^{3+}, and Ga^{3+}, all three ions can be codeposited on the surface of the substrate (cathode) when the potential is sufficiently negative. Interestingly, Se also codeposits with Cu, In, and Ga under certain conditions even though it is expected to deposit on the counter-electrode (anode). The conditions of electrodeposition of CIGS thin films are related to the individual electrochemical reactions that correspond with the following Nernst equation, leading to the deposition of Cu, In, Ga, and Se. In this case, the electrodeposition of CIGS films on the cathode is most probably caused by the combination of electrochemical and chemical reactions as follows:

$$M^{n+} + ne^- \rightarrow M \tag{23}$$

$$Cu^{2+} + 2e^- \rightarrow Cu(s) \tag{24}$$

$$E = E^0_{Cu} + RT/2F\ln[Cu^{2+}] = 0.337 + 0.0295\log[Cu^{2+}] \quad (25)$$

$$In^{3+} + 3e^- \rightarrow In(s) \quad (26)$$

$$E = E^0_{In} + RT/3F\ln[In^{3+}] = -0.342 + 0.0197\log[In^{3+}] \quad (27)$$

$$Ga^{3+}\ 3e^- \rightarrow Ga(s) \quad (28)$$

$$E = E^0_{Ga} + RT/3F\ln[Ga^{3+}] = -0.529 + 0.0197\log[Ga^{3+}] \quad (29)$$

$$H_2SeO_3 + 4H^+ + 4e^- \rightarrow Se + 3H_2O \quad (30)$$

$$E = 0.741 - 0.0591pH + 0.0148\log[H_2SeO_3] \quad (31)$$

$$HSeO_3^- + 5H^+ + 4e^- \rightarrow Se + 3H_2O \quad (32)$$

$$E = 0.778 - 0.0739pH + 0.0148\log[HSeO_3^-] \quad (33)$$

$$SeO_3^{2-} + 6H^+ + 4e \rightarrow Se + 3H_2O \quad (34)$$

$$E = 0.875 - 0.0886pH + 0.0148\log[SeO_3^{2-}] \quad (35)$$

$$xM + ySe \rightarrow M_xSe_y \quad (36)$$

where E is the electrode equilibrium potential with respect to the standard hydrogen electrode (SHE), and E^0_{Cu}, E^0_{In} and E^0_{Ga} are the standard electrode potentials of Cu, In, and Ga, respectively. F is Faraday's constant equal to 96485 Coulomb/mole. The concentration of H^+ is converted to pH scale and shown in the equations. The logarithmic of concentrations of different chemical species shifts the electrode equilibrium potential at the specific concentration.

The electrodeposition of direct selenide compounds [Cu–Se, In_2Se_3, Ga_2Se_3, $CuInSe_2$, $Cu(InGa)Se_2$] is also possible because many metallic selenide compounds have large formation energies, ΔG (e.g., −386 kJ/mol for In_2Se_3, −418 kJ/mol for Ga_2Se_3, and −104 kJ/mol for Cu_2Se) as referred below. The redox potentials of the metal selenide films, therefore, move in a favorable direction. The redox potential, for example, of the reaction:

$$2In^{3+} + 3Se + 6e \rightarrow In_2Se_3 \quad (37)$$

will shift by an amount of $\Delta G/6F$ = +0.65 V with respect to the deposition potential of metallic indium.[77] This means that the deposition of indium selenide can take place below 0.35 V. A similar mechanism (Kroger's mechanism) takes place for Cu–Se, Ga_2Se_3, $CuInSe_2$, and $Cu(InGa)Se_2$.

In 1983, Bhattacharya[63] reported for the first time the electrodeposition of a CIS film. At NREL currently, CIGS films are prepared from a modified acidic bath containing 0.02–0.05 M $CuCl_2$, 0.04–0.06 M $InCl_3$, 0.01–0.03 M H_2SeO_3, and 0.08–0.1 M $GaCl_3$. The deposition baths are prepared using a base solution of 10 g of LiCl and 20 g of pH hydrion buffer 3 dissolved in 1000 mL of solution. The resulting pH of the deposition bath solution is between 2 and

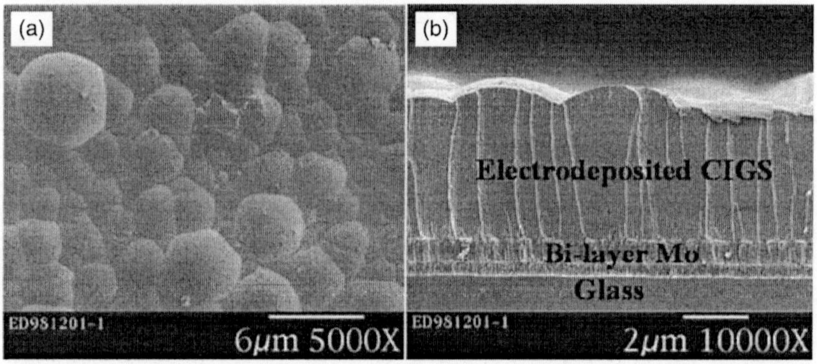

Figure 7.6. SEM of the electrodeposited CIGS precursor film: (a) surface morphology and (b) cross section.

3. Cu, In, Ga, and Se are codeposited from the solution at room temperature in a three-electrode cell configuration, where the reference electrode is a platinum pseudo-reference, the counter electrode is platinum gauze, and the working electrode is the substrate. The substrates typically used are glass, DC-sputtered with about 1 μm of Mo. In all experiments, the applied potential is –1.0 V versus the Pt pseudo-reference electrode. The corresponding current density range for the deposition is 5 to 7 mA/cm^2.

Good morphology of the precursor films is essential to obtain high-efficiency devices. Figures 7.6a and 7.6b show scanning electron microscope (SEM) morphology of the surface and cross section of a film with acceptable morphology. As the micrograph reveals, the as-deposited precursor films are densely packed with descent particle size. The stoichiometric or indium-rich precursor films tend to have poor morphologies that do not produce good devices.

The electrodeposited precursor films prepared in our laboratory that produced high-efficiency devices were Cu-rich films. These precursor films required additional In, Ga, and Se, deposited by PVD, to adjust their final composition to $CuIn_{1-x}Ga_xSe_2$. During this second step, the substrate temperature was maintained at 560 °C ± 10 °C. Figure 7.7 presents the Auger analysis of the final absorber and shows nonuniform distribution of Ga in the absorber and more Ga near the surface. This result is primarily from the second-stage PVD addition. The Ga hump is not helpful for hole collection. The device efficiencies are expected to increase by optimizing the Ga distribution in the absorber layers. The optimized layers should have less Ga in the front and more Ga on the back, which facilitates hole collection.

The photovoltaic devices were then completed with a 50-nm layer of chemically deposited CdS, 50 nm of radio frequency (RF) sputtered intrinsic ZnO, and 350 nm of Al-doped ZnO and bilayer Ni/Al top contacts deposited by e-beam. Finally, a 100-nm layer of MgF_2 is deposited by e-beam to minimize

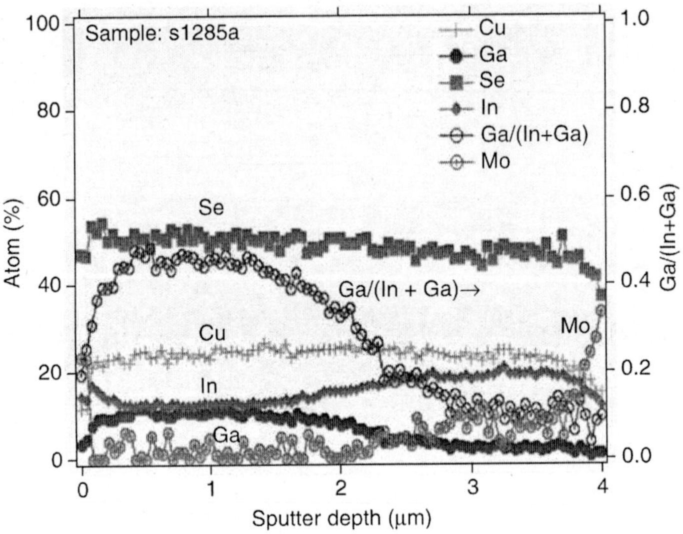

Figure 7.7. Auger analysis of the CIGS absorber layer. [Reproduced with permission from Ref. 78(b). Copyright 2000 Elsevier.]

reflection. The best device obtained from the electrodeposited precursor film has an efficiency of 15.4%.[78]

Electrodeposition of CIS films has also been studied by several other research groups. Qui et al.[79] reported an active-area efficiency of 7% using direct electrodeposited polycrystalline CIS films that were annealed in Ar atmosphere at about ~400 °C. Guillemoles et al.[80] reported a total-area efficiency of 6.5% using electrodeposited CIS thin films. The as-deposited films were annealed at 450 °C under selenium partial-pressure. Sudo et al.[81] and Matsuoka et al.[82] studied electrodeposited CIS and CIGS films extensively using Raman scattering measurements. Kapur et al.[83] reported a total-area efficiency of 8.43% using electrodeposited stacked Cu and In layers that were subsequently annealed in a H_2Se atmosphere. Oliveira et al.[84] prepared an electrodeposited CIS film from an acidic electrolyte containing citrate ions as a complexing agent. Kemell et al.[85] reported electrodeposition of CIS thin films from SCN^- baths.

Guillen et al.[86] combined electrodeposited Cu–Se alloy and sputtered In precursors, followed by annealing in a Se atmosphere, to obtain CIS thin films with Cu/In composition ratios varying between 0.25 and 2.75. De Silva et al.[87] characterized electrodeposited CIS films using XPS, X-ray diffraction (XRD), X-ray fluorescence (XRF), glow-discharge optical-emission spectroscopy (GDOES), and SEM for structural, stoichiometric, and morphological properties. Wenyi et al.[88] prepared electrodeposited CIS films at room temperature in the potentiostatic mode from a bath containing CuCl, $InCl_3{\cdot}4H_2O$, and SeO_2 at pH 1. Kampmann and et al.[89] reported a 4.8%-efficient Mo/CIS/CdS/ZnO device using an electrodeposited CIS layer and also a buffer layer. Bhattacharya et al.[9,90] reported the electrodepostion of CuInX (X = Se, S, Te). Friedfeld et al.[91] electrodeposited CIGS polycrystalline thin films with various Ga-to-In ratios using a two-step electrodeposition process. The process involves the electrodeposition of a Cu–Ga precursor film, followed by the electrodeposition of a Cu–In–Se thin film. The resulting CuGa/CuInSe bilayer was then annealed at 600 °C for 60 min in flowing Ar to form a CIGS thin film. The Cu–Ga electrodeposition was performed from a $CuSO_4$, $Ga_2(SO_4)_3$, and NaOH solution based on the work of Mehlin et al.[92] Ugarte et al.[93] electrodeposited CIS thin films in a glycine acid medium. Lincot et al.[77] reported an efficiency of 11.3% for an electrodeposited CIGS absorber with a band gap of 1.47 eV.

7.3.2 Electroless Deposition of CIGS

The electroless deposition methods for fabricating CIGS films for photovoltaic applications are also potentially attractive techniques. A device fabricated using electroless precursor films has already resulted in a solar cell efficiency of 13.4%.[94] The electroless method is based on short-circuiting the conducting substrate to an easily oxidizable redox component in the electrolyte bath. This method, unlike its closely related electrochemical counterpart, has the advantage of simplicity in that it does not require instrumentation for potentiostatic or galvanostatic control. Therefore, it is particularly promising for routine and large-scale fabrication of large-area thin films. The major drawback of electroless deposition is the potential contamination of the deposited layers from the dissolved anodic materials.

Codeposition of Cu–In–Ga–Se by the electroless process was performed from a bath containing $CuCl_2$, $InCl_3$, $GaCl_3$, and LiCl dissolved in deionized water at room temperature (24 °C) and without stirring. The pH of the bath was adjusted to 2.4 by adding incremental amounts of dilute HCl. Electroless precursors were prepared by short-circuiting the Mo substrate to a Fe, Zn, or Al counter-electrode. The electroless process for preparing CIGS precursor films is accomplished by the combination of electrochemical and chemical reactions. The counter-electrode (Fe) initiates the electrochemical reaction. The electroless deposition of CIGS most likely occurs as follows:

	E° vs SHE	
$Fe \rightarrow Fe^{2+} + 2e$	−0.447	(38)
$Zn \rightarrow Zn^{2+} + 2e$	−0.763	(39)
$Al \rightarrow Al^{3+} + 3e$	−1.662	(40)
$Cu^{+} + e \rightarrow Cu$	0.521	(41)
$Cu^{2+} + 2e \rightarrow Cu$	0.342	(42)

$$In^{3+} + 3e \rightarrow In \qquad -0.338 \qquad (43)$$

$$Ga^{3+} + 3e \rightarrow Ga \qquad -0.549 \qquad (44)$$

$$SeO_3^{2-} + 3H_2O + 4e \rightarrow Se + 6OH^- \qquad -0.366 \qquad (45)$$

$$Cu, In, Ga, Se \rightarrow Cu_xIn_yGa_zSe_n \qquad \text{Chemical reaction} \qquad (46)$$

The deposition potential is composed of the equilibrium reduction potentials (E_{eq}), overpotential, and ohmic potential drop (IR_s) in the solution. The applied potential (E) during codeposition of Cu–In–Ga–Se using an Fe electrode is 0.447 + rest potential (the electrode potential of the unpolarized cell), which probably makes the Ga deposition possible. The composition of the as-deposited CIGS precursor as analyzed by ICP was $CuIn_{0.35}Ga_{0.02}Se_{2.3}$ (Cu and Se-rich). The as-deposited films are amorphous or polycrystalline in nature and needed to be annealed/crystallized for making an active absorber layer. Additional In, Ga, and Se were added to the electroless precursor film by PVD to adjust the final composition to $CuIn_{1-x}Ga_xSe_2$. The substrate (precursor film) temperature during the PVD step was 550 °C ± 10 °C. The films were also selenized by exposure to selenium vapor during the cool-down time (~40 °C/min).

The addition of In and Ga and selenization at high temperature were crucial in obtaining the 13.4%-efficient device. Photovoltaic device fabrication was completed by a CBD of about 500 Å of CdS, followed by RF sputtering of 500 Å of intrinsic ZnO and 3500 Å of Al_2O_3-doped conducting ZnO. Bilayer Ni/Al top contacts were deposited in an e-beam system. The final step in the fabrication sequence is the deposition of 100 nm of MgF_2 as an antireflection coating.

Bhattacharya et al.[95] conducted deep-level transient spectroscopy (DLTS) measurements to estimate trap levels in CIGS thin-film solar cells prepared from both electrodeposited and electroless-deposited thin films. Deep-level impurities in CIGS solar cells grown by physical vapor deposition, electroplating (EP), and electroless or autoplating (AP) were investigated by using capacitance-voltage (C-V) and DLTS.[96] C-V profiling was used to determine the net carrier-concentration profile.

The net carrier concentration, shown in Fig. 7.8, was obtained at a frequency of 100 kHz. DLTS spectra were recorded using reverse- and forward-bias modes in the temperature range of 80–350 K. In the reverse-bias mode, the devices were reverse biased from −1.2 V to −0.2 V, with a pulse width of 1 ms. Two hole (majority-carrier) trap levels were found in all the devices. These levels were designated as H_1 at E_v+0.26 and H_2, for which an activation energy could not be resolved. Upon minority-carrier injection (forward-bias mode), DLTS showed two additional electron (minority-carrier) traps, which are labeled E_1 (E_c-0.1 eV) and E_2 (E_c-0.83 eV) in Table 7.1. The spectra were measured at an emission time of 465.2 s^{-1}, and the width of the

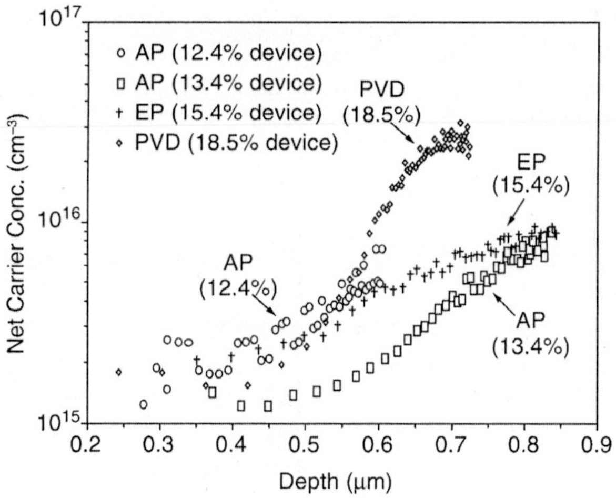

Figure 7.8. The net carrier concentration versus depth for the devices measured by C-V (DLTS measurements). The levels are designated for AP = autoplated (electroless deposition); EP = electroplated (electrodeposition); and PVD = physical vapor deposition. [Reproduced with permission from Ref. 95. Copyright 2001 Elsevier.]

TABLE 7.1. Trap Concentration and Activation Energy of Each Trap Measured by DLTS

Devices	Trap Level	Energy Level (eV)	Trap Concentration (cm^{-3})
PVD	E_1	$E_c - 0.1$	2.1×10^{14}
	H_1	$E_v + 0.26$	2.7×10^{14}
EP-15.4%	E_1	$E_c - 0.09$	8.4×10^{14}
	H_1	$E_v + 0.27$	1.2×10^{13}
EP-13.4%	E_1	$E_c - 0.09$	1.2×10^{15}
	H_1	$E_v + 0.26$	1.5×10^{14}
AP-12.4%	E_1	Not resolved	6.6×10^{14}
	H_1	$E_v + 0.26$	3.2×10^{14}
	E_2	$E_c - 0.83$	2.1×10^{14}

filling pulse was 1 ms. Table 7.1 summarizes the trap concentration and activation energy of each trap. The E_1 trap level was detected in all devices, whereas the E_2 level was observed only in the AP device that had a lower efficiency of 12.4%. It was suggested that the E_2 trap level is an effective recombination center. The calculated Shockley–Hall–Read (SHR) lifetime of the E_2 trap level, using the appropriate thermal velocity, which captures the cross section and the trap density, is about 0.5 ns. Previously, several research groups reported the H_1 and E_1 trap levels in CIS, but they could not detect the E_2 trap level.[97–100]

7.4 ELECTRODEPOSITION OF OXIDE SUPERCONDUCTORS

Electrodeposition techniques have also been applied to obtain high-current-carrying epitaxial Tl-oxide and Bi-oxide superconductor films after postannealing of the precursor films.[101–109] The availability of Bi-2223 high-temperature-superconductor (HTS) powder-in-tube (PIT) tape, with acceptable performance for long lengths, has provided the ability to construct a wide range of HTS electric power components. As a result, there are major worldwide projects in developing HTS electric power components for demonstration in a utility environment. The prospects for HTS use in power applications would be enhanced if HTS conductors with lower cost and current-carrying performance above 10^5 A/cm^2 at 77 K in magnetic fields of 3–5 T were available. Biaxially textured Tl-oxide-based superconductors could conceivably provide a lower cost and improved performance over Bi-2223 if serious attention were to be directed toward the development of long lengths. The single-layer Tl-1223 phase, with Bi or Pb and Sr substitution, offers improved coupling for the Cu–O planes and a higher temperature irreversibility line. Polycrystalline development, moreover, is more forgiving in this system than in either the Y- or Bi-oxides because of the ability to form uniaxially textured films with a "colony microstructure" that provides a percolative path for transport current. The grain boundaries in the Tl-oxide-based superconductors are also intrinsically clean, and the superconductivity (SC) behavior is markedly less sensitive to O_2 stoichiometry than YBCO.

Nonvapor methods such as electrodeposition are low-cost, nonvacuum techniques that have successfully allowed the preparation of superconducting oxide precursor films at thicknesses up to 5 μm in a few minutes and have been reacted in less than an hour to form high-quality superconducting films. A significant advantage of the electrodeposition technique is the mixture of the constituent elements that occurs at an atomic scale, which significantly decreases the annealing time and improves the transport properties of superconducting films. The critical current density of 1.1×10^6 A/cm^2 at 77 K in zero field for the electrodeposited Tl-oxide films are the highest ever reported for a processing technique that does not involve a vapor transport method, such as pulsed-laser deposition, sputtering, or e-beam. These latter methods have demonstrated typical current densities for Tl-oxide epitaxial films on single crystal substrates of around 10^6 A/cm^2 at 77 K in zero field, which represents the highest value that is also obtained by the electrodeposition technique.

7.4.1 Electrodeposition of Tl–Bi–Sr–Ba–Ca–Cu–O

The electrodeposited $(TlBi)_{1.1}Sr_{1.6}Ba_{0.4}Ca_2Cu_4O_x$ (TBSBCCO) precursor films were obtained by co-electrodeposition of the constituent metals using nitrate salts dissolved in dimethyl sulfoxide (DMSO) solvent. A typical electrolyte-bath composition for the TBSBCCO films consisted of 1-g $TlNO_3$, 1-g $Bi(NO_3)_3{\cdot}5H_2O$, 12.6-g $Sr(NO_3)_2$, 11.5-g $Ba(NO_3)_2$, 6.8-g $Ca(NO_3)_2{\cdot}4H_2O$, and 2.3-g $Cu(NO_3)_2{\cdot}6H_2O$ dissolved in 390-mL DMSO solvent. The substrates were single-crystal $LaAlO_3$ (LAO) coated with 300 Å of Ag.

The final film thicknesses were increased either by increasing the deposition time or by multilayer deposition. Although the film thickness increases with longer deposition time, the film morphology tends to be poor, as shown in Fig. 7.9. A two-layer technique that used two layers of TBSBCCO films, with an intermediate layer of Ag, improved the film uniformity and morphology, as shown in Fig. 7.10. The deposition process for the multiple layer is as follows: (1) Single-crystal substrates are coated with 300 Å of Ag; (2) TBSBCCO films (0.4–1.3 μm) are prepared by electrodeposition (ED) on Ag/LAO; (3) 300 Å of Ag are deposited on ED-TBSBCCO/Ag/LAO; (4) the second layer of TBSBCCO is electrodeposited (0.4–1.3 μm) on Ag/ED-TBSBCCO/Ag/LAO; and (5) the complete two-layer system is reacted.

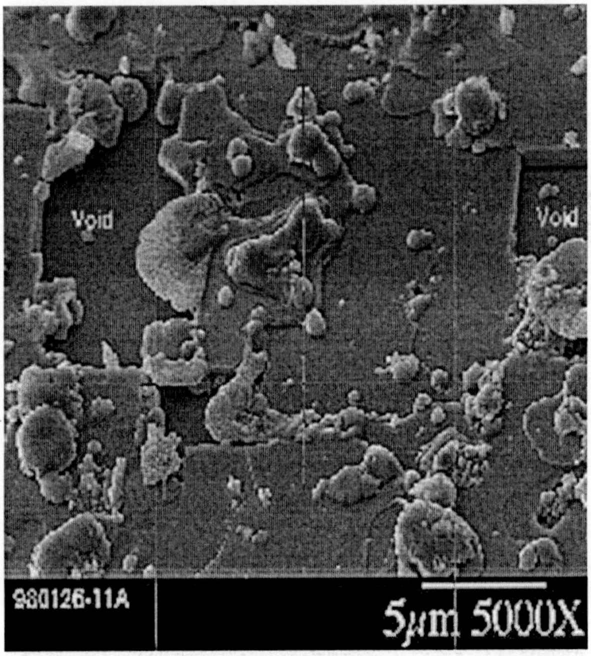

Figure 7.9. SEM of single-layer 0.8-μm electrodeposited TBSBCCO. [Reproduced with permission from Ref. 106. Copyright 2000 Elsevier and Ref. 103. Copyright 2002 IOP Publishing Limited.]

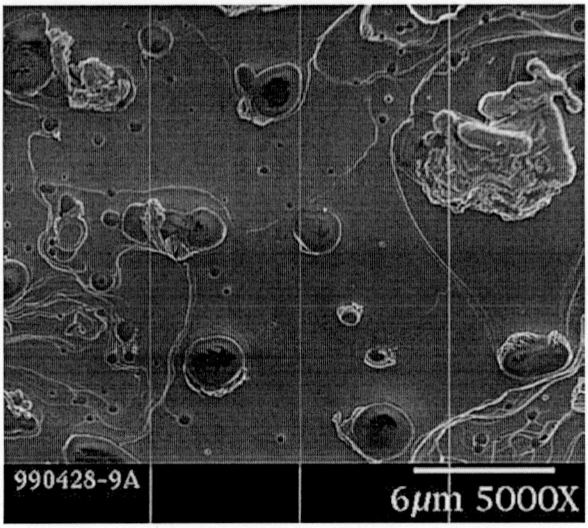

Figure 7.10. SEM of two-layer 0.8-μm electrodeposited TBSBCCO. [Reproduced with permission from Ref. 106. Copyright 2000 Elsevier and Ref. 103. Copyright 2002 IOP Publishing Limited.]

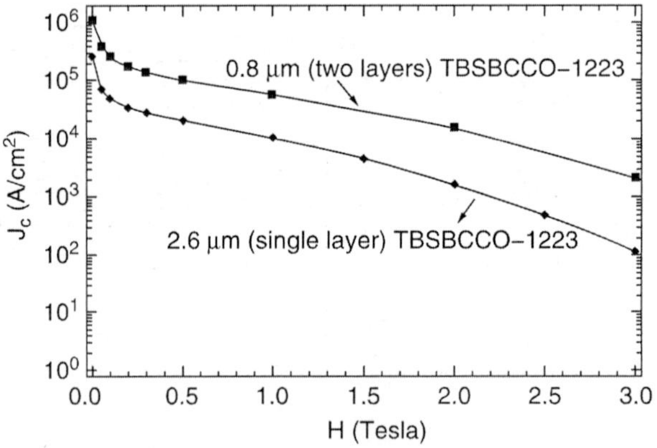

Figure 7.11. The magnetic field dependencies of transport J_c at 77 K (H // c) for a two-layer 0.8-μm ED-TBSBCCO/Ag/ED-TBSBCCO/Ag/LAO and a single-layer 2.6-μm ED-TBSBCCO/Ag/LAO film. [Reproduced with permission from Ref. 106. Copyright 2000 Elsevier.]

The electrodeposited precursor films, annealed in air at 870 °C in the presence of a TBSBCCO pellet, produce a biaxial textured Tl-1223 phase, as confirmed by an XRD pole-figure measurement. The omega and phi scans indicate full-width at half-maximum (FWHM) values of only 0.92 ° and 0.6 °, respectively, which indicates a very high-quality film. The superconductive transition temperature of the Tl-1223 film, determined resistively, was about 110 K. Figure 7.11

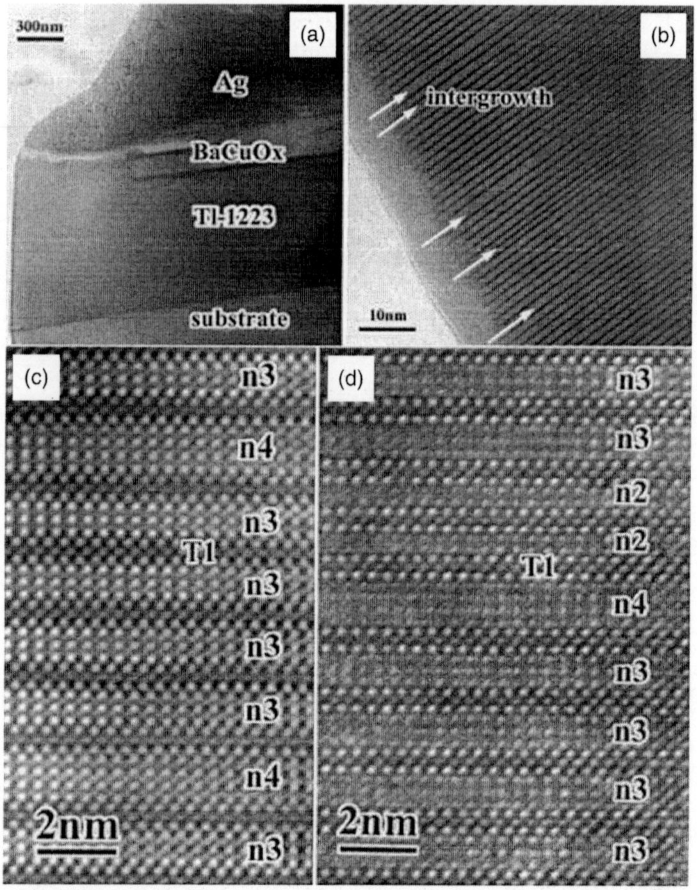

Figure 7.12. TEM images of (a) TBSBCCO-1223 film (1.6 μm, two-layer) in low magnification; (b) TBSBCCO-1223 film (1.6 μm, two-layer) in medium magnification (arrows indicate intergrowth); and (c) and (d) are HRTEM images. [Reproduced with permission from Ref. 103. Copyright 2002 IOP Publishing Limited.]

shows the critical current density versus magnetic field values at 77 K of a two-layer 0.8-μm film and a single-layer 2.6-μm film. At 77 K and no magnetic field, the critical current-density value of the two-layer, 0.8-μm-thick film is 1.1×10^6 A/cm^2 using the field criterion of 1 μV/cm. The superconducting current density of two-layer ED-TBSBCCO improved significantly over single-layer TBSBCCO films.

The transmission electron microscopy (TEM) images of a two-layer Tl-1223 film are shown in Fig. 7.12, which confirms the epitaxial nature of the annealed electrodeposited films. All films showed a significant amount of intergrowth, as shown by a high-resolution transmission electron microscopic (HRTEM) measurement in Figs. 7.12c and d of a representative two-layer

1.6-μm-thick TBSBCCO film. In Figs. 7.12c and d, the $Tl_{1-a}(Bi)_aO$ layer is represented by T1 and the n-layers represent the number of infinite $Ca_{n-1}Cu_nO_y$ layers between two $Ba_{1-b}(Sr)_bO$ layers. Figures 7.12c and d show only one type of $Tl_{1-a}(Bi)_aO$ layer (single layer) but several types of n-layers. In these figures, n2 represents a $CaCu_2O_y$ layer between two $Ba_{1-b}(Sr)_bO$ layers, which results into Tl-1212 in combination with T1; n3 represents a $Ca_2Cu_3O_y$ layer between two $Ba_{1-b}(Sr)_bO$ layers, which results in Tl-1223 in combination with T1; and n4 represents a $Ca_3Cu_4O_2$ layer between two $Ba_{1-b}(Sr)_bO$ layers, which results in Tl-1234 in combination with T1. In summary, Tl-1212 and Tl-1234 film growth was observed along with Tl-1223 in the electrodeposited films. Ag was detected only on the surface of the processed films, not in the substrate interface (Fig. 7.12a), which indicates that even though Ag may initially help to enhance the melt growth phase formation, it eventually migrates to the surface of the partial-melting Tl-oxide superconducting oxide films. In almost all of these films, $BaCuO_x$ phases were also observed as shown in Fig. 7.12a. Figure 7.12b shows clearly the magnitude of intergrowths in diffraction contrast images. An X-ray energy-dispersive spectrometer attached to the microscope performed the qualitative analysis of Ag and $BaCuO_x$.

7.4.2 Electrodeposition of Bi–Sr–Ca–Cu–O

The electrodeposited $Bi_2Sr_2Ca_1Cu_2O_X$ (BSCCO) precursor films were obtained by co-electrodeposition of the constituent metals using nitrate salts dissolved in DMSO solvent. The electrodeposition was performed in a closed-cell configuration at room temperature (~24°C). The cation ratios of the electrodeposition bath were adjusted systematically to obtain BSCCO precursor compositions. A typical electrolyte-bath composition for the BSCCO films consisted of 2.0-g $Bi(NO_3)_3{\cdot}5H_2O$, 1.0-g $Sr(NO_3)_2$, 0.6-g $Ca(NO_3)_2{\cdot}4H_2O$, and 0.9-g $Cu(NO_3)_2{\cdot}6H_2O$ dissolved in 400 mL of DMSO solvent. The substrates were single-crystal LAO coated with 300 Å of Ag.

The electrodeposited precursor films were processed by the melt-quench/melt-growth processing technique. The comparison of omega-scan and phi-scan values obtained from ED Bi-2212 films and pulsed-laser-deposited (PLD) Bi-2212 on LAO substrate showed a similar value or a slightly better value for the ED film compared with the PLD film on LAO. The TEM analysis (Fig. 7.13) of the ED Bi-2212 films on LAO showed phase-pure Bi-2212 film development. The interface shown in Fig. 7.13 between the LAO and the Bi-2212 phase of the processed film is also very clean. Only about 3% by volume stacking faults were observed in the film. At 4.2 K and no magnetic field, the reported critical current-density value of a 1-μm-thick film was 5×10^6 A/cm^2. The electrodeposition process, therefore, produced high-quality biaxially textured Bi-2212 films with promising critical current-density characteristics.

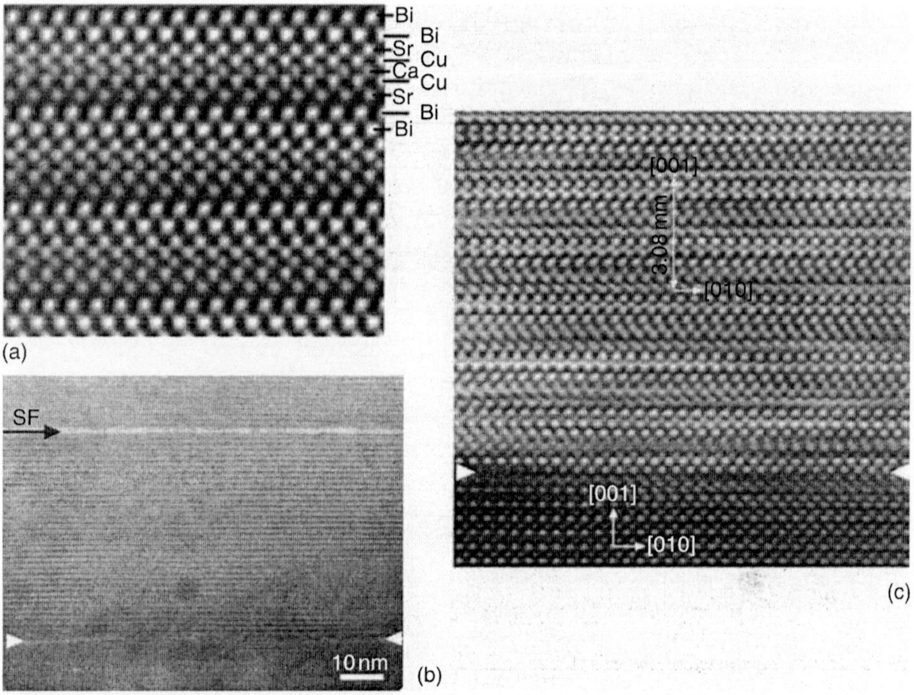

Figure 7.13 TEM analysis of the ED Bi-2212 films on LAO: (a) HRTEM, (b) clean interface, and (c) stacking fault (SF). [Reproduced with permission from Ref. 101. Copyright 2004 IOP Publishing Limited.]

7.5 ELECTRODEPOSITION OF CERIUM OXIDE FILMS

Doped and undoped cerium oxide films were fabricated successfully by electrodeposition.[110,111] Cerium oxide (commonly called ceria) has applications in several key areas of coating technology. Ceria film is a major component in a catalyst used for automobile exhaust pollution control,[112,113] which results from the ease of converting between Ce(+III) and Ce(+IV) oxidation states, thus allowing the release or uptake of oxygen under reducing or oxidizing conditions. In the automobile industry, ceria and doped ceria films are also potential materials for oxygen sensors.[114,115] Based on their excellent chemical stability and transparency, ceria-based films are used as a counter-electrode in electrochromic devices.[116,117] These devices are based on optically active films that can change their refractive index n when subjected to an electrochemical reduction or oxidation reaction.

Epitaxial ceria films grown on (001) silicon substrates are also of interest for microelectronics applications,[118,119] at least in part because the fluorite crystal structure of CeO_2 (space group Fm$\bar{}$3m, cubic lattice parameter a = 5.411 Å) closely matches that of silicon (a = 5.43088 Å). The misfit coefficient ($\Delta a/a$) is only 0.35%. Samarium- and gadolinium-doped ceria have attracted attention as a potential substitute for the solid electrolyte in the yttrium-stabilized zirconium (YSZ) oxide used in solid-oxide fuel-cell (SOFC) applications.[120–122] The high oxygen–ion conductivity and low interfacial losses with the electrodes would allow operating SOFCs at lower temperatures (T = 600 °C–800 °C) than are used in conventional YSZ-based SOFCs. The chemical stability, as well as the excellent lattice match with the second-generation $YBa_2Cu_3O_{7-\delta}$ (YBCO) superconductor, also promotes ceria as an excellent buffer-layer candidate for fabricating microwave devices on single-crystal substrates[123,124] and for manufacturing YBCO-coated conductors on RABiTS (rolling-assisted biaxially textured substrate) Ni-based substrates[125,126] or high-performance alloys such as Hastelloy.[127,128]

$CeO_{2-\delta}$ films have been electrodeposited from an aqueous electrolyte mixture containing cerium halide (0.5 g) in 200 mL of deionized water. $Ce_{1-x}Sm_xO_{2-\delta}$ thin films, with $x = 0.08$ and 0.2, were deposited from an aqueous mixture containing 0.5 g of cerium halide and 0.01 and 0.02 g of samarium halides, respectively. The average pH of the electrolyte was about 4. Electrodeposition was performed at a current density of 0.17 mA/cm^2 and under constant stirring in a vertical two-electrode cell configuration. The average rate of deposition was about 10 nm/min.

As-deposited films analyzed by XRD scans are amorphous/polycrystalline in nature. Postannealing of the precursor films at 910 °C to 980 °C produced biaxially textured single-crystal-like films. The optimum gas mixture that yielded the most-crystalline $CeO_{2-\delta}$ films with ~37-nm particle size was Ar/0.5% H_2. Stoichiometric CeO_2 crystallized in a cubic fluorite-type structure with a lattice parameter a = 5.4115 Å. $Ce_{0.92}Sm_{0.08}O_{2-\delta}$ films were annealed at temperatures ranging from 910 °C to 980 °C for 2 h in a flowing gas mixture of Ar/0.5% H_2. The $Ce_{0.92}Sm_{0.08}O_{2-\delta}$ was mostly ⟨001⟩ oriented, with about 0.5% ⟨111⟩ orientation. The relative intensity of the (002) reflection peak increased with increasing processing temperature. This suggested that the grain size of ceria particles increases with the annealing temperature. The estimation of the grain size calculated by the Scherrer formula as a function of thermal-treatment temperature is shown in Fig. 7.14. The grain size increased from about 18 nm at 910 °C to 35 nm at 980 °C.

The pure ceria films were textured with out-of-plane ($\Delta\omega$) and in-plane (Δ) mosaic dispersions as low as 3.8° (Fig. 7.15a) and 6.1° (Fig. 7.15b), respectively. Although the out-of-plane angular dispersion increased from 3.8° to 5.1° when Sm doping was increased, a significant improvement was achieved in both out-of-plane and in-plane texture when compared with the Ni-based substrates; i.e., $\Delta\omega$ = 7.9° and Δ = 7.6° (Fig. 7.16).

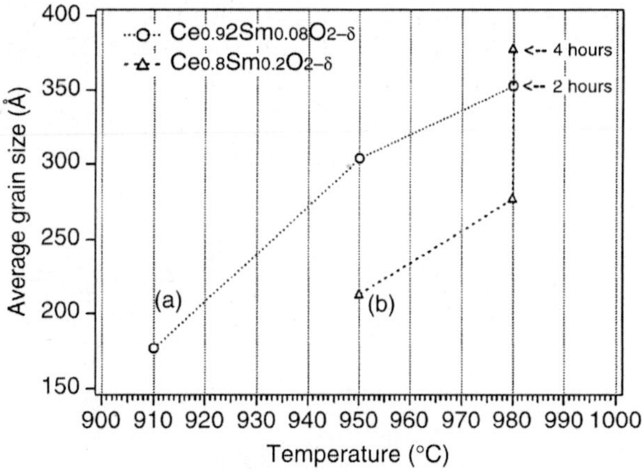

Figure 7.14. The estimation of the CeO_2 grain size calculated by Scherrer formula as a function of temperature. [Reproduced with permission from Ref. 110. Copyright 2006 Wiley-VCH.]

7.6 ELECTRODEPOSITION OF $Gd_2Zr_2O_7$

Electrodeposition was used to prepare a biaxially textured $Gd_2Zr_2O_7$ (GZO) buffer layer on Ni-W substrates.[129] Buffer layers provide chemically inert, continuous, and smooth bases for the growth of the superconductor oxide films. They also prevent both the diffusion of metal to the high-temperature superconductor (HTS) layer and the oxidation of the metal substrate when superconductor oxide films are processed at high temperature (~800 °C) in an oxygen atmosphere (100 ppm or more).

GZO films were electrodeposited from an electrodeposition bath containing 0.29-g gadolinium halide and 0.1-g zirconium halide dissolved in a 150-mL electrolyte solution. Electrodeposition was performed at a current density of 1 mA/cm^2 and under constant stirring in a vertical two-electrode cell configuration. The average rate of deposition was about 25 nm/min.

As-deposited films were amorphous in nature. Postannealing of the precursor films at 980 °C to 1050 °C produced biaxially textured single-crystal-like films. A representative XRD θ/2θ scan of an annealed GZO film is shown in Fig. 7.17 and indicates that only the c-axis-oriented (004) GZO reflection was identified. The omega scan on the (004) reflection (Fig. 7.18) shows that the dispersion along the c-axis of the GZO grains is as low as 5.7°. The omega-scan value of GZO is much better compared with the value of 7.4° of the base Ni–W substrate. The representative pole-figure scans of GZO (Fig. 7.19) show that GZO is biaxially textured. An average in-plane mosaic spread obtained from the phi scan of the GZO film is about 7.2°, as shown in Fig. 7.20, which is also slightly improved compared with the phi-scan value (7.6°) of the base

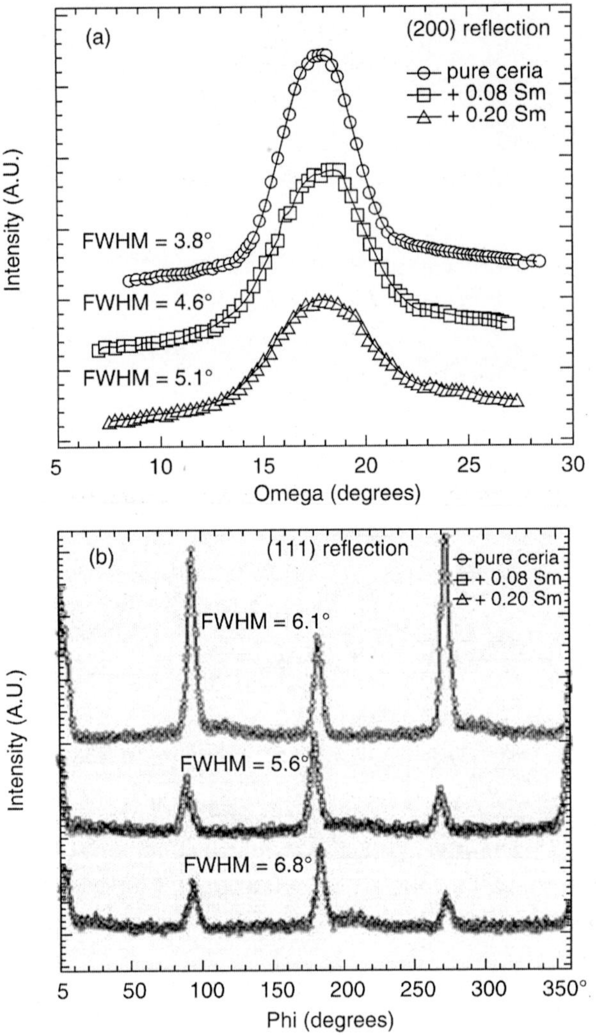

Figure 7.15. (a) XRD out-of-plane (Δω) and (b) XRD in-plane (Δφ) mosaic dispersions of $CeO_{2-\delta}$, $Ce_{0.92}Sm_{0.08}O_{2-\delta}$, and $Ce_{0.8}Sm_{0.2}O_{2-\delta}$. [Reproduced with permission from Ref. 110. Copyright 2006, Wiley-VCH.]

Ni–W substrate. The improved values of omega are attributed to the relatively small grain size of the electrodeposited GZO precursor layers, which showed better grain alignment after annealing. Atomic force microscopy (AFM) images of the GZO layer, shown in Fig. 7.21, reveal that the average layer roughness is about 3 nm and that the film is free of cracks. High-performance YBCO films have been demonstrated on these types of electrodeposited buffer layers. A 300-nm-thick YBCO film showed a high critical current density (J_c) of 3.3 MA/cm^2 at 77 K, in a self-field.

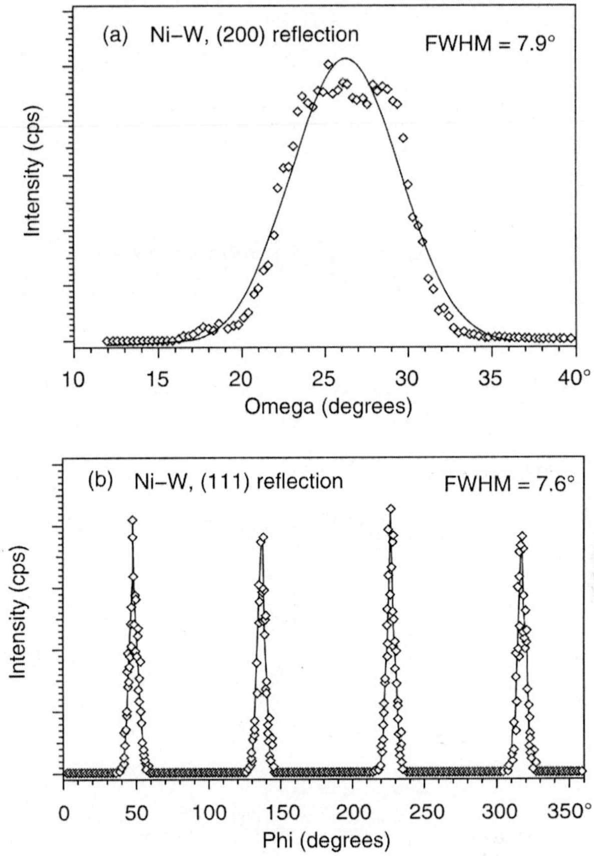

Figure 7.16. (a) XRD out-of-plane (Δω) and (b) XRD in-plane (Δφ) mosaic dispersions of Ni-based substrates. [Reproduced with permission from Ref. 110. Copyright 2006 Wiley-VCH.]

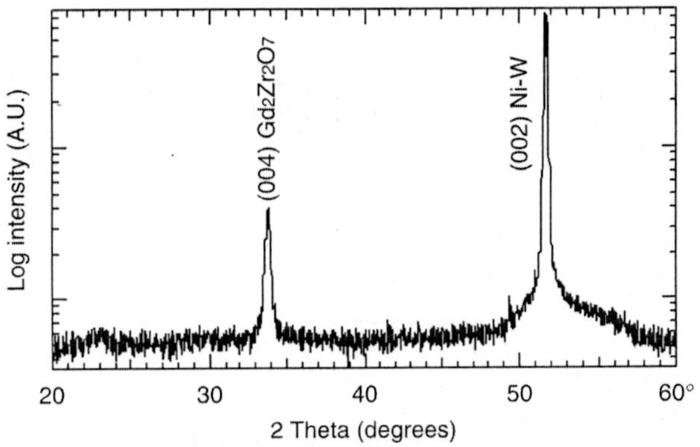

Figure 7.17. X-ray diffractographs of an annealed electrodeposited GZO layer on a Ni–W substrate. [Reproduced with permission from Ref. 129. Copyright 2007 IEEE.]

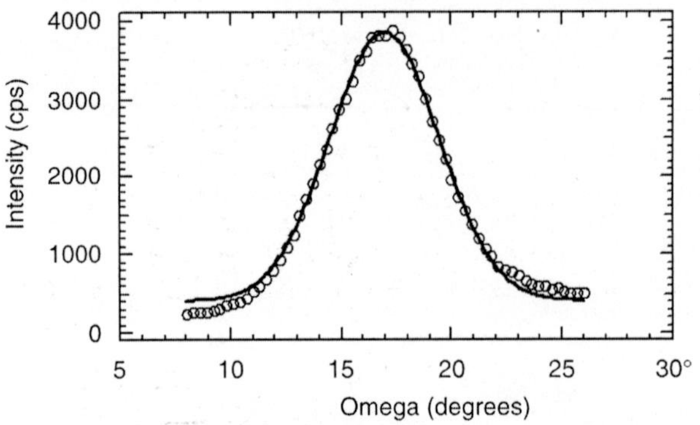

Figure 7.18. XRD omega scan of an annealed electrodeposited GZO layer on a Ni–W substrate. [Reproduced with permission from Ref. 129. Copyright 2007 IEEE.]

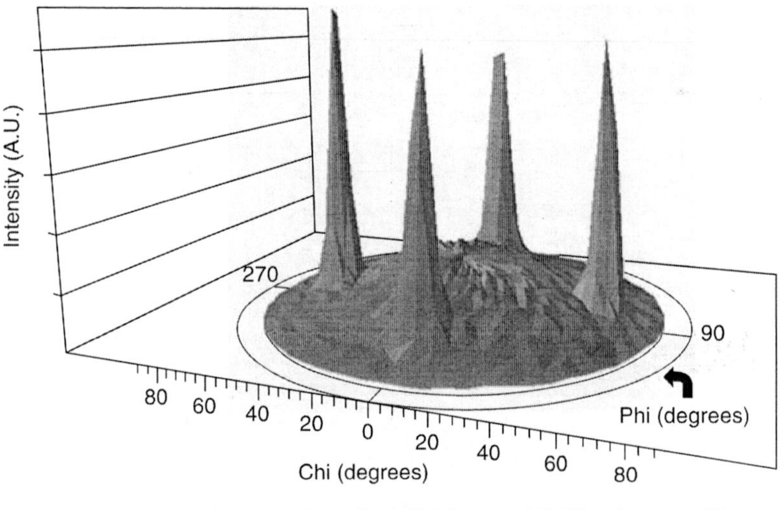

Figure 7.19. Pole figure of electrodeposited GZO on a Ni–W substrate. [Reproduced with permission from Ref. 129. Copyright 2007 IEEE.]

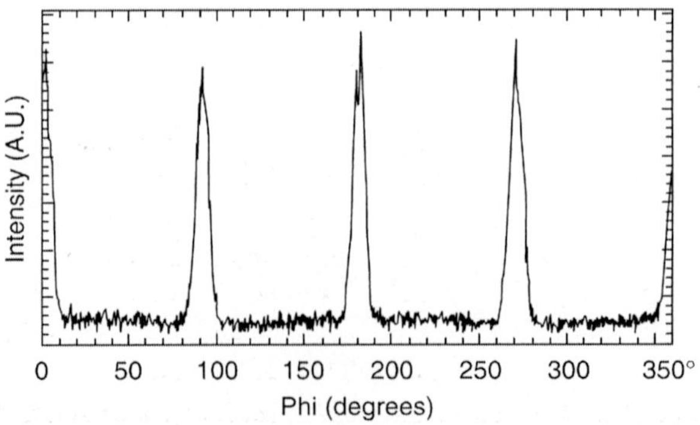

Figure 7.20. XRD phi scans of an annealed electrodeposited GZO layer on a Ni–W substrate. [Reproduced with permission from Ref. 129. Copyright 2007 IEEE.]

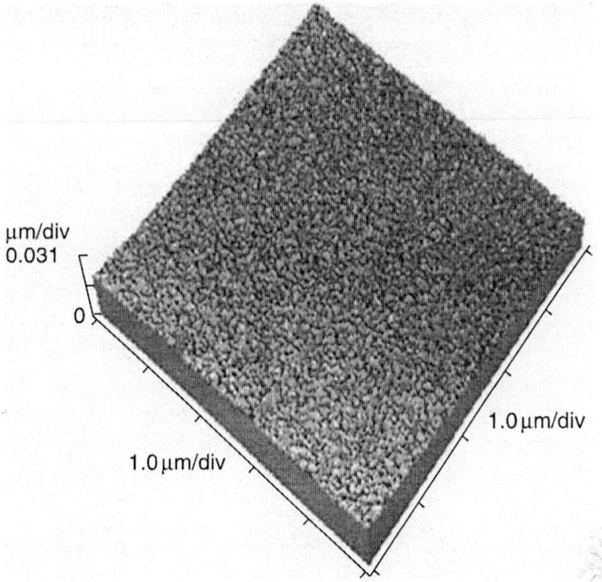

Figure 7.21. AFM micrograph indicating roughness of only 3 nm for an electrodeposited annealed GZO film on a Ni–W substrate. [Reproduced with permission from Ref. 129. Copyright 2007 IEEE.]

REFERENCES

1. Bruckman, G. 1933. Preparation and properties of thin lead sulfide films with special reference to their detector action. *Kolloid-z* 65:1–11.
2. Kitaev, G. A.; Uritskaya, A. A.; Moksushin, S. G. 1965. Conditions for the chemical deposition of thin films of cadmium sulfide on a solid surface. *Russ. J. Phys. Chem.* 39:1101–1102.
3. Chopra, K. L.; Das, S. R. 1983. Solution growth process. In *Thin Film Solar Cells*. Plenum Press, New York. pp. 221–572.
4. Kaur, I.; Pandya, D. K.; Chopra, K. L. 1980. Growth kinetics and polymorphism of chemically deposited CdS films. *J. Electrochem. Soc.* 127:943–948.
5. Pramanik, P.; Bhattacharya, R. N. 1980. A chemical method for deposition of thin film of Bi_2S_3. *J. Electrochem. Soc.* 127:2087.
6. Bhattacharya, R. N.; Pramanik, P. 1981. New chemical methods for the deposition of $Cu_{1.8}S$ and TlSe thin film. *Bull. Mater. Sci.* 3:403–408.
7. Bhattacharya, R. N.; Pramanik, P. 1982. Semiconductor liquid junction solar cell based on chemically deposited Bi_2S_3 thin film and some semiconducting properties of bismuth chalcogenides. *J. Electrochem. Soc.* 129:332–335.
8. Acharya, H. N.; Bose, H. N. 1971. A chemical method of preparing photoconducting lead sulphide films. *Phys. Stat. Sol.* 6(a):K43–K44.
9. Bhattacharya, R. N.; Cahen, D.; Hodes, G. 1984. Electrodeposition of $CuInS_2$ layers and their photoelectrochemical characterization. *Solar Energy Mater.* 10:41–45.

10. Bhattacharya, R. N. 1983. Solution growth and electrodeposited $CuInSe_2$ thin films. *J. Electrochem. Soc.* 130:2040–2042.

11. Pramanik, P.; Bhattacharya, R. N. 1982. Chemical methods for the deposition of thin films of Sb_2Se_3. *J. Solid State Chem.* 44:425.

12. Pramanik, P.; Bhattacharya, R. N.; Mondal, A. 1980. A chemical method for the deposition of thin films of Bi_2Se_3. *J. Electrochem. Soc.* 127:1857–1858.

13. Kessler, J.; Wennerberg, J.; Bodegard, M.; Stolt, L. 2003. Highly efficient Cu(In,Ga)Se_2 mini-modules. *Sol. Energy Mater. Solar Cells* 75:35–46.

14. Basol, B. M.; Kapur, V. K.; Leidholm, C. R.; Halani, A. 1996. Flexible and light weight copper indium diselenide solar cells. *Conference Record of the Twenty-Fifth IEEE Photovoltaic Specialists Conference* (NREL Report No. TP-410-21091). pp. 157–162.

15. Abou-Ras, D.; Kostorz, G.; Romeo, A.; Rudmann, D.; Tiwari, A. N. 2005. Structural and chemical investigations of CBD- and PVD-CdS buffer layers and interfaces in Cu(In,Ga)Se_2-based thin film solar cells. *Thin Solid Films* 480–481:118–123.

16. Kessler, J.; Velthaus, K. O.; Ruckh, M.; Laichinger, R.; Schock, H. W.; Lincot, D.; Ortega, R.; Vedel, J. 1992. Chemical bath deposition of CdS on $CuInSe_2$, etching effects and growth kinetics. *Proceedings of the Sixth International Photovoltaic Science and Engineering Conference (New Delhi, India; Feb. 10–14).* pp. 1005–1010.

17. Perkins, C. L.; Hasoon, F. S.; Al-Thani, H. A.; Asher, S. E.; Sheldon, P. 2005. XPS and UPS investigation of NH_4OH-exposed Cu(In,Ga)Se_2 thin films. *Proceedings of the 31st IEEE Photovoltaics Specialists Conference (Lake Buena Vista, FL; Jan. 3–7).* IEEE, New York. pp. 255–258.

18. Kronik, L.; Burstein, L.; Leibovich, M.; Shapira, Y.; Gal, D.; Moons, E.; Beier, J.; Hodes, G.; Cahen, D.; Hariskos, D.; Klenk, R.; Schock, H.-W. 1995. Band diagram of the polycrystalline CdS/Cu(In,Ga)Se_2 heterojunction. *Appl. Phys. Lett.* 67:1405–1407.

19. Wada, T.; Nishitani, M.; Negami, T.; Kohara, N.; Ikeda, M.; Terauchi, M. 1994. Microstructural characterization of substrate-type and superstrate-type $CuInSe_2$ thin film solar cells. *Proc. 12th European Photovoltaic Solar Energy Conf.* pp. 1542–1545.

20. Furlong, M. J.; Froment, M.; Benard, M. C.; Cortes, R.; Tiwari, A. N.; Krejci, M.; Zogg, H.; Lincot, D. 1998. Aqueous solution epitaxy of CdS layers on $CuInSe_2$. *J. Crystal Growth* 193:114–122.

21. Kitaev, G. A.; Mokrushin, S. G.; Uritskaya, A. A. 1965. Experimental studies of laminar systems. *Colloid. J. USSR* 27:38–41.

22. Froment, M.; Lincot, D. 1995. Phase formation processes in solution at the atomic level: Metal chalcogenide semiconductors. *Electrochim. Acta.* 40:1293–1303.

23. Parfitt, G. D. 1976. Surface chemistry of oxides. *Pure Appl. Chem.* 48:415–418.

24. Nielsen, A. E. 1964. Diffusion controlled growth. *Kinetics of Precipitation.* Pergamon Press, Oxford, UK. p. 34.

25. Lide, D. R. 1993–1994. Solubility product constants. In *CRC Handbook of Chemistry and Physics*, 74th edition. CRC Press, Boca Raton, FL. pp. 8–49.

26. Kylner, A.; Lindgren, J.; Stolt, L. 1996. Impurities in chemical bath deposited CdS films for Cu(In,Ga)Se_2 solar cells and their stability. *J. Electrochem. Soc.* 143:2662–2669.

27. Webb, J. D.; Rose, D. H.; Niles, D. W.; Swartzlander, A.; Al-Jassim, M. M. 1997. FTIR, EPMA, Auger, and XPS analysis of impurity precipitates in CdS films. *Conference Record of the Twenty-Sixth IEEE Photovoltaic Specialists Conference*, (Cat. No.97CB36026). pp. 399–402.

28. Hashimoto, Y.; Nakanishi, T.; Andoh, T.; Ito, K. 1995. CdS thin film deposited in iodide-containing chemical bath. *Jpn. J. Appl. Phys.* 34:L382–L385.

29. Arias-Carbajal Readigos, A.; Garcia, V. M.; Gomezdaza, O.; Campos, J.; Nair, M. T. S.; Nair, P. K. 2000. Substrate spacing and thin-film yield in chemical bath deposition of semiconductor thin films. *Semicond. Sci. Technol.* 15: 1022–1029.

30. Popescu, V.; Pica, E. M.; Pop, I.; Grecu, R. 1999. Optical properties of cadmium sulfide thin films, chemically deposited from baths containing surfactants. *Thin Solid Films* 349:67–70.

31. Froment, M.; Bernard, M. C.; Cortes, R.; Lincot, D.; Ortega-Borges, R.; Mokili, B. Edited by Paunovic, M. 1995. Study of CdS epitaxial films chemically deposited on InP single crystals. *Proceedings of the Second Symposium on Electrochemically Deposited Thin Films*. pp. 121–135.

32. Mereu, B.; Sarau, G.; Pentia, E.; Draghici, V.; Lisca, M.; Botila, T.; Pintilie, L. 2004. Field-effect transistor based on nanometric thin CdS films. *Materials Sci. & Engineer. B* 109:260–263.

33. Meth, J. S.; Zane, S. G.; Sharp, K. G.; Agrawal, S. 2003. Patterned thin film transistors incorporating chemical bath deposited cadmium sulfide as the active layer. *Thin Solid Films* 444:227–234.

34. O'Brien, P.; McAleese, J. 1998. Deposition and characterization of cadmium sulfide thin films by chemical bath deposition. *J. Mater. Chem.* 8:2309–2314.

35. Dona, J. M.; Herrero, J. 1994. Process and film characterization of chemical-bath-deposited ZnS thin films. *J. Electrochem. Soc.* 141:205–210.

36. Vincente-Perez, S.; Durand, J. S.; Alvarez, M. D. 1992. Limitations of complexes logarithmic diagrams as a function of the ligand concentration-diagrams of conditioned variable. *An. Quim.* 88(7–8):683–688.

37. Nakada, T.; Mizutani, M.; Hagiwara, Y.; Kunioka, A. 2001. High-efficiency Cu(In,Ga)Se_2 thin-film solar cells with a CBD-ZnS buffer layer. *Solar Energy Mater. Solar Cells* 67:255–260.

38. (a) Bhattacharya, R. N.; Contreras, M. A.; Teeter, G. 2004. 18.5% CIGS device using single-layer, chemical-bath-deposited ZnS(O,OH). *Jpn. J. Appl. Phys.* 43:L1475–L1476 and (b) Bhattacharya, R. N.; Ramanathan, K.; Gedvilas, L.; Keyes, B. 2005. Cu(In,Ga)Se_2 thin-film solar cells with ZnS(O,OH), Zn–Cd–S(O,OH), and CdS buffer layers. *J. Phys. Chem. Solids* 66:1862–1864.

39. Marcotrigiano, G.; Peyronel, G.; Battistuzzi, R. 1972. Kinetics of desulfuration of S-35-labeled thiourea in sodium hydroxide studied by chromatographic methods. *J. Chem. Soc., Perkin Trans.* 2:1539–1541.

40. Bhattacharya, R. N.; Contreras, M. A.; Egaas, B.; Noufi, R. N.; Kanevce, A.; Sites, J. R. 2006. High efficiency thin-film $CuIn_{1-x}Ga_xSe_2$ photovoltaic cells using a $Cd_{1-x}Zn_xS$ buffer layer. *Appl. Phys. Lett.* 89:253503(1–2).

41. (a) Turnbull, D. 1953. The kinetics of precipitation of barium sulfate from aqueous solution. *Acta Met.* 1:684–691; and (b) O'Rourke, J. D.; Johnson, R. A. 1955. Kinetics and mechanism in formation of slightly soluble ionic precipitates. *Anal. Chem.* 27:1699–1704.

42. Yamada, A.; Miyazaki, H.; Chiba, Y.; Konagai, M. 2005. High-efficiency $Cu(InGa)Se_2$ solar cells with a zinc-based buffer layer. *Thin Solid Films* 480–481:503–508.

43. Mikami, R.; Miyazaki, H.; Abe, T.; Yamada, A.; Konagai, M. 2003. Chemical bath deposited (CBD)-ZnO buffer layer for CIGS solar cells. Edited by Kurokawa, K.; Kazmerski, L. L.; McNelis, B.; Yamaguchi, M.; Wronski, C.; Sinke, W. C. *Proceedings of 3rd World Conference on Photovoltaic Energy Conversion* (IEEE Cat. No. 03CH37497). P1 Vol. 1. pp. 519–522.

44. Hariskos, D.; Ruckh, M.; Ruehle, U.; Walter, T.; Schock, H. W.; Hedstroem, J.; Stolt, L. 1996. A novel cadmium free buffer layer for $Cu(In,Ga)Se_2$ based solar cells. *Solar Energy Mater. Solar Cells* 41–42:345–353.

45. Kim, K. H.; Larina, L.; Yoon, K. H.; Konagai, M.; Ahn, B. T. 2005. Growth of an $In_x(OOH,S)_y$ buffer layer and its application to $Cu(In,Ga)(Se,S)_2$ solar cells. *Mater. Sci. Forum.* 475–479:1681–1684.

46. Larina, L.; Kim, K. H; Yoon, K. H; Konagai, M.; Ahn, B. T. 2003. Thin film CIGS-based solar cells with an In-based buffer layer fabricated by chemical bath deposition. Edited by Kurokawa, K.; Kazmerski, L. L.; McNelis, B.; Yamaguchi, M.; Wronski, C.; Sinke, W. C. *Proceedings of 3rd World Conference on Photovoltaic Energy Conversion* (IEEE Cat. No.03CH37497), Vol. 1. pp. 531–534.

47. Tokita, Y.; Chaisitsak, S.; Miyazaki, H.; Mikami, R.; Yamada, A.; Konagai, M. 2002. Novel $In(OH)_3:Zn^{2+}$ buffer layer for $Cu(InGa)Se_2$ based solar cells. *Jpn. J. Appl. Phys.* 41:7407–7412.

48. Gangopadhyay, U.; Kim, K.; Mangalaraj, D.; Junsin, Y. 2004. Low cost CBD ZnS antireflection coating on large area commercial mono-crystalline silicon solar cells. *Appl. Surf. Sci.* 230:364–370.

49. Ennaoui, A.; Weber, M.; Saad, M.; Harneit, W.; Lux-Steiner, M. Ch.; Karg, F. 2000. Chemical bath deposited $Zn(Se,OH)_x$ on $Cu(In,Ga)(S,Se)_2$ for high efficiency thin film solar cells: Growth kinetics, electronic properties, device performance and loss analysis. *Thin Solid Films* 361–362:450–453.

50. Yamaguchi, T.; Yamamoto, Y.; Tanaka, T.; Yoshida, A. 1999. Preparation and characterization of (Cd,Zn)S thin films by chemical bath deposition for photovoltaic devices. *Thin Solid Films* 343–344:516–519.

51. Rajebnhonsale, M. R.; Pawar, S. H. 1982. Growth and structural properties of $Cd_{1-x}Zn_xS$ films formed by chemical bath deposition technique. *Indian J. Pure and Appl. Phys.* 20(8):652–653.

52. Cheng, J.; Fan, D.; Wang, H.; Liu, B.; Zhang, Y.; Yan, H. 2003. Chemical bath deposition of crystalline ZnS thin films. *Semicond. Sci. Technol.* 18:676–679.

53. Yamaguchi, K.; Yoshida, T.; Lincot, D.; Minoura, H. 2003. Mechanistic study of chemical deposition of ZnS thin films from aqueous solutions containing zinc acetate and thioacetamide by comparison with homogeneous precipitation. *J. Phys. Chem.* B 107:387–397.

54. Aguilar, J. O.; Gomez-Daza, O.; Brito, L. A.; Nair, M. T. S.; Nair, P. K. 2005. Optical and mechanical characteristics of clear and solar control laminated glass using zinc sulphide and copper sulphide thin films. *Surface & Coatings Technol.* 200:2557–2565.

55. Lipowsky, P.; Jia, S.; Hoffmann, R. C.; Jin-Phillipp, N. Y.; Bill, J.; Ruhle, M. 2006. Thin film formation by oriented attachment of polymer-capped nanocrystalline ZnO. *Zeitschrift fur Metallkunde* 97(5):607–613.

56. Hankare, P. P.; Chate, P. A.; Delekar, S. D.; Bhuse, V. M.; Asabe, M. R.; Jadhav, B. V.; Garadkar, K. M. 2006. Structural and opto-electrical properties of molybdenum diselenide thin films deposited by chemical bath method. *J. Cryst. Growth* 291:40–44.

57. Mane, R. S.; Lee, W. J.; Pathan, H. M.; Han, S. H. 2005. Nanocrystalline TiO_2/ZnO thin films: Fabrication and application to dye-sensitized solar cells. *J. Physical Chem.* B 109:24254–24259.

58. Cheng, H.-C.; Chen, C.-F.; Lee, C.-C. 2006. Thin-film transistors with active layers of zinc oxide (ZnO) fabricated by low-temperature chemical bath method. *Thin Solid Films* 498:142–145.

59. Kale, R. B.; Lokhande, C. D. 2005. Influence of air annealing on the structural, morphological, optical and electrical properties of chemically deposited ZnSe thin films. *Appl. Surf. Sci.* 252:929–938.

60. Lokhande, C. D.; Ennaoui, A.; Patil, P. S.; Giersig, M.; Muller, M.; Diesner, K.; Tributsch, H. 1998. Process and characterisation of chemical bath deposited manganese sulphide (MnS) thin films. *Thin Solid Films* 330:70–75.

61. Gumus, C.; Ulutas, C.; Esen, R.; Ozkendir, O. M.; Ufuktepe, Y. 2005. Preparation and characterization of crystalline MnS thin films by chemical bath deposition. *Thin Solid Films* 492:1–5.

62. Fan, D. B; Wang, H.; Zhang, Y. C.; Cheng, J.; Yan, H.; Yoshimura, M. 2004. Direct fabrication of oriented MnS thin films by chemical bath deposition. *Surf. Rev. and Lett.* 11:27–31.

63. Bhattacharya, R. N. 1983. Solution growth and electrodeposited $CuInSe_2$ thin films. *J. Electrochem. Soc.* 130:2040–2042.

64. Pillai, P. K. V.; Vijayakumar, K. P.; Mukherjee, P. S. 1994. Room temperature deposition of $CuInSe_2$ thin films by a chemical method. *J. Mater. Sci. Letters* 13:1725–1726.

65. Valyomana, A. G.; Sajeev, T. P. 1991. Photoconductivity studies of $CuInSe_2$ thin films prepared by the chemical bath deposition technique. *Physica Status Solidi A* 127:K113–K116.

66. Murali, K. R. 1988. Preparation and characterization of chemically deposited $CuInSe_2$ films. *Thin Solid Films* 167:L19–L22.

67. Zeenath, N. A.; Pillai, P. K. V.; Bindu, K.; Lakshmy, M.; Vijakakumar, K. P. 2000. Study of trap levels by electrical techniques in p-type $CuInSe_2$ thin films prepared using chemical bath deposition. *J. Mater. Sci.* 35:2619–2624.

68. Pathan, H. M.; Lokhande, C. D. 2005. Chemical deposition and characterization of copper indium diselenide (CISe) thin films. *Appl. Surf. Sci.* 245:328–334.

69. Dhanam, M.; Balasundaraprabhu, R.; Jayakumar, S.; Gopalakrishnan, P.; Kannan, M. D. 2002. Preparation and study of structural and optical properties of chemical

bath deposited copper indium diselenide thin films. *Physica Status Solidi A* 191: 149–160.

70. Cao, B.; Cai, W.; Li, Y.; Sun, F.; Zhang, L. 2005. Ultraviolet-light-emitting ZnO nanosheets prepared by a chemical bath deposition method. *Nanotechnol.* 16:1734–1738.
71. Ganchev, M.; Stratieva, N.; Tzvetkova, E.; Gadjov, I. 2003. Kinetics of the chemical bath deposition of ZnSe films. *J. Mater. Sci.: Materials in Electronics* 14:847–848.
72. Bindu, K.; Lakshmi, M.; Bini, S.; Sudha Kartha, C.; Vijayakumar, K. P.; Abe, T.; Kashiwaba, Y. 2002. Amorphous selenium thin films prepared using chemical bath deposition: Optimization of the deposition process and characterization. *Semicond. Sci. Technol.* 17:270–274.
73. Bini, S.; Bindu, K.; Lakshmi, M.; Sudha Kartha, C.; Vijayakumar, K. P.; Kashiwaba, Y.; Abe, T. 2000. Preparation of $CuInS_2$ thin films using CBD Cu_xS films. *Renewable Energy* 20:405–413.
74. Nair, P. K.; Nair, M. T. S.; Garcia, V. M.; Arenas, O. L.; Pena, Y.; Castillo, A.; Ayala, I. T.; Gomezdaza, O.; Sanchez, A.; Campos, J.; Hu, H.; Suarez, R.; Rincon, M. E. 1998. Semiconductor thin films by chemical bath deposition for solar energy related applications. *Solar Energy Mater. Solar Cells* 52:313–344.
75. Ito, K.; Nakamura, K. 1996. Preparation of ZnO thin films using the flowing liquid film method. *Thin Solid Films* 286:35–36.
76. Ullal, H. S.; Zweibel, K.; Roedern, B. V. 1997. Current status of polycrystalline thin-film PV technologies. *Proceedings of the 26th IEEE PVSC (Anaheim, CA).* pp. 301–305.
77. Lincot, D.; Guillemoles, J. F.; Taunier, S.; Guimard, D.; Sicx-Kurdi, J.; Chaumont, A.; Roussel, O.; Ramdani, O.; Hubert, C.; Fauvarque, J. P.; Bodereau, N.; Parissi, L.; Panheleux, P.; Fanouillere, P.; Naghavi, N.; Grand, P. P.; Benfarah, M.; Mogensen, P.; Kerrec, O. 2004. Chalcopyrite thin film solar cells by electrodeposition. *Solar Energy* 77:725–737.
78. (a) Bhattacharya, R. N.; Batchelor, W.; Hiltner, J. F.; Sites, J. R. 1999. Thin-film $CuIn_{1-x}Ga_xSe_2$ photovoltaic cells from solution-based precursor layers. *Appl. Phys. Lett.* 75:1431–1433 and (b) Bhattacharya, R. N.; Batchelor, W.; Ramanathan, K.; Contreras, M. A.; Moriarty, T. 2000. The performance of $CuIn_{1-x}Ga_xSe_2$-based photovoltaic cells prepared from low-cost precursor films. *Solar Energy Mater. Solar Cells* 63:367–374.
79. Qui, S. N.; Li, L.; Qui, C. X.; Shih, I.; Champness, C. H. 1995. Diffusion length measurements on electrodeposited $CuInSe_2$ cells. *Solar Energy Mater. Solar Cells* 37:389–393.
80. Guillemoles, J. F.; Cowache, P.; Lusson, A.; Fezzaa, K.; Boisivon, F.; Vedel, J.; Lincot, D. 1996. One step electrodeposition of $CuInSe_2$: Improved structural, electronic, and photovoltaic properties by annealing under high selenium pressure. *J. Appl. Phys.* 79:7293–7302.
81. Sudo, Y.; Endo, S.; Irie, T. 1993. Preparation and characterization of electrodeposited $CuInSe_2$ thin films. *Jpn. J. Appl. Phys.* 32:1562–1567.
82. Matsuoka, T.; Nagahori, Y.; Endo, S. 1994. Preparation and characterization of electrodeposited $CuGa_xIn_{1-x}Se_2$ thin films. *Jpn. J. Appl. Phys.* 33:6105–6110.

83. Kapur, V. K.; Basol, B. M.; Tseng, E. S. 1987. Low cost methods for the production of semiconductor films for $CuInSe_2$/CdS solar cells. *Solar Cells* 21:65–72.

84. Oliveira, M. C. F.; Azevedo, M.; Cunha, A. 2002. A voltammetric study of the electrodeposition of $CuInSe_2$ in a citrate electrolyte. *Thin Solid Films* 405: 129–134.

85. Kemell, M.; Saloniemi, H.; Ritala, M.; Leskela, M. 2001. Electrochemical quartz crystal microbalance study of the electrodeposition mechanisms of $CuInSe_2$ thin films. *J. Electrochem. Soc.* 148:C110–C118.

86. Guillen, C.; Martinez, M. A.; Herrero, J. 2000. $CuInSe_2$ thin films obtained by a novel electrodeposition and sputtering combined method. *Vacuum* 58:594–601.

87. De Silva, K. T. L.; Priyantha, W. A. A.; Jayanetti, J. K. D. S.; Chithrani, B. D.; Siripala, W.; Blake, K.; Dharmadasa, I. M. 2001. Electrodeposition and characterisation of $CuInSe_2$ for applications in thin film solar cells. *Thin Solid Films* 382:158–163.

88. Wenyi, L.; Jian, Y.; Xuebin, Y.; Qiulong, C.; Xun, C. 2000. Preparation of $CuInSe_2$ film with electrodeposition. *Proceedings of the SPIE—The International Society for Optical Engineering*, Vol. 4086. pp. 217–220.

89. Kampmann, A.; Sittinger, V.; Rechid, J.; Reineke-Koch, R. 2000. Large area electrodeposition of $Cu(In,Ga)Se_2$. *Thin Solid Films* 361–362:309–313.

90. Bhattacharya, R. N.; Rajeshwar, K. 1986. Electrodeposition of CuInX (X = Se, Te) thin films. *Solar Cells* 16:237–243.

91. Friedfeld, R.; Raffaelle, R. P.; Mantovani, J. G. 1999. Electrodeposition of $CuIn_xGa_{1-x}Se_2$ thin films. *Solar Energy Mater. Solar Cells* 58:375–385.

92. Mehlin, M.; Rammasch, J.; Fritz, H. P. 1994. Preparation of $CuGaSe_2$ thin-film solar-cells comprising an electrochemical gallium deposition step. *Zeitschrift Fur. Naturforschung* 49b:1597–1605.

93. Ugarte, R.; Schrebler, R.; Cordova, R.; Dalchiele, E. A.; Gomez, H. 1999. Electrodeposition of $CuInSe_2$ thin films in a glycine acid medium. *Thin Solid Films* 340:117–124.

94. Bhattacharya, R. N.; Batchelor, W.; Ramanathan, K.; Contreras, M. A.; Moriarty, T. 2000. The performance of $CuIn_{1-x}Ga_xSe_2$-based photovoltaic cells prepared from low-cost precursor films. *Solar Energy Mater. Solar Cells* 63:367–374.

95. Bhattacharya, R. N.; Balcioglu, A.; Ramanathan, K. 2001. Deep-level transient spectroscopy (DLTS) of $CdS/CuIn_{1-x}Ga_xSe_2$-based solar cells prepared from electroplated and auto-plated precursors, and by physical vapor deposition. *Thin Solid Films* 384:65–68.

96. Lang, D. V. 1974. Deep-level transient spectroscopy: A new method to characterize traps in semiconductors. *J. Appl. Phys.* 45:3023–3032.

97. Abou-Elfotouh, F. A.; Moutinho, H.; Bakry, A.; Coutts, T. J.; Kazmerski, L. L. 1991. Characterization of the defect levels in copper indium diselenide. *Solar Cells* 30:151–160.

98. Ahrenkiel, R. K. 1989. The effect of deep states on the photovoltaic performance of $CdZnS/CuInSe_2$ thin film devices. *Solar Cells* 16:549–565.

99. Herberholz, R.; Igalson, M.; Schock, H. W. 1998. Distinction between bulk and interface states in $CuInSe_2$/CdS/ZnO by space charge spectroscopy. *J. Appl. Phys.* 83:318–325.

100. Kuranouchi, S.; Konagai, M. 1995. Characterization of $ZnO/CdS/CuInSe_2$ thin-film solar cells by deep-level transient spectroscopy. *Jpn. J. Appl. Phys.* 34:2350–2351.

101. Bhattacharya, R. N.; Chen, J.; Spagnol, P.; Huang, J. Y.; Ren, Z. F. 2004. Superconductor Bi-oxide films via an electrodeposition process. *Supercond. Sci. Technol.* 17:120–124.

102. Chen, J.; Bhattacharya, R. N. 2003. Growth of 1–2 μm thick biaxially textured Bi-2212 films on (1 0 0) $LaAlO_3$ single crystal substrates by electrodeposition. *Physics C* 399:171–177.

103. Bhattacharya, R. N.; Banerjee, D.; Wen, J. G.; Padmanabhan, R.; Wang, Y. T.; Chen, J.; Ren, Z. F.; Hermann, A. M.; Blaugher, R. D. 2002. Structural studies of electrodeposited and sprayed thallium-oxide films. *Supercond. Sci. Technol.* 15:1288–1294.

104. Blaugher, R. D.; Bhattacharya, R. N.; Chen, J.; Padmanabhan, R. 2002. Alternative HTS coated conductors. *Physica* C 382:72–79.

105. Bhattacharya, R. N.; Xing, Z.; Wu, J. Z.; Chen, J.; Yang, S. X.; Ren, Z. F.; Blaugher, R. D. 2002. Superconducting thallium oxide and mercury oxide films. *Physica* C 377:327–332.

106. Bhattacharya, R. N.; Wu, H. L.; Wang, Y.-T.; Blaugher, R. D.; Yang, S. X.; Wang, D. Z.; Ren, Z. F.; Tu, Y.; Verebelyi, D.T.; Christen, D. K. 2000. Improved electrodeposition process for the preparation of superconducting thallium oxide films. Improved electrodeposition process for the preparation of superconducting thallium oxide films. *Physica* C 333:59–64.

107. Bhattacharya, R. N.; Feldmann, M.; Larbalestier, D.; Blaugher, R. D. 2001. Electrodeposition process for the preparation of superconducting thallium oxide films. *IEEE Trans. Appl. Supercond.* 11:3102–3105.

108. Bhattacharya, R. N.; Blaugher, R. D.; Ren, Z. F.; Li, W.; Wang, J. H.; Paranthaman, M.; Verebelyi, D. T.; Christen, D. K. 1998. Superconducting thallium oxide films by electrodeposition method. *Physica* C 304:55–65.

109. Bhattacharya, R. N.; Blaugher, R. D.; Ren, Z. F.; Li, W.; Wang, J. H.; Paranthaman, M.; Verebelyi, D. T.; Christen, D. K. 1998. Superconducting epitaxial $(TlBi)_{0.9}Sr_{1.6}Ba_{0.4}Ca_2Cu_3Ag_{0.2}O_x$ film from an electrodeposited precursor. *Electrochemi. Solid-State Lett.* 1:165–167.

110. Phok, S.; Bhattacharya, R. N. 2006. Effect of samarium doping on electrodeposited CeO_2 thin film. *Physica Status Solidi (a)* 203:3734–3742.

111. Bhattacharya, R. N.; Phok, S.; Spagnol, P.; Chaudhuri, T. 2006. Electrodeposited biaxially textured buffer layer for $YBa_2Cu_3O_{7-\delta}$ (YBCO) superconductor oxide films. *J. Electrochem. Soc.* 153:C273–C276.

112. Trovarelli, A. 1996. Catalytic properties of CeO_2 and CeO_2-containing materials. *Catal. Rev. Sci. Eng.* 38:439–520.

113. Rodriguez, J. A.; Jirsak, T.; Freitag, A.; Hanson, J. C.; Larese, J. Z.; Chaturvedi, S. 1999. Interaction of SO_2 with CeO_2 and Cu/CeO_2 catalysts: Photoemission, XANES and TPD studies. *Catal. Lett.* 62:113–119.

114. Izu, N.; Murayama, N.; Shin, W.; Matsubara, I.; Kanzaki, S. 2004. Resistive oxygen sensors using cerium oxide thin films prepared by metal organic chemical vapor deposition and sputtering. *Jpn. J. Appl. Phys.* 43:6920–6924.

115. Beie, H. J.; Gnorich, A. 1991. Oxygen gas sensors based on CeO_2 thick and thin films. *Sens. Actuators B* 4:393–399.

116. Porqueras, I.; Person, C.; Corbella, C.; Vives, M.; Pinyol, A.; Bertran, E. 2003. Characteristics of e-beam deposited electrochromic CeO_2 thin films. *Solid State Ionics* 165:131–137.

117. Ozer, N. 2001. Optical properties and electrochromic characterization of sol-gel deposited ceria films. *Solar Energy Mater. Solar Cells* 68:391–400.

118. Inoue, T.; Yamamoto, Y.; Koyama, S.; Suzuki, S.; Ueda, Y. 1990. Epitaxial growth of CeO_2 layers on silicon. *Appl. Phys. Lett.* 56:1332–1333.

119. Nishikawa, Y.; Yamaguchi, T.; Yoshiki, M.; Satake, H.; Fukushima, N. 2002. Interfacial properties of single-crystalline CeO_2 high-k gate dielectrics directly grown on Si (111). *Appl. Phys. Lett.* 81:4386–4388.

120. Yahiro, H.; Eguchi, Y.; Eguchi, K.; Arai, H. 1988. Oxygen ion conductivity of the ceria samarium oxide system with fluorite structure. *J. Appl. Electrochem.* 18:527–531.

121. Inaba, H.; Tagawa, H. 1996. Ceria-based solid electrolytes. *Solid State Ionics* 83:1–16.

122. Peng, R.; Xia, C.; Peng, D.; Meng, G. 2004. Effect of powder preparation on $(CeO_2)_{0.8}(Sm_2O_3)_{0.1}$ thin film properties by screen-printing. *Mater. Lett.* 58:604–608.

123. Schneider, R.; Aidam, R.; Zaitsev, A.; Geerk, J.; Linker, G.; Ratzel, F.; Smithey, R. 2001. Resonators and filters made of YBaCuO thin films on sapphire wafers. *Physica* C 351:21–24.

124. Lorenz, M.; Hochmuth, H.; Natusch, D.; Lippold, G.; Svetchnikov, V. L.; Kaiser, T.; Hein, M. A.; Schwab, R.; Heidinger, R. 1999. Ag-doped double-sided PLD-YBCO thin films for passive microwave devices in future communication systems. *IEEE Trans. Appl. Supercond.* 9:1936–1939.

125. Kim, S.-M.; Song, S. C.; Lee, S. Y. 2001. Effect of CeO_2, $BaTiO_3$ and $CeO_2/BaTiO_3$ double buffer layers on the superconducting properties of $Y_1Ba_2Cu_3O_{7-x}$ grown on metallic substrates by pulsed laser deposition. *Physica* C 351:379–385.

126. Thiele, K.; Jooss, Ch.; Hoffmann, J.; Kautschor, L. O.; Dzick, J.; Freyhardt, H. C. 2001. Grain boundaries in $YBa_2Cu_3O_{7-\delta}$ films grown on bicrystalline Ni substrates. *Physica* C 355:203–210.

127. Develos-Bagarinao, K.; Yamasaki, H.; Nakago, Y.; Obara, H.; Yamada, H. 2003. Microcrack-free thick $YBCO/CeO_2/Al_2O_3$ films prepared by a large-area pulsed laser deposition system. *Physica C* 392–396:1229–1235.

128. Takahashi, Y.; Araki, T.; Yamagiwa, K.; Yamada, Y.; Kim, S. B.; Iijima, Y.; Takeda, K.; Hirabayashi, I. 2001. Preparation of YBCO films on CeO_2 buffered metallic substrates by the TFA-MOD method. *Physica C* 357–360:1003–1006.

129. Bhattacharya, R. N.; Phok, S.; Xu, Y.; Bhattacharya, R. 2007. Electrodeposited biaxially textured buffer layers for YBCO superconductors. *IEEE Transaction of ASC, Seattle*, 17. pp. 3321–3324.

CHAPTER 8

Successive Ionic Layer Adsorption and Reaction (SILAR) and Related Sequential Solution-Phase Deposition Techniques

SEPPO LINDROOS and MARKKU LESKELÄ
Department of Chemistry, University of Helsinki, Helsinki, Finland

8.1 INTRODUCTION

Thin films for electronic applications have been grown by solution-phase techniques for many years. The solution-phase techniques are characterized by simple setups, low price, and low temperature. There are disadvantages as well, however, including the environmental impact of the solvents employed in the process and the generally lower film quality, compared with analogous films made at higher temperatures. The oldest and even today most commonly used solution growth technique is chemical bath deposition (CBD). In the CBD technique, the thin film is grown in a reaction vessel containing simultaneously all the precursors needed for the deposition (see Chapter 7). The successive ionic layer adsorption and reaction (SILAR) technique can be regarded as a modification of CBD. In SILAR, the thin-film growth occurs sequentially. The precursor solutions are placed in different reaction vessels and the substrate is immersed in turn in the vessel containing the cation precursor and the vessel containing the anion precursor. In between the cation and anion immersions, the substrate is thoroughly rinsed with purified solvent (i.e., normally water). Through repetition of these cycles, thin films are grown layer by layer.

The complete SILAR process was first described by Nicolau in 1985.[1–3] Since then it has mostly been used to grow oxide and chalcogenide thin films.

Several review articles have been published about SILAR-grown films.[4–7] The SILAR technique, including its advantages and disanvantages and the equipment employed, is presented in Section 8.2. Materials that have been prepared by SILAR are reviewed in Section 8.3. Short descriptions of the related ILGAR, ECALE, and other sequential solution-phase techniques follow in Sections 8.4– 8.6.

8.2 SILAR

8.2.1 Basic Principles of SILAR

Film growth in the SILAR technique proceeds by alternate immersion of the substrate in the cation and anion precursor solutions, with intermediate rinsing of the substrate. Thus, one growth cycle consists of four steps, as described in Fig. 8.1.

In the first step, the substrate is immersed in the cation solution and the ions of the precursor solution are adsorbed onto the surface of the substrate. There is an oxide layer on the surface of most substrate materials, including metal oxide substrates and glass or silicon. This oxide layer and the surface anion layer in metal sulphides is effectively negatively charged in a water solution with pH above ~2.[8] Upon immersion in the cation solution, a double layer is formed where cations form an inner layer and the anions of the cation precursor provide a charge-balancing outer layer. In the following step, the substrate is rinsed with purified water so that only the immobile double layer, which is tightly stuck on the surface, remains on the substrate surface. In the next reaction step, the substrate is immersed in the anion precursor solution.

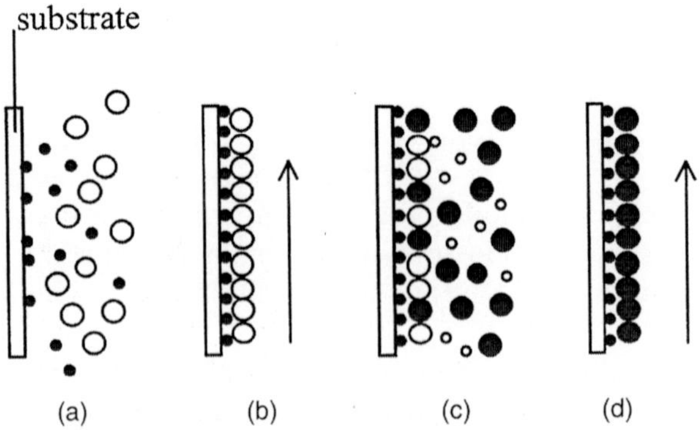

Figure 8.1. Schematic view of thin-film growth during one SILAR cycle. (a) cation adsorption: • cation ○ anion; (b) rinsing; (c) anion reaction ● anion ∘ cation; and (d) rinsing. Reprinted with permission from Lindroos 1997.

The ions of the solution are diffused into the surface and react with the adsorbed cations of the inner layer. The first solid thin-film layer has now been formed on the substrate surface. Proceeding in this way, the growth of the thin film occurs only on the surface of the substrate. In a final step, the substrate is rinsed to remove the ions from the diffusion layer. A second growth cycle can now commence. The overall reaction occuring during one SILAR cycle can be presented as follows:

$$mM^{n+}(aq) + nA^{m-}(aq) \rightarrow M_mA_n(s) \quad (1)$$

The factors affecting thin-film growth in the SILAR process include the concentrations, pH, and counter-ions of the precursor solutions. In the deposition of ZnS films, for example, an increase in the zinc ion concentration has been observed to increase the growth rate up to a saturation point where total coverage of the surface is achieved. Use of sulfate instead of chloride as the counter-ion for zinc clearly reduced the growth rate. The reason is that at pH 5.5, which was used in the ZnS deposition, the concentrations of monovalent zinc species in the $ZnCl_2$ solution, $ZnCl^+$ and $Zn(OH)^+$, are higher than the concentration of $Zn(OH)^+$ in the $ZnSO_4$ solution. Monovalent cations have been shown to adsorb more readily than divalent cations.[1]

The cation precursor can also be complexed by an organic ligand, which enables the choice of cation species by adjusting the pH of the solution.[9,10] In complexation, the metal ions are ideally situated in one chemical environment. Therefore, the deposition reaction is in better control than in uncomplexed reactions, where several different chemical species may exist. Also, the size of the counter-ion of the cation precursor may have an effect on the growth rate, as was observed in the deposition of PbS.[10] Proper pretreatment of the substrate has a notable role for good attachment of the thin film to substrate as has been observed in deposition of ZnS and PbS films by SILAR.[9,10] Temperature may also have a role in film growth because of the higher mobility of ions in the precursor solutions and a more efficient rinsing step.

8.2.2 Advantages and Disadvantages of SILAR

The possibility to grow good-quality thin films at room temperature and normal pressure is the main advantage of SILAR relative to the gas-phase techniques. Moreover, since vacuum systems are not required, SILAR deposition equipment is simple and inexpensive. Similarly, toxic chemicals, such as selenium compounds, which are easier and safer to handle as solutions than as gases, can be more conveniently employed in SILAR. From the environmental point of view, a notable advantage of SILAR is that the system is totally closed and all the chemicals that are used are recyclable. Compared with other solution-phase methods, especially with CBD, an important advantage of SILAR is the facile control over film thickness using the number of deposition cycles. In principle, relatively thick films may be grown by SILAR without a change in the morphology or other properties of the film. However, most likely the industrial use of SILAR will be concentrated on growing well-defined very thin films, from some nanometers to tens of nanometers.

There are also several disadvantages of the SILAR process relative to other deposition techniques, including, most notably, that it is slower than most vapor-phase techniques. The typical increment of film thickness in one deposition cycle is 1–3 Å, which is higher than normally achieved in atomic layer deposition (ALD), the gas-phase equivalent of SILAR. However, the much longer cycle time, because of the long rinsing step needed, means that films are much slower to prepare by SILAR than by ALD. On the other hand, since SILAR is not yet used industrially, the process times have not been fully optimized. To some extent, the slow growth rate of SILAR can be compensated by using large-area substrates and simultaneously a large number of substrates. Another disadvantage of the SILAR process has been the fairly high oxygen contamination of chalcogenide films (e.g., ZnS) deposited using the process, owing to the aqueous deposition conditions. The instability of some important substrate materials (Si, Al_2O_3) toward the aqueous and often alkaline precursor solutions has also been a drawback of the SILAR technique.

8.2.3 SILAR Deposition Equipment

The first type of equipment for SILAR was that described by Nicolau,[1] who arranged alternating reaction and rinsing vessels in a circle (Fig. 8.2). During the deposition, the vessels and the substrate are placed under a hood in a nitrogen atmosphere and the substrate is lifted and rotated from one vessel to another by two electric motors. The rinsing water is circulated from a reservoir to the rinsing vessels with the aid of a peristaltic pump.[1,9]

The second type of SILAR equipment was developed by Kanniainen and Lindroos.[9,10] The equipment is controlled by a computer, with the substrate being moved from reaction to rinsing vessel by a laboratory xyz robot. The installation also allows the movement of the substrate in the rinsing and reaction vessels, which promotes film growth and rinsing. A schematic drawing of the equipment is shown in Fig. 8.3.[9,10]

Ultrasonic–assisted rinsing has recently been introduced to SILAR to shorten the rinsing time and enhance the effectiveness of the rinsing step. A commercial ultrasonic bath with a frequency of 50 kHz and a power of 100 W has been used.[11]

A method in which the precursor solutions are successively injected into a cell containing the substrate and rinsed in between has been used to analyze the morphology of SILAR-grown films by atomic force microscopy (AFM).[10] Recently, this approach has been applied to the growth of core/shell nanocrystals by Li et al.[12]

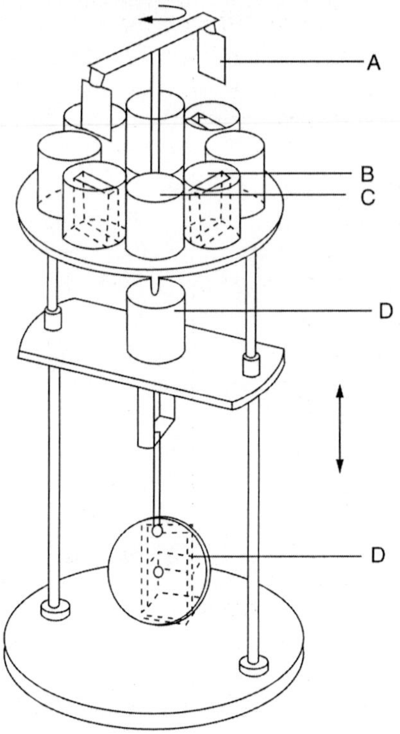

Figure 8.2. First version of the SILAR deposition equipment: (a) substrate, (b) rinsing vessel, (c) precursor vessel, and (d) electric motors to rotate and lift the substrate. Reprinted with permission from Lindroos 1997.

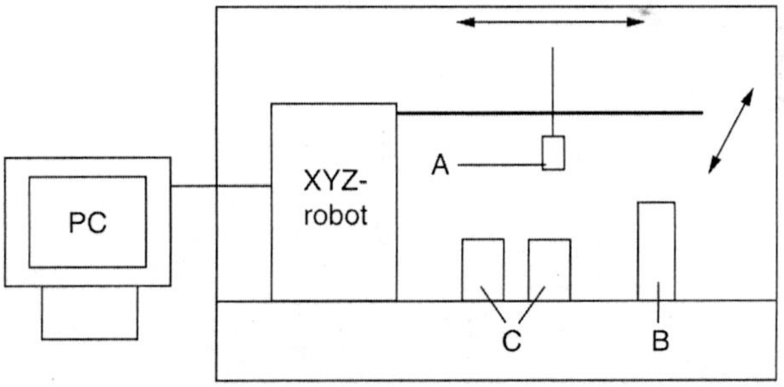

Figure 8.3. Schematic drawing of modified equipment for SILAR deposition: (a) substrate, (b) rinsing vessel, and (c) reaction vessels. Reprinted with permission from Lindroos 1997.

8.2.4 Mechanism of Film Growth in SILAR

The growth mechanisms of two typical materials, ZnS and PbS, have been studied by AFM by both *in situ* and *ex situ* techniques.[9,10,13–17] An AFM liquid cell was used for the *in situ* studies for comparison with films grown with the SILAR equipment.

The *in situ* measurements revealed that ZnS film growth on a mica substrate is three-dimensional from the beginning, with the rms roughness of the film increasing steadily from the first cycles onward.[15] ZnS SILAR growth on a soda lime glass substrate is different.[13] A glass surface is much rougher than a mica surface, and the deposition of ZnS occurs in the valleys of the glass surface. This was clearly evident in the *in situ* AFM images, with the rms roughness value also clearly decreasing with ZnS deposition. The *ex situ* images on soda lime glass substrate showed the ZnS film to be very flat up to 50 cycles. After this number of cycles, the rms roughness increased rapidly, indicating a change of the film growth mode from two-dimensional to island or Stranski–Krastanov type.[10,13] After 200 cycles, the ZnS film began to flatten from the coalescence of the islands or from more favorable growth of ZnS in the valleys between the tops of the ZnS islands. At this point in the film growth, the whole surface of the substrate was covered by zinc sulfide and there was no uncovered glass surface on which the growth of ZnS could be less favorable than on the already deposited ZnS. The SILAR growth of ZnS on GaAs was different.[16] In the early stages, the growth was three-dimensional, possibly because of the slight dissolution of the substrate material. However, after this point, the growth was mainly two-dimensional and the film roughness remained low. On (100)Si, in turn, the rms roughness increased steadily, reaching 14 nm after 100 cycles (Fig. 8.4).[17]

Atomic force microscopy has also been applied to study the development of PbS film morphology on a glass substrate.[10,14] The rms roughness increased rapidly during the first cycles. Judging from contact force AFM and scanning electron microscopy (SEM) images, the film was formed as separate islands up to a thickness of 30–40 nm. However, lateral force microscopy (LFM) and force modulation (FM) revealed film formations after just 50 cycles (~2-nm nominal thickness), which indicates full surface coverage. The growth of PbS thus follows the Stranski–Krastanov mode.[10,14]

8.3 MATERIALS GROWN BY SILAR

Oxide and chalcogenide films are the primary materials that have been produced by the SILAR technique. A few metal films and films of other materials have also been produced.

8.3.1 Oxide Films

Two basic methods have been used to grow metal oxide thin films by the SILAR technique (see Table 8.1). The more common of these methods consists of the adsorption of metal hydroxide ions on the substrate surface followed by thermal treatment to convert hydroxide to an oxide. Another way to produce metal oxide films is to use hydrogen peroxide as the anion precursor and then to convert the formed metal peroxide film to an oxide film. Several examples of each approach are discussed in more detail below.

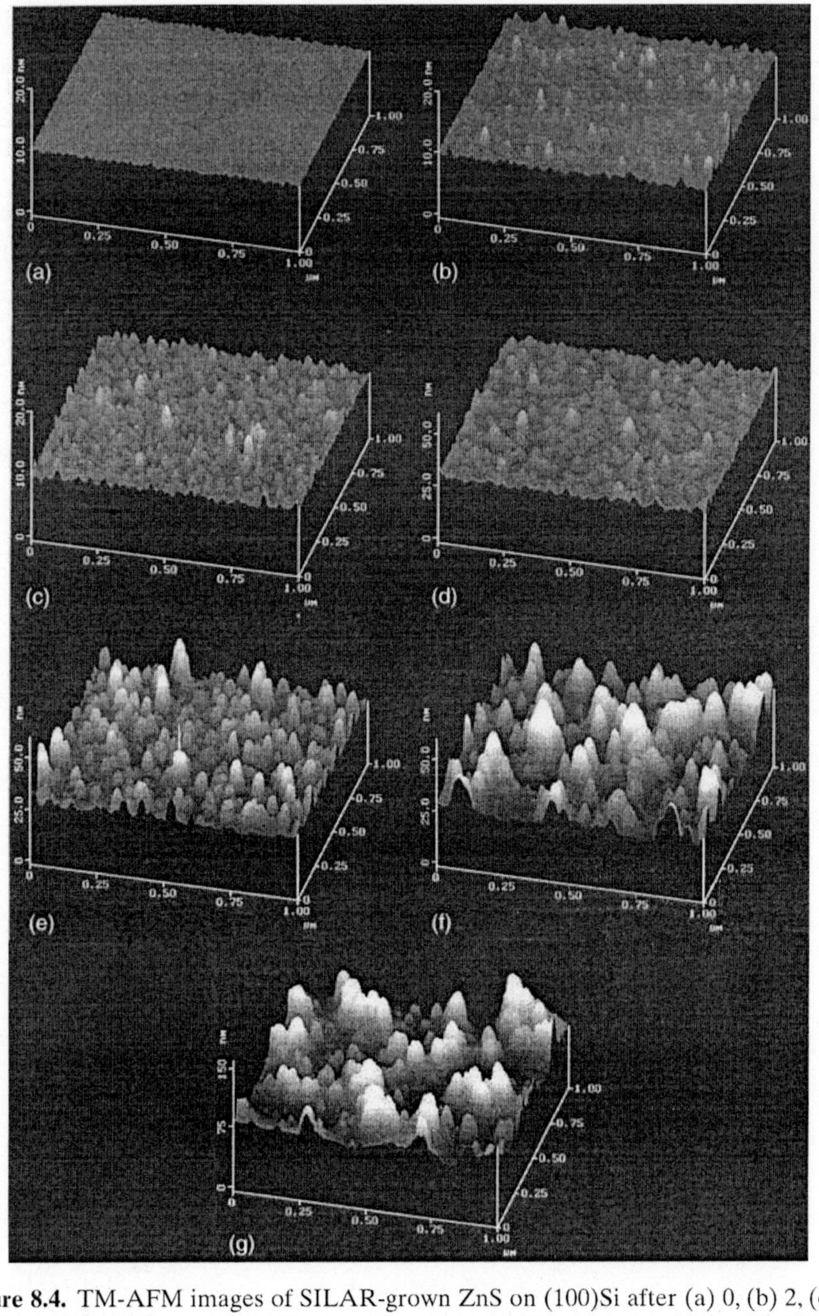

Figure 8.4. TM-AFM images of SILAR-grown ZnS on (100)Si after (a) 0, (b) 2, (c) 5, (d) 10, (e) 20, (f) 50, and (g) 100 cycles. Depth scale: 10 nm (a)–(c), 25 nm (d)–(f), and 75 nm (g) from black to white. Reprinted from Valkonen, Lindroos, Resch, Leskelä, Friedbacher and Grasserbauer. 1998. Applied Surface Science 136, Copyright (1998) with permission from Elsevier.

TABLE 8.1. Oxide Films Grown by SILAR

Oxide	Precursors	Characterization	Reference
ZnO	$ZnSO_4 + NH_4OH$ H_2O (95–100 °C)	XRD, UV, XRF, conductivity, photoresponse	18–21
ZnO	$ZnSO_4 + NH_3$ H_2O (95 °C)	XRD, SEM, TEM, electron diffraction, AFM, photoluminescence	22, 24–27
ZnO	$ZnSO_4$ + ethylenediamine Na_2S (+ oxidation at 400 °C)	XRD, SEM, UV, photoluminescence	23
ZnO:Sn	$ZnSO_4 + NaOH + Na_2SnO_3$ H_2O (95 °C)	SEM, EDX, resistivity, gas (NO_2) sensing	29
ZnO	Zn(II)acetate + NH_3 H_2O (hot)	SEM, UV, XRD, TEM, cathodoluminescence	28
ZnO	$ZnCl_2$ + ethylenediamine $H_2O_2 + NH_3$	UV, XRD, SEM	30, 31
Cu_2O	$CuSO_4 + Na_2S_2O_3$ NaOH (70 °C)	XRD, UV, electrical properties, photocurrent response	33
CuO_2	$Cu(NO_3)_2$, $Cu(CH_3COO)_2$ $H_2O_2 + NH_3$	UV	34
CuO_2/BaO_2	$Cu(NO_3)_2$, $Ba(NO_3)_2$ $Cu(CH_3COO)_2$, $Ba(CH_3COO)_2$ $H_2O_2 + NH_3$	Ellipsometry	35
CuO, Fe_2O_3, CoO, MnO_2	$Fe(NH_3)_2SO_4$, $CoSO_4$, $Cu(CH_3COO)_2$, $MnCl_2$ + $KMnO_4$	XRD, SEM, gas (CO, H_2, O_3) response	37
SnO_2	SnF_2 H_2O_2 + NaOH	Ellipsometry	38
TiO_2	$(NH_4)_2TiO(C_2O_4)_2$ NaOH	XRD, AFM, photocatalytic activity, synchotron XRD	40
TiO_2	$TiCl_3$ NH_4OH	XPS, GIXRD, AFM, SEM, HRTEM, UV, ellipsometry, SEM, XRD, TEM, EDX, UV, photoelectrochemical	41

TABLE 8.1. *Continued*

Oxide	Precursors	Characterization	Reference
TiO_2	Ti(IV) isopropoxide H_2O	SEM, XRD, UV, resistivity	42
CeO_2	$Ce(CH_3COO)_3$ $H_2O_2+NH_3$	UV, FTIR	44
CeO_2, CeO_2:Sm	$Ce(CH_3COO)_3$, $Ce(NO_3)_3$, $Sm(NO_3)_3$ H_2O_2	XRD, XPS, SEM	43
PbO_2	$Pb(NO_3)_2$ + $Na(CH_3COO)$ $KMnO_4$ + KOH	UV, FTIR, ellipsometry, XRD	45
MnO_2	$Mn(NO_3)_2$, $Mn(CH_3COO)_2$, $MnCl_2$ $KMnO_4$	FTIR, XRD, XPS, ellipsometry	46
ZrO_2, $ZnSiO_4$, MnO_2	Zr^{4+}, Zn^{2+}, Mn^{2+} OH^-, SiO_4^{4-}, H_2O_2 + OH^-	XRD	47
ZrO_2	$Zr(SO_4)_2$ NaOH	AFM, XPS, RBS, friction measurements	48
Tl_2O_3 SnO_2 CuO_x La_yNdO_z	$TlNO_3$, $La(CH_3COO)_3$, Nd_2O_3 + H_2O_2, H_2O_2 + KOH	Ellipsometry, FTIR, XPS	49
Co_3MnO_x $LiCo_3MnO_x$	Co salt, Mn salt, $LiNO_3$	XPS, FTIR, XRD	50
Y_2O_3, La_2O_3, Eu_2O_3	Y, La and Eu nitrate H_2O_2 + NaOH	Ellipsometry	51
Ag_xMnO_2	$AgNO_3$ $Mn(CH_3COO)_2$	XPS, XRD, FTIR, reflectance	52
$Zr_{2.6}(OH)_x$ $PMo_{8.0}Sn_{1.9}O_y$	$Zr(CH_3COO)_4$ $SnCl_2$ $H_3PMo_{12}O_{40}$	FTIR, UV, XPS	53

8.3.1.1 ZnO. Zinc oxide is one of the most widely considered oxide materials with respect to SILAR. ZnO films were first grown using the SILAR technique by Jiménez-González et al. in the mid-1990s.[18–20] The films were grown using ammonia-complexed zinc(II) sulfate as a zinc precursor, with hot water (96 °C) being used to convert the zinc hydroxide to zinc oxide. The rate of film growth was relatively high, 2.5 nm/cycle. Characterization by X-ray diffraction (XRD) indicated that the films were hexagonal and highly (002) oriented. This orientation was partially lost when the films were annealed at

350 °C. The optical spectra of the SILAR-grown ZnO film showed high transmittance (80–90%) in the visible region, and the band gap was 3.42 eV for a 30-nm-thick film. The band gap decreased when the film thickness increased, and for a 60-nm-thick film, the band gap was 3.38 eV. The photoresponse was studied after application of different annealing conditions. Heat treatment in air improved the dark current and photocurrent, but treatment in oxygen had no specific effect. The best results were obtained when the films were treated in hydrogen or in hydrogen followed by oxygen, after which the dark currents were 2.8×10^{-7} A and 1.8×10^{-6} A, respectively.[18–21]

Doping of zinc oxide with nickel, copper, and cadmium was also studied by Jiménez-González.[18] Doping had no effect on the crystalline properties of the as-grown ZnO films. However, after annealing, the degree of crystallographic orientation of the films was normally reduced. Only films doped with nickel retained the orientation after annealing, which had an effect on film conductivity. After annealing, the dark conductivity was increased when nickel and cadmium were used as dopants, but it was decreased when copper was the dopant.[18]

Zinc oxide has been deposited by the "normal" SILAR system and recently also by ultrasonic irradiated SILAR. The latter method was introduced by Gao et al.,[22] who employed ammonia complexed zinc (0.02 M $[Zn(NH_3)_4]^{2+}$ with zinc/ammonia ratio 1/10). The deposition cycle consisted of immersion of the substrate in the zinc precursor solution, immersion of the substrate in a water beaker with ultrasonic irradiation (formation of $Zn(OH)_2$ precipitate), and immersion of the substrate in a hot water bath (to convert $Zn(OH)_2$ to ZnO). The films were crystalline and (002) oriented as indicated by selective-area electron diffraction. The films were formed of large ZnO particles (200–300 nm) and of nanocrystals (3–10 nm), and they were stoichiometric; i.e., the Zn/O ratio was 51%/49%. When the films were annealed at 400 °C, the intensity of the (100) XRD reflection clearly increased while there was a slight decrease in the (002) reflection. The transmittance in the UV-Vis spectrum increased after annealing, possibly because of the decrease in porosity. The preferred orientation along the (002) plane of ZnO was observed in films grown on a silicon substrate (Si(100)), with more abundant orientation than for films grown on glass. AFM measurements revealed an rms roughness of 47.5 nm for a 300-nm-thick film. The growth rate was as high as 30 nm/growth cycle, indicating that the growth was not of pure SILAR type, but that some homogeneous precipitation occurred during film formation (Fig. 8.5).[11,22–26]

Zinc oxide films have also been grown by thermal oxidation of SILAR-grown ZnS thin films.[27] The oxidation was performed both in air and in an oxygen–argon mixture at 400 °C for 2 h. Oxidation in air gave better quality films, with particle sizes of 20–50 nm and (100) preferred orientation in the XRD pattern. The band gaps of the ZnO films were lower, 3.16 eV and 3.22 eV, respectively (for the different annealing atmospheres), than the bulk value of ZnO (3.37 eV).[27]

A SILAR-grown ZnO thin film has also been used as a seed layer for CBD-grown nanostructures.[28] The seed layer was grown from an ammine-com-

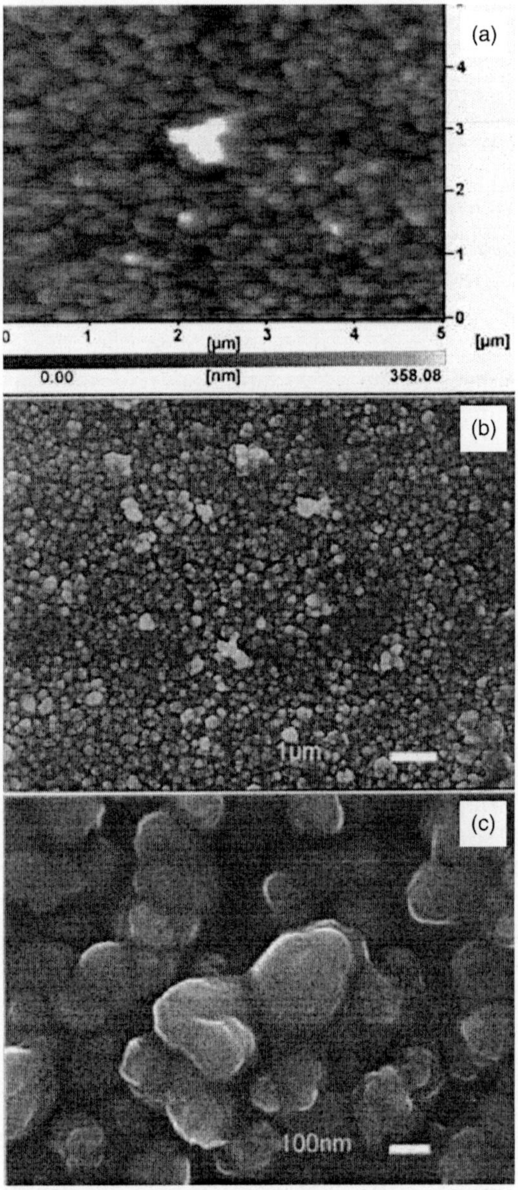

Figure 8.5. Morphology of SILAR-grown ZnO films on Si(100) substrate: (a) AFM image, (b) SEM image, and (c) image of individual ZnO particles. Reprinted from Gao, Li and Yu. 2005. Materials Research Bulletin 40, Copyright (2005) with permission from Elsevier.

plexed zinc precursor with ultrasonic-assisted rinsing and followed by normal hot water treatment to convert the zinc hydroxide film to zinc oxide. The substrate (indium tin oxide, ITO) pretreatment and hot water temperature were varied to study the morphology of the ZnO film. The best high-density ZnO seed layer was obtained with electrochemical pretreatment of the ITO-substrate by applying an anodic oxidation potential at room temperature for 3600 s and with a 95 °C hot water bath. The ZnO reflections could be observed in XRD after hydrothermal annealing.[28]

SILAR-grown ZnO films have been tested for gas sensor applications.[29] The ZnO films, doped with tin for this purpose, were grown from a mixture of dilute zinc sulfate, sodium hydroxide, and sodium tin(IV)oxide solutions. The final step, resulting in the oxide film, was treatment of the substrate and film in a nearly boiling water bath. The NO_2 gas sensing properties were tested for films doped with Al, Cu, Pd, and Sn, but only the film doped with tin exhibited sensitivity toward NO_2. The sensitivity of the ZnO:Sn film was 5%/ppm after rapid photothermal processing (RPP). The best sensitivity was obtained when the tin concentration was 5–10%.[29]

Another route to grow ZnO thin films by the SILAR method is to use hydrogen peroxide as an oxygen precursor. The as-grown zinc peroxide film was homogeneous and polycrystalline, showing only the cubic XRD reflections of zinc peroxide. After annealing, during which the zinc peroxide decomposed to zinc oxide, the films were amorphous. The band gap of the ZnO film was 3.7 eV, and the transmittance in UV-VIS was over 90%.[30–32]

8.3.1.2 Cu_2O. Copper sulfate complexed with thiosulfate has been used together with hot sodium hydroxide to deposit copper(I) oxide films. When the films were annealed at 350 °C in air, they converted to copper(II) oxide, but annealing at 400 °C in nitrogen did not change the composition of the original copper(I) oxide film. The electrical conductivity of the CuO films was high.[33] Copper oxide films have also been grown via the peroxide route by Tolstoi et al., who used a copper ammine solution and basic hydrogen peroxide.[34–36] The effect of the concentration of the precursors was investigated by ellipsometric and spectrophotometric methods.

Copper, iron, cobalt, and manganese oxide nanoclusters have been grown by the SILAR method on SnO_2 substrates to investigate their gas sensing properties.[37] As-grown metal hydroxide films were annealed at 100–400 °C to convert them to metal oxides. The metal oxide nanoclusters could not be observed by SEM, but after annealing at 800 °C, thicker films (50–60 nm) showed typical polycrystalline XRD reflections of the metal oxides. The crystallite size of the oxides was small (i.e., 4–6 nm).[37]

8.3.1.3 TiO_2. Titanium oxide (TiO_2) is a technologically important material, used in photocatalytic and solar cells, electronic and optical devices, air purification, and many biological and medical applications. Photocatalytically active TiO_2 films have been grown using aqueous solutions of 0.06 M

$(NH_4)_2TiO(C_2O_4)_2 \cdot H_2O$ (pH 3.76) for the titanium precursor and 0.06 M NaOH (pH 12.69) for the oxygen precursor. The as-grown films were polycrystalline with no preferred orientation. After annealing, the particle size ranged from ten to a few hundred nanometers. TiO_2 films hydrothermally treated at 200 °C exhibited a higher photocatalytic activity than films annealed in a furnace at 600 °C. They also had better mechanical stability under photocatalytic conditions.[39]

The same precursors, 0.005 M ammonium titanyl oxalate monohydrate ($(NH_4)_2TiO(C_2O_4)_2 \cdot H_2O$, pH 3.81) and 0.01 M sodium hydroxide (NaOH, pH 11.57), were used to grow anatase films on Ta sheets. The growth rate was high, 6.4 nm/growth cycle, with a linear relationship between the relative film thickness and the number of growth cycles. The O/Ti ratio of the as-grown TiO_2 films, analyzed by XPS, was 2.49 on the surface of the films, and after 10 s of sputtering (i.e., deeper in the films), the value was 2.04. After annealing for 16 h at 393 K, the values were correspondingly 2.46 (surface) and 2.08 (after sputtering). The as-grown films did not show any diffraction rings in high-resolution transmission electron microscopy (HRTEM). However, the annealed films showed four rings that could be attributed to anatase crystal planes. The average crystal diameter was 6 ± 1 nm as revealed by grazing incidence XRD. The roughness of the as-grown films (75 nm thickness) was 16.2 nm, and it dropped to 11.8 nm after annealing of the same film. Change in the morphology of the films could be observed in AFM and SEM images.[40]

TiO_2 films have also been grown from 10 to 13 wt.% $TiCl_3$ and NH_4OH solutions. The growth rate was 0.6 nm/cycle. According to the XRD pattern, the as-deposited film consisted of a mixture of TiO_2 and Ti_3O_5. However, after annealing at 450 °C, only TiO_2 was observed. Also, the annealed films showed better photo-activity than the as-deposited films.[41]

Titanium isopropoxide has been used as an alternative Ti^{4+} source, with distilled water at 333 K as the oxygen precursor. The films were amorphous after annealing at 673 K for 2 h. According to SEM images, the film consisted of spherical grains of random size. The absorption edge of the annealed film shifted to the longer wavelength side and the resistivity decreased by one order of magnitude, possibly because of structural changes.[42]

8.3.1.4 CeO_2 and $(Ce,Sm)O_2$. CeO_2 and $(Ce,Sm)O_2$ films were grown on silicon using 0.06 M cerium acetate or cerium nitrate and a basic solution of hydrogen peroxide (3–30%) as precursors. Doping with samarium was successful only when cerium nitrate and samarium nitrate were used. Doping was achieved by adjusting the concentration ratio of the precursor solutions. A film thickness of 60 nm was achieved after 200 cycles. However, the thickness could not be increased beyond this point because of dissolution of the hydroxide film to the precursor solution at low pH. The films were annealed in air at 878 K for 50 min or in a reactor at 378 K for 15 h. After annealing, the film consisted of particles with diameters near 20 nm. The stoichiometry of the material was CeO_2.[43] The Fourier transform infrared (FTIR) spectrum showed water in the as-grown films, and according to the UV-VIS spectra, the oxidation state of cerium was +4.[44]

8.3.1.5 MnO_2. Manganese(IV) oxide films have been grown using different Mn(II) solutions and the $KMnO_4$ solution as precursors. Although the as-deposited films were amorphous and contained some water, according to XPS analysis, the films were MnO_2.[46] Also the peroxide route has been used to obtain MnO_2 films by the SILAR technique. The as-deposited films were amorphous. However, after annealing at 473 K, the films were crystalline MnO_2. Further annealing to 773 K resulted in loss of oxygen and a composition change to Mn_2O_3.[47]

8.3.1.6 ZrO_2. Zirconium sulfate and sodium hydroxide were used to grow ZrO_2 films. Although the growth rate remained low during the first cycles, after 10 cycles, it increased to 0.62 nm/cycle. The pH of the zirconium solution had a marked effect on the growth rate. When diluted sulfuric acid was added to the zirconium precursor solution, the growth rate dropped to 0.1 nm/cycle. The Zr/O ratio of the films was 1/2.6 as determined by Rutherford backscattering spectroscopy (RBS). The film roughness increased in parallel with the number of deposition cycles. The ZrO_2 films were amorphous, and after annealing at 473 K in air, they became crystalline monoclinic ZrO_2. Direct annealing at 923 K produced oxygen-deficient tetragonal ZrO_2.[47,48]

8.3.1.7 Other Oxide Films. Thallium(III) nitrate and hydrogen peroxide were used to grow Tl_2O_3 films on silicon. The growth rate was 0.5 nm/cycle.[49] Thin films with mixed metal oxide composition, Co_3MnO_x and $LiCo_3MnO_x$, have also been produced by SILAR for lithium battery cathode applications. The films were amorphous, and infrared (IR) spectroscopy revealed hydroxide groups in the films. The composition of the films was confirmed by X-ray photoelectron spectroscopy.[50] Also, a green emitting phosphor Zn_2GeO_4:Mn has been grown on a flexible polyimide substrate.[47] Nanolayers of Y, La, and Eu hydroxyperoxides have been grown by SILAR from the corresponding metal nitrates and a basic solution of hydrogen peroxide. The thickness of the films, both as-grown and after annealing, was studied.[51] SILAR-grown Ag_xMnO_2 oxide material has been found by XPS to consist of metallic silver and MnO_2. The composition was confirmed by FTIR, reflectance spectra, and XRD.[52] Zr–P–Mo–Sn–O monolayers have also been grown by SILAR on a silica surface. The material was found to have the composition $Zr_{2.6}(OH)_xPMo_{8.0}Sn_{1.9}O_y \cdot nH_2O$.[53]

8.3.2 Chalcogenide Films

The first deposition experiments using SILAR were performed to grow zinc and cadmium chalcogenide films.[1] Metal salt solutions and sodium sulfide were used as precursors in most experiments. Table 8.2 provides a more complete

TABLE 8.2. Chalcogenide Films Grown by SILAR

Chalcogenide	Precursors	Characterization	Reference
ZnS CdS $Zn_{1-x}Cd_xS$	$ZnCl_2$,$ZnSO_4$ $CdCl_2$ Na_2S	UV, XRD, XRF, SEM, IR, RBS, Auger, SIMS, resistivity, cathodoluminescence, RHEED	1, 54–56
ZnS CdS $Zn_{1-x}Cd_xS$	$Cd(NO_3)_2$,$ZnSO_4$ Na_2S	SEM, electron diffraction	57
ZnS CdS CdS/ZnS $Cd_xZn_{1-x}S$	$ZnCl_2$ $CdCl_2$ Na_2S	UV, XRD, RBS, nuclear reaction analysis, SEM, AFM, EXAFS, stress	9, 10, 58–60, 62, 64–70, 75, 76, 86, 87, 89
ZnS:Mn	$ZnCl_2$,$MnCl_2$ Na_2S	XRD, UV, RBS, nuclear reaction analysis, SEM	9, 61
ZnS	$ZnCl_2$ + triethanol amine, ethylenediamine Na_2S	XRD, SEM, EDX, EXAFS, UV	9, 63
ZnS	$ZnSO_4$ + NH_3 Na_2S	XRD, HRTEM, EDX, UV, resistivity	71
ZnS	$ZnSO_4$ + ethylenediamine Na_2S	XRD, SEM, AFM, UV	11
ZnSe	$Zn(CH_3COO)_2$ + tartaric acid Na_2SeSO_3	XRD, SEM, EDX, UV, resistivity	73
ZnTe	$ZnSO_4$ Na_2Te	XRD, UV, SEM, EDX	74
CdS	CdF_2, $Cd(ClO_4)_2$, $CdCl_2$, CdI_2, $Cd(CH_3COO)_2$, $Cd(HCOO)_2$ + 2-mercaptoethylamine hydrochloride, cysteine, ethylenediamine, triethanolamine, monoethanolamine, nitrilotriacetic acid trisodium salt Na_2S	XRD, UV, AFM, Auger, XPS	77, 78
CdS	$Cd(CH_3COO)_2$ Na_2S	XRD, SEM, EDX, TEM, RBS	79
CdS	$Cd(NO_3)_2$ Na_2S	UV, I-V cell characterization, SEM, EDX, TEM, ultraviolet photoemission spectroscopy	80

TABLE 8.2. *Continued*

Chalcogenide	Precursors	Characterization	Reference
CdS	CdO + oleic acid + octadecane S + octadecane	UV, photoluminescence, TEM, HRTEM, XRD, XPS, photooxidation, EDX	82, 88
CdS/ZnS	CdO, ZnO + oleic acid + octadecane S + octadecane	UV, photoluminescence, TEM, HRTEM,	72, 90
CdSe	$CdSO_4$ + tartaric acid Na_2SeSO_3	HRTEM, EDX, XRD, UV, resistivity	81
CdSe	$Cd(CH_3COO)_2$ + tartaric acid, triethanolamine Na_2SeSO_3	XRD, SEM, EDX, HRTEM, AFM, Raman, UV, resistivity	83
PbS	$Pb(NO_3)_2$, $Pb(CH_3COO)_2$ Na_2S, thioacetamide	XRD, SEM	10, 92
PbS	$Pb(CH_3COO)_2$ + triethanolamine thioacetamide	XRD, SEM, RBS, nuclear reaction analysis, AFM, stress	10, 91, 93–99, 101
PbS/CdS	$Pb(CH_3COO)_2$, $Pb(CH_3COO)_2$ + triethanolamine, $Pb(NO3)_2$, $CdCl_2$ thioacetamide, Na_2S	XRD, XPS, AFM	100
PbS/In_2S_3	$Pb(CH_3COO)_2$ $InCl_3$ Na_2S	Surface photovoltage spectra, I-V, quantum efficiency, SEM, ERDA	101, 102
PbSe	$Pb(CH_3COO)_2$ + triethanolamine Na_2SeSO_3 + hydrazine hydrate	XRD, RBS, nuclear reaction analysis, SEM	103
HgS	$Hg(CH_3COO)_2$ Na_2S	XRD, SEM, UV	104
Cu_xS	$CuSO_4$ + triethanolamine thiourea	XRD, UV	105
CuS CdS/CuS	$Cu(CH_3COO)_2$, $CdCl_2$ Na_2S	XRD, SEM, EDX, UV	106
Cu_2S	$CuSO_4$ + triethanolamine + hydrazine hydrate Na_2S	XRD, SEM, EDS, RBS, HRTEM, UV, resistivity	107
As_2S_3	As_2O_3 + HCl + EDTA $Na_2S_2O_3$	XRD, UV, reisitivity	108

TABLE 8.2. *Continued*

Chalcogenide	Precursors	Characterization	Reference
Sb_2S_3	Sb_2O_3 $Na_2S_2O_3$	XRD, UV, resistivity	109
Sb_2S_3	KSb(tartrate) Na_2SeSO_3	XRD, SEM, UV, reisitivity	110
Bi_2S_3	$Bi(NO_3)_3$ + triethanolamine thioacetamide + hydrazine hydrate	XRD, UV, SEM, resistivity	112, 114
CdS-Bi_2S_3	$Cd(CH_3COO)_2$ $Bi(NO_3)_3$ Na_2S Thioacetamide	XRD, UV, I-V characteristics, photoresponse	112
Bi_2Se_3	$Bi(NO_3)_3$ Na_2SeSO_3	XRD, SEM, AFM, HRTEM, EDX, RBS, I-V characteristics, photovoltaic characteristics, C-V characteristics	114
Bi_2Se_3	$Bi(NO_3)_3$ + triethanolamine Na_2SeSO_3	XRD, SEM, UV, resistivity	113
Bi_2Se_3 Sb_2Se_3 Bi_2Se_3-Sb_2Se_3	$Bi(NO_3)_3$ + triethanolamine $KSbO(C_4H_4O_6)$ Na_2SeSO_3	XRD, SEM, AFM, HRTEM, EDX	111
In_2S_3	$InCl_3$ Na_2S	XRD, SEM, UV, resistivity	115
In_2S_3	$InCl_3$ + triethanolamine + hydrazine hydrate Na_2S	XRD, UV, SEM, RBS,	116
In_2S_3	$InCl_3$ Na_2S	XRD, XPS, UV	117
SnS_2	$SnCl_2$ Na_2S	XRD, EDX, SEM, AFM, UV, photoluminescence I-V characteristics	118
CoS	$CoSO_4$ Na_2S	XRD, UV, resistivity	119
$CuInS_2$	$CuCl_2$ $InCl_3$ Na_2S	XRD, SEM, XPS, UV, conductivity, Hall mobility	121, 122
Ag_7SbS_6	Ag salt Na_3SbS_4	UV, IR, XPS, XRD	123

list of chalcogenide films deposited by SILAR, with a more detailed discussion of selected examples given in the following sections.

8.3.2.1 ZnS. Zinc sulphide is an important semiconductor material with uses in electronic, photonic, and cathodoluminescence devices. Zinc sulfide was first grown using SILAR by Nicolau,[1] and since then, several articles on ZnS SILAR deposition have appeared.[1,8–10,54–71] The growth rate of ZnS has been found to be proportional to concentrations of the precursor solutions. The maximum growth rate was 0.31 nm/cycle, with a saturated zinc precursor solution, and the mean surface coverage was then 90–100%. The growth rate has also been found to depend on substrate material. On silicon, the growth rate was much lower, 0.02 nm/cycle, at the beginning of the growth process.[17] The growth rate on polymer surfaces, 0.16 nm/cycle, was higher than on glass.[59,60] Complexing the zinc precursor has been shown to increase the growth rate of ZnS.[9,63]

As studied by XRD, the ZnS films grown on glass were apparently amorphous up to a nominal thickness of 30 nm. Afterward, the films showed a polycrystalline cubic structure with (111) preferential orientation.[1,9,54,58] An increase of the cubic (111) reflection coincided with a decrease in rms roughness. On a (100) GaAs substrate, a polycrystalline ZnS thin film with weak cubic reflections was observed after a 50-nm-thick amorphous phase.[64–66] The SILAR-grown ZnS thin films on polymer substrates showed poorer crystallinity than those on glass.[9,59,60] The refractive indices of SILAR-grown ZnS were 2.07–2.25 on a glass substrate, 2.15–2.30 on an ITO substrate, and 1.75–2.10 on GaAs, with the bulk value being 2.30.[9,54,64–66] Complexing of the zinc precursor resulted in films with a slightly lower refractive index.[9,63] According to RBS measurements, the ZnS films were stoichiometric or contained an excess of zinc (i.e., the S/Zn ratio was 0.8). Oxygen and hydrogen impurities were detected by RBS, XPS, and nuclear reaction analysis (NRA). Although IR studies revealed water in films grown on Ge substrates, no hydroxide was found by IR, RBS, or Auger measurements.[1,9,54,58] On the other hand, EXAFS measurements indicated the presence of hydroxide in the surface layer of the films.[62,63] HRTEM images revealed random orientations of nanoparticles with dimensions of 50–60 nm in a 250-nm-thick ZnS film. The band gap of SILAR-grown ZnS films is 3.30 eV, and the resistivity is on the order of $10^6\,\Omega$ cm.[71] Measurement of the stress properties of ZnS films on GaAs revealed that increase in the thickness of the film was accompanied by change in the stress from tensile to compressive.[64–66] ZnS films have been doped with Mn using either $MnCl_2$ or $KMnO_4$ solutions as manganese source.[54,61]

Ultrasonic-assisted rinsing has been tested for the growth of ZnS thin films. The growth rate clearly increased, and the morphology of the films became smoother, whereas the grains became smaller. The band gap of the ZnS films was 3.67 eV.[11]

A zinc sulfide layer in a CdSe/ZnS/CdSe quantum system structure has also been grown by SILAR from ZnO and sulfur precursors in octadecane.[72]

Through change in the thickness of the ZnS layer, the two quantum systems were controlled to be either electronically coupled or decoupled. The photoluminescence quantum efficiency was as high as 30% at room temperature.[72]

8.3.2.2 ZnSe, ZnTe. Both pure zinc acetate and zinc acetate complexed with tartrate, together with sodium selenosulfate, have been used as precursors for ZnSe films. The as-deposited films were poorly crystalline or amorphous. After annealing at 200 °C, however, they showed a cubic structure, with the crystallite size increasing from 6 nm to 13 nm upon annealing. The stoichiometry of the films was Zn/Se 52/48. The band gap was 2.8 eV, and the electrical resistivity was on the order of $10^7\,\Omega$ cm at room temperature.[73] Zinc sulfate and sodium telluride were the precursors for the deposition of ZnTe films on a glass substrate.[74] The films were nanocrystalline and perhaps slightly zinc rich, with a Zn/Te composition of 53/47. The optical band gap of a 75-nm-thick film was 2.75 eV, compared with the bulk value of 2.3 eV for ZnTe.[74]

8.3.2.3 CdS. Diluted cadmium chloride and sodium sulfide solutions have been the most common precursors in the growth of cadmium sulfide films by SILAR. The growth rate of CdS was much lower than that of ZnS. The maximum growth rate was 0.14 nm/cycle, and the mean surface coverage was 0.3 when the precursor solutions were saturated. The grain size in the CdS films was larger (30–60 nm) than that in the ZnS films and the surface structure was rougher. The films were either stoichiometric or, like the ZnS films, contained some oxygen as an impurity, as analyzed by RBS, EDX, and Auger. Other significant impurities, as determined by secondary ion mass spectrometry (SIMS) were sodium and iron. The CdS films have been generally polycrystalline, the main reflection being either the (111) reflection of the cubic form or the (002) reflection of the hexagonal form. Epitaxial CdS films have been grown on (111) InP. The refractive index was 2.15–2.35, with a room temperature resistivity of $3 \times 10^5\,\Omega$ cm.[66,67,75,76]

The use of chelating reagents in CdS SILAR deposition has been thoroughly investigated. CdF_2, $CdCl_2$, CdI_2, $Cd(ClO_4)_2$, $Cd(CH_3COO)_2$, and Cd(HCOO) have been used as precursors together with 2-mercaptoethylamine, cysteine, ethylenediamine, triethanolamine, or monoethanolamine as chelating reagents, in a 1:1 molar ratio.[77,78] The growth rate increased when a complexed cadmium precursor was used. The highest quality CdS films were achieved with triethanolamine as a chelating agent. The surface structure of the film was then smooth, and the S/Cd ratio was the same as for the films grown with uncomplexed $CdCl_2$ as a precursor.

Ethylene glycol has been used instead of water as solvent in the deposition of CdS films from cadmium acetate and sodium sulfide precursors. The films were amorphous and showed some excess cadmium (Cd/S atomic percentage ratio was 52.3/47.7) in EDX analysis.[79]

A nanostructured TiO_2/CdS/CuSCN solar cell has been fabricated, where the CdS layer was grown by the basic SILAR technique. The cell efficiency was 1.3%. The CdS coating completely covered the TiO_2 particles, and dispersed CdS particles were observed by SEM and TEM.[80]

8.3.2.4 CdSe, CdTe. CdSe films have been grown from complexed (with tartaric acid and triethanolamine) cadmium acetate or cadmium sulfate solutions and sodium selenosulfate.[12,72,81–84] The films were amorphous or nanocrystalline with an average crystallite size of 6 nm. The optical band gap was 1.8–2.1 eV, and the electrical resistivity was of the order 10^4–10^6 Ω cm.

A modified SILAR system has been used to grow CdSe in CdS/CdSe core shell semiconductor nanocrystals.[12] A cadmium precursor solution, with CdO dissolved with oleic acid in octadecane, was injected onto the substrate, and the Se solution (Se powder dissolved with tributylphosphine in octadecane) was similarly injected. The temperature of the reaction solution was 185 °C. A CdS outer layer in the CdS/CdSe/CdS colloidal quantum wells was deposited by alternating injections of cadmium and sulfur both in octadecane solutions at 230–240 °C. These structures showed high PL quantum yields (20–40%), relatively narrow emission bands, and tunable emission colors from about 520 to 650 nm depending on the number of CdSe monolayers.

Cadmium acetate and sodium tellurite solutions have similarly been used to grow CdTe films by SILAR. The films were polycrystalline (hexagonal), and the grain size was on the order of 22 nm (film thickness 272 nm). The optical band gap was 1.41 eV.[85]

8.3.2.5 $Cd_xZn_{1-x}S$ Solid Solution Films. There is a small (7%) difference between the lattice constants of ZnS and CdS. $Cd_xZn_{1-x}S$ thin films have been grown by two methods: from mixed-cation precursor solutions and from separate cadmium and zinc precursors.[54–57,64,67,86,87] The thin films were rich in cadmium if equal numbers of cadmium and zinc growth cycles were used in the growth process. The chemical compositions of $Cd_xZn_{1-x}S$ (x = 0–1) films have been characterized by chemical analysis and EDX analysis, with the peak position in the XRD spectra tracking the composition of the solid solution film (Fig. 8.6). CdSe nanocrystals covered by the SILAR-grown multilayer structure $CdS/Zn_{0.5}Cd_{0.5}S/ZnS$ (Fig. 8.7) showed a high fluorescence quantum yield of 70–85%.[88] The SILAR technique allows a gradual change in the shell composition from CdS to ZnS, so that the desirable properties of both materials are present in the structure. The lattice strain is small, and the particles show high crystallinity. Additionally the band offsets are high, and the particles are electronically well passivated.

8.3.2.6 CdS/ZnS Multilayer Thin Films. The low deposition temperature of SILAR allows the growth of very thin layers to achieve multilayer structures. CdS/ZnS multilayer films have been grown from separate cadmium, zinc, and sulfur precursor solutions. Multilayer structures with layer thicknesses of 2–5 nm have been fabricated, and the separate layers could be seen by SEM. RBS measurements revealed that the layers were separated with only

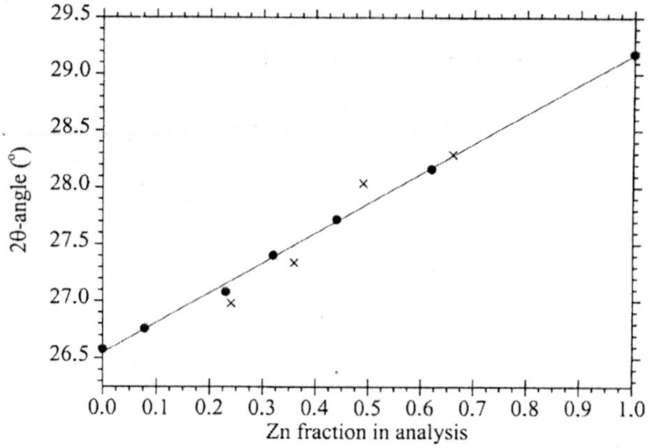

Figure 8.6. The position of XRD reflection versus zinc content for $Cd_xZn_{1-x}S$ thin films grown by SILAR on ITO (circle) and glass (cross) substrates. Reprinted from Valkonen, Lindroos, and Leskelä. 1998. Applied Surface Science 134, Copyright (1998) with permission from Elsevier.

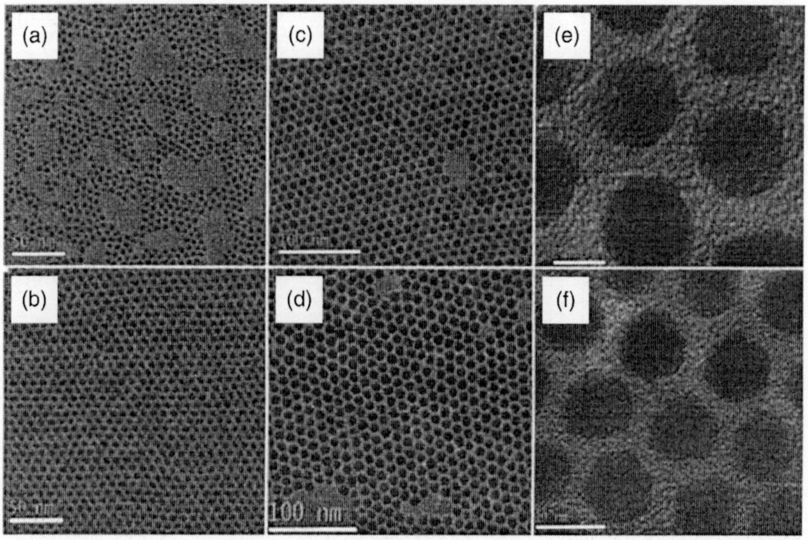

Figure 8.7. TEM images of SILAR-grown $CdS/Zn_{0.5}Cd_{0.5}S/ZnS$ multishell structure: (a) TEM image of CdSe-cores, scale bar is 50 nm; (b) two monolayers CdS on CdSe-cores (a), scale bar is 50 nm; (c) and (e) 3.5 monolayers $Zn_{0.5}Cd_{0.5}S$ on CdS (b), scale bar is 100 nm (c) and 6 nm (e); (d) and (f) 2.5 monolayers ZnS on $Zn_{0.5}Cd_{0.5}S$ (c), scale bar is 100 nm (d) and 10 nm (f). Reprinted with permission from Xie, Kolb, Li, Basché and Mews 2005 Journal of the American Chemical Society 127. Copyright 2005 American Chemical Society.

slight intermixing. The intermixing is mainly from the roughness of the film, not from the chemical mixing by diffusion.[67,68,76,89]

SILAR has been used for the synthesis of CdS/ZnS coatings for CdSe quantum dots. The precursor solutions were prepared by dissolving CdO, ZnO, and S in oleic acid and octadecane. The final coating consisted of three layers of CdS and three additional layers of ZnS. The photonic band structure of the photonic crystal had a modifying influence on the photoluminescence of the embedded quantum dots.[90]

8.3.2.7 PbS, PbSe. Lead sulfide has been used in IR detectors for decades. Lead sulfide was first deposited using the SILAR method by Kanniainen et al.[10,91,92] Subsequently, several papers on SILAR-grown PbS on silicon substrates were published.[93–99] The choice of lead precursor has been found to have an effect on growth rate, particle size, and orientation of the PbS film. The growth rate was slowest with lead nitrate solution. Use of lead acetate solution as a precursor increased the growth rate up to 0.06 nm/cycle, and its complexation with triethanolamine resulted in a growth rate of 0.12 nm/cycle. The complexation of lead acetate had no effect on the grain size. Use of a nitrate solution, on the other hand, resulted in a reduced grain size. The PbS thin films grown by SILAR were polycrystalline (i.e., cubic) with different degrees of preferred orientation. The most preferred orientation occurred when a complexed lead acetate solution was used as a precursor (Fig. 8.8). According to RBS analysis, the PbS thin film grown with the triethanolamine-complexed lead acetate solution was stoichiometric. In addition, 6 to 8 at.% of light impurities (such as H, C, N, and O) were detected.[10,91–99]

Multilayer CdS–PbS films have been grown using different cycles for CdS and PbS. The XRD patterns showed that the double-layer components were

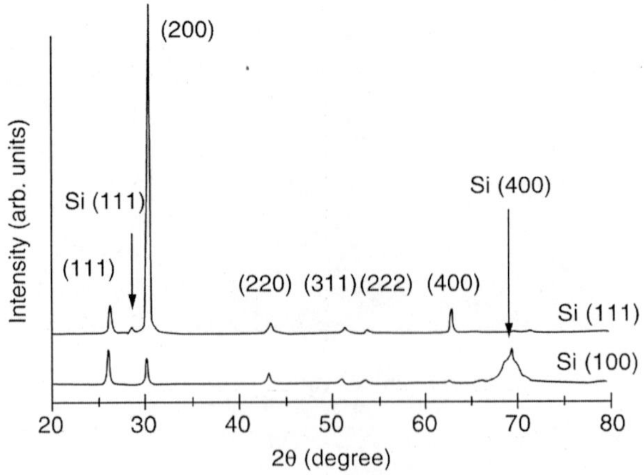

Figure 8.8. XRD spectrum of a SILAR-grown PbS thin film on silicon. Reprinted from Puišo, Tamulevicius, Laukaitis, Lindroos, Leskelä, and Snitka. 2002. Thin Solid Films 403–404. Copyright (2002) with permission from Elsevier.

polycrystalline. CdS was hexagonal, and PbS showed high cubic (200) orientation. The roughness of the layer structure was lowest (i.e, 37.4 nm for a 100-nm PbS + 100-nm CdS structure) when lead acetate complexed with triethanolamine was used as the lead precursor. Without complexation, a lead acetate precursor produced very rough films in an analogous structure (i.e., ~143.9 nm).[100]

An ultra-thin heterostructure (35 cycles of In_2S_3 and 15 cycles of PbS) was grown by SILAR on Mo-coated glass and on SnO_2:F (FTO) substrates. The film composition was analyzed by elastic recoil detection analysis (ERDA), and the films were essentially stoichiometric but contained 3 at.% oxygen and 13–14 at.% hydrogen. The layer structure was preserved after annealing at 250 °C. The photoelectrical properties of a solar cell structure were examined as a function of annealing temperature. A solar cell efficiency of 3–5% seems to be realistic with a SnO_2:F/TiO_2/$In(OH)_xS_y$/PbS(O)/PEDOT:PSS/C structure.[101,102]

The lead precursor in PbSe film deposition was lead acetate complexed with triethanolamine. The growth rate for PbSe was 0.18–0.16 nm/cycle. The films were polycrystalline (i.e., cubic) without preferred orientation. The stoichiometry of SILAR-grown PbSe was found to be 1:1 within the limits of the RBS technique. Impurities detected were 5 at.% of oxygen and 8 at.% of hydrogen.[103]

8.3.2.8 CuS and Cu_2S. Copper(II) sulfate complexed with triethanolamine or copper(II) chloride has been the copper precursor for CuS films.[105–107] The resulting films were hexagonal (covellite) with a band gap of 2.36 eV.[105–107] Annealing of the films up to 500 °C changed the composition to Cu_xS (x = 1.96–1.86). Cu_2S films have been directly grown from copper(II) sulfate complexed with a mixture of triethanolamine and hydrazine hydrate and sodium sulfide as the sulphur source.[105,107] The Cu_2S films exhibited a hexagonal crystal structure and were composed of particles 24 nm in size. RBS studies showed the presence of some oxygen. The resistivity of the films was on the order of $10^{-2}\,\Omega$ cm.[105,107]

8.3.2.9 Sb_2S_3, Sb_2Se_3. Antimony oxide with tartaric acid served as a cation precursor, and sodium thiosulfate acted as an anion precursor for Sb_2S_3 films. The resulting amorphous films yielded a band gap of 1.8 eV and a resistivity of $10^7\,\Omega$ cm. Antimony potassium tartrate at pH 3 was the cationic precursor for Sb_2Se_3 films, with sodium selenosulfate serving as the anionic precursor, yielding a growth rate of 0.56 nm/cycle. The average grain size was 3 nm after 150 deposition cycles (84 nm thickness). Study by AFM showed the films to be rough. The band gap of the films was 1.3 eV, with a resistivity of $10^5\,\Omega$ cm.[109–111]

8.3.2.10 Bi_2S_3, Bi_2Se_3. Bismuth nitrate complexed with triethanolamine or EDTA and thioacetamide, with the addition of hydrazine hydrate, have been used as Bi_2S_3 film precursors. The as-grown, 140-nm-thick film was amorphous but became polycrystalline after annealing in air at 200 °C for 2 h. The average grain size was 3–4 nm. The optical band gap was 1.78 eV. Bismuth nitrate complexed with triethanolamine and sodium selenosulfite have been used to grow Bi_2Se_3 films. Although the films did not show any defined peaks in XRD when deposited on glass substrate, crystallinity was improved on silicon. The average crystallite size on silicon was 50–60 nm, as determined by AFM and SEM. Severe selenium loss was observed when the films were annealed up to 400 °C. The band gap of an as-grown film was 0.38 eV, with a resistivity of $3.54 \times 10^5\,\Omega$ cm. When the same precursors as for Bi_2Se_3 and Sb_2Se_3 were used together, a solid solution Bi_2Se_3–Sb_2Se_3 was grown. The film morphology of the nominally amorphous film was rougher than that for the separate Bi_2Se_3 and Sb_2Se_3 films. The optical band gap of a film grown using precursor solutions with Bi:Sb concentration ratio 1:1 was 1.57 eV, and the resistivity was $10^4\,\Omega$ cm.[111–114]

8.3.2.11 In_2S_3. Indium sulfide films have been grown using solutions of indium(III) chloride and indium(III) sulfate complexed with triethanolamine and hydrazine hydrate.[115–117] The films grown from the indium chloride precursor were amorphous or nanocrystalline (i.e., cubic structure), but those grown from complexed indium precursor were rhombohedral, possibly because of the complexation. Annealing the films grown from indium chloride at 400 °C improved the crystallinity of the films, and a tetragonal In_2S_3 phase was detected, along with secondary indium-rich phases like InS and In_6S_7.[117] The band gap varied from 2.3 to 2.7 eV, and it was reduced to 1.4–1.5 eV after annealing at 400 °C.[115–117]

8.3.2.12 SnS_2. Weakly crystalline SnS_2 films have been grown from tin(II) chloride and sodium sulfide precursors. EDX analysis gave a Sn/S ratio of 1/2.02. The surface roughness of a 290-nm-thick film was 32 nm. The band gap was 2.22 eV, and the photoluminescence spectra, using a 325 nm excitation source, showed peaks at 549 nm and 700 nm.[118]

8.3.2.13 MoS_2. SILAR-grown molybdenum sulfide and zirconium oxide films have been tested as coatings for MEMS silicon test devices. The MoS_2 and ZrO_2 coatings decreased the coefficient of friction relative to that of uncoated devices. The SILAR-grown coatings showed high wear.[120]

8.3.2.14 $CuInS_2$, $CuInSe_2$. $CuInS_2$ (CIS), films have been grown from mixed copper(II) chloride, indium(III) chloride cation precursor, and sodium sulfide anion precursor solutions.[121,122] XPS and XRD analyses revealed that, when the copper/indium concentration ratio in the solution was 1.25, a stoichiometric CIS film could be grown. The electrical parameters obtained with different copper/indium concentration ratios have been investigated.[121]

Copper(II) chloride, indium(III) chloride, and sodium selenosulfite were applied in separate growth cycles to obtain $CuInSe_2$ chalcopyrite films. After annealing at 400 °C, the CuSe XRD reflections disappeared and the reaction leading to the formation of $CuInSe_2$ was complete. The films were stoichiometric after annealing, as detected by XPS. The particle size increased with annealing and reached the value 200–400 nm. The cross-sectional SEM images showed good adherence to the glass substrate. Resistivity, carrier concentration, and Hall mobility of the annealed films were 0.58 Ωcm, $7.8 \times 10^{16}\,cm^{-3}$ and $5.2\,cm^2/Vs$, respectively.[121,122]

8.3.3 Films of Metals and Other Materials

See Table 8.3.

8.3.3.1 Cu, Ag. Metallic copper films have been grown by SILAR from a diluted solution of copper(II) acetate and a basic solution of formaldehyde as a reducing agent. Separate copper islands were formed on the substrate during the first growth cycles, and these islands coalesced when the film grew thicker, giving rise to a columnar structure. The films were slightly porous (Fig. 8.9) and polycrystalline, even on polymer substrates.[124]

Silver nitrate complexed with ethylenediamine and glucose and other reducing agents have been used to grow silver films by SILAR. After annealing at 300 °C the films showed cubic structure and the I–V curves exhibited linear behavior, whereas the as-grown films showed nonlinear behavior.[125]

8.3.3.2 CuI and CuSCN. Both CuI and CuSCN require copper in the monovalent state. This was achieved by complexing Cu^{2+} ions with thiosulfate, which also reduces Cu^{2+} to Cu^+. Potassium iodide or thiocyanate served as an

TABLE 8.3. Metal and Miscellaneous Other Films Grown by SILAR

Material	Precursors	Characterization	Reference
Cu	$Cu(CH_3COO)_2$ HCHO + NaOH	XRD, resistance, SEM, AFM, XPS	124
Ag	$Ag(NO_3)$ + ethylenediamine Glucose	AFM, I-V, UV, XRD	125
CuI CuSCN	$CuSO_4 + Na_2S_2O_3$ KI KSCN	XRD, SEM, AFM, UV, quantum efficiency	126, 127
$Zn_xZr(OH)_yF_z$	K_2ZrF_6 $Zn(CH_3COO)_2$ + $NH_4(CH_3COO) + NH_4OH$	XPS, FTIR	128
ScF_3 LaF_3	$Sc(NO_3)_3$ $La(NO_3)_3$, $LaCl_3$ HF, NH_4F	Ellipsometry, FTIR, XPS	129

Figure 8.9. SEM image of a SILAR-grown copper thin film. The scale bar is 2 μm. Reprinted from Lindroos, Ruuskanen, Ritala and Leskelä 2004. Thin Solid Films 460, Copyright (2004) with permission from Elsevier.

anion precursor. On glass and FTO, the polycrystalline CuI films were cubic in structure. The corresponding CuSCN films on glass exhibited a rhombohedral crystal structure. In contrast to the films on glass, CuSCN films on an FTO substrate were orthorhombic. The average crystallite size was 50–100 nm for a 300-nm-thick CuI film on glass and 500–600 nm for a 500-nm-thick CuSCN film grown on glass. The band-gap values were 2.97 and 3.72 eV for CuI and CuSCN, respectively.[32,126,127]

8.3.3.3 $Zn_xZr_yF_z$. A fluoro complex of zirconium, ZrF_6^{2-}, and an ammine complex of zinc, $Zn(NH_3)_4^{2+}$, have been used to grow $Zn_xZr_yF_z$ films on silicon. The Zn/Zr/F ratio was 0.08/1/0.32, and the films contained hydroxyl groups and physisorbed water.[128]

8.3.3.4 ScF_3 and LaF_3. Both separate and multilayer scandium and lanthanum fluoride films have been grown from nitrates and lanthanum chloride and sodium and ammonium fluorides. XPS analysis revealed that the films were stoichiometric and that the films contained both hydroxide groups and water.[129]

8.4 ILGAR

8.4.1 Basic Principles of ILGAR

Ion layer gas reaction (ILGAR) is an alternative sequential chemical deposition technique for which a metal-containing precursor solution is first transferred to the substrate by dipping (dip-ILGAR) or spraying (spray-ILGAR)

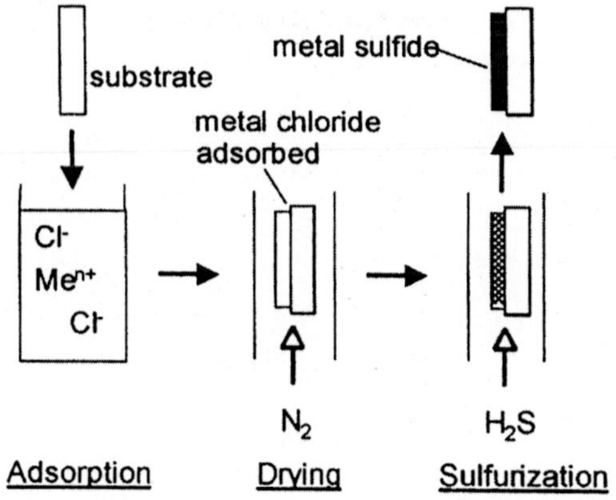

Figure 8.10. Schematic drawing of the basic principle of ILGAR. Reprinted from Möller, Fischer, Muffler, Könenkamp, Kaiser, Kelch and Lux-Steiner 2000. Thin Solid Films 361–362, Copyright (2000) with permission from Elsevier.

and dried. The dry solid precursor layer is then allowed to react with a hydrogen chalcogenide or ammonia and water gas. Figure 8.10 shows a schematic drawing of the ILGAR cycle. This cycle can be repeated several times to achieve the desired film thickness. The spray-ILGAR process more rapidly provides thicker films compared with the dip-ILGAR approach.[130–132]

The advantages of the ILGAR technique are the same as for SILAR. The process and equipment are simple, and vacuum or high-temperature conditions are not necessarily needed. Additionally, the film thickness can be controlled by the number of deposition cycles. An advantage of ILGAR compared with SILAR is the fact that the reaction takes place at a solid–gas interface, where the solid precursor layer is chemically converted to the final film. The quality of the final film is thus only dependent on a metal precursor and on the completeness of the solid–gas reaction. In contrast to solution-based reactions, solvent incorporation in the film is not typically a factor affecting film quality. The growth rate is also generally higher for the ILGAR than for the SILAR technique—i.e., a film 20–50 nm thick can be grown in 10 mins using ILGAR. Furthermore, a reagent gas more easily penetrates narrow holes and pores compared with liquids.[130–132]

8.4.2 Materials Grown by ILGAR

ILGAR has been used to grow materials for solar cells, namely CdS, Cu_2S, and In_2S_3 to form $CuInS_2$ and oxide films, most commonly ZnO. A more complete list of materials deposited by ILGAR is given in Table 8.4.

The CdS films may be used as a buffer layer between absorber and front contact in a solar cell. The CdS films deposited using the ILGAR technique

TABLE 8.4. Films Grown by ILGAR, ECALE, and Other Sequential Solution-Phase Deposition Methods

Material	Precursors	Characterization	Reference
ILGAR			
CdS Cu_2S In_2S_3 $CuInS_2$	$CdCl_2$ CuCl CuI $InCl_3$ H_2S	XRD, SEM, EDX, UV	130–133
$CuInS_2$	CuCl $InCl_3$ H_2S	XRD, XPS, SEM, UV, conductivity, Hall mobility, DSC, TG	135–137
$CuInS_2$	$Cu(CH_3COO)_2$ $InCl_3$	XRD, EDX, HRTEM, SAED	134
ZnO	$Zn(ClO_4)_2$ NH_3	XRD, EDX, UV, XPS, inverse photoemission	138–143
ECALE			
ZnS, ZnSe CdS, CdSe CdTe $Cd_xZn_{1-x}S$ $Cd_xZn_{1-x}Se$ CdS_xSe_{1-x}	$CdSO_4$ $ZnSO_4$ Na_2S Na_2SeO_3 SeO_2 TeO_2 Te	XPS, differntial pulse anodic stripping, AMF, photoelectrochemical characterization, EXAFS, surface-XRD, X-ray photoelectron diffraction (XPD), electron probe microanal., SEM, XRD, LEED, Auger, STM	145–157
PbS PbSe PbSe/PbTe	$Pb(NO_3)_2$ + EDTA $PbClO_4$ Na_2S SeO_2 TeO	XRD, AFM, EPMA, UV, IR	158–161
Bi_2S_3 Bi_2Te_3	$Bi(NO_3)_3$ Na_2S TeO_2	XRD, SEM, electron probe microanalysis (EPMA), FESEM	162–166
Sb_2Te_3	Sb_2O_3 TeO_2	XRD, FESEM, EDX, FTIR	167
In_2Se_3	$In_2(SO_4)_3$ SeO_2	XRD, AFM, UV, photoelectrochemical spectrum	168
GaAs InAs InAs/InSb	$Ga_2(SO_4)_3$ $In_2(SO_4)_3$ $HAsO_2$ Sb_2O_3	LEED, STM, Auger, coulometry, XRD, AFM, XPS, differential pulse anodic stripping	169–172

TABLE 8.4. *Continued*

Other sequential methods			
Material	Precursors	Characterization	Reference
ZnO	Zn salt H_2O (95–100 °C)	XRD	173
ZnO	Zn salt H_2O (95–100 °C)		174–178
Cu_2O	$CuSO_4 + Na_2S_2O_3$ NaOH (70 °C)	XRD	179
$Ti(HPO_4)_2$	$TiSO_4$ Na_2HPO_4 NaH_2PO_4	XPS, FTIR, SEM	180

have been homogeneous, and the XRD results showed both hexagonal and cubic crystal structures.[130,131] The broad XRD peaks indicated the presence of very small crystallites. EDX analysis of as-grown CdS films showed the presence of high chlorine impurity levels—from 8.3 to 21.7 at.%. The band gap of the as-grown CdS films was 2.78 eV. However, after annealing at 375 °C, the band gap shifted to 2.45 eV, which is near to the band-gap energy of CBD grown CdS, 2.43 eV.[130,131]

Also, indium(III) sulfide can be used as a solar cell buffer layer. In_2S_3 has been grown by the spray-ILGAR method using an elevated substrate temperature of between 100 and 400 °C.[130,132,133] At 200 °C, the film was very compact and the grains were barely observable, but at 300 °C and above, the film became rough and crystallites could be clearly observed. According to XRD analysis, the films were cubic and no significant preferred orientation could be observed. ERDA and X-ray fluorescence (XRF) analysis showed large residual chlorine content, from 3 to 17 at.%, depending on the substrate temperature. The lowest chlorine impurity level was observed for a substrate temperature of 350 °C.[130,132,133]

The $CuInS_2$ absorber layer of a solar cell has been grown by ILGAR using two methods—either by depositing sequentially the binary sulfides Cu_2S and In_2S_3 and annealing the film after deposition or by using a mixed precursor solution of CuI or CuCl and In_2Cl_3.[134–137] Using the latter method, it has been easier to get a stoichiometric $CuInS_2$ film. The as-grown films were nanocrystalline in nature, but after annealing at 500 °C in an Ar/H_2S atmosphere, the XRD peaks were narrower indicating crystal growth. Using an equimolar precursor solution, a copper-rich film was formed on glass and TiO_2 substrates and Cu_2S could be found by XRD. On porous silica, however, only $CuInS_2$ was observed by XRD. The $CuInS_2$ film, grown from a precursor solution with $[Cu^+]/[In^{3+}]$ ratio 1.55, was analyzed by XPS and it had the Cu/In ratio 0.99 and S/(Cu + In) 0.98. The morphology of $CuInS_2$ films has been observed to be dependent on indium concentration in the precursor solution. When the concentration was smaller than 0.05 M, the films were smooth and uniform, but a higher concentration resulted in rougher films. The band gap of the $CuInS_2$ films was 1.48–1.56 eV.[134–137]

ZnO films, for a solar cell buffer layer, have been grown by dip-ILGAR using a zinc perchlorate solution and a gaseous mixture of ammonia and water as precursors.[138–143] As the last step of the growth cycle, the sample was heated, or alternatively, the sample was heated only at the end of the process, under air at temperatures of between 100 °C and 300 °C. However, the best films were obtained when a process employing warm ammonia/water gas was used. The films were hexagonal in crystal structure with a (002) preferred orientation. The band gap of the ZnO was 3.38 eV. Solar cells with buffer layer and absorber layer parts of the materials grown by ILGAR have been found to have efficiencies of as high as 10.7%.[138–143]

8.5 ECALE

8.5.1 Basic Principles of ECALE

The electrochemical atomic layer epitaxy (ECALE) technique, also known as electrochemical atomic layer deposition (EC-ALD), is based on layer-by-layer electrodeposition. Each constituent of the thin film are deposited separately using underpotential deposition (UPD) of that element. UPD is a process wherein an atomic layer of one element is deposited on the surface of a different element at a potential under that needed to deposit the element on itself. ECALE has been used to grow mainly II–VI and III–V compounds. A thorough review of ECALE research has been published by Stickney.[144] A summary of the materials deposited using ECALE are given in Table 8.4, with a more detailed discussion for a few select examples given below.

8.5.2 Materials Grown by ECALE

Zinc chalcogenide thin films have been grown by ECALE using zinc sulphate as metal source and sodium sulphide and sodium selenite as chalcogenide precursors.[145–148] The formation of the first layers of ZnS on (111)Au has been analyzed by STM and XPS.[145] HRSEM images showed that the film surface was very flat, even at an atomic level. On the other hand, thicker ZnS films were formed of well-separated crystal nuclei. The stoichiometry of a thicker ZnS film showed a slight excess of sulphur, with a Zn:S ratio of 1:1.2. The band gap of a thicker film (deposition time 12 h) was 3.60 eV.[147]

Cadmium chalcogenide thin films have been grown by ECALE using cadmium sulphate as metal source and sodium sulphide, sodium selenite, selenium oxide, and tellurium oxide as chalcogenide precursors.[149–154] The electrochemical growth processes have been widely studied. The films have been found to be stoichiometric by XPS and electrochemical stripping measurements. The thin films of CdS had the hexagonal wurtzite structure.[149–151] AFM studies of the CdS films showed large clusters homogeneously distributed in the films, with the size and density of these clusters not changing with an increasing number of deposition cycles.[149] EXAFS studies showed the presence of a tetragonal structure in the CdSe films.[153] Cadmium sulphide selenides, CdS_xSe_{1-x}, and ternary cadmium and zinc sulphides and selenides, $Cd_xZn_{1-x}S$, $Cd_xZn_{1-x}Se$, have also been grown by ECALE.[155–157] The ratio of cation and anion was close to 1:1 as determined by XPS. AFM revealed that the growth was both by two- and three-dimensional growth.[156,157] The band gap was linear, following the composition of the films. CdTe films have been grown epitaxially on Au.[154] The films were produced by the alternate reductive underpotential deposition of Te and Cd. The epitaxial growth was confirmed by STM.[154]

Lead nitrate complexed with EDTA and lead perchlorate and sodium sulphide have been used for PbS ECALE-deposition.[158,159] The films were cubic and highly (200) oriented, and AFM images showed the same cubic structure.[158,159] PbSe films were also cubic, and the band gap of a film after 50 deposition cycles was $8000\,cm^{-1}$.[160] PbSe/PbTe superlattices, with 4.2-nm and 7.0-nm periods, have been grown by ECALE.[161] The (111) reflection in the XRD pattern showed a first-order satellite peak and one second-order peak, indicating the formation of the superlattice. AFM images of the superlattice structure showed a small amount of three-dimensional growth.[161]

Bismuth nitrate together with sodium sulphide and tellurium oxide have been used to grow bismuth chalcogenide films.[162–166] The stoichiometric ratio 2:3 has been confirmed by XPS, EDX, and XRD.[162–166] The Bi_2Te_3 films were rough and consisted of particles with a diameter of 30–100 nm, and electron probe microanalysis showed a worm-like network structure.[164–166]

Antimony telluride films have been grown from antimony(III) and tellurium(IV) oxides.[167] Antimony telluride films were stoichimetric and consisted of nanoscale particles of the size 100 nm. The films had a good crystallinity.[167] Indium selenide films were grown from indium sulphate and selenium oxide precursors.[168] The films consisted of large particles, 70 to 200 nm in diameter. The band gap was 1.73 eV.[168]

Oxidative underpotential deposition (UPD) of As from a $HAsO_2$ solution and reductive UPD of Ga from a $Ga_2(SO_4)_3$ solution have been used to deposit GaAs on a gold substrate.[169,170] InAs films grown by ECALE were stoichiometric, and the surface roughness did not increase from the substrate roughness during film deposition.[171] InAs/InSb superlattice structures, with periods of one monolayer of InAs and half a monolayer of InSb, have been grown.[172] AFM showed that the sample with a superlattice structure was much rougher than the InAs film.[172]

8.6 OTHER SEQUENTIAL SOLUTION-PHASE DEPOSITION TECHNIQUES

Ristov et al. was the first to prepare ZnO films with a sequential deposition process.[173] The process used resembles SILAR but lacks the rinsing steps, which may lead to a less-defined film morphology compared with SILAR grown films, because the reaction occurs also in the diffusion layer instead of only on the surface (as in SILAR). ZnO films were grown using an ammonia-complexed zinc precursor. The films were also doped with tin(II) ions. A similar method has been used by Mitra et al.[174–178] to grow ZnO films. The as-deposited films consisted of zinc hydroxide and zinc oxide. However, after annealing at 150 °C, only ZnO was found by XRD. The films could be used as gas sensors, with the best response being found toward H_2.[176–178]

Nearly stoichiometric copper(I) oxide thin films have been grown using a similar stepwise technique without rinsing.[179] Ultra-thin titanium phosphate films have also been grown using a stepwise growth process from $Ti(SO_4)_2$ and sodium hydrogen phosphate and sodium dihydrogen phosphate precursors. The composition $Ti(HPO_4)_2$ was confirmed by XPS and FTIR.[180]

REFERENCES

1. Nicolau, Y. F. 1985. Solution deposition on thin solid compound films by a successive ionic-layer adsorption and reaction process. *Appl. Surf. Sci.* 22–23: 1061–1074.
2. Nicolau, Y. F. 1986. Method and apparatus for deposition on a substrate of a thin film containing cationic and anionic constituents. French Patent No. 2569427.
3. Nicolau, Y. F. 1987. Method and apparatus for deposition on a substrate of a thin film containing cationic and anionic constituents. US Patent No. 4675207.
4. Tolstoi, V. P. 1993. Synthesis of thin-layer structures by the ionic layer deposition method. *Russ. Chem. Rev.* 62:237–242.
5. Niesen, T. P.; De Guire, M. R. 2001. Review: Deposition of ceramic thin films at low temperatures from aqueous solutions. *J. Electroceram.* 6:169–207.
6. Niesen, T. P.; De Guire, M. R. 2002. Review: Deposition of ceramic thin films at low temperatures from aqueous solutions. *Solid State Ionics* 151:61–68.
7. Pathan, H. M.; Lockhande, C. D. 2004. Deposition of metal chalcogenide thin films by successive ionic layer adsorption and reaction (SILAR) method. *Bull. Mater. Sci.* 27:85–111.
8. Nicolau, Y. F.; Menard, J. C. 1992. An electrokinetic study of ZnS and CdS surface chemistry. *J. Colloid Interface Sci.* 148:551–570.
9. Lindroos, S. 1997. The successive ionic layer adsorption and reaction (SILAR) growth and characterization of ZnS and ZnS:Mn thin films. Ph.D. thesis. University of Helsinki, Helsinki, Finland.

10. Kanniainen, T. 2001. Studies of zinc and lead chalcogenide thin films grown by SILAR (successive ionic layer adsorption and reaction) technique. Ph.D. thesis. University of Helsinki, Helsinki, Finland.

11. Gao, X. D.; Li, X. M.; Yu, W. D. 2004. Morphology and optical properties of amorphous ZnS films deposited by ultrasonic-assisted successive ionic layer adsorption and reaction method. *Thin Solid Films* 468:43–47.

12. Li, J. J.; Wang, Y. A.; Guo, W.; Keay, J. C.; Mishima, T. D.; Johnson, M. B.; Peng, X. 2003. Large-scale synthesis of nearly monodisperse CdSe/CdS core/shell nanocrystals using air-stable reagents via successive ion layer adsorption and reaction. *J. Am. Chem. Soc.* 125:12567–12575.

13. Kanniainen, T.; Lindroos, S.; Prohaska, T.; Friedbacher, G.; Leskelä, M.; Grasserbauer, M.; Niinistö, L. 1995. Growth of zinc sulfide thin films with the successive ionic layer adsorption and reaction method as studied by atomic force microscopy. *J. Mater. Chem.* 5:985–9.

14. Resch, R.; Friedbacher, G.; Grasserbauer, M.; Kanniainen, T.; Lindroos, S.; Leskelä, M.; Niinistö, L. 1997. Lateral force microscopy and force modulation microscopy on SILAR-grown lead sulfide samples. *Appl. Surf. Sci.* 120:51–57.

15. Resch, R.; Prohaska, T.; Friedbacher, G.; Grasserbauer, M.; Kanniainen, T.; Lindroos, S.; Leskelä, M.; Niinistö, L.; Broekaert, J. A. C. 1998. In-situ investigation of ZnS deposition on mica by successive ionic layer adsorption and reaction method as studied with atomic force microscopy. *Fresenius' J. Anal. Chem.* 353:772–777.

16. Valkonen, M. P.; Lindroos, S.; Kanniainen, T.; Leskelä, M.; Resch, R.; Friedbacher, G.; Grasserbauer, M. 1998. Atomic force microscopy studies of ZnS films grown on (100) GaAs by the successive ionic layer adsorption and reaction method. *J. Mater. Res.* 13:1688–1692.

17. Valkonen, M.; Lindroos, S.; Resch, R.; Leskelä, M.; Friedbacher, G.; Grasserbauer, M. 1998. Growth of zinc sulfide thin films on (100)Si with the successive ionic layer adsorption and reaction method studied by atomic force microscopy. *Appl. Surf. Sci.* 136:131–136.

18. Jiménez-González, A. E. 1997. Modification of ZnO thin films by Ni, Cu, and Cd doping. *J. Solid State Chem.* 128:176–180.

19. Jiménez-González, A.; Nair, P. K. 1995. Photosensitive ZnO thin films prepared by the chemical deposition method SILAR. *Semicond. Sci. Technol.* 10: 1277–1281.

20. Jiménez-González, A.; Suárez-Parra, R. 1996. Effect of heat treatment on the properties of ZnO thin films prepared by successive ion layer adsorption and reaction (SILAR). *J. Cryst. Growth* 167:649–655.

21. Robles, M.; Tagueena-Martinez, J.; del Rio, J. A. 1997. Effective conductivity of chemically deposited ZnO thin films. *Thin Solid Films* 293:320–326.

22. Gao, X. D.; Li, X. M.; Yu, W. D. 2004. Synthesis and optical properties of ZnO nanocluster porous films deposited by modified SILAR method. *Appl. Surf. Sci.* 229:275–281.

23. Gao, X. D.; Li, X. M.; Yu, W. D.; Li, L.; Qiu, J. J.; Peng, F. 2007. Low-temperature deposition of transparent ZnO films by the ultrasonic-mediated stepwise method. *Sol. Energy Mater. Sol. Cells* 91:467–473.

24. Gao, X. D.; Li, X. M.; Yu, W. D. 2005. Rapid preparation, characterization, and photoluminescence of ZnO films by a novel chemical method. *Mater. Res. Bull.* 40:1104–1111.

25. Gao, X.-D.; Li, X.-M.; Yu, W.-D.; Li, L.; Peng, F.; Zhang, C.-Y. 2006. Microstructure analysis and formation mechanism of ZnO nanoporous film via the ultrasonic irradiation mediated SILAR method. *J. Cryst. Growth* 291:175–182.

26. Gao, X. D.; Li, X. M.; Yu, W. D.; Li, L.; Qiu, J. J. 2007. Seed layer-free synthesis and characterization of vertically grown ZnO nanorod array via the stepwise solution route. *Appl. Surf. Sci.* 253:4060–4065.

27. Gao, X. D.; Li, X. M.; Yu, W. D. 2004. Structure and UV photoluminescence of nanocrystalline ZnO films prepared by thermal oxidation of ZnS films. *Mater. Sci. Eng. B* 113:274–278.

28. Ku, C.-H.; Wu, J.-J. 2006. Aqueous solution route to high-aspect-ratio zinc oxide nanostructures on indium tin oxide substrates. *J. Phys. Chem. B* 110: 12981–12985.

29. Shishiyanu, S. T.; Shishiyanu, T. S.; Lupan, O. I. 2005. Sensing characteristics of tin-doped ZnO thin films as NO_2 gas sensor. *Sens. Actuators B* 107:379–386.

30. Lindroos, S.; Valkonen, M.; Leskelä, M. 1999. Growth of ZnO_2/ZnO thin films by successive ionic layer adsorption and reaction—SILAR—technique. *Adv. Sci. Technol.* 20:43–50.

31. Lindroos, S.; Leskelä, M. 2000. Growth of zinc peroxide (ZnO_2) and zinc oxide (ZnO) thin films by the successive ionic layer adsorption and reaction—SILAR—technique. *Inter. J. Inorg. Mater.* 2:197–201.

32. Sankapal, B. R.; Sartale, S. D.; Lokhande, C. D.; Ennaoui, A. 2004. Chemical synthesis of Cd-free wide band gap materials for solar cells. *Sol. Energy Mater. Sol. Cells* 83:447–458.

33. Nair, M. T. S.; Guerrero, L.; Arenas, O. L.; Nair, P. K. 1999. Chemically deposited copper oxide thin films: Structural, optical and electrical characteristics. *Appl. Surf. Sci.* 150:143–151.

34. Tolstoi, V. P.; Molotkina, E. V. 1994. The synthesis of copper-peroxide nanolayers on silicon and fused-quartz surfaces by the ionic deposition technique.. *Russ J. Inorg. Chem.* 39:372–374.

35. Tolstoi, V. P.; Tolstobrov, E. V. 1994. CuO_2/BaO_2 superlattices grown by ionic layering. *Inorg. Mater.* 30:875–878.

36. Tolstoy, V. P. 1997. The peroxide route of the successive ionic layer deposition procedure for synthesizing nanolayers of metal oxides, hydroxides and peroxides. *Thin Solid Films* 307:10–13.

37. Korotcenkov, G.; Macsanov, V.; Brinzari, V.; Tolstoy, V.; Schwank, J.; Cornet, A.; Morante, J. 2004. Influence of Cu-, Fe-, Co-, and Mn-oxide nanoclusters on sensing behavior of SnO_2 films. *Thin Solid Films* 467:209–214.

38. Tolstoi, V. P. 1993. Synthesis of ultrathin layers of $SnO_2 \cdot nH_2O$ on a silicon surface by ion layer deposition. *Russ. J. Inorg. Chem.* 38:1063–1065.

39. Park, S.; DiMasi, E.; Kim, Y.-I.; Han, W.; Woodward, P. M.; Vogt, T. 2002. The preparation and characterization of photocatalytically active TiO_2 thin films and nanoparticles using successive-ionic-layer-adsorption-and-reaction. *Thin Solid Films* 515:1250–1254.

40. Johnson, S. E.; Burgoon, M. W. P.; Wang, Q.; White, J. M. 2006. Low-temperature preparation of anatase thin films on tantalum. *Langmuir* 22:6570–6577.

41. Pathan, H. M.; Min, S.-K.; Desai, J. D.; Jung, K.-D.; Joo, O.-S. 2006. Preparation and characterization of titanium dioxide thin films by SILAR method. *Mater. Chem. Phys.* 97:5–9.

42. Kale, S. S.; Mane, R. S.; Chung, H.; Yoon, M.-Y.; Lokhande, C. D.; Han, S.-H. 2006. Use of successive ionic layer adsorption and reaction (SILAR) method for amorphous titanium dioxide thin films growth. *Appl. Surf. Sci.* 253:421–424.

43. Park, S.; Herman, G. S.; Kezler, D. A. 2003. Oxide films: Low temperature deposition and crytallization. *J. Solid State Chem.* 175:84–87.

44. Tolstoy, V. P.; Ehrlich, A. G. 1997. The synthesis of $CeO_{2+n} \cdot n\ H_2O$ nanolayers on silicon and fused-quartz surfaces by the successive ionic layer deposition technique. *Thin Solid Films* 307:60–64.

45. Tolstoi, V. P.; Tolstobrov, E. V. 2002. Synthesis of highly oriented α-PbO_2 layers on the surfaces of single-crystal silicon and quartz by successive ionic layer deposition. *Russ. J. Appl. Chem.* 75:1529–1531.

46. Tolstoy, V. P.; Murin, I. V.; Reller, A. 1997. The synthesis of Mn(IV) oxide nanolayers by the successive ionic layer deposition method. *Appl. Surf. Sci.* 112: 255–257.

47. Park, S.; Clark, B. L.; Keszler, D. A.; Bender, J. P.; Wager, J. F.; Reynolds, T. A.; Herman, G. S. 2002. Low-temperature thin-film deposition and crystallization. *Science* 297:65.

48. Liu, J.-F.; Nistorica, C.; Gory, I.; Skidmore, G.; Mantiziba, F. M.; Gnade, B. E. 2005. Layer-by-layer deposition of zirconium oxide films from aqueous solutions for friction reduction in silicon-based microelectromechanical system devices. *Thin Solid Films* 492:6–12.

49. Tolstoi, V. P. 1995. Syntheisis of $Tl_2O_3 \cdot H_2O$ nanolayers on a silica surface by ion deposition. *Russ. J. Inorg. Chem.* 40:208–210.

50. Gulina, L. B.; Tolstoi, V. P.; Murin, I. V. 2001. Study of Co—Mn—O— and Li—Co—Mn—O—containing layers synthesized by successive ionic layer deposition on silica surface. *Russ. J. Appl. Chem.* 74:1955–1957.

51. Tolstoi, V. P.; Molotilkina, E. V. 1994. Synthesis of nanolayers of Y, La, and Eu hydroperoxides by ionic layering on the silicon surface. *Inorg. Mater.* 30:201–203.

52. Tolstoi, V. P.; Tolstobrov, E. V. 2004. Synthesis of "hybrid" $Ag_x^0 \cdot MnO_2 \cdot nH_2O$ metal oxide nanolayers by ionic deposition. *Russ. J. Gen. Chem.* 74:323–326.

53. Gulina, L. B.; Tolstoi, V. P. 2004. Synthesis on silica surface by the ionic deposition technique of nanolayers of heteropolycompounds on the basis of phosphomolybdic acid. *Russ. J. Gen. Chem.* 74:327–330.

54. Nicolau, Y. F.; Dupuy, M.; Brunel, M. 1990. ZnS, CdS and $Zn_{1-x}Cd_xS$ thin films deposited by the successive ionic layer adsorption and reaction process. *J. Electrochem. Soc.* 137:2915–2924.

55. Nicolau, Y. F.; Menard, J. C. 1988. Solution growth of zinc sulfide, cadmium sulfide and zinc cadmium sulfide thin films by the successive ionic-layer adsorption and reaction process; growth mechanism. *J. Cryst. Growth* 92:128–42.

56. Nicolau, Y. F.; Menard, J. C. 1990. Procedure for solution growth of zinc sulfide, cadmium sulfide, and zinc cadmium sulfide ($Zn_{1-x}Cd_xS$) thin films by successive ionic-layer adsorptions and reactions. *J. Appl. Electrochem.* 20:1063–1066.

57. Klechkovskaya, N. N.; Maslov, V. N.; Muradov, M. B. 1989. Growth and structure of semiconducting films of CdS, ZnS, and solid solutions based on them, obtained by the mechanisms of chemisorption from solutions of electrolytes. *Soviet Physics Crystallogr.* 34:105–107.

58. Lindroos, S.; Kanniainen, T.; Leskelä, M. 1994. Growth of ZnS thin films by liquid phase atomic layer epitaxy (LPALE). *Appl. Surf. Sci.* 75:70–74.

59. Lindroos, S.; Kanniainen, T.; Leskelä, M. 1996. Zinc sulfide thin films grown by SILAR on poly(vinyl chloride) and polycarbonate substrates. *J. Mater. Chem.* 6:1497–1500.

60. Lindroos, S.; Kanniainen, T.; Leskelä, M. 1997. Growth of zinc sulfide thin films by successive ionic layer adsorption and reaction—(SILAR)—method on polyester substrates. *Mater. Res. Bull.* 32:1631–1636.

61. Lindroos, S.; Kanniainen, T.; Leskelä, M.; Rauhala, E. 1995. Deposition of manganese-doped zinc sulfide thin films by the successive ionic layer adsorption and reaction (SILAR) method. *Thin Solid Films* 263:79–84.

62. Lindroos, S.; Charreire, Y.; Kanniainen, T.; Leskelä, M.; Benazeth, S. 1997. Zn K-edge EXAFS study of SILAR-grown zinc sulfide thin films. *J. Mater. Chem.* 7:741–745.

63. Lindroos, S.; Charreire, Y.; Bonnin, D.; Leskelä, M. 1998. Growth and characterization of zinc sulfide thin films deposited by the successive ionic layer adsorption and reaction (SILAR) method using complexed zinc ions as the cation precursor. *Mater. Res. Bull.* 33:453–459.

64. Laukaitis, G. 2000. Stress and surface analysis of SILAR grown CdS, ZnS and $Cd_xZn_{1-x}S$ thin films on (100)GaAs. Ph.D. thesis. Kaunas Technical University, Kaunas, Lithuania.

65. Laukaitis, G.; Lindroos, S.; Tamulevicius, S.; Leskelä, M.; Raukaitis, M. 2000. Stress and surface studies of SILAR grown ZnS thin films on (100)GaAs substrates. *Mater. Sci. Eng. A* 288:223–230.

66. Laukaitis, G.; Lindroos, S.; Tamulevicius, S.; Leskelä, M. 2001. Stress and morphological development of CdS and ZnS thin films during the SILAR growth on (100)GaAs. *Appl. Surf. Sci.* 185:134–139.

67. Valkonen, M. 1998. Preparation and characterization of CdS, ZnS thin films and CdS/ZnS multilayer thin films grown by the SILAR technique. Ph.D. thesis. University of Helsinki, Helsinki, Finland.

68. Valkonen, M. P.; Kanniainen, T.; Lindroos, S.; Leskelä, M.; Rauhala, E. 1997. Growth of ZnS, CdS and multilayer ZnS/CdS thin films by SILAR technique. *Appl. Surf. Sci.* 115:386–392.

69. Resch, R.; Friedbacher, G.; Grasserbauer, M.; Kanniainen, T.; Lindroos, S.; Leskelä, M.; Niinistö, L. 1997. In situ investigations on the SILAR-growth of ZnS films as studied by tapping mode atomic force microscopy (TM-AFM). *Fresenius' J. Anal. Chem.* 358:80–84.

70. Resch, R.; Friedbacher, G.; Grasserbauer, M.; Lindroos, S.; Kanniainen, T.; Valkonen, M. P.; Leskelä, M. 1998. Investigation of ZnS thin films on Si(100) by

phase detection imaging and Young's modulus microscopy. *Fresenius' J. Anal. Chem.* 361:613–617.

71. Lokhande, C. D.; Pathan, H. M.; Giersig, M.; Tributsch, H. 2002. Preparation of $Zn_x(O,S)_y$ thin films using modified chemical bath deposition method. *Appl. Surf. Sci.* 187:101–107.

72. Battaglia, D.; Blackman, B.; Peng, X. 2005. Coupled and decoupled dual quantum systems in one semiconductor nanocrystal. *J. Amer. Chem. Soc.* 127:10889–10897.

73. Kale, R. B.; Lokhande, C. D. 2004. Room temperature deposition of ZnSe thin films by successive ionic layer adsorption and reaction (SILAR) method. *Mater. Res. Bull.* 39:1829–1839.

74. Kale, S. S.; Mane, R. S.; Pathan, H. M.; Shaikh, A. V.; Joo, O.-S.; Han, S.-H. 2007. Preparation and characterization of ZnTe thin films by SILAR method. *Appl. Surf. Sci.* 253:4335–4337.

75. Tamulevicius, S.; Valkonen, M. P.; Laukaitis, G.; Lindroos, S.; Leskelä, M. 1999. Stress and surface studies of SILAR grown CdS thin films on (100)GaAs. *Thin Solid Films* 355–356:430–434.

76. Valkonen, M. P.; Lindroos, S.; Leskelä, M. 1998. $Cd_xZn_{1-x}S$ solid solution thin films, CdS thin films and CdS/ZnS thin films grown by the SILAR technique. *Appl. Surf. Sci.* 134:283–291.

77. Sasagawa, M.; Nosaka, Y. 2001. Studies on the effects of Cd ion sources and chelating reagents on atomic layer CdS deposition by successive ionic layer adsorption and reaction (SILAR) method. *Phys. Chem. Chem. Phys.* 3:3371–3376.

78. Sasagawa, M.; Nosaka, Y. 2002. Electrochemical evaluation of the roles of chelating reagents in Cd ion adsorption on CdS surface for the successive ionic layer adsorption and reaction (SILAR) deposition. *J. Electroanal. Chem.* 536: 141–144.

79. Lokhande, C. D.; Sankapal, B. R.; Pathan, H. M.; Muller, M.; Giersig, M.; Tributsch, H. 2001. Some structural studies on successive ionic layer adsorption and reaction (SILAR)—deposited CdS thin films. *Appl. Surf. Sci.* 181:277–282.

80. Larramona, G.; Choné, C.; Jacob, A.; Sakakura, D.; Delatouche, B.; Péré, D.; Cieren, X.; Nagino, M.; Bayón, R. 2006. Nanostructured photovoltaic cell of the type titanium dioxide, cadmium sulfide thin coating, and copper thiocyanate showing high quantum efficiency. *Chem. Mater.* 18:1688–1696.

81. Pathan, H. M.; Sankapal, B. R.; Desai, J. D.; Lokhande, C. D. 2003. Preparation and characterization of nanocrystalline CdSe thin films deposited by SILAR method. *Mater. Chem. Phys.* 78:11–14.

82. Battaglia, D.; Li, J. J.; Wang, Y.; Peng, X. 2003. Colloidal two-dimensional systems: CdSe quantum shells and wells. *Angew. Chem. Int. Ed.* 42:5035–5039.

83. Kale, R. B.; Sartale, S. D.; Chougule, B. K.; Lokhande, C. D. 2004. Growth and characterization of nanocrystalline CdSe thin films deposited by the successive ionic layer adsorption and reaction method. *Semicond. Sci. Technol.* 19: 980–986.

84. Dias, E. A.; Sewall, S. L.; Kambhampati, P. 2007. Light harvesting and carrier transport in core/barrier/shell semiconductor nanocrystals. *J. Phys. Chem. C* 111:708–713.

85. Ubale, A. U.; Dhokne, R. J.; Chikhlikar, P. S.; Sangawar, V. S.; Kulkanari, K. 2006. Characterization of nanocrystalline cadmium telluride thin films grown by successive ionic layer adsorption and reaction (SILAR) method. *Bull. Mater. Sci.* 29:165–168.

86. Valkonen, M. P.; Ritala, M.; Lindroos, S.; Leskelä, M. 1998. Solid solution $Cd_xZn_{1-x}S$ thin films grown by atomic layer epitaxy and successive ionic layer adsorption and reaction techniques. *Mater. Sci. Forum* 287–288:367–370.

87. Laukaitis, G.; Lindroos, S.; Tamulevicius, S.; Leskelä, M.; Rackaitis, M. 2000. SILAR deposition of $Cd_xZn_{1-x}S$ thin films. *Appl. Surf. Sci.* 161:396–405.

88. Xie, R.; Kolb, U.; Li, J.; Basché, T.; Mews, A. 2005. Synthesis and characterization of highly luminescent CdSe-core $CdS/Zn_{0.5}Cd_{0.5}S/ZnS$ multishell nanocrystals. *J. Amer. Chem. Soc.* 127:7480–7488.

89. Valkonen, M. P.; Lindroos, S.; Kanniainen, T.; Leskelä, M.; Tapper, U.; Kauppinen, E. 1997. Thin multilayer CdS/ZnS films grown by SILAR technique. *Appl. Surf. Sci.* 120:58–64.

90. Fleischhaker, F.; Zentel, R. 2005. Photonic crystals from core-shell colloids with incorporated highly fluorescent quantum dots. *Chem. Mater.* 17:1346–1351.

91. Kanniainen, T.; Lindroos, S.; Ihanus, J.; Leskelä, M. 1996. Growth of strongly orientated lead sulfide thin films by successive ionic layer adsorption and reaction (SILAR) technique. *J. Mater. Chem.* 6:161–164.

92. Kanniainen, T.; Lindroos, S.; Leskelä, M. 1995. Deposition of lead sulfide thin films by successive ionic layer and reaction-technique. *Adv. Sci. Technol.* 5:291–298.

93. Kanniainen, T.; Lindroos, S.; Resch, R.; Leskelä, M.; Friedbacher, G.; Grasserbauer, M. 2000. Structural and topographical studies of SILAR-grown highly oriented PbS thin films. *Mater. Res. Bull.* 35:1045–1051.

94. Puišo, J. 2004. Growth kinetics and properties of lead sulfide thin films deposited on crystalline silicon using successive ionic layer adsorption and reaction method. Ph.D. thesis. Kaunas University of Technology, Kaunas, Lithuania.

95. Puišo, J.; Tamulevicius, S.; Lindroos, S.; Leskelä, M.; Snitka, V.; Pucëta, M. 2001. Growth kinetics of thin PbS films deposited by SILAR. *Lithuanian Journal of Physics* 41:361–366.

96. Puišo, J.; Tamulevicius, S.; Laukaitis, G.; Lindroos, S.; Leskelä, M.; Snitka, V. 2002. Growth of PbS thin films on silicon substrate by SILAR technique. *Thin Solid Films* 403–404:457–461.

97. Puišo, J.; Lindroos, S.; Tamulevicius, S.; Leskelä, M.; Snitka, V. 2003. Growth of ultra thin PbS films by SILAR technique. *Thin Solid Films* 428:223–226.

98. Puišo, J.; Lindroos, S.; Tamulevicius, S.; Leskelä, M.; Snitka, V. 2003. The influence of the pretreatment of Si substrate on the growth of PbS thin films in the SILAR technique. *Solid State Phenomena* 94:261–264.

99. Puišo, J.; Tamulevicius, S.; Laukaitis, G.; Lindroos, S.; Leskelä, M.; Pucëta, M. 2002. Growth of PbS thin films on silicon substrate by SILAR technique. *Thin Solid Films* 403–404:457–461.

100. Lindroos, S.; Puišo, J.; Tamulevicius, S.; Leskelä, M. 2004. CdS-PbS multilayer thin films grown by the SILAR method. *Solid State Phenomena* 99–100: 243–246.

101. Gavrilov, S.; Oja, I.; Lim, B.; Belaidi, A.; Bohne, W.; Strub, E.; Röhrich, J.; Lux-Steiner, M.-Ch.; Dittrich, Th. 2006. Charge selective contact on ultra-thin $In(OH)_xS_y/Pb(OH)_xS_y$ heterostructure prepared by SILAR. *Phys. Status Solidi A* 203:1024–1029.

102. Oja, I.; Belaidi, A.; Dloczik, L.; Lux-Steiner, M-Ch.; Dittrich, Th. 2006. Photoelectrical properties of $In(OH)_xS_y/PbS(O)$ structures deposited by SILAR on TiO_2. *Semicond. Sci. Technol.* 21:520–526.

103. Kanniainen, T.; Lindroos, S.; Ihanus, J.; Leskelä, M. 1996. Growth of lead selenide thin films by successive ionic layer adsorption and reaction (SILAR) technique. *J. Mater. Chem.* 6:983–986.

104. Patil, R. S.; Lokhande, C. D.; Mane, R. S.; Pathan, H. M.; Oh-Shim, J.; Sung-Hwan, H. 2006. Successive ionic layer adsorption and reaction (SILAR) trend for nanocrystalline mercury sulfide thin films growth. *Mater. Sci. Eng. B* 129:59–63.

105. Sartale, S. D.; Lockhande, C. D. 2001. Effect of annealing on the structural and optical properties of SILAR grown Cu_xS thin films. *Indian J. Phys.* 75A: 375–378.

106. Lindroos, S.; Arnold, A.; Leskelä, M. 2000. Growth of CuS thin films by the successive ionic layer adsorption and reaction method. *Appl. Surf. Sci.* 158:75–80.

107. Pathan, H. M.; Desai, J. D.; Lokhande, C. D. 2002. Modified chemical deposition and physico-chemical properties of copper sulphide (Cu_2S) thin films. *Appl. Surf. Sci.* 202:47–56.

108. Sartale, S. D.; Lokhande, C. D. 2000. Preparation and characterization of As_2S_3 thin films deposited using successive ionic layer adsorption and reaction (SILAR) method. *Mater. Res. Bull.* 35:1345–1353.

109. Sankapal, B. R.; Ganesan, V.; Lokhande, C. D. 2000. Studies on deposition of antimony triselenide thin films by chemical method: SILAR. *Indian J. Pure Appl. Phys.* 38:606–610.

110. Sankapal, B. R.; Mane, R. S.; Lokhande, C. D. 1999. Preparation and characterization of Sb_2S_3 thin films using a successive ionic layer adsorption and reaction (SILAR) method. *J. Mater. Sci. Lett.* 18:1453–1455.

111. Lokhande, C. D.; Sankapal, B. R.; Sartale, S. D.; Pathan, H. M.; Giersig, M.; Ganesan, V. A. 2001. A novel method for the deposition of nanocrystalline Bi_2Se_3, Sb_2Se_3 and Bi_2Se_3—Sb_2Se_3 thin films—SILAR. *Appl. Surf. Sci.* 182:413–417.

112. Ahire, R. R.; Sankapal, B. R.; Lokhande, C. D. 2001. Preparation and characterization of Bi_2S_3 thin films using modified chemical bath deposition method. *Mater. Res. Bull.* 36:199–210.

113. Sankapal, B. R.; Mane, R. S.; Lokhande, C. D. 2000. Preparation and characterization of Bi_2Se_3 thin films deposited by successive ionic layer adsorption and reaction (SILAR) method. *Mater. Chem. Phys.* 63:230–234.

114. Sankapal, B. R.; Mane, R. S.; Pathan, H. M.; Muller, M.; Giersig, M.; Tributsch, H.; Ganeshan, V. 2002. Structural characterization of chemically deposited Bi_2S_3 and Bi_2Se_3 thin films. *Appl. Surf. Sci.* 187:108–115.

115. Mane, R. S.; Lokhande, C. D. 2003. Studies on structural, optical and electrical properties of indium sulfide thin films. *Mater. Chem. Phys.* 78:15–17.

116. Pathan, H. M.; Lokhande, C. D.; Kulkarni, S. S.; Amalnerkar, D. P.; Seth, T.; Han, S.-H. 2005. Some studies on successive ionic layer adsorption and reaction (SILAR) grown indium sulphide thin films. *Mater. Res. Bull.* 40:1018–1023.

117. Ranjith, R.; John T. T.; Kartha C. S.; Vijayakumar K. P.; Abe T.; Kashiwaba Y. 2007. Post-deposition annealing effect on In_2S_3 thin films deposited using SILAR technique. *Mater. Sci. Semicond. Process.* 10:49–55.

118. Deshpande, N. G.; Sagade, A. A.; Gudage, C. D.; Lokhande, C. D. 2007. Growth and characterization of tin disulfide (SnS_2) thin film deposited by successive ionic layer adsorption and reaction (SILAR) technique. *J. Alloys Compd.* 436: 421–426.

119. Sartale, S. D.; Lokhande, C. D. 2000. Deposition of cobalt sulphide thin films by successive ionic layer adsorption and reaction (SILAR) method and their characterization. *Indian J. Pure Appl. Phys.* 38:48–52.

120. Nistorica, C.; Liu, J.-F.; Gory, I.; Skidmore, G. D.; Mantiziba, F. M.; Gnade, B. E.; Kim, J. 2005. Tribological and wear studies of coatings fabricated by atomic layer deposition and by successive ionic layer adsorption and reaction for microelectromechanical devices. *J. Vacuum Sci. Technol. A.* 23:836–840.

121. Shi, Y.; Jin, Z.; Li, C.; An, H.; Qiu, J. 2006. Effect of [Cu]/[In] ratio on properties of $CuInS_2$ thin films prepared by successive ionic layer absorption and reaction method. *Appl. Surf. Sci.* 252:3737–3743.

122. Shi, Y.; Jin, Z.; Li, C.; An, H.; Qiu, J. 2007. Effects of post-heat treatment on the characteristics of chalcopyrite $CuInSe_2$ film deposited by successive ionic layer absorption and reaction method. *Thin Solid Films* 515:3339–3343.

123. Gulina, L. B.; Tolstoi, V. P. 2002. Synthesis of Ag_7SbS_6 nanolayers on the silica surface by ionic layer deposition. *Russ. J. Gen. Chem.* 72:840–843.

124. Lindroos, S.; Ruuskanen, T.; Ritala, M.; Leskelä, M. 2004. Growth of Cu thin films by the successive ionic layer adsorption and reaction (SILAR) method. *Thin Solid Films* 460:36–40.

125. Wang, L.; Kang, S.; Mu, J. 2006. Electrical properties of Ag thin films deposited by the improved SILAR method. *J. Dispersion Sci. Technol.* 27:393–397.

126. Sankapal, B. R.; Ennaoui, A.; Guminskaya, T.; Dittrich, Th.; Bohne, W.; Röhrich, J.; Strub, E.; Lux-Steiner, M. Ch. 2005. Characterization of p-CuI prepared by the SILAR technique on Cu-tape/n-$CuInS_2$ for solar cells. *Thin Solid Films* 480–481:142–146.

127. Sankapal, B. R.; Goncalves, E.; Ennaoui, A.; Lux-Steiner, M. Ch. 2004. Wide band gap p-type windows by CBD and SILAR methods. *Thin Solid Films* 451–452:128–132.

128. Tolstoy, V. P.; Altangerel, B. 2007. A new "fluoride" synthesis route for successive ionic layer deposition of the $Zn_xZr(OH)_yF_z \cdot nH_2O$ nanolayers. *Mater. Lett.* 61:123–125.

129. Zhuchkov, B. S.; Tolstoy, V. P.; Murin, I. V. 1997. Synthesis of ScF_3, LaF_3 nanolayers and $nLaF_3$-$mScF_3$ multinanolayers at the surface of silicon by successive ionic layer deposition method. *Solid State Ionics* 101–103:165–170.

130. Muffler, H.-J.; Fischer, Ch.-H.; Diesner, K.; Lux-Steiner, M. C. 2001. ILGAR—A novel thin-film technology for sulfides. *Sol. Energy Mater. Sol. Cells* 67: 121–127.

131. Möller, J.; Fischer, Ch.-H.; Muffler, H.-J.; Könenkamp, R.; Kaiser, I.; Kelch, C.; Lux-Steiner, M. C. 2000. A novel deposition technique for compound semiconductors on highly porous substrates: ILGAR. *Thin Solid Films* 361–362: 113–117.

132. Allsop, N. A.; Schönmann, A.; Belaidi, A.; Muffler, H.-J.; Mertesacker, B.; Bohne, W.; Strub, E.; Röhrich, J.; Lux-Steiner, M. C.; Fischer, Ch.-H. 2006. Indium sulfide thin films deposited by the spray ion layer gas reaction technique. *Thin Solid Films* 513:52–56.

133. Allsop, N. A.; Hänsel, A.; Visbeck, S.; Niesen, T. P.; Lux-Steiner, M. C.; Fischer, Ch.-H. 2006. The dry and damp heat stability of chalcopyrite solar cells prepared with an indium sulphide buffer deposited by the spray-ILGAR technique. *Thin Solid Films* 511–512:55–59.

134. Das, K.; Panda, S. K.; Chaudhuri, S. 2007. Fabrication of nanostructured $CuInS_2$ thin films by ion layer gas reaction method. *Appl. Surf. Sci.* 253:5166–5172.

135. Qiu, J.; Jin, Z.; Qian, J.; Shi, Y.; Wu, W. 2005. Influence of post-heat treatment on the properties of $CuInS_2$ thin films deposited by an ion layer gas reaction (ILGAR). *J. Cryst. Growth* 282:421–428.

136. Qiu, J.; Jin, Z.; Qian, J.; Shi, Y.; Wu, W. 2005. ILGAR $CuInS_2$ films from various preparation conditions. *Mater. Lett.* 59:2735–2740.

137. Qiu, J.; Jin, Z.; Wu, W.; Xiao, L.-X. 2006. Characterization of $CuInS_2$ thin films prepared by ion layer gas reaction method. *Thin Solid Films* 510:1–5.

138. Bär, M.; Fischer, Ch.-H.; Muffler, H.-J.; Zweigart, S.; Karg, F.; Lux-Steiner, M. C. 2003. Replacement of the CBD–CdS buffer and the sputtered i-ZnO layer by an ILGAR-ZnO WEL: Optimization of the WEL deposition. *Sol. Energy Mater. Sol. Cells* 75:101–107.

139. Bär, M.; Muffler, H.-J.; Fischer, Ch.-H.; Lux-Steiner, M. C. 2001. ILGAR technology IV: ILGAR thin film technology extended to metal oxides. *Sol. Energy Mater. Sol. Cells* 67:113–120.

140. Bär, M.; Weinhardt, L.; Heske, C.; Muffler, H.-J.; Umbach, E.; Lux-Steiner, M. Ch.; Niesen, Th. P.; Karg, F.; Fischer, Ch.-H. 2006. Chemical insights into the Cd^{2+}/NH_3 treatment—An approach to explain the formation of Cd-compounds on $Cu(In,Ga)(S,Se)_2$ absorbers. *Sol. Energy Mater. Sol. Cells* 90:3151–3157.

141. Fischer, Ch.-H.; Bär, M.; Glatzel, Th.; Lauermann, I.; Lux-Steiner, M. C. 2006. Interface engineering in chalcopyrite thin film solar devices. *Sol. Energy Mater. Sol. Cells* 90:1471–1485.

142. Hariskos, D.; Spiering, S.; Powalla M. 2005. Buffer layers in $Cu(In,Ga)Se_2$ solar cells and modules. *Thin Solid Films* 480–481:99–109.

143. Muffler, H-J.; Bär, M.; Lauermann, I.; Rahne, K.; Schröder, M.; Lux-Steiner, M. C.; Fischer, C.-H.; Niesen, T. P.; Karg, F. 2006. Colloid attachment by ILGAR-layers: Creating fluorescing layers to increase quantum efficiency of solar cells. *Sol. Energy Mater. Sol. Cells* 90:3143–3150.

144. Stickney, J. L. 1999. Electrochemical atomic layer epitaxy. *Electroanalytical Chem.* 21:75–209.

145. Gichuhi, A.; Shannon, C.; Perry, S. 1999. A scanning tunneling microscopy and X-ray photoelectron spectroscopy study of electrochemically grown ZnS monolayers on Au(111). *Langmuir* 15:5654–5661.

146. Torimoto, T.; Obayashi, A.; Kuwabata, S.; Yasuda, H.; Mori, H.; Yoneyama, H. 2000. Preparation of size-quantized ZnS thin films using electrochemical atomic layer epitaxy and their photoelectrochemical properties. *Langmuir* 16:5820–5824.

147. Öznülüer, T.; Erdogan, I.; Demir, Ü. 2006. Electrochemically induced atom-by-atom growth of ZnS thin films: A new approach for ZnS co-deposition. *Langmuir* 22:4415–4419.

148. Pezzatini, G.; Caporali, S.; Innocenti, M.; Foresti, M. 1999. Formation of ZnSe on Ag(111) by electrochemical atomic layer epitaxy. *J. Electroanal. Chem.* 475: 164–170.

149. Innocenti, M.; Cattarin, S.; Cavallini, M.; Loglio, F.; Foresti, M. 2002. Characterisation of thin films of CdS deposited on Ag(111) by ECALE. A morphological and photoelectrochemical investigation. *J. Electroanal. Chem.* 532:219–225.

150. Cecconi, T.; Atrei, A.; Bardi, U.; Forni, F.; Innocenti, M.; Loglio F.; Foresti M.; Rovida G. 2001. X-ray photoelectron diffraction (XPD) study of the atomic structure of the ultrathin CdS phase deposited on Ag(111) by electrochemical atomic layer epitaxy (ECALE). *J. Electron Spectrosc. Relat. Phenom.* 114–116: 563–568.

151. Colletti, L.; Flowers, Jr. B.; Stickney, J. 1998. Formation of thin films of CdTe, CdSe, and CdS by electrochemical atomic layer epitaxy. *J. Electrochem. Soc.* 1998:1442–1449.

152. Lister, T. E.; Stickney, J. L. 1996. Formation of the first monolayer of CdSe on Au(111) by electrochemical ALE. *Appl. Surf. Sci.* 107:153–160.

153. Loglio, F.; Innocenti, M.; D'Acapito, F.; Felici, R.; Pezzatini, G.; Salvietti, E.; Foresti, M. L. 2005. Cadmium selenide electrodeposited by ECALE: electrochemical characterization and preliminary results by EXAFS. *J. Electroanal. Chem.* 575:161–167.

154. Hayden, B. E.; Nandhakumar, I. S. 1998. In situ STM study of CdTe ECALE bilayers on gold. *J. Phys. Chem. B* 102:4897–4905.

155. Foresti, M. L.; Milani, S.; Loglio, F.; Innocenti, M.; Pezzatini, G.; Cattarin. S. 2005. Ternary CdS_xSe_{1-x} deposited on Ag(111) by ECALE: synthesis and characterization. *Langmuir* 21:6900–6907.

156. Innocenti, M.; Cattarin, S.; Loglio, F.; Cecconi, T.; Seravalli, G.; Foresti, M. L. 2004. Ternary cadmium and zinc sulfides: Composition, morphology and photoelectrochemistry. *Electrochim. Acta* 49:1327–1337.

157. Loglio, F.; Innocenti, M.; Pezzatini, G.; Foresti, M. L. 2004. Ternary cadmium and zinc sulfides and selenides: electrodeposition by ECALE and electrochemical characterization. *J. Electroanal. Chem.* 562:117–125.

158. Öznülüer, T.; Erdogan, I.; Sisman, I.; Demir, Ü. 2005. Electrochemical atom-by-atom growth of PbS by modified ECALE method. *Chem. Mater.* 17:935–937.

159. Tsukasa, T. A.; Takabayashi, S.; Mori, H.; Kuwabata, S. 2002. Photoelectrochemical activities of ultrathin lead sulfide films prepared by electrochemical atomic layer epitaxy. *J. Electroanal. Chem.* 522:33–39.

160. Vaidyanathan, R.; Stickney, J. L.; Happek, U. 2004. Quantum confinement in PbSe thin films electrodeposited by electrochemical atomic layer epitaxy (EC-ALE). *Electrochim. Acta* 49:1321–1326.

161. Vaidyanathan, R.; Cox, S. M.; Happek, U. 2006. Preliminary studies in the electrodeposition of PbSe/PbTe superlattice thin films via electrochemical atomic layer deposition (ALD). *Langmuir* 22:10590–10595.

162. Öznülüer, T.; Demir, Ü. 2002. Formation of Bi_2S_3 thin films on Au(111) by electrochemical atomic layer epitaxy: Kinetics of structural changes in the initial monolayers. *J. Electroanal. Chem.* 529:34–42.

163. Yang, J.; Zhu, W.; Gao, X.; Bao, S.; Fan, X. 2005. Electrochemical aspects of the formation of Bi_2Te_3 thin film via the route of ECALE. *J. Electroanal. Chem.* 577:117–123.

164. Zhu, W.; Yang, J. Y. Hou, J.; Gao, X. H.; Bao, S. Q.; Fan X. A. 2005. Optimization of the formation of bismuth telluride thin film by using ECALE. *J. Electroanal. Chem.* 585:83–88.

165. Zhu, W.; Yang, J. Y.; Gao, X. H.; Bao, S. Q.; Fan, X. A.; Zhang, T. J.; Cui, K. 2005. Effect of potential on bismuth telluride thin film growth by electrochemical atomic layer epitaxy. *Electrochim. Acta* 50:4041–4047.

166. Zhu, W.; Yang, J.; Gao, X.; Hou, J.; Zhang, T.; Cui, K. 2005. Growth of bismuth telluride thin film on Pt by electrochemical atomic layer epitaxy. *Trans. Nonferrous Metals Society of China* 15:404–409.

167. Yang, J. Y.; Zhu, W.; Gao, X. H.; Fan, X. A.; Bao, S. Q.; Duan, X. K. 2007. Electrochemical aspects of depositing Sb_2Te_3 compound on Au substrate by ECALE. *Electrochim. Acta* 52:3035–3039.

168. Vaidyanathan, R.; Stickney, J. L.; Cox, S. M. 2003. Formation of In_2Se_3 thin films and nanostructures using electrochemical atomic layer epitaxy. *J. Electroanal. Chem.* 559:55–61.

169. Villegas, I.; Stickney, J. L. 1992. Gallium arsenide deposition on the (100) and (110) planes of gold by electrochemical atomic layer epitaxy: A low-energy electron diffraction, Auger electron spectroscopy, and scanning tunneling microscopy study. *J. Vac. Sci. Technol. A* 10:3032–3038.

170. Villegas, I.; Stickney, J. L. 1992. Preliminary studies of gallium arsenide deposition on gold (100), (110), and (111) surfaces by electrochemical atomic layer epitaxy. *J. Electrochem. Soc.* 139:686–94.

171. Innocenti, M.; Forni, F.; Pezzatini, G.; Raiteri, R.; Loglio, F.; Foresti, M. L. 2001. Electrochemical behavior of As on silver single crystals and experimental conditions for InAs growth by ECALE. *J. Electroanal. Chem.* 514:75–82.

172. Wade, T. L.; Vaidyanathan, R.; Happek, U.; Stickney, J. L. 2001. Electrochemical formation of a III-V compound semiconductor superlattice: InAs/InSb. *J. Electroanal. Chem.* 500:322–332.

173. Ristov, M.; Sinadinovski, G.; Grozdanov, I.; Mitreski, M. 1987. Chemical deposition of ZnO films. *Thin Solid Films* 149:65–71.

174. Mitra, P.; Chatterjee, A. P.; Maiti, H. S. 1998. ZnO thin film sensor. *Mater. Lett.* 35:33–38.

175. Mitra, P.; Chatterjee, A. P.; Maiti, H. S. 1998. Chemical deposition of ZnO films for gas sensor. *J. Mater. Sci. Mater. Electron.* 9:441–445.

176. Mitra, P.; Khan, J. 2006. Chemical deposition of ZnO films from ammonium zincate bath. *Mater. Chem. Phys.* 98:279–284.

177. Mitra, P.; Maiti, H. S. 2004. A wet-chemical process to form palladium oxide sensitiser layer on thin film zinc oxide based LPG sensor. *Sens. Actuators B* 97:49–58.

178. Chatterjee, A. P.; Mitra, P.; Mukhopadhyay, K. 1999. Chemically deposited zinc oxide thin film gas sensor. *J. Mater. Sci.* 34:4225–4231.

179. Ristov, M.; Sinadinovski, G.; Grozdanov, I. P. 1985. Chemical deposition of Cu_2O thin films. *Thin Solid Films* 123:63–67.

180. Wang, Q.; Zhong, L.; Sun, J.; Shen, J. 2005. A facile layer-by-layer adsorption and reaction method to the preparation of titanium phosphate. *Chem. Mater.* 17:3563–3569.

CHAPTER 9

Evaporation-Induced Self-Assembly for the Preparation of Porous Metal Oxide Films

BERND SMARSLY* and DINA FATTAKHOVA-ROHLFING**

*Institute of Physical Chemistry, Justus-Liebig-University of Giessen, Giessen, Germany
**Ludwig-Maximilians-Universität München, Department of Physical Chemistry and Biochemistry, Munich, Germany

9.1 INTRODUCTION

Porous thin films are of great interest because of their potential technological impact in applications using heterogeneous interface processes (e.g., interfacial electron/ion transfer and diffusion), such as sensors, selective membranes, photovoltaics and energy storage devices, electrochromic coatings, and barrier layers. The introduction of patterned porosity into a material drastically increases its interfacial surface area and surface-to-bulk ratio, thus inducing additional properties complementary to the intrinsic physico-chemical properties of the bulk. Several property variations can be further extended by functionalization of the surface with organic or inorganic moieties, or by incorporation of guest species into the pore structure. One of the key challenges in this area is the problem of controlling, ordering, and combining of pore structure over different length scales. Evaporation-induced self-assembly (EISA) is nowadays one of the most popular techniques enabling manufacturing of films with periodically organized porosity via solution processing.

Invented less than a decade ago,[1] EISA has rapidly developed into a universal technique for fabrication of organized porous and patterned nanocomposite materials, ranging from metal oxides and chalcogenides to carbons, polymers, and metals.[2,3] In addition to generating ordered mesoporous films, this technique may also be used to incorporate functional molecules and nanoparticles into the mesostructured films, with the formation of hybrid structures.

The number of publications concerning utilization of the EISA process for fabrication of different structured materials is counted in the hundreds, which is far beyond the possibilities of this chapter to review in depth. Rather, we intend to provide a brief introduction into EISA and its application to the fabrication of functional thin films for electronic applications (e.g., electrochromic layers and solar cells), with a special focus on fabrication of crystalline mesoporous films of metal oxides. Attention will also be given to techniques used to evaluate the pore structure of the thin films. For the other aspects of the EISA process, for example its mechanism,[4] strategies for preparation of crystalline porous metal oxides,[5] mesoporous nanohybrid materials,[6] periodic organic silica materials,[7,8] or postgrafting functionalization of mesoporous framework,[9] we kindly recommend the reader to refer to the referenced comprehensive reviews.

9.2 THE EISA PROCESS

EISA can be described as a solution deposition technique, which combines sol-gel synthesis with self-assembly and microphase separation. EISA is actually an extension of more general templating approaches, for which preparation of porous materials is achieved via spontaneous self-assembly of precursor and template species with formation of supramolecular periodic structures.[10–12] As templates for preparation of mesoporous materials, amphiphilic molecules are most often used. These molecules possess mutually immiscible hydrophilic and hydrophobic parts, which therefore segregate in appropriate solvents with formation of micelles of various shape. The driving force of micelle formation is microphase separation in order to minimize the total surface energy. The micelles can further aggregate in periodic superstructures, so-called lyotropic liquid-crystalline phases, which can serve as organic templates for controlled assembly of the inorganic precursor added to the surfactant solution.[13] Such a liquid-crystal templating (LCT) mechanism (Fig. 9.1) is also referred to also as "nanocasting," where a stabilized surfactant mesophase is assigned as a robust template for condensation of the inorganic phase. Another proposed mechanism is called cooperative self-assembly (CSA), in which a surfactant and an inorganic precursor form a combined intermediate phase, which behaves as an independent surfactant species and builds up a hybrid structure. The nature of the process is strongly dependent on the chemical composition of the system, but in most cases, the process is likely to follow a combination of both mechanisms.[4]

The prerequisite for templating is the presence of weak noncovalent interactions between the hydrophilic micelle interface and the precursor, so that the total interfacial energies of the resulting composite are lowered. Too strong interactions, as for example strong electrostatic ones, can lead to the

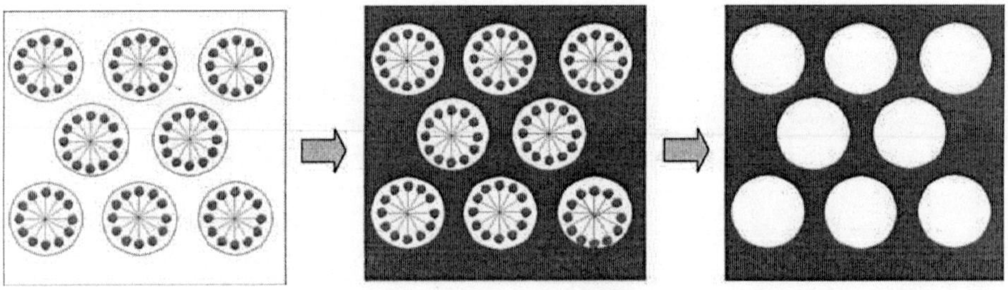

Figure 9.1. Scheme of micelle templating. The circular objects correspond to micelles and the gray color to the condensed inorganic oxide network.

formation of surfactant-stabilized single particles instead of continuous inorganic networks, whereas, in contrast, too weak or no interaction usually causes the phase separation of organic and inorganic phases.

The classically used templates are low-weight surfactants, which usually contain the long-chain alkanes (oligoethylene) as a hydrophobic part. Depending on the nature of the hydrophilic group, they are divided into ionic (e.g., cationic such as cetyltrimetylammonium bromide (CTAB) or anionic such as sodium lauryl sulphate) and nonionic (e.g., containing oligo(ethyleneoxide) as a head group). The surfactants rely mostly on water or water mixtures as solvent, where they self-organize after exceeding their critical micellar concentration (cmc) into various periodic structures (e.g., cubic, hexagonal, or lamellar) depending on their concentration and temperature. Such a rich library of accessible periodic organic structures enables preparation of different mesoporous inorganic materials, whose mesostructure can be roughly predicted on the basis of phase diagrams of corresponding surfactants and can be controlled by the surfactant concentration.[14]

The pore diameter that can be obtained using the low-weight surfactants is, however, restricted to a relatively small size of 2–5 nm. The available pore size can be substantially extended when block copolymers (amphiphilic polymers) are used as templates. Their hydrophilic block is in most cases poly(ethyleneoxide) (PE) or, more seldom, poly(vinylpyridine).[15] The variety of available hydrophobic blocks is broader and is represented by poly(propylene oxide) (PP),[16] polystyrene (PS),[17] polybutadiene (PB),[18] and poly(ethylene-co-butylene) (KL).[19] The most often used block of copolymer templates are a commercially available family of Pluronic triblock-copolymers with a general formula $PE_xPP_yPE_x$ and Brij diblock-copolymers with a formula of $CH_3(CH_2)_xPE_yOH$. The block copolymers form periodic micellar structures also in less polar solvents such as ethanol, tetrahydrofuran, ethers, and for PS-PEO,[20] even in toluene, which substantially increases the amount of used precursor and correspondingly the inorganic materials that can be prepared on their base. An additional advantage of block copolymers as templates is their higher molecular weight and rich variability in size and composition,

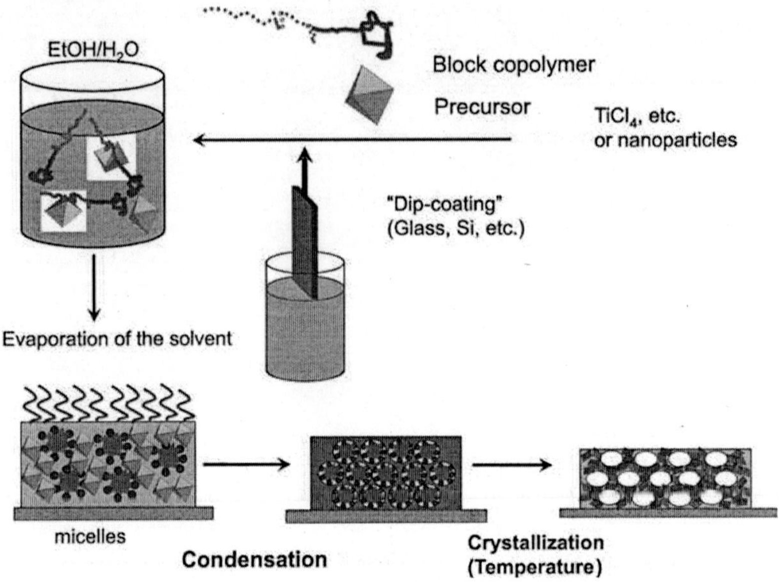

Figure 9.2. Scheme of the EISA process.

which provides broader possibilities for manufacturing of diverse mesoporous structures and increases the range of available pore sizes.

The peculiarity of the EISA process is that it deals with diluted precursor solutions (Fig. 9.2). The low viscosity of such solutions makes them especially suitable for manufacturing of thin films of high optical quality via different deposition techniques such as dip- or spin-coating, spraying, or casting. The surfactant concentration in such a solution is, however, below the cmc, so no micelles are initially present for most amphiphiles. Organization of the surfactant molecules into micelles takes place after evaporation of some part of the solvent, thus triggering self-assembly of template and precursor species into periodically organized structures. This implies that evaporation-induced self-assembly is a complex dynamic process, which can be influenced by practically all processing parameters. The existing mechanistic studies provide the fundamental understanding of the processes taking place during evaporation-induced self-assembly, thus providing a solid and reliable platform for planning of synthetic strategies for the manufacture of desirable mesoporous structures. The complexity and variability of the EISA process and the number of parameters influencing the total equilibrium make, however, theoretical forecasting practically impossible and one must rely, to a certain extent, on a trial-and-error approach.

Formation of organized mesoporous structures in the EISA process is generally governed by three major parameters: (1) the nature and composition of the precursor solution, which predefines the chemical composition and, to a great extent, the mesostructure of the final film; (2) the deposition process, which drives the self-assembly of precursor and template moieties by evaporation of the volatile solvent, thus directing formation of the periodically structured inorganic–organic composite material; and (3) the postdeposition treatment, which is aimed at condensation and, in some cases, crystallization of a continuous inorganic framework and, if required, removal of the template with liberation of free pores.

The precursors for the EISA-processed thin films are homogeneous solutions of inorganic precursor and template in concentrations below cmc in some volatile solvent. Alcohols and, especially, ethanol are the most commonly used solvents, as they are compatible with most precursors and templates and have good wetting abilities with respect to hydrophilic substrates. Water is typically another important reagent in the hydrolysis reaction. The amount of water controls the rate and degree of hydrolysis/condensation of the inorganic precursor, influencing the chemical composition and the quality of meso-organization of the final film. On the other hand, water is one of the components of the phase diagram of the structure-directing agent. So by variation of the solvent/template/water ratio, it is possible to obtain different mesostructures. The precursor solution can also contain the catalysts of the hydrolysis and condensation of inorganic precursor, which is usually hydronium or hydroxide ions. The catalysts can be also formed *in situ*. The most typical example of this situation is the formation of hydrochloric acid as a product of hydrolysis, when metal chloride is used as the inorganic precursor.[21]

The nature of the compound employed as inorganic precursor plays a decisive role in the chemical composition of the obtained inorganic matrix and, to a great extent, in the type of mesostructure that is finally formed. The precursor species should be able to assemble around the hydrophilic part of the micelles with the formation of a continuous inorganic network. Therefore, they have to possess a certain affinity to the polar micelle interface and enough mobility to follow the formation of a hybrid organic–inorganic structure in a cooperative self-assembly process. This is mostly achieved by using hydrolyzed species as inorganic building blocks, obtained from easily hydrolyzable inorganic compounds. The most typical precursors for formation of mesoporous metal oxides are metal alkoxides and metal chlorides, but hydroxides, salts, and complexes with labile ligands are also employed. The degree of hydrolysis and condensation and the surface charge of inorganic species are extremely important parameters, determining the mobility of these species, the strength of their interaction with the micelle interface and hence their suitability to self-assembly, and the rate of inorganic network condensation, which should not take place until the formation of the organized templated structure is basically completed. These parameters are controlled to a great extent by the pH of the solution, presence of condensation catalysts/inhibitors, amount of water present in the reaction media and the reaction atmosphere, and the duration of the hydrolysis reaction (i.e., precursor aging). The most suitable inorganic building blocks are slightly condensed low-weight inorganic oligomers, which can stay stable in solution for a long time and can be triggered to condense when necessary by application of some external parameter, such as temperature, humidity, ultraviolet (UV) irradiation, or condensation catalyst.

Finally, the ratio of the inorganic component to the organic template is of a great importance for the formation of the desired mesoporous structure. This parameter is usually calculated as the volume fraction of organic and inorganic components, which as a rule varies in a range of 20–80 volume % of the organic part. Variation in the fraction of the template often enables obtaining different mesoporous structures.

Self-assembly of the precursor components in the EISA process starts after evaporation of part of the volatile solvent, which progressively enriches the solution with respect to surfactant, precursor, and water. When their concentration in the deposited film achieves a certain critical level, mesostructure formation takes place. The latter is in equilibrium with the processing atmosphere. So the relative humidity, as well as the temperature during film deposition, represent some of the most important parameters influencing the mesostructure formation.

Kinetic control is another important issue. Assembly of the inorganic building blocks with an organic template always competes with condensation of the inorganic part. When the rate of the latter process is faster than that of the former one, self-assembly can be "frozen" before the formation of a periodic structure is completed. When cooperative assembly of the inorganic precursor and the templating agent is not thermodynamically forbidden or kinetically impeded, the EISA process results in formation of an organic–inorganic periodic mesophase exhibiting liquid crystalline behavior. Formation of a stable final structure often passes through a critical state called "tunable steady state," [4] where the structure is still labile and can be influenced and modified by the equilibration with atmosphere humidity.

The last stage of the EISA process, namely postdeposition treatment, is extremely important for the completion of the mesophase obtained after the film deposition. The objectives that can be achieved during this stage are multifold. First, stabilization and additional condensation of the hydrolyzed inorganic framework are usually required, which can be achieved by different means depending on the selected system—most typically by a thermal treatment. The other possible ways of matrix stabilization include UV treatment, exposure to condensation catalyst atmosphere (thus, treatment in a NH_3 atmosphere can be used for condensation of a silica matrix), and drying. Second, modification of the inorganic scaffold into a phase with desired properties can be achieved, such as crystallization of the amorphous inorganic component. This concept will be described in more detail in Section 9.4, which is devoted to preparation of crystalline mesoporous oxides. Also, very important, removal of the organic component with liberation of free pores is generally needed, which can be achieved via thermal treatment, UV or plasma decomposition of the organic template, or its extraction with organic solvent or supercritical CO_2. Finally, the chemical functionality of the films' surface can be adjusted in this stage by substitution of the surface hydroxyl groups with the desired functional moieties.[9]

9.3 CHARACTERIZATION OF SELF-ASSEMBLED FILMS

Compared with the corresponding bulk materials, the structural characterization of thin, self-assembled films prepared by EISA imposes certain complications. The analysis of structural parameters such as the pore size distribution, the pore volume, and the type of mesostructure (e.g., lamellae and cubic) is complicated mainly by the low amount of material but also by factors resulting from the film geometry itself. For instance, self-assembled films often show a preferred orientation of the mesostructure with respect to the substrate, which can only be discerned by appropriate techniques adapted to films, especially grazing incidence X-ray analysis. In particular, major insights into the self-aggregation process can be achieved by *in situ* methods following the structure formation in real time. Among other methods, 2D-GISAXS (grazing incidence small-angle X-ray scattering), SRSAXS (SAXS in symmetric reflection), X-ray reflectometry, wide-angle X-ray scattering (WAXS), ellipsoporosimetric porosimetry, (high-resolution) transmission electron microscopy, time-resolved infrared spectroscopy, surface acoustic waves (SAW), and optically polarized xenon nuclear magnetic resonance (NMR) have proven to be highly relevant for this purpose. Since the advances in the preparation of self-assembled thin films is strongly based on the parallel progress in appropriate characterization techniques, in the following sections, several of these novel analytical methods are described, which were specifically designed for thin-film analysis.

9.3.1 Positron Annihilation Lifetime Spectroscopy (PALS)

PALS is sensitive to different kinds of defects, such as dislocations, grain boundaries, voids, and pores. Similar to most scattering techniques, PALS is a noninvasive technique, thereby allowing for the detection of inaccessible pores, and it is sufficiently sensitive to determine the porosity in thin films. PALS is based on the decay of positrons into two photons ("annihilation").[22] Entering the sample, the positrons (e.g., created by a ^{22}Na radioactive source) lose their energy by inelastic collisions with electrons. These "thermalized" positrons have energies on the order of a few meV and form positroniums (Ps, the electron–positron bound state), which diffuse through the solid until annihilation after their lifetime. The natural lifetime of Ps (i.e., 142 ns) is reduced by annihilation with electrons during collisions and, thus, becomes longer when a positron or positronium is localized at spaces with lower electron density, such as voids. Therefore, positrons can be used as a probe to investigate the average sizes of the free volume, size distribution, and the free volume concentration by measuring their lifetimes.[23] Since the determination of pore size distributions is based on the trapping of positrons in voids of varying size, PALS was reported to lead to a single, average lifetime in the case of interconnected pore systems. PALS has been successfully applied to the investigation of porosity in various types of porous materials, such as thin siliceous films.[24,25]

9.3.2 Gas Physisorption

The fundamental principle of all physisorption techniques is the dependence on the pore size of gas condensation into the corresponding fluid, as expressed, for example, by the Kelvin equation. The gases have to be chosen according to the requirement that sorption/condensation in mesopores can be detected at reasonable pressures and temperatures, and that the interaction with the solid can be described entirely on the basis of physical, but not chemical, sorption mechanisms.[26] Therefore, for this reason, nitrogen and argon are usually used for the characterization of porous solids. Another reason to employ these gases is the advance in the theoretical understanding of sorption phenomena for such gases.[26] Simplistically, in a typical physisorption experiment, the volume of adsorbed gas is measured as a function of the pressure of the gas in relation to its saturation pressure. Physisorption represents a powerful methodology, as various structural parameters can be obtained, such as the pore size distribution, surface area (e.g., by the BET formalism), and the overall meso- and micropore volume. However, although there is no difference in the theory between physisorption in porous bulk materials and thin films, the standard setups and experimental procedures introduced for the bulk are not applicable to films. As a main problem, physisorption analysis on thin films has to face the low quantities of material and the fact that the films are prepared on a substrate, which impedes the analysis for several reasons. Standard setups for the characterization of powders, using nitrogen or argon, require at least a surface area of several meters squared, which demands at least ca. 10 mg for any valid analysis. Evidently, the amount of material in thin films can be orders of magnitude less. Scratching off material from a large number of samples is not desirable because of potential damage to the mesostructure. Therefore, significant effort has been dedicated in the past years to develop strategies to allow for physisorption experiments on a single specimen of thin film, i.e., typically with a specimen area of the order of 1 cm^2 only. Three main approaches have emerged to fill the gap of routine physisorption measurements on thin films.

9.3.2.1 Krypton Sorption. Volumetric adsorption using gases with low saturated vapor pressure has been found to be an effective technique to gain detailed structural information for small quantities of porous materials, especially using krypton (Kr).[27] The substitution of nitrogen by Kr reduces significantly the amount of unadsorbed molecules in the dead volume, allows for the characterization of small surface areas, and is thus ideal for mesoporous

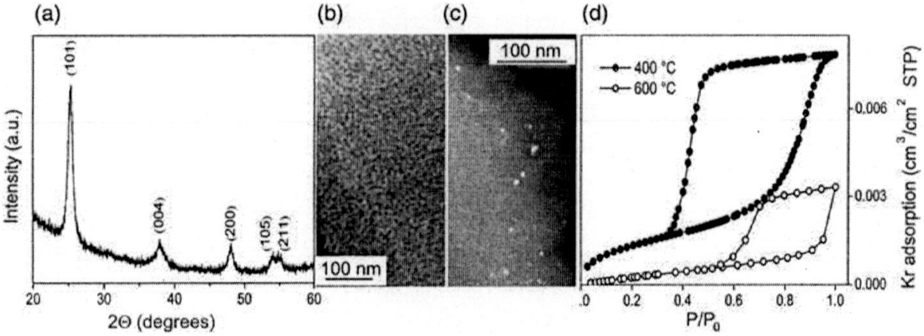

Figure 9.3. Characterization of mesoporous TiO_2 films templated by Pluronics block copolymers using diverse characterization techniques: XRD pattern (a), transmission electron microscope (TEM) image (b), dark-field TEM image (c), and isotherms of Kr adsorption (d). The Pluronic-templated TiO_2 films were calcined at 400 °C (solid points) and 600 °C (open points). The films were prepared according to Alberius et al. (Ref. 14).

films.[27] However, a certain experimental inconvenience is associated with the necessity to perform the measurements at higher temperatures than that of liquid nitrogen (77 K). In a recent study by Zukal, it was demonstrated that the pore condensation of krypton occurs only in mesopores that are smaller than 6 nm.[27] Thus, liquid argon (87 K) represents an ideal means to provide a suitable temperature for the detection of larger mesopores, and Kr was successfully applied to characterize ordered mesoporous films.[28] Figure 9.3 shows an example of a krypton isotherm of a porous silica sample with 3-nm mesopores.[29] It can be expected that krypton physisorption will develop into a standard characterization tool for thin porous films, as for nitrogen sorption for bulk material, which however could be affected by the price of liquid argon.

9.3.2.2 Surface Acoustic Waves (SAWs). The basic idea of this technique is to use the dependence of the frequency and propagation of surface acoustic waves on mass loading in a film. The porous film has to be deposited on a piezoelectric substrate (quartz), which is then placed into a physisorption setup to condense nitrogen at 77 K. Adsorption and condensation of N_2 result in a shift of the oscillation frequency, and thus measurements of the oscillation frequency as a function of N_2 partial pressure provide an adsorption–desorption isotherm.[30] Although the technique has proven to provide a concise characterization of porous films,[29,30] the requirement for the deposition directly onto the SAW piezoelectric substrate represents a certain restriction.

9.3.2.3 Spectroscopic Ellipsometry Porosimetry (EP). In general, ellipsometry takes advantage of the change of polarization of a polarized light beam after reflection from a surface. From the parameters (Ψ and Δ), obtained from the Fresnel coefficients of the reflected beam, diverse information can be obtained, especially the film thickness and refractive index. Thus, classic ellipsometry allows for the determination of the overall porosity of films. Very recently, ellipsometric measurements were coupled with physisorption of gases into the film under ambient conditions.[31,32] In such experiments, the change in the refractive index is plotted against the relative pressure of the gas (e.g., ethanol and toluene). Such isotherms provide the pore volume of the film and a pore size distribution. This novel technique has several advantages, including the possibility to use small pieces of specimens. Also, the method is sensitive enough to enable the study of sorption phenomena in films at realistic conditions regarding applications, namely room temperature and gases with other than purely physical interaction. However, a main drawback of this technique is the lack of suitable theoretical approaches to model this interaction and, thus, to extract the exact porosity parameters.

9.3.3 Small-Angle X-Ray Scattering (SAXS)

9.3.3.1 2D SAXS/GISAXS. One of the main challenges in the characterization of self-assembled films is the determination of the exact mesostructure (e.g., lamellae and cubic), the periodicity (unit cell parameters), and its relative orientation with respect to the substrate. For instance, ordered packings [body-centered (bcc) or face-centered cubic (fcc)] of spherical mesopores can adopt different orientations such as [100] or [110] with respect to the substrate. Small-angle x-ray scattering using 2D detectors has emerged as the main tool to elucidate both the mesostructure and its orientation, and numerous studies have reported appropriate experimental and theoretical 2D SAXS methods (e.g., see Fig. 9.4). Two main experimental approaches must be distinguished. Experiments performed in "grazing incidence" geometry (GISAXS) take advantage of the fact that an x-ray beam undergoes total reflections at small angles of incidence, thus minimizing absorption by the sample and maximizing the desired signals of the mesostructure. Thereby, GISAXS has turned out to be an invaluable tool for time-resolved measurements at high-flux synchrotron setups (e.g., during the mesostructure formation and the stabilization, which significantly contributed to an advanced understanding of the self-assembly in thin films and the subsequent sintering).[33–35] Since the main hurdle for widespread application is the nonavailability of synchrotron facilities, recent efforts have focused on the development of approaches using laboratory 2D SAXS setups. Indeed, it was demonstrated that 2D SAXS measurements are feasible even with laboratory instruments using thin substrates.[36,37] In contrast to GISAXS, these studies also demonstrated that 2D SAXS patterns acquired at different angles of incidence between 0° and 90° offer the opportunity to obtain more information on the degree of preferred orientation and disorder.[38]

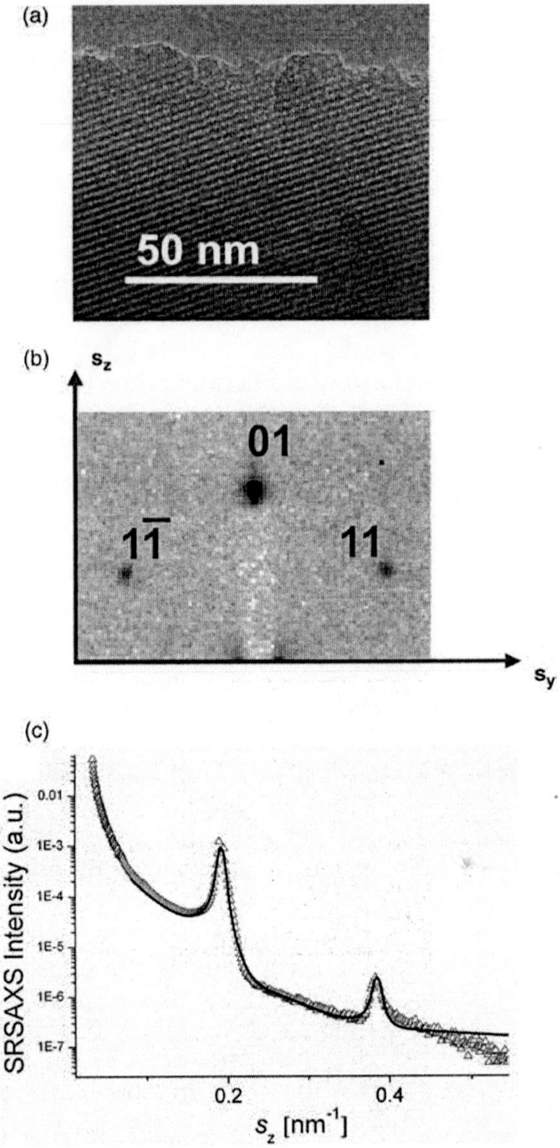

Figure 9.4. Characterization of mesoporous SiO_2 films with cylindrical mesopores (ca. 3 nm in diameter) templated using Brij 58 surfactant: TEM image a), 2D GISAXS pattern with crystallographic indexation b), and SRSAXS/XRR analysis c). The experimental data in c) thus correspond to a detailed scan along the s_z axis in b), using a suitable diffractometer. The films were prepared according to Ref. 39 and analyzed by the methods described therein.

9.3.3.2 X-Ray Reflectometry (XRR)/SAXS in Symmetric Reflection (SRSAXS). Although 2D detectors are needed to account for an unambiguous determination of the type of mesostructure, recently advanced analytical and experimental approaches were developed to extract precise values for the periodicity and particularly the pore size and even disorder parameters from X-ray scattering data. The main idea of X-ray scattering measurements in reflectivity mode or "symmetric reflection" is the precise measurement of the scattering intensity along the specular reflection, which is close to and above the critical angle.[39–42] Evidently, such measurements only provide structural information perpendicular to the substrate and hence should be complemented by 2D SAXS analysis. However, suitable evaluation approaches provide various structural parameters with excellent accuracy, such as the film thickness, the pore diameter, and wall thickness, and even translational disorder.[39,40] Measurements of the critical angle even allow for the determination of the absolute porosity, because the density of the film is related to this angle.[41,42] In essence, both XRR and SRSAXS analyses are based on the fitting of the entire scattering curve using various structural parameters as fit parameters. These approaches were successfully applied to different types of self-assembled films[39,40,42] and can be expected to establish an interesting alternative to physisorption experiments, because they can be performed on laboratory setups.

9.4 GENERATION OF MESOPOROUS CRYSTALLINE METAL OXIDE FILMS VIA EVAPORATION-INDUCED SELF-ASSEMBLY

Crystallinity is often a critical issue in electronic applications based on the semiconducting properties of metal oxides. This requirement is imposed by applications involving electron transport processes initiated by heterogeneous electron transfer at an interface, such as sensorics, photovoltaics, or electrochromics. Metal oxide films for such kinds of applications should combine a very high interface surface area and a well-developed accessible pore structure, with a high degree of crystallinity. Despite the seeming simplicity, a simultaneous combination of these properties in one material is a nontrivial task. Use of molecular precursors, which gives EISA an advantage in flexibility and variability of the obtainable structures, turns out to be an Achilles' heel of this method when crystalline materials have to be produced. Condensation of the hydrolyzed precursor results as a rule in amorphous inorganic polymer networks containing a large amount of free hydroxyl groups. Their solidification to more dense metal oxides leads unavoidably to the shrinkage of the formed skeleton, which causes deformation of the formed mesostructure. The deformation problems become much more significant when further crystallization of the solidified inorganic mesophase has to be achieved, whereas uncontrollable nucleation and crystal growth can lead to a complete collapse of a thermodynamically metastable periodic structure.

Most studies concerning EISA processing of crystalline mesoporous metal oxides deal with titanium dioxide. Intensive research on this material results from its technologically important applications in photocatalysis[43] and dye-sensitized solar cells (Graetzel cells),[44] which require high surface area and crystallinity of the titania layer for efficient performance. The classic EISA processing of mesoporous titania films uses titanium alkoxides or titanium tetrachloride as the titania precursor, Pluronic polymer as a templating agent, and ethanol as a volatile solvent.[14,45] Thermal treatment of the deposited layers at temperatures above 400 °C leads to titania films with cubic (e.g., Im3m), hexagonal (e.g., P6m) or worm-like mesostructures consisting typically of anatase nanocrystals as the major component and often some proportion of layered TiO_2 (B).[46,47] The phase composition can be altered by the choice of precursor. Thus, Chen et al. demonstrated that TiO_2 films with varying composition of anatase and rutile phases can be prepared, when a mixture of $Ti(OBu)_4$ and $TiCl_4$ is used as precursor.[48]

Titania films prepared by the methods described above are, however, just partially crystalline. Although WAXS patterns indicate formation of anatase crystals of ca. 10–12 nm in size (Fig. 9.3a), the electron microscopy study demonstrates that the elongated crystals are actually embedded into an amorphous mesoporous matrix (Fig. 9.3c). The degree of crystallinity for such films usually does not exceed 60%; attempts to increase it by calcination at higher temperatures cause uncontrolled crystal growth, which leads to collapse of mesoporosity and a drastic decrease in the surface area (Fig. 9.3d).

The main problem of mesostructure stability during crystallization is, therefore, the fast growth of crystalline seeds caused by diffuse sintering. To prevent this, the preparation procedure should be carried out in such a way that the nucleation is triggered at once, so that a large number of nuclei of the same size are formed simultaneously. The necessary condition for simultaneous nuclei formation is a structural homogeneity of the amorphous inorganic network before the onset of crystallization or, in other words, the absence of large density fluctuations that could serve as nucleation centers. It implies that a complete condensation and dehydration of the inorganic precursor should precede the nucleation process. As was shown by Grosso et al., this can be achieved by a long-term treatment of the deposited film at temperatures slightly below the crystallization temperature of the metal oxide, T_c, followed by a fast ramp to temperatures above T_c—a process that is also called the delayed rapid crystallization (DRC) mechanism.[34,49] The condensation has to be supported by the organic template, which should withstand to a certain extent the temperature treatment. In the case of Pluronic-templated titania films, such a treatment leads to one-dimensional shrinkage of the initially formed bcc cubic pore structure, with merging of the pores, resulting in a periodic grid-like pore structure with completely crystalline titania walls. The DRC technique was extended to a large number of other metal oxides, enabling preparation of crystalline grid-like ZrO_2, Nb_2O_5, or even ternary transition metal oxides.[50–52]

Another way to complete condensation of an inorganic matrix is a treatment of the deposited films at different humidity in combination with slightly elevated temperatures. This approach, called delayed humidity treatment (DHT), enables controlled mesoorganization and condensation of the metal oxides (e.g., SnO_2) with a low condensation rate.[53,54] The thermal stability of such mesostructures is, however, low. So, in most cases, just semicrystalline metal oxide films can be obtained.

The thermal stability of mesoporous frameworks substantially increases with an increase in the wall thickness and pore size, which can be varied even for the same template by changing the processing conditions. Ozin et al.[55] developed a way to prepare crystalline titania films with a 2D-hexagonal architecture by replacement of ethanol in the Pluronic-containing precursor solution with more hydrophobic butanol-1. The latter promotes phase separation at low surfactant-to-titania ratios, resulting in thicker pore walls, which are more compatible with the crystal growth during subsequent calcination.

Fabrication of the mesoporous crystalline structures is greatly facilitated by the use of a more robust template, which is able to withstand the conditions of crystallization in a broader temperature range. This was demonstrated by introduction of two families of block copolymer templates: KLE (KL–PEO)[19] and PIB–b–PEO,[18] which substantially extend the applicability of EISA toward preparation of different classes of crystalline metal oxides with developed mesoporous architechture. Both families of polymers feature large hydrophobic/hydrophilic contrast because of a large size and the different polarities of hydrophilic and hydrophobic blocks. This large contrast makes their self-assembly less sensitive to the character of the surroundings, ensuring fast micellization and mesoorganization in different precursor solutions. The larger size of the micelles and their enhanced thermal stability makes the mesostructure more robust and better adapted to accommodation of the growing crystals.

Using KLE as a template, preparation of various crystalline metal oxides with highly periodic mesoporous structure becomes possible using a relatively simple synthetic procedure.[56] Besides TiO_2 (see Fig. 9.5)[21] and other transition metal oxides, such as SnO_2,[36] WO_3,[57] α-Fe_2O_3 and α-FeOOH,[58] indium tin oxide (ITO),[59] and many others, even oxides with very high crystallization temperatures such as Al_2O_3[60] and HfO_2,[61] were prepared without loss of mesoporosity. The necessary condition for successful preparation is stabilization of the inorganic skeleton before onset of crystallization, which is achieved by a prolonged treatment below the crystallization temperature. Crystallization, which is started by heating to higher temperatures, is actually a solid–solid transformation of a completely condensed amorphous network into the crystalline phase. The metal oxides templated with the KLE polymer possess structural similarity from a close packing of the spherical KLE micelles, leading to formation of bcc cubic periodic structures after their self-assembly. Condensation at elevated temperatures leads to the unidirectional shrinkage of spherical micelles in the direction perpendicular to a substrate plane,

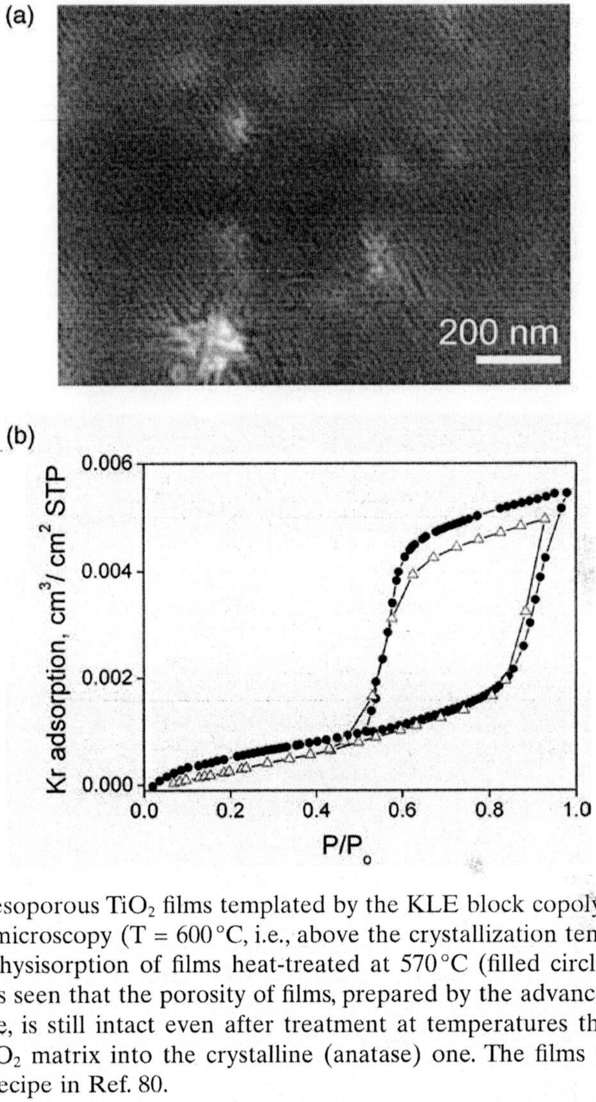

Figure 9.5. Mesoporous TiO_2 films templated by the KLE block copolymer. (a) Scanning electron microscopy (T = 600 °C, i.e., above the crystallization temperature) and (b) Krypton physisorption of films heat-treated at 570 °C (filled circles) and 650 °C (triangles). It is seen that the porosity of films, prepared by the advanced block copolymer template, is still intact even after treatment at temperatures that convert the amorphous TiO_2 matrix into the crystalline (anatase) one. The films were prepared based on the recipe in Ref. 80.

resulting in formation of ellipsoidal pores with a size of 12 to 80 nm along their long axis (Fig. 9.6). The degree of shrinkage can be very significant, achieving up to 50% depending on calcination temperature. This is one of the disadvantages of the EISA process, which is practically unavoidable when starting from a liquid molecular precursor.

The problem of structure shrinkage can be solved, when pre-formed nanocrystalline particles are used as the starting structural elements. This was successfully demonstrated for SnO_2[62] and CeO_2[63] nanoparticles, which were assembled in highly periodic cubic crystalline structures using KLE as a

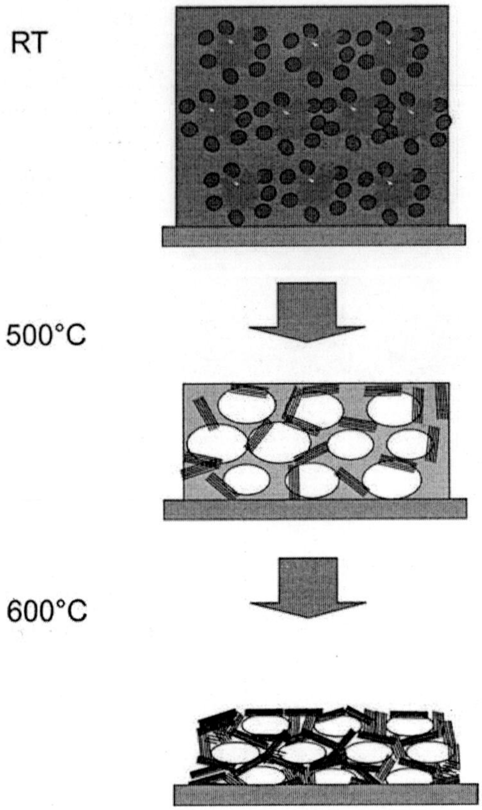

Figure 9.6. Illustration of the crystallization process (the layered stacks symbolize the nanocrystals) and shrinkage of the mesostructure in thin films upon heat treatment.

structure-directing agent. The use of crystalline particles as building blocks implies, however, certain requirements to their size and solubility. The particles should be perfectly dispersible in the starting solution, and their size should be compatible with the micelle size of the structure-directing templates, which limits their diameter to ca. 5–7 nm. This so far is a limiting factor for widespread use of nanoparticles in the EISA process, whereas a majority of the published syntheses mostly provide irreversible agglomerates of very small crystallites.

To sum up, EISA is a versatile technique for fabrication of thin films of metal oxides with different degrees of crystallinity and periodic nanostructures, which are often beyond the reach of other deposition techniques. Probably no other method enables controlled preparation of highly organized porous films with a long-scale periodicity in a simple wet deposition process. The films processed by EISA feature very high optical quality, low surface roughness, and very good thickness homogeneity, which makes them especially suitable for optical applications. This technique is, however, limited to relatively thin films; the thicknesses of films prepared by a conventional EISA approach are usually in a submicron range and can be increased to a maximum of a few microns by repeated deposition. This makes EISA a method of choice when homogeneous thin coating have to be prepared, but it renders it unsuitable for generation of very thick films. Use of molecular precursors that can be intimately mixed on a molecular level gives EISA an advantage for preparation of mixed metal oxides. This was successfully applied for fabrication of crystalline mixed oxides and even such complex materials as perovskites, e.g., $BaTiO_3$ and $SrTiO_2$.[56] The processing parameters should, however, be adjusted in order to equilibrate the hydrolysis and condensation rates of different precursors. This can be achieved by different strategies, for example, by stabilization of the faster hydrolizable atom with some ligand[64] by prehydrolysis of the more reactive component. The template-directed crystallization of metal oxides can in some cases lead to mesoporous crystalline films with aligned orientation of single crystallites (so-called "soft epitaxy"), which was observed for KLE-templated Nb_2O_5.[65] By a choice of suitable template and processing conditions, highly crystalline mesoporous films of various metal oxides can be generated. The size of the crystallites in such films is, however, small and usually does not exceed 15 nm; additional crystal growth leads in most cases to the collapse of periodic mesoporosity. The limited crystal size of EISA-generated crystalline films should be kept in mind when those films are used in different applications, since it can substantially influence the physico-chemical properties of the materials.

9.5 ELECTRONIC APPLICATIONS

The 3D-architechture of mesoporous films featuring interpenetrating interfaces of solid material and open voids provides ideal conditions for heterogeneous reactions and bulk mass transport processes. It is therefore not surprising that mesoporous films attract great interest in various electronic applications, where both phenomena play a key role. Depending on the electric conductivity of the inorganic scaffold, we divide the EISA-prepared metal oxide layers tested so far in electronic applications into two major groups, namely those with nonconducting (insulating) and semiconducting framework.

9.5.1 Mesoporous Films with Insulating Framework

This category includes a large variety of silica, zirconia, and alumina mesoporous films. Although the inorganic scaffold of such layers does not transport electric current, the pore architecture, which can be also used as a host matrix for incorporation of functional molecules, can alter electron transport to and from the conducting surface, thus influencing electronic properties of the complete system.

Thus, silica films have been applied as electrode modification layers in electrochemical applications. Etienne et al.[66] demonstrated that the pore structure of silica films has a profound effect on permeability of those layers toward different electrochemical probes. For the same pore size, the rate of diffusion changes drastically for different pore morphologies, being the fastest for the cubic Pm3n pore structure and decreasing for the 3D hexagonal P6(3)/mmc and further for 2D hexagonal P6m pore structures. The electrochemical response is also dramatically influenced by the charge of the molecular probe and the solution pH, leading to either preconcentration or permselective behavior of silica electrode coatings. The variety of permselective properties of mesoporous silica layers, which are very interesting for electrosensing and electrochemical applications, can be further extended because of modification of the pore surface with different functional groups. This was demonstrated in numerous processes, for example by Walcarius et al.[67–69] for porous silicates grafted with amine or thiol groups, which were successfully applied as electrode layers for applications in electroanalysis, including preconcentration associated with voltammetric detection. Fattakhova-Rohlfing et al. described application of mesoporous silica films grafted with amino groups as permselective electrode layers, whose selectivity and permeability can be reversibly controlled by changing pH.[70,71] Because of the surface charge being controlled by the degree of protonation of amino groups at different pH levels, such layers completely block or enhance the flux of ionic species through them depending on their charge. At the same time, because of the high porosity, such layers remain permeable for the neutral species in the whole range of pH. The high porosity of the EISA-prepared films ensures as well fast penetration rates of the analytes and very fast response to the changes in pH levels. The properties introduced by the developed porosity in the mesoporous range make electrode layers based on mesoporous silica films unique compared with those of typical compact layers, thus opening an additional perspective for their use in permselective, analytical, electrocatalitic, and sensor applications.

Because of the suitable pore size, pore connectivity, and rich surface chemistry, the mesoporous silica layers can be used as a structure-directing matrix for incorporation of electroactive species, enabling preparation of modified electrode layers. Despite the insulating character of the silica scaffold, electron transport through such layers takes place because of electron hopping between the adjacent electroactive species. This point was demonstrated by Fattakhova-Rohlfing et al.,[72] who used mesoporous silica films for ionic or covalent immobilization of descrete electroactive moieties, such as polyoxometalate $PMo_{12}O_{40}^{3-}$, hexacyanoferrate(III), and ferrocene. The resulting transparent homogeneous films featured enhanced electrochemical activity. The amount of the molecules involved in the charge transfer sequence is proportional to the film thickness and is comparable with the total amount of the embedded guests, which demonstrates that essentially the whole bulk volume of the modified silica films is electrochemically accessible.

Sel et al.[29] demonstrated that the pore shape and pore connectivity play the crucial role in the electrical conductivity of the functionalized silica channels. A sufficient connection between the mesopores is indispensable to allow sufficient functionalization, a facilitated electron hopping in the matrix, and a better accessibility of the electrode surface.

9.5.2 Mesoporous Films with a Semiconducting Framework

Semiconducting metal oxides are the active layers of various electronic systems, such as energy storage devices, capacitors, electrochromic coatings, sensors, or electrooptic switches. Charge flux through such layers is provided by the electron transport from electrode to the active sites in the metal oxide, which is always accompanied by the transfer of counter ions from the solution in the opposite direction in order to maintain electroneutrality. The amount of transported charge is defined by the number of available active sites (mostly metal ions), whereas the kinetics of charge transport is determined by the rates of ion diffusion toward the semiconductor interface, an interfacial electron transfer, and finally, coupled solid-state electron/ion transport, which as a rule is the rate-limiting step. For thick layers, which are mostly needed for sufficient efficiency, the kinetic limitations become a very significant problem, hindering the total performance of the electronic device. Porosity is therefore expected to enhance significantly the performance of metal oxide layers, particularly because of facilitation of interfacial charge transfer and solid-state diffusion processes.

The fast response times are very desirable for electrochromic layers, for which the change in the color should respond immediately to the change in the electrode potential. The influence of the porosity on the switching times was studied by Sallard et al., who prepared electrochromic WO_3 layers with regular cubic arrangement of the pores, using KLE as a template.[57,73] As expected, the electrochromic response times for both the coloration and the bleaching steps were significantly shortened as compared with nonporous WO_3 layers prepared by similar sol-gel methodology. Moreover, the used synthesis procedure enabled preparation of films with an adjustable degree of crystallinity while maintaining the same pore structure. This made possible a study of the influence of both crystallinity and mesoporosity on the electrochromic performance of WO_3 films. It was shown that increasing crystallinity decreases the charge capacity of the films, but it drastically improves the stability of the films toward multiple Li^+ or H^+ insertion/extraction. The porous completely crystalline films demonstrate excellent long-term cycling stability, whereas partially crystalline or amorphous layers degrade after repetitive cycling. For the best performance, the developed porosity should be combined with the high crystallinity of the inorganic matrix.

Another electrochromic system based on mesoporous EISA-produced films was realized by Jheong et al.,[74] who fabricated Pluronic-templated TiO_2

films with viologen molecules anchored in the pores. The efficiency of this system is based on interfacial electron transfer between the semiconductor and the attached viologen moieties acting as electrochromophores. The mesoporous TiO_2 structure was shown to be much more efficient than the ordinary nanoparticulate TiO_2 layer both in the coloration efficiency and switching response. The observed adsorbance of mesoporous layers was three times higher compared with that of a particulate layer of similar thickness, whereas the response time was shortened to 1 s compared with 3 s for the nonporous films. The advantageous performance of mesoporous films was attributed to the high interfacial surface area of such films enabling adsorption of a much higher amount of electrochromophores and, additionally, to the uniform shape, size, and connectivity of the pores promoting nonhindered diffusion of electrolyte to the interface. Moreover, the solid-state electron transport through the mesoporous TiO_2 scaffold was proposed to be more efficient because of the decreased diffusion path length.

Similar electrochemical processes, namely coupled electron/ion charge transfer, take place also in energy storage systems. Therefore, homogeneous mesoporosity is expected also in this case to have a positive impact on the performance of such systems. This was demonstrated for example by the group of Kavan et al.[75–79] and Fattakhova-Rohlfing et al.[80] for mesoporous TiO_2 layers of varying degree of crystallinity templated by block copolymer templates. All mesoporous films demonstrate high insertion capacity approaching the theoretically maximum value, which is apparently from the better accessibility of the active sites for the electrode reaction in mesoporous material. The common feature of all mesoporous titania layers is the very fast rate of Li^+ insertion and extraction, in good agreement with theoretical expectations. As for electrochromic WO_3 layers, the increased crystallinity of mesoporous films has a positive influence on the multiple insertion/extraction stability.[80] The high insertion capacity, good reversibility, and very fast charge/discharge rates make mesoporous crystalline titania layers very promising candidates for thin-layer charge storage devices, which can be especially advantageous for miniaturized systems.

Very recently, even transparent conducting oxides (TCOs), such as indium-tin-oxide (ITO), have been prepared using suitable KLE templates.[59] As one potential application, such porous TCOs (ZnO, etc.) are interesting for use in dye-sensitized solar cells. In general, such porous electrodes cover a variety of potential electro-optical applications, because they are both conducting and transparent.

The investigation of possible utilization of EISA-manufactured layers in electronic applications has started relatively recently, but the already performed studies demonstrate a very high potential of mesoporous films for technologies using interfacial and bulk charge transport. The advantages of the EISA-prepared layers become especially evident when the interfacial charge transfer from the species attached to the interface plays the key role in system performance.

9.6 MESOPOROUS FILMS IN DYE-SENSITIZED SOLAR CELLS

Because of the increasing need for renewable sources of energy, substantial effort is dedicated to the development of novel types of solar cells. Silicon-based cells have reached large-scale production and can be operated under varying environmental conditions. However, costly and elaborate production technology has spurred the search for alternative concepts beyond the silicon-based technology. In particular, it would be desirable to substitute costly semiconductors by cheap and abundant compounds, such as metal oxide films, which can be obtained by evaporation-induced self-assembly in combination with sol-gel chemistry or the deposition of mesoporous powder (e.g., by doctor-blading). Dye-sensitized solar cells (DSSCs) have already been shown to allow for promising light-to-electricity conversion efficiencies of up to 10% for certain types.[44] The classic "Graetzel"-type cell uses TiO_2 as an n-type semiconductor and a suitable dye and electrolyte, but also other metal oxides are discussed to serve as the charge carrier.[81] In general, the oxide has to be in a highly crystalline state in order to fulfill the requirements for a semiconductor with sufficient conductivity and chemical stability. In short, the "classic" Graetzel-type DSSC works as follows (see the review of L. M. Peter[81] for an overview): Photocurrents in the DSSC are created when light is absorbed by the sensitizer dye, leading to electron injection into the conduction band of TiO_2. The electrons move through the TiO_2 network by a random walk process until they reach the electrode. The dye is regenerated by electron transfer from iodide before it has time to undergo bleaching. The I_3- ions produced in the regeneration step diffuse to the cathode (usually covered with platinum), where they are reduced to iodide ions completing the regenerative cycle. Alternative concepts use mesoporous TiO_2 in combination with conducting polymers. Up until some years ago, the TiO_2 used in DSSCs was mainly of coarse-grain morphology, i.e., with particle sizes on the scale of >100 nm. Although the performance of DSSCs is regarded as being dependent on various parameters (e.g., electrolyte and dye), it is generally assumed that modification of the semiconductor morphology still has the potential to lead to a significant enhancement of the efficiency. Since the harvesting of the light correlates with the amount of dye, the principle idea of the usage of well-defined mesoporous metal oxide films is the increase in accessible surface area, thus resulting in an increased interface between dye and TiO_2. The mesoporous TiO_2 film therefore serves as a "host" for the dye, or polymers in the case of solid electrolyte solar cells. As another advantage, the EISA process is supposed to yield films with highly uniform thickness. Additionally, EISA can produce films without cracks, which represents an important prerequisite for the performance of DSSCs. Hence, the major difficulty in using mesoporous metal oxide films is the impregnation of the mesopore surface by the dye or the filling by the hole-transporting organic material.

Although the idea of applying mesoporous thin films of TiO_2 (or other suitable oxides) prepared by self-assembly strategies in combination with typical coating procedures (dip-coating or spin-coating) seems to be straightforward, surprisingly the literature describing the photovoltaic properties of such films is not extensive. Only very recently, the number of publications covering this potential application has been continuously increasing. In particular, one has to differentiate between two procedures, which are reported in the literature. Films of excellent overall quality can only be obtained by the EISA process as described in the next part as "method 1." However, still most publications describe the fabrication of coatings obtained by the deposition (e.g., by doctor-blading or spray deposition) of a ground powder of mesoporous TiO_2 (referred to in the following section as "method 2").

In general, the reports describing the performance of DSSCs based on mesoporous TiO_2 by either method are consistent regarding the positive effect of applying nanostructured TiO_2 films.

Method 1: The number of publications on metal oxide–based DSSCs prepared by EISA is still surprisingly small.[82–85] Several studies report promising results regarding the incident-photon-to-current-efficiency (IPCE) and the ability to incorporate conducting polymers and dyes. In a systematic study, Sanchez et al. studied the influence of sintering and mesoporosity on the performance of mesoporous TiO_2 thin films containing polymeric hole conductors, which were introduced into the porous matrix by impregnation.[82] It was clearly demonstrated that highly crystalline TiO_2 is needed for reasonable energy conversion and that encouraging values for the fill factor can be achieved by such films. At too high sintering temperatures, a reduction in the fill factor was found, which is attributable to the loss of porosity from sintering. A general weakness of EISA-derived films used in photovoltaics is the low thickness of the films. Because of the need to start from a dilute solution, the EISA process does not allow for the generation of coatings thicker than ca. 500 nm in one step, as pointed out by several groups. An interesting approach was presented by Ahn et al., using a 320-nm mesoporous TiO_2 layer (obtained by EISA) as an interlayer between the electrode and a micrometer-thick, porous TiO_2 film.[85] It could be demonstrated that the introduction of such a film of defined mesoporosity had a positive effect on the general performance of a Graetzel-type cell.

To overcome the limitation of the thin films, in a modified version of the EISA process Somani et al. prepared films with macroporous TiO_2 by a two-step procedure, using poly(styrene) colloidal particles as templates. First, films of self-assembled mono-dispersed poly(styrene)beads were obtained using a solvent-induced self-assembly technique.[86] Second, these films were soaked in an aqueous solution of a Ti-compound, annealed and calcined to yield porous TiO_2 films of several microns in thickness with pores of ca. 300 nm in diameter. Such films showed a better light conversion efficiency, which is attributed to the excellent connectivity of the pores and the intimate contact of the dye with the TiO_2 matrix. In a different approach, multiple dip-coating (applying EISA) was applied to yield thicker films.

These experiments indicated that an increase of solar conversion efficiency on the order of 50% can be achieved for cells containing TiO_2 mesoporous films with a thickness ≥1 μm.[5]

Method 2: The majority of films of templated mesoporous TiO_2 are still prepared by the deposition of preformed mesoporous TiO_2 powder using suitable coating techniques, e.g., doctor-blading. Ngamsinlapasathian et al. report a study on the performance of a DSSC based on such TiO_2, deposited on a nonporous TiO_2 underlayer.[87] By such a procedure, light conversion efficiencies of up to 8% could be achieved. In addition, most of the fundamental understanding and advances in the application of porous TiO_2 films used in DSSCs was achieved with porous powders. Menzies and coworkers reported that the post-treatment of mesoporous TiO_2 films by infiltration with indium (III) and zirconium (IV) oxide precursor solutions led to an improved efficiency.[88]

In conclusion, so far, templated mesoporous TiO_2 films show certain promise, but they have not reached the performance of the nanocrystalline TiO_2 films used in the classic "Graetzel" cell, in which the porosity (void between the nanoparticles) is disordered. This fact is somehow at variance with the assumption that an ordered mesoscopic structure should be superior to a nonorganized one because of, for example, the better homogeneity and good connection between the nanocrystals. Aside from the film thickness, one of the problems encountered with templated mesoporous films certainly is the accessibility of the dye and hole-conducting polymers (in the case of solid-state cells); since dyes and polymers possess a relatively low mobility, any constraints will impede an intimate covering of the pore walls throughout the sample. A loose packing of nanoparticles (a "disordered" mesopore structures) can be expected to be advantageous in this respect. Indeed, León et al. describe that photoinduced charge-transfer processes take place efficiently between the dye and the substrate (nanocrystalline anatase).[89] Still, there is a general lack of fundamental understanding of how the morphology on the nanometers scale influences the solar-cell performance.

Progress in the usage of templated mesoporous TiO_2 for DSSCs has been limited, and thus, novel concepts are needed, for instance by using different oxides than those of titanium. Very recently, new solar-cell concepts were reported (Fig. 9.7), taking advantage of different oxides. Fattakhova-Rohlfing et al. described the deposition of semiconducting metal oxides within mesoporous indium-tin-oxide (ITO), which was itself obtained by EISA/sol-gel-templating. Finally, a metal oxide semiconductor was deposited within the mesoporous ITO, and a dye was incorporated by soaking; thus, it could be demonstrated that a photocurrent was measurable.[59]

As a main feature, the usage of a conducting matrix can be expected to help to decrease the resistance and to increase the mean life of the generated electrons. Also, a porous conducting electrode enables a more intimate contact between the electrode and the semiconductor. Despite the prelimi-

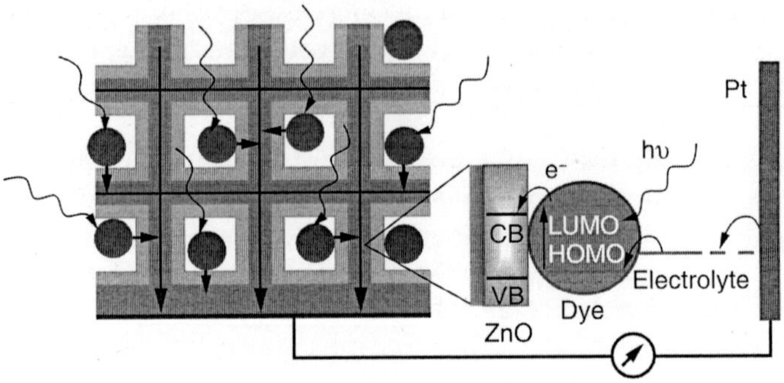

Figure 9.7. Illustration of the usage of mesoporous films of transparent conducting oxides for novel types of solar cells. The dark gray areas correspond to ITO, the brighter ones to an oxide deposited onto the TCO matrix. The sphere symbolizes a dye. For instance, such films can be used as porous electrodes to include dyes and to deposit semiconductors such as ZnO.

nary character of such approaches, evidently more intensive research is needed to bring mesoporous metal oxide films to applicability in the field of light harvesting.

9.7 CONCLUSIONS

So far, self-assembled, templated films with an ordered mesoscopic structure have been only regarded as exotic model materials, since no large-scale application has been established. However, the enormous efforts to unravel the formation mechanism and structure of such thin, mesostructured films have now provided the base to design such coatings with the desired type of porosity, especially with higher complexity. For instance, it can be envisaged that in the near future, films will be prepared with alternating layers of mesoporous oxides with different chemical composition, thus allowing for the buildup of coatings with arbitrary functionality. In particular, progress in the preparation of mesoporous films with high crystallinity and mesoporosity can be expected to spur the development of novel types of films in electro-optical and electrochemical devices, thus also creating economic benefit and probably contributing to the solution of environmental problems.

REFERENCES

1. Brinker, C. J.; Lu, Y. F.; Sellinger, A.; Fan, H. Y. 1999. Evaporation-induced self-assembly: Nanostructures made easy. *Adv. Mater.* 11:579–585.
2. Huang, Y.; Cai, H. Q.; Yu, T.; Sun, X. L.; Tu, B.; Zhao, D. Y. 2007. Highly ordered mesoporous carbonaceous frameworks from a template of a mixed amphiphilic

triblock-copolymer system of PEO-PPO-PEO and reverse PPO-PEO-PPO. *Chem.—Asian J.* 2:1282–1289.

3. Tanaka, S.; Katayama, Y.; Tate, M. P.; Hillhouse, H. W.; Miyake, Y. 2007. Fabrication of continuous mesoporous carbon films with face-centered orthorhombic symmetry through a soft templating pathway. *J. Mater. Chem.* 17:3639–3645.
4. Grosso, D.; Cagnol, F.; Soler-Illia, G.; Crepaldi, E. L.; Amenitsch, H.; Brunet-Bruneau, A.; Bourgeois, A.; Sanchez, C. 2004. Fundamentals of mesostructuring through evaporation-induced self-assembly. *Adv. Funct. Mater.* 14:309–322.
5. Smarsly, B.; Antonietti, M. 2006. Block copolymer assemblies as templates for the generation of mesoporous inorganic materials and crystalline films. *Eur. J. Inorg. Chem.* 1111–1119.
6. Soler-Illia, G.; Innocenzi, P. 2006. Mesoporous hybrid thin films: The physics and chemistry beneath. *Chem.—Eur. J.* 12:4478–4494.
7. Sayari, A.; Hamoudi, S. 2001. Periodic mesoporous silica-based organic—inorganic nanocomposite materials. *Chem. Mater.* 13:3151–3168.
8. Hoffmann, F.; Cornelius, M.; Morell, J.; Froba, M. 2006. Periodic mesoporous organosilicas (PMOs): Past, present, and future. *J. Nanosc. Nanotech.* 6:265–288.
9. Vinu, A.; Hossain, K. Z.; Ariga, K. 2005. Recent advances in functionalization of mesoporous silica. *J. Nanosc. Nanotech.* 5:347–371.
10. Soler-illia, G. J. D.; Sanchez, C.; Lebeau, B.; Patarin, J. 2002. Chemical strategies to design textured materials: From microporous and mesoporous oxides to nanonetworks and hierarchical structures. *Chem. Rev.* 102:4093–4138.
11. Yang, P. D.; Zhao, D. Y.; Margolese, D. I.; Chmelka, B. F.; Stucky, G. D. 1998. Generalized syntheses of large-pore mesoporous metal oxides with semicrystalline frameworks. *Nature* 396:152–155.
12. Lu, A. H.; Schuth, F. 2006. Nanocasting: A versatile strategy for creating nanostructured porous materials. *Adv. Mater.* 18:1793–1805.
13. Kresge, C. T.; Leonowicz, M. E.; Roth, W. J.; Vartuli, J. C.; Beck, J. S. 1992. Ordered mesoporous molecular-sieves synthesized by a liquid-crystal template mechanism. *Nature* 359:710–712.
14. Alberius, P. C. A.; Frindell, K. L.; Hayward, R. C.; Kramer, E. J.; Stucky, G. D.; Chmelka, B. F. 2002. General predictive syntheses of cubic, hexagonal, and lamellar silica and titania mesostructured thin films. *Chem. Mater.* 14: 3284–3294.
15. Kramer, E.; Forster, S.; Goltner, C.; Antonietti, M. 1998. Synthesis of nanoporous silica with new pore morphologies by templating the assemblies of ionic block copolymers. *Langmuir* 14:2027–2031.
16. Zhao, D. Y.; Feng, J. L.; Huo, Q. S.; Melosh, N.; Fredrickson, G. H.; Chmelka, B. F.; Stucky, G. D. 1998. Triblock copolymer syntheses of mesoporous silica with periodic 50 to 300 angstrom pores. *Science* 279:548–552.
17. Goltner, C. G.; Henke, S.; Weissenberger, M. C.; Antonietti, M. 1998. Mesoporous silica from lyotropic liquid crystal polymer templates. *Angew. Chem. Int. Ed.* 37: 613–616.
18. Goltner, C. G.; Berton, B.; Kramer, E.; Antonietti, M. 1999. Nanoporous silicas by casting the aggregates of amphiphilic block copolymers: The transition from cylinders to lamellae and vesicles. *Adv. Mater.* 11:395–398.

19. Thomas, A.; Schlaad, H.; Smarsly, B.; Antonietti, M. 2003. Replication of lyotropic block copolymer mesophases into porous silica by nanocasting: Learning about finer details of polymer self-assembly. *Langmuir* 19:4455–4459.

20. Freer, E. M.; Krupp, L. E.; Hinsberg, W. D.; Rice, P. M.; Hedrick, J. L.; Cha, J. N.; Miller, R. D.; Kim, H. C. 2005. Oriented mesoporous organosilicate thin films. *Nano Lett.* 5:2014–2018.

21. Smarsly, B.; Grosso, D.; Brezesinski, T.; Pinna, N.; Boissiere, C.; Antonietti, M.; Sanchez, C. 2004. Highly crystalline cubic mesoporous TiO_2 with 10-nm pore diameter made with a new block copolymer template. *Chem. Mater.* 16:2948–2952.

22. Schrader, D. M.; Jean, Y. C. 1988. *Positron and Positronium Chemistry*. Elsevier, Amsterdam, The Netherlands.

23. Gidley, D. W.; Frieze, W. E.; Dull, T. L.; Yee, A. F.; Ryan, E. T.; Ho, H. M. 1999. Positronium annihilation in mesoporous thin films. *Phys. Rev.* B 60: R5157–R5160.

24. Petkov, M. P.; Weber, M. H.; Lynn, K. G.; Rodbell, K. P. 2001. Porosity characterization by beam-based three-photon positron annihilation spectroscopy. *Appl. Phys. Lett.* 79:3884–3886.

25. Dull, T. L.; Frieze, W. E.; Gidley, D. W.; Sun, J. N.; Yee, A. F. 2001. Determination of pore size in mesoporous thin films from the annihilation lifetime of positronium. *J. Phys. Chem. B.* 105:4657–4662.

26. Thommes, M. 2004. Physical adsorption characterization of ordered and amorphous mesoporous materials. In *Nanoporous Materials: Science and Engineering*, edited by Lu, G. Q.; Zhao, X. S. Imperial College Press, London. pp. 317–364.

27. Zukal, A. 2006. Adsorption and pore condensation of krypton on mesoporous silicas at 77 K. *Microporous Mesoporous Mater.* 92:220–226.

28. Chiu, C. Y.; Chiang, A. S. T.; Chao, K. J. 2006. Mesoporous silica powders and films—Pore size characterization by krypton adsorption. *Microporous Mesoporous Mater.* 91:244–253.

29. Sel, O.; Sallard, S.; Brezesinski, T.; Rathousky, J.; Dunphy, D. R.; Collord, A.; Smarsly, B. M. 2007. Periodically ordered meso- and macroporous SiO_2 thin films and their induced electrochemical activity as a function of pore hierarchy. *Adv. Funct. Mater.* 17:3241–3250.

30. Glaves, C. L.; Frye, G. C.; Smith, D. M.; Brinker, C. J.; Datye, A.; Ricco, A. J.; Martin, S. J. 1989. Pore structure characterization of porous films. *Langmuir* 5:459–466.

31. Boissiere, C.; Grosso, D.; Lepoutre, S.; Nicole, L.; Bruneau, A. B.; Sanchez, C. 2005. Porosity and mechanical properties of mesoporous thin films assessed by environmental ellipsometric porosimetry. *Langmuir* 21:12362–12371.

32. Fuertes, M. C.; Lopez-Alcaraz, F. J.; Marchi, M. C.; Troiani, H. E.; Luca, V.; Miguez, H.; Soler-Illia, G. 2007. Photonic crystals from ordered mesoporous thin-film functional building blocks. *Adv. Funct. Mater.* 17:1247–1254.

33. Grosso, D.; Babonneau, F.; Albouy, P. A.; Amenitsch, H.; Balkenende, A. R.; Brunet-Bruneau, A.; Rivory, J. 2002. An in situ study of mesostructured CTAB-silica film formation during dip coating using time-resolved SAXS and interferometry measurements. *Chem. Mater.* 14:931–939.

34. Grosso, D.; Soler-Illia, G.; Crepaldi, E. L.; Cagnol, F.; Sinturel, C.; Bourgeois, A.; Brunet-Bruneau, A.; Amenitsch, H.; Albouy, P. A.; Sanchez, C. 2003. Highly porous TiO_2 anatase optical thin films with cubic mesostructure stabilized at 700 degrees C. *Chem. Mater.* 15:4562–4570.

35. Gibaud, A.; Grosso, D.; Smarsly, B.; Baptiste, A.; Bardeau, J. F.; Babonneau, F.; Doshi, D. A.; Chen, Z.; Brinker, C. J.; Sanchez, C. 2003. Evaporation-controlled self-assembly of silica surfactant mesophases. *J. Phys. Chem. B* 107:6114–6118.

36. Brezesinski, T.; Fischer, A.; Iimura, K.; Sanchez, C.; Grosso, D.; Antonietti, M.; Smarsly, B. M. 2006. Generation of self-assembled 3D mesostructured SnO_2 thin films with highly crystalline frameworks. *Adv. Funct. Mater.* 16:1433–1440.

37. Tate, M. P.; Urade, V. N.; Kowalski, J. D.; Wei, T. C.; Hamilton, B. D.; Eggiman, B. W.; Hillhouse, H. W. 2006. Simulation and interpretation of 2D diffraction patterns from self-assembled nanostructured films at arbitrary angles of incidence: From grazing incidence (above the critical angle) to transmission perpendicular to the substrate. *J. Phys. Chem. B* 110:9882–9892.

38. Ruland, W.; Smarsly, B. M. 2007. 2D SAXS of self-assembled nanocomposite films with oriented arrays of spheres: Determination of lattice type, preferred orientation, deformation and imperfection. *J. Appl. Cryst.* 40:409–417.

39. Ruland, W.; Smarsly, B. 2005. SAXS of self-assembled nanocomposite films with oriented two-dimensional cylinder arrays: An advanced method of evaluation. *J. Appl. Cryst.* 38:78–86.

40. Ruland, W.; Smarsly, B. 2004. SAXS of self-assembled oriented lamellar nanocomposite films: An advanced method of evaluation. *J. Appl. Cryst.* 37:575–584.

41. Dourdain, S.; Bardeau, J. F.; Colas, M.; Smarsly, B.; Mehdi, A.; Ocko, B. M.; Gibaud, A. 2005. Determination by x-ray reflectivity and small angle x-ray scattering of the porous properties of mesoporous silica thin films. *Appl. Phys. Lett.* 86:113108.

42. Dourdain, S.; Gibaud, A. 2005. On the capillary condensation of water in mesoporous silica films measured by x-ray reflectivity. *Appl. Phys. Lett.* 87:223105.

43. Fujihira, M.; Satoh, Y.; Osa, T. 1981. Heterogeneous photocatalytic oxidation of aromatic-compounds on TiO_2. *Nature* 293:206–208.

44. Gratzel, M. 2001. Photoelectrochemical cells. *Nature* 414:338–344.

45. Yang, P. D.; Zhao, D. Y.; Margolese, D. I.; Chmelka, B. F.; Stucky, G. D. 1999. Block copolymer templating syntheses of mesoporous metal oxides with large ordering lengths and semicrystalline framework. *Chem. Mater.* 11:2813–2826.

46. Bosc, F.; Ayral, A.; Albouy, P. A.; Guizard, C. 2003. A simple route for low-temperature synthesis of mesoporous and nanocrystalline anatase thin films. *Chem. Mater.* 15:2463–2468.

47. Bosc, F.; Ayral, A.; Albouy, P. A.; Datas, L.; Guizard, C. 2004. Mesostructure of anatase thin films prepared by mesophase templating. *Chem. Mater.* 16:2208–2214.

48. Chen, L.; Yao, B.; Cao, Y.; Fan, K. 2007. Synthesis of well-ordered mesoporous titania with tunable phase content and high photoactivity. *J. Phys. Chem. C* 111:11849–11853.

49. Crepaldi, E. L.; Soler-Illia, G.; Grosso, D.; Cagnol, F.; Ribot, F.; Sanchez, C. 2003. Controlled formation of highly organized mesoporous titania thin films: From

mesostructured hybrids to mesoporous nanoanatase TiO_2. *J. Amer. Chem. Soc.* 125:9770–9786.

50. Grosso, D.; Boissiere, C.; Nicole, L.; Sanchez, C. 2006. Preparation, treatment and characterisation of nanocrystalline mesoporous ordered layers. *J. Sol-Gel Sci. Technol.* 40:141–154.
51. Crepaldi, E. L.; Soler-Illia, G.; Grosso, D.; Sanchez, M. 2003. Nanocrystallised titania and zirconia mesoporous thin films exhibiting enhanced thermal stability. *New J. Chem.* 27:9–13.
52. de Zarate, D. O.; Boissiere, C.; Grosso, D.; Albouy, P. A.; Amenitsch, H.; Amoros, P.; Sanchez, C. 2005. Preparation of multi-nanocrystalline transition metal oxide (TiO_2-$NiTiO_3$) mesoporous thin films. *New J. Chem.* 29:141–144.
53. Urade, V. N.; Hillhouse, H. W. 2005. Synthesis of thermally stable highly ordered nanoporous tin oxide thin films with a 3D face-centered orthorhombic nanostructure. *J. Phys. Chem. B* 109:10538–10541.
54. Pan, J. H.; Chai, S. Y.; Lee, C.; Park, S. E.; Lee, W. I. 2007. Controlled formation of highly crystallized cubic and hexagonal mesoporous SnO_2 thin films. *J. Phys. Chem. C* 111:5582–5587.
55. Choi, S. Y.; Mamak, M.; Coombs, N.; Chopra, N.; Ozin, G. A. 2004. Thermally stable two-dimensional hexagonal mesoporous nanocrystalline anatase, meso-nc-TiO_2: Bulk and crack-free thin film morphologies. *Adv. Funct. Mater.* 14:335–344.
56. Grosso, D.; Boissiere, C.; Smarsly, B.; Brezesinski, T.; Pinna, N.; Albouy, P. A.; Amenitsch, H.; Antonietti, M.; Sanchez, C. 2004. Periodically ordered nanoscale islands and mesoporous films composed of nanocrystalline multimetallic oxides. *Nature Mater.* 3:787–792.
57. Brezesinski, T.; Fattakhova-Rohlfing, D.; Sallard, S.; Antonietti, M.; Smarsly, B. M. 2006. Highly crystalline WO_3 thin films with ordered 3D mesoporosity and improved electrochromic performance. *Small* 2:1203–1211.
58. Brezesinski, T.; Groenewolt, M.; Antonietti, M.; Smarsly, B. 2006. Crystal-to-crystal phase transition in self-assembled mesoporous iron oxide films. *Angew. Chem. Int. Ed.* 45:781–784.
59. Fattakhova-Rohfing, D.; Brezesinski, T.; Rathousky, J.; Feldhoff, A.; Oekermann, T.; Wark, M.; Smarsly, B. 2006. Transparent conducting films of indium tin oxide with 3D mesopore architecture. *Adv. Mater.* 18:2980–2983.
60. Kuemmel, M.; Grosso, D.; Boissiere, U.; Smarsly, B.; Brezesinski, T.; Albouy, P. A.; Amenitsch, H.; Sanchez, C. 2005. Thermally stable nanocrystalline gamma-alumina layers with highly ordered 3D mesoporosity. *Angew. Chem. Int. Ed.* 44: 4589–4592.
61. Brezesinski, T.; Smarsly, B.; Iimura, K.; Grosso, D.; Boissiere, C.; Amenitsch, H.; Antonietti, M.; Sanchez, C. 2005. Self-assembly and crystallization behavior of mesoporous, crystalline HfO_2 thin films: A model system for the generation of mesostructured transition-metal oxides. *Small* 1:889–898.
62. Ba, J. H.; Polleux, J.; Antonietti, M.; Niederberger, M. 2005. Non-aqueous synthesis of tin oxide nanocrystals and their assembly into ordered porous mesostructures. *Adv. Mater.* 17:2509–2512.

63. Deshpande, A. S.; Pinna, N.; Smarsly, B.; Antonietti, M.; Niederberger, M. 2005. Controlled assembly of preformed ceria nanocrystals into highly ordered 3D nanostructures. *Small* 1:313–316.

64. Supplit, R; Husing, N.; Gross, S.; Bernstorff, S.; Puchberger, M. 2007. Hafnium oxide doped mesostructured silica films. *Eur. J. Inorg. Chem.* 2797–2802.

65. Brezesinski, T.; Antonietti, M.; Smarsly, B. M. 2007. Self-assembled metal oxide bilayer films with "single-crystalline" overlayer mesopore structure. *Adv. Mater.* 19:1074–1078.

66. Etienne, M.; Quach, A.; Grosso, D.; Nicole, L.; Sanchez, C.; Walcarius, A. 2007. Molecular transport into mesostructured silica thin films: Electrochemical monitoring and comparison between p6m, P6(3)/mmc, and Pm3n structures. *Chem. Mater.* 19:844–856.

67. Walcarius, A. 2001. Electrochemical applications of silica-based organic-inorganic hybrid materials. *Chem. Mater.* 13:3351–3372.

68. Etienne, M.; Walcarius, A. 2005. Evaporation induced self-assembly of templated silica and organosilica thin films on various electrode surfaces. *Electrochem. Comm.* 7:1449–1456.

69. Etienne, M.; Cortot, J.; Walcarius, A. 2007. Preconcentration electroanalysis at surfactant-templated thiol-functionalized silica thin films. *Electroanalysis* 19:129–138.

70. Fattakhova-Rohlfing, D.; Wark, M.; Rathousky, J. 2007. Electrode layers for electrochemical applications based on functionalized mesoporous silica films. *Sens. Actuat. B-Chem.* 126:78–81.

71. Fattakhova-Rohlfing, D.; Wark, M.; Rathousky, J. 2007. Ion-permselective pH-switchable mesoporous silica thin layers. *Chem. Mater.* 19:1640–1647.

72. Fattakhova-Rohlfing, D.; Rathousky, J.; Rohlfing, Y.; Bartels, O.; Wark, M. 2005. Functionalized mesoporous silica films as a matrix for anchoring electrochemically active guests. *Langmuir* 21:11320–11329.

73. Sallard, S.; Brezesinski, T.; Smarsly, B. M. 2007. Electrochromic stability of WO_3 thin films with nanometer-scale periodicity and varying degrees of crystallinity. *J. Phys. Chem. C* 111:7200–7206.

74. Jheong, H. K.; Kim, Y. J.; Pan, J. H.; Won, T. Y.; Lee, W. I. 2006. Electrochromic property of the viologen-anchored mesoporous TiO_2 films. *J. Electroceramics* 17:929–932.

75. Attia, A.; Zukalova, M.; Pospisil, L.; Kavan, L. 2007. Electrochemical impedance spectroscopy of mesoporous Al-stabilized TiO_2 (anatase) in aprotic medium. *J. Solid State Electrochem.* 11:1163–1169.

76. Attia, A.; Zukalova, M.; Rathousky, J.; Zukal, A.; Kavan, L. 2005. Mesoporous electrode material from alumina-stabilized anatase TiO_2 for lithium ion batteries. *J. Solid State Electrochem.* 9:138–145.

77. Kavan, L.; Rathousky, J.; Gratzel, M.; Shklover, V.; Zukal, A. 2001. Mesoporous thin film TiO_2 electrodes. *Microporous Mesoporous Mater.* 44:653–659.

78. Kavan, L.; Rathousky, J.; Gratzel, M.; Shklover, V.; Zukal, A. 2000. Surfactant-templated TiO_2 (anatase): Characteristic features of lithium insertion electrochemistry in organized nanostructures. *J. Phys. Chem. B* 104:12012–12020.

79. Kavan, L.; Attia, A.; Lenzmann, F.; Elder, S. H.; Gratzel, M. 2000. Lithium insertion into zirconia-stabilized mesoscopic TiO_2 (anatase). *J. Electrochem. Soc.* 147:2897–2902.
80. Fattakhova-Rohlfing, D.; Wark, M.; Brezesinski, T.; Smarsly, B. M.; Rathousky, J. 2007. Highly organized mesoporous TiO_2 films with controlled crystallinity: A Li-insertion study. *Adv. Funct. Mater.* 17:123–132.
81. Peter, L. M. 2007. Dye-sensitized nanocrystalline solar cells. *Phys. Chem. Chem. Phys.* 9:2630–2642.
82. Lancelle-Beltran, E.; Prene, P.; Boscher, C.; Belleville, P.; Buvat, P.; Lambert, S.; Guillet, F.; Boissiere, C.; Grosso, D.; Sanchez, C. 2006. Nanostructured hybrid solar cells based on self-assembled mesoporous titania thin films. *Chem. Mater.* 18:6152–6156.
83. Malfatti, L.; Falcaro, P.; Amenitsch, H.; Caramori, S.; Argazzi, R.; Bignozzi, C. A.; Enzo, S.; Maggini, M.; Innocenzi, P. 2006. Mesostructured self-assembled titania films for photovoltaic applications. *Microporous Mesoporous Mater.* 88:304–311.
84. Zukalova, M.; Zukal, A.; Kavan, L.; Nazeeruddin, M. K.; Liska, P.; Gratzel, M. 2005. Organized mesoporous TiO_2 films exhibiting greatly enhanced performance in dye-sensitized solar cells. *Nano Lett.* 5:1789–1792.
85. Ahn, K. S.; Kang, M. S.; Lee, J. W.; Kang, Y. S. 2007. Effects of a surfactant-templated nanoporous TiO_2 interlayer on dye-sensitized solar cells. 2007. *J. Appl. Phys.* 101:084312.
86. Somani, P. R.; Dionigi, C.; Murgia, M.; Palles, D.; Nozar, P.; Ruani, G. 2005. Solid-state dye PV cells using inverse opal TiO_2 films. *Solar Energy Mater. Solar Cells* 87:513–519.
87. Ngamsinlapasathian, S.; Pavasupree, S.; Suzuki, Y.; Yoshikawa, S. 2006. Dye-sensitized solar cell made of mesoporous titania by surfactant-assisted templating method. *Solar Energy Mater. Solar Cells* 90:3187–3192.
88. Menzies, D. B.; Dai, Q.; Bourgeois, L.; Caruso, R. A.; Cheng, Y. B.; Simon, G. P.; Spiccia, L. 2007. Modification of mesoporous TiO_2 electrodes by surface treatment with titanium(IV), indium(III) and zirconium(IV) oxide precursors: Preparation, characterization and photovoltaic performance in dye-sensitized nanocrystalline solar cells. *Nanotechnology* 18:125608.
89. Leon, C. P.; Kador, L.; Peng, B.; Thelakkat, M. 2006. Characterization of the adsorption of ru-bpy dyes on mesoporous TiO_2 films with UV-Vis, Raman, and FTIR spectroscopies. *J. Phys. Chem. B* 110:8723–8730.

CHAPTER 10

Engineered Nanomaterials as Soluble Precursors for Inorganic Films

DMITRI V. TALAPIN

The University of Chicago, Department of Chemistry, Chicago, IL

10.1 INTRODUCTION

Crystalline inorganic metals and semiconductors offer superior sets of electronic and optical characteristics relative to organic systems, but they are difficult to form by low-cost processes. Crystallization of inorganic semiconductors requires high-temperature treatments that force trade-offs between device performance and compatibility with inexpensive and flexible plastic substrates. The development of applications ranging from displays, photovoltaics, and light-emitting devices to "smart cards", radio frequency tags, and sensors could be accelerated by introducing lower cost alternatives to conventional silicon technology.[1,2]

Solution-based processes such as spin coating, dip coating, or inkjet printing can offer substantial cost reductions for fabrication of electronic and optoelectronic devices when combined with properly engineered nanoscale building blocks like clusters,[3] nanocrystals,[4] nanorods,[5] nanowires,[6] and nanotubes.[7] Nowadays many inorganic materials can be synthesized in these forms. Following this proposed classification, *clusters* can be described as large molecules with an atomically defined structure, having structural units of inorganic solids and typically built of tens to hundreds of atoms—e.g., $[Cd_{17}S_4(SC_6H_5)_{28}]^{2-}$ (Fig. 10.1a).[8] Clusters have closed structural shells and correspond to the pronounced minima of the free energy versus cluster size dependence.[3] Larger particles whose crystalline core consists of ~10^2–10^6 atoms fall into the category of "*nanocrystals*" (Fig. 10.1b). Here, nearly continuous tunability of the particle size becomes possible, as an addition or removal of an atom cell or unit cell leads to only a small variation of the

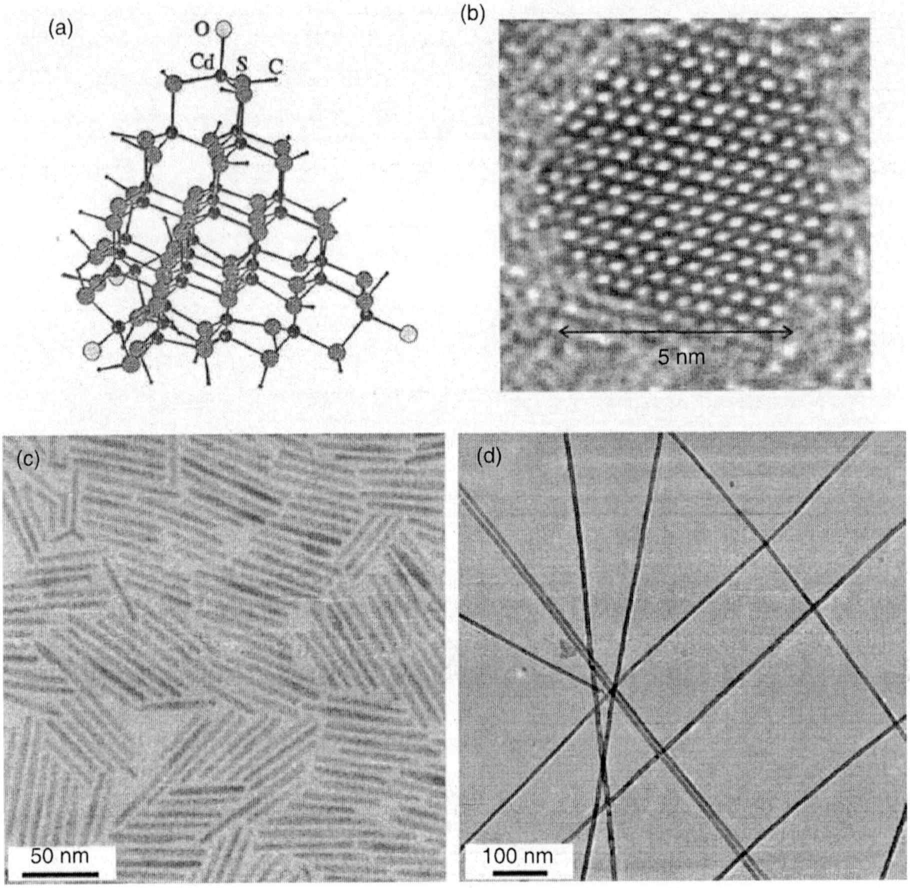

Figure 10.1. (a) Structure of a $[Cd_{32}S_{14}(SCH_2CH(OH)CH_3)_{36}]\cdot(H_2O)_4$ nanocluster, as derived from the single-crystal X-ray data. Reproduced from Ref. 8, Copyright 1995, with permission from the American Chemical Society. (b) A high-resolution transmission electron microscope (TEM) image of a CdSe nanocrystal. (c,d) TEM images of CdS nanorods and PbSe nanowires, respectively.

nanocrystal free energy. Modern synthetic techniques allow precise tailoring of both size and shape of metal and semiconductor nanocrystals (Fig. 10.1c).[9] *Nanowires* can be defined as structures that have a lateral size constrained to tens of nanometers or less and an unconstrained longitudinal size (Fig. 10.1d). Individual semiconductor nanowires show very good electronic performance, in some cases exceeding that of respective bulk materials.[10] *Carbon nanotubes* represent another very interesting family of engineered nanoscale building blocks for solution-processed electronic devices and sensors. Individual single-wall nanotubes demonstrate very high chemical stability and outstanding room-temperature carrier mobilities, outperforming known inorganic semiconductors.[7,11] It is also worth mentioning hybrid solution-processed inorganic–organic nanostructures, which potentially allow combining advantages of both inorganic and organic materials.[12] In this chapter we provide an overview of different concepts to designing inorganic materials for electronic, optoelectronic, and sensing applications using nano-entities as the soluble building blocks.

10.2 SYNTHESIS OF INORGANIC NANOMATERIALS

The approaches used for preparation of inorganic nanomaterials can be divided into two broad categories: solution-phase colloidal synthesis and gas-phase synthesis. Metal and semiconductor nanoparticles are usually synthesized via solution-phase colloidal techniques,[4,9,13] whereas high-temperature gas-phase processes like chemical vapor deposition (CVD), pulsed laser deposition (PLD), and vapor transfer are widely used for synthesis of high-quality semiconductor nanowires and carbon nanotubes.[6,7] Such division reflects only the current research bias, as promising routes to metallic nanoparticles are also available based on vapor condensation[14] and colloidal syntheses of high-quality semiconductor nanowires.[15]

The colloidal synthesis of nanocrystalline particles generally involves several consecutive stages: nucleation from an initially homogeneous solution, growth of the preformed nuclei, isolation of particles reaching the desired size from the reaction mixture, and post-preparative size fractionation.[4,16] As a rule, temporal separation of the nucleation event from the growth of the nuclei is required for narrow size distribution.[4,17] As the nucleation rate is temperature-sensitive, it is possible to achieve such separation by controlling the temperature profile during nanoparticle synthesis. During the earlier stages of nanoparticle research, most work was performed on II–VI compounds such as CdS and CdSe. The initial concept of "keeping particles small" involved the strategy of arrested precipitation, which makes use of water pools of inverse micelles as nanometer-sized reactors.[18] This was followed by employing stabilizing (also called capping or passivating) agents (ligands), which prevent uncontrollable growth and aggregation of nanoparticles.[17]

The choice of proper stabilizing agents can be considered a key point in an advanced colloidal synthesis of uniform crystalline nanoparticles. Stabilizing agents regulate the growth rate and the size of nanoparticles, prevent them (to a smaller or larger extent) from oxidation, and provide passivation of dangling bonds at the particle surface, thus at least partially eliminating surface traps. Importantly, stabilizers, consisting of organic molecules with free functional groups, make nanoparticles "soluble" in different solvents and allow their further handling using solution-based processing like spin-coating[19] and inkjet printing.[20]

Metal nanoparticles are synthesized by reduction of metal ions using reducing agents such as borohydride, amines, and 1,2-diols in the presence of stabilizing agents, typically long-chain alkyl thiols (e.g., dodecanethiol), amines, or fatty acids. This approach works well for noble metals (Au,[21–23] Ag,[22,23] Pd[24]), transition metals (Co,[25,26] Ni,[26] Fe[27]), and their alloys (FePt,[28] $CoPt_3$[29]). Noble metal nanoparticles are often synthesized at room temperature, whereas colloidal synthesis of transition metal and alloy nanocrystals is performed at high temperatures, up to 300 °C, to allow the annealing of crystalline defects in the particles.[25–29] Examples of reactions used for synthesis of metal nanocrystals are given below:

$$AuCl_3 \xrightarrow{25°C,\ NaBH_4,\ \text{didodecyldimethylammonium bromide}} \text{Au nanocrystals} \quad (1)$$

$$Fe(CO)_5 + Pt(\text{acetylacetonate})_2 \xrightarrow{300°C,\ \text{oleic acid–oleyl amine}} \text{FePt nanocrystals} \quad (2)$$

The reactions used for synthesis of II–VI (CdSe, CdTe), III–V (InP, InAs), and IV–VI (PbS, PbSe, PbTe) *semiconductor nanocrystals* are outlined by Schemes 3–5.[4,17,30–32] The syntheses are carried out at high temperatures, and in the presence of long-chain alkylphosphines (trioctylphosphine, TOP), alkylphosphine oxides (trioctylphosphine oxide, TOPO), alkylamines (hexadecylamine, HDA), and alkylphosphonic acids as the stabilizing agents.

$$Cd(CH_3)_2 + (C_8H_{17})_3PSe \xrightarrow{300°C,\ \text{HDA–TOPO–TOP}} \text{CdSe nanocrystals} \quad (3)$$

$$InCl_3 + [(CH_3)_3Si]_3P \xrightarrow{260°C,\ \text{TOP}} \text{InP nanocrystals} \quad (4)$$

$$Pb(CH_3COO)_2 + (C_8H_{17})_3PSe \xrightarrow{180°C,\ \text{oleic acid}} \text{PbSe nanocrystals} \quad (5)$$

The nanocrystal size can be tuned by adjusting the concentrations of stabilizing agents, reaction temperature, and duration of the particle growth. It is also possible to vary the nanocrystal shape by using multicomponent mixtures of stabilizing agents that provide differences in growth rates in the different crystallographic directions. For example, depending on the concentration of alkylphosphonic acid and the heating regime, rod-, arrow-, rise-, teardrop-, and tetrapod-shaped CdSe nanocrystals can be selectively synthesized.[5,33]

Semiconductor nanowires can be synthesized via seeded vapor–liquid–solid (VLS) growth, using CVD or PLD techniques.[6,34,35] In this approach, metal nanoparticles are used as catalysts to initiate and direct the nanowire growth as schematically shown in Figure 10.2. To date, many group IV, II–VI, III–V and IV–VI semiconducting materials have been prepared in the form of high-quality crystalline nanowires.[6] Excellent diameter uniformity and low concentration of structural defects can be considered as major advantages of VLS synthesis. On the other hand, this technique is difficult to scale up. Another limitation is associated with the incompatibility of Au nanoparticles, typically used as the catalyst (Fig. 10.2), with an industrial CMOS process. This last limitation can be overcome for oxides and some chalcogenide semiconductors by growing catalyst-free, one-dimensional nanostructures (so-called nanobelts or nanoribbons) by vapor-transfer techniques.[36]

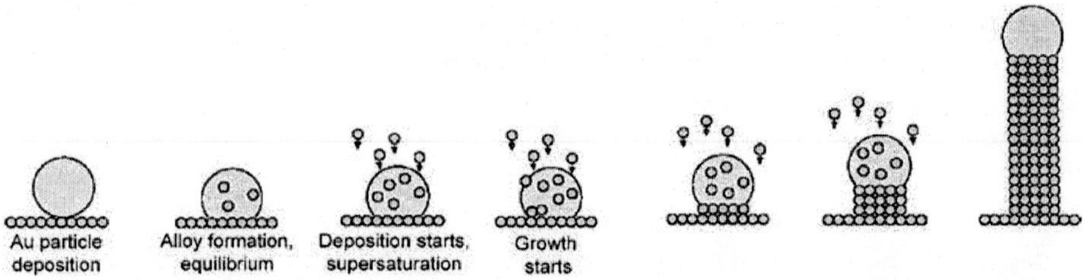

Figure 10.2. Vapor–liquid–solid growth of semiconductor nanowires.

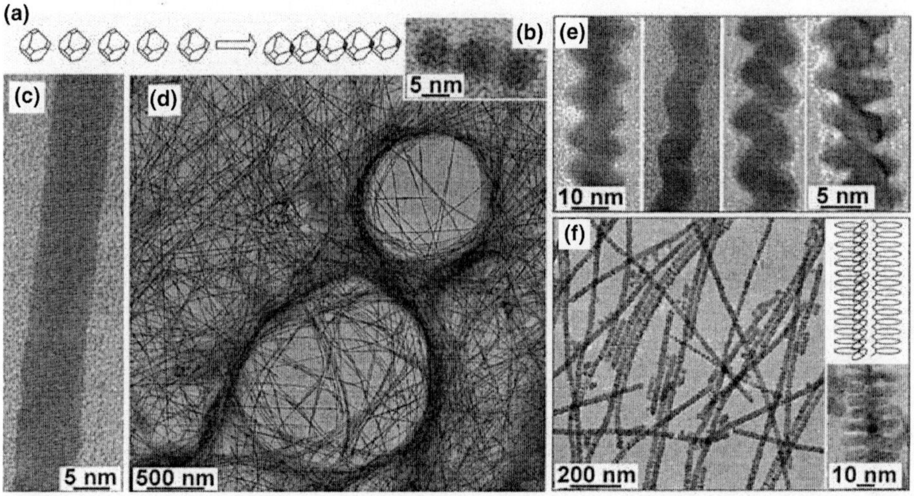

Figure 10.3. PbSe nanowires synthesized by the oriented attachment and fusion of nanoparticles. (a) Schematic representation of the oriented attachment process. (b) High-resolution TEM image of a "trimer" formed at the early stage of the reaction. (c) High-resolution and (d) overview TEM images of straight PbSe nanowires. (e) TEM images of zigzag and helical PbSe nanowires. (f) TEM images of branched PbSe nanowires.

Some important applications (e.g., solar cells, displays, printable electronics, and thermoelectrics) require large quantities of nanowires at a reasonably low price. Solution-phase synthetic methods, such as solution–liquid–solid (SLS) growth[37] and oriented attachment,[15,38] are now extensively explored to provide low-cost routes to the preparation of single-crystalline semiconductor nanowires. The SLS synthesis is similar to VLS, but nanowires grow in solution in the presence of surfactant molecules, which prevents aggregation and precipitation of nanowires. Nanoparticles of low-melting-point metals, like Bi or In, are used as the catalysts for nanowire growth.[37] The other approach to solution-phase synthesis of nanowires is based on the oriented attachment of nanoparticle building blocks. For example, the chaining of PbSe nanoparticles followed by their fusion is a method for fabricating high-quality, single-crystalline PbSe nanowires[15] (Fig. 10.3). Synthesis through oriented attachment allows for controlling the nanowire morphology. In addition to straight wires, zigzag, helical, branched, and tapered nanowires can all be prepared by adjustment of the reaction conditions (Fig. 10.3e,f). The different nanowire morphologies may be advantageous depending on the applications. Straight nanowires with minimal surface roughness provide the high carrier mobilities necessary for high-performance FETs.[10] On the other hand, the performance of nanowire-based sensors[39,40] and photovoltaic devices[41] should improve by increasing the wire surface area; i.e., highly branched nanowires may best match the requirements for these applications. The performance of nanowire-based thermoelectric devices[42] may substantially benefit from the multiple scattering of acoustic phonons in zigzag and helical nanowires.

Several techniques are available to produce single- and multiwall carbon nanotubes (CNTs), including arc discharge, laser ablation, high-pressure carbon monoxide decomposition (HiPco), and chemical vapor deposition (CVD).[7,43,44] In the arc discharge process, the carbon contained in the negative electrode sublimates because of the high temperatures caused by the discharge. In the laser ablation process, a pulsed laser vaporizes a graphite target, which condenses in the form of CNTs. This method produces primarily single-walled carbon nanotubes with a controllable diameter determined by the reaction temperature. In CVD synthesis, the substrate is covered with a layer of metal catalyst particles, most commonly nickel, cobalt, or iron, and it is heated to approximately 700 °C in an atmosphere of a carbon-containing gas (such as acetylene, ethylene, ethanol, and methane).[45] Of the various means for nanotube synthesis, CVD shows the most promise for industrial-scale deposition in terms of its price/unit ratio. The development of inexpensive synthesis techniques is vital to the future of carbon nanotechnology and commercial-scale applications.

10.3 NANOPARTICLES AS SOLUBLE BUILDING BLOCKS FOR INORGANIC FILMS

Colloidal nanocrystals are the logical building blocks for solution-processed electronic and optoelectronic devices because they are solution-processable fragments of crystalline inorganic metals and semiconductors. Furthermore, tuning the particle size can provide an additional degree of freedom for tailoring electronic and optical properties of the material, because the properties of individual semiconductor nanocrystals can differ substantially from the respective bulk materials. Quantum confinement in semiconductor nanocrystals gives rise to discrete electron and hole states that can be precisely tuned by varying particle size and shape (Fig. 10.4).[46,47] Efforts to harness the quantum tunability of semiconductor nanocrystals have led to many successes in optical and optoelectronic applications such as light emitting devices[48] and biological tags.[49] However, realizing solid-state electronic applications (e.g., field-effect

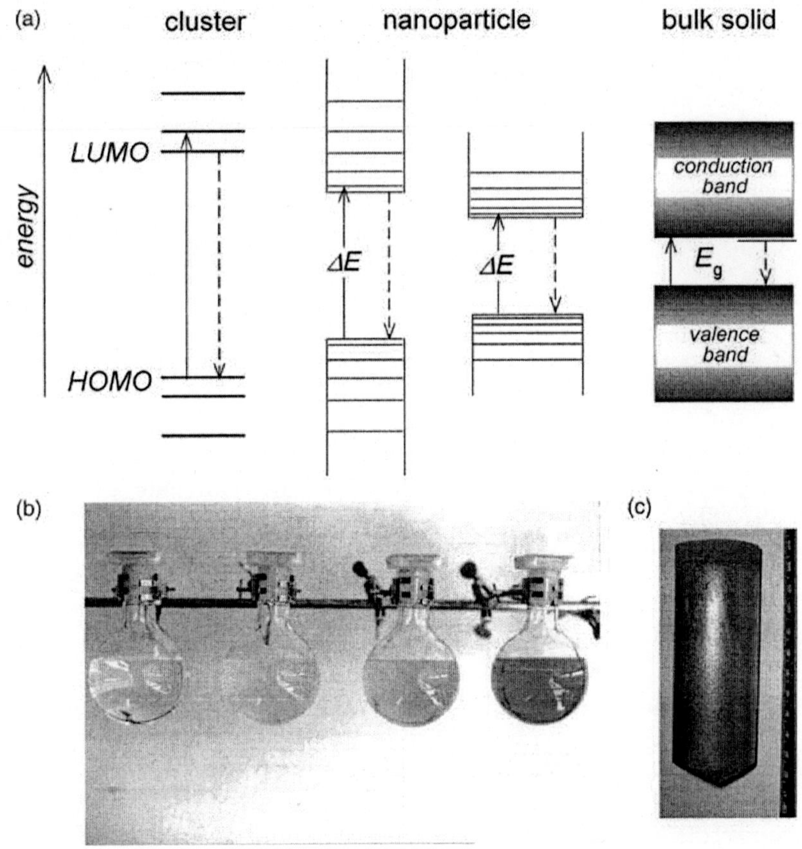

Figure 10.4. (a) Schematic energy diagrams of clusters, nanoparticles, and bulk semiconductors. (b) Manifestation of the size quantization effect as a color change of aqueous colloidal solutions of CdSe nanoparticles (courtesy of A. Rogach). The particle size changes from left to right from ~1.5 to ~4.5 nm. (c) Bulk CdSe crystal. (*See color insert.*)

devices) of these nanoscale building blocks has been more challenging. The confinement of carriers inside the nanocrystals imparts their fascinating size-tunable properties, but, until very recently, it has frustrated the efforts to contact and integrate them efficiently into devices that switch with useful speeds.

10.3.1 Sintering Metal and Semiconductor Nanoparticles into Continuous Polycrystalline Films

Nanoparticles can be deposited onto a substrate by spin coating or inkjet printing a colloidal solution and can be sintered into a continuous conductive film during the subsequent annealing step. Coalescence of individual

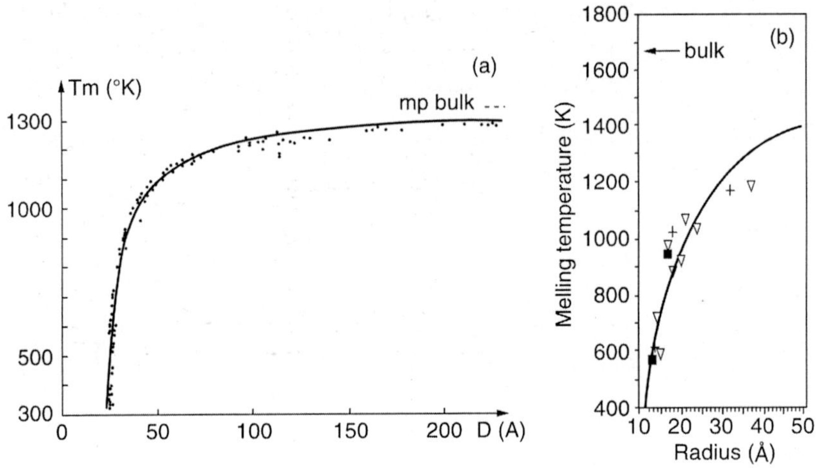

Figure 10.5. Size dependence of the melting temperature for (a) Au and (b) CdS nanoparticles. Reproduced from Ref. 51, Copyright 1976, with permission from the American Institute of Physics, and from Ref. 52, Copyright 1992, with permission from the AAAS, respectively.

nanoparticles yields polycrystalline solids with grains considerably larger than the initial nanoparticles.[50] Sintering individual nanocrystals is facilitated by significant reduction of the melting temperature with decreasing particle size.[51,52] Size-dependent melting-point depression is a general phenomenon observed across a broad class of materials, including technologically important metals and semiconductors (Fig. 10.5). The reduction of the melting point can be described by the effect of surface tension, which plays an increasingly important role when particle size decreases below 10 nm.[52] Changes in the melting temperature (T_m) of the particle of size R_{sol} compared to that of the bulk (T_b) are given by[51,52]

$$T_m - T_b = \frac{2T_m}{L\rho_{sol}R_{sol}}\left[\gamma_{sol} - \gamma_{liq}\left(\frac{\rho_{sol}}{\rho_{liq}}\right)^{2/3}\right] \quad (6)$$

where L is the molar latent heat of fusion and γ and ρ are the surface tension and density of the solid (*sol*) and liquid (*liq*) phases, respectively. The size-dependent suppression of the melting temperature of metals and semiconductors can exceed 1000 °C, which allows for the sintering of very small (~2 nm) nanocrystals at temperatures below 300 °C (Fig. 10.5 and 10.6).[51,52]

The ability of nanocrystals to fuse at relatively modest temperatures provides a low-temperature route to thin-film growth. This approach has been successfully used for formation of metallic (Au,[53] Ag[54]) and semiconducting (CdS,[52] CdSe,[55] HgTe[56]) thin films with good electrical conductivity. However, a serious limitation of this approach is associated with the presence of organic

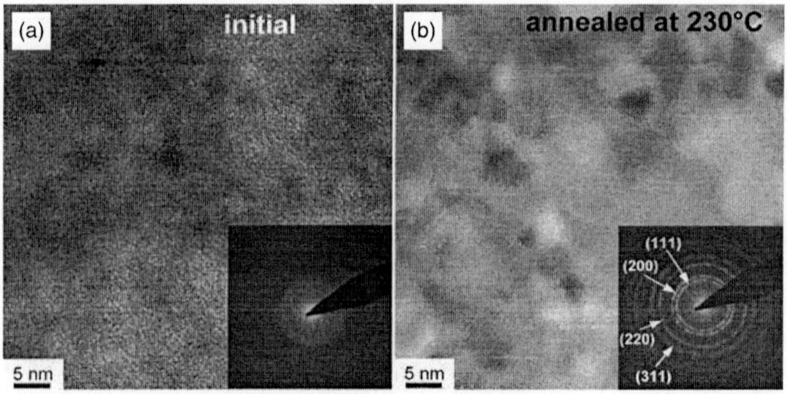

Figure 10.6. (a) High-resolution TEM image of a film of CdS nanoparticles stabilized with 5-aminotetrazole and (b) high-resolution TEM image of the same film after annealing at 230 °C. The insets show selected area electron diffraction patterns (a) before and (b) after film annealing.

ligands at the nanoparticle surface, which is necessary to stabilize particles in solution.[4] These ligand molecules often partially decompose and carbonize during the sintering process, contaminating semiconducting materials and reducing the size of crystalline grains. As a result, carrier mobility in semiconducting films obtained by sintering of corresponding nanoparticles is much lower than the mobility of CVD-grown thin films. For example, reported electron mobility in CdSe thin films derived from 2-nm nanoparticles sintered at 350 °C was 1 $cm^2V^{-1}s^{-1}$, more than two orders of magnitude lower than the mobilities in vapor-deposited CdSe films.[55] Continued improvement should be possible by designing special surface ligands, which can be reliably removed from the nanoparticle surface without leaving any solid residue behind. Tetrazole derivatives are promising candidates for removable stabilizing ligands because they can be controllably destroyed by mild thermal treatments.[57] Tetrazoles are five-membered heterocyclic compounds containing four nitrogen atoms in the cycle and having the highest content of nitrogen among all organic substances (for example, 82.3 wt.% for 5-aminotetrazole). Tetrazoles can thermally decompose at relatively low temperatures (400–500 K) with formation of only gaseous products. CdS nanoparticles capped with 5-aminotetrazole can be successfully sintered into a continuous film at temperatures as low as 230 °C (Fig. 10.6).

Colloidal semiconductor nanocrystals are attracting growing attention as the building blocks for inexpensive, large-area, solution-processed solar cells. The advantages here are the scalable and controlled synthesis, an ability to be processed in solution, the broadband absorption, and the superior transport properties of traditional photovoltaic semiconductors. Solar cells that rely exclusively on colloidal nanocrystals have been anticipated theoretically[58] and

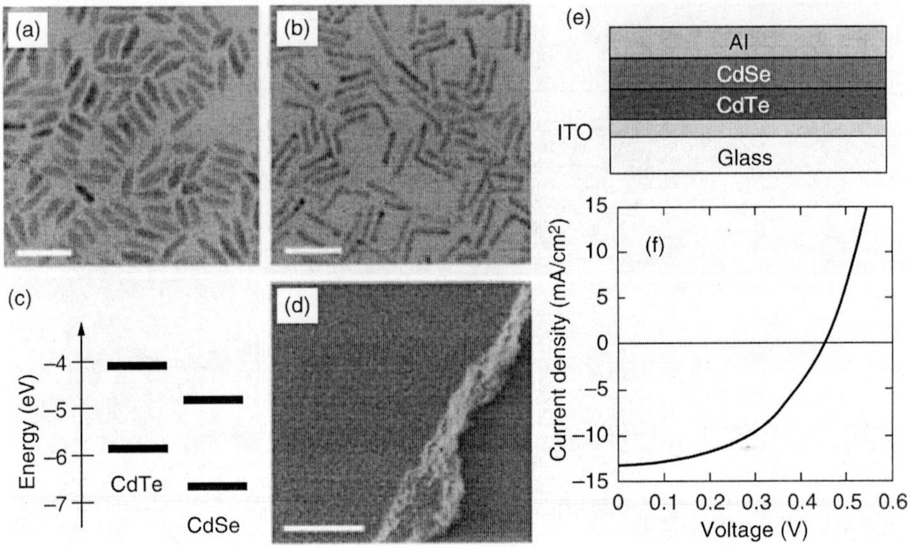

Figure 10.7. Transmission electron micrographs of (a) CdSe and (b) CdTe nanocrystals. Scale bar, 40 nm. (c) An energy diagram of valence and conduction band levels for CdTe and CdSe illustrates the type II charge-transfer junction formed between the two materials. (d) A typical spin-cast film of colloidal nanocrystals imaged by scanning electron microscopy (SEM). Scale bar, 1 μm. (e) Schematic structure of an ITO/100-nm CdTe/100-nm CdSe/Al bilayer nanocrystal solar cell. (f) I–V characteristics of the sintered nanocrystal solar-cell under simulated AM1.5G illumination. Reproduced from Ref. 59, Copyright 2005, with permission from AAAS.

recently demonstrated by the Alivisatos group using rod-shaped CdSe and CdTe nanocrystals[59] (Fig. 10.7a,b). In fabricating devices, the nanocrystals were spin-cast from a pyridine solution, which created densely packed homogeneous nanocrystal films over large areas (Fig. 10.7d). Planar heterojunctions were fabricated by sequentially spin-casting films of CdTe and then CdSe on indium tin oxide/glass coated with 2-Å alumina. Thermally deposited aluminum was used as a top contact (Fig. 10.7e).

The photovoltaic conversion in the nanocrystal bilayer devices is based on donor–acceptor charge transfer.[59] A schematic energy diagram (Fig. 10.7c) illustrates the staggered band alignment of this nanocrystalline donor–acceptor pair. After absorption and charge transfer, majority holes in the CdTe diffuse into the ITO but are blocked from moving through the CdSe toward the Al electrode. Likewise, majority electrons in the CdSe can diffuse only toward the Al, and not through the CdTe to the ITO.

Annealing of the devices at 400 °C sinters the CdSe and CdTe nanorods, concurrently improving carrier transport. To facilitate the sintering process, nanocrystal thin films are exposed to a saturated solution of $CdCl_2$ in methanol before the annealing. After sintering, films of CdSe and CdTe remain insulating in the dark, but they show about two orders of magnitude enhancement in photoconductivity. Optimized nanocrystal solar cell devices demonstrate a short-circuit current of 13.2 mA/cm^2, an open-circuit voltage of 0.45 V, and a fill factor of 0.49, which corresponds to a power conversion efficiency of 2.9% under simulated AM1.5G illumination (Fig. 10.7f). Sintered nanocrystalline solar cells have exhibited very good stability against photo-oxidation and represent a promising way toward inexpensive large-area photovoltaic devices.

10.3.2 Electronic Materials Based on Nanoparticle Assemblies

Sintering individual nanocrystals into a polycrystalline film increases film conductance but often generates high concentrations of structural defects that limit carrier lifetime and mobility.[55] An alternative approach suggests using ordered close-packed nanoparticle arrays for electronic and optoelectronic applications. Monodisperse colloidal nanocrystals tend to pack into ordered superlattices during slow evaporation of the carrier solvent[4] or gentle destabilization of the colloidal solution[16] (Fig. 10.8). The lateral size of long-range ordered domains can approach hundreds of microns.[60,61] Among a variety of

Figure 10.8. Transmission electron micrographs of superlattices of (a) $CoPt_3$ and (b) CdSe nanocrystals. (c,d) High-resolution images of [100] and [110] projections of the CdSe superlattice with the corresponding Fourier transformations. Reproduced from Ref. 16, Copyright 2002, with permission from Wiley.

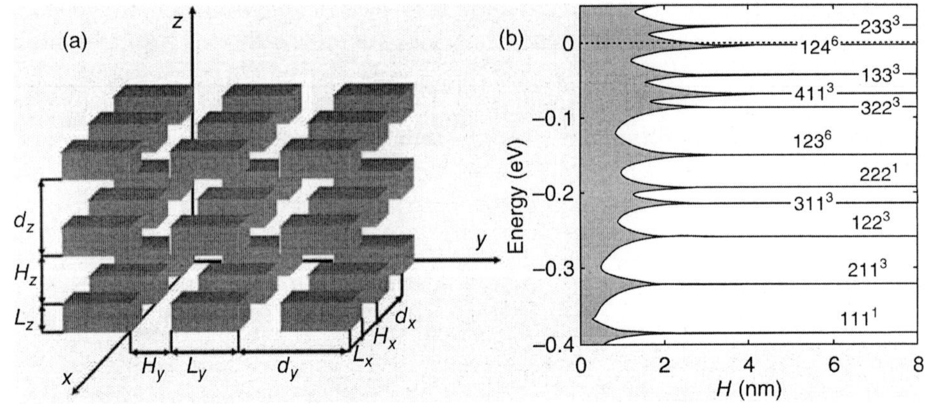

Figure 10.9. (a) Schematic structure of a silicon quantum dot crystal and (b) its calculated electronic structure as a function of interparticle distance H. The size of the nanoparticles is $L = 6.5\,\text{nm}$. At small H, a splitting of the quantized energy levels of single dots results in the formation of three-dimensional minibands. Reproduced from Ref. 64, Copyright 2001, with permission from the American Institute of Physics.

packing symmetries, the close-packed, face-centered cubic (*fcc*) lattice was predicted to be most thermodynamically stable.[62] However, formation of nanoparticle superlattices with hexagonal-close-packed (*hcp*) and simple hexagonal (*sh*) structures is common in the case of both semiconductor (CdSe, PbSe) and magnetic (Fe_2O_3, $CoPt_3$) nanoparticles.[63] *Hcp* and *fcc* superlattices often form simultaneously.

The presence of long-range translational ordering in nanoparticle assemblies makes them different from common amorphous and polycrystalline solids. The coupling among ordered quantum dots can lead to a splitting of the quantized carrier energy levels of single particles and can result in the formation of three-dimensional minibands (Fig. 10.9).[64] By changing the size of the nanocrystals, interparticle distances, barrier height, and regimentation, one can control the electronic band structure of these artificial *quantum dot crystals*.[65] Calculations show that the carrier density of states, effective mass tensor, and other properties of quantum dot crystals are different from those of bulk solids and quantum well superlattices.[64] Generally speaking, nanocrystal assemblies can be considered as a novel type of condensed matter, whose behavior depends both on the properties of the individual building blocks and on the many-body exchange interactions. The engineering of three-dimensional minibands in quantum dot crystals allows one to fine-tune electronic and optical properties of such nanostructures, providing exciting possibilities for low-cost electronic, photovoltaic, and thermoelectric devices.

Optical characteristics of nanocrystals and nanocrystal assemblies have been extensively studied in the past,[66,67] whereas their electronic properties have attracted considerable attention only recently.[68–70] Charge transport in an array of nanocrystals separated by energy barriers (a so-called "nanocrystal solid") depends on matching the energy levels of neighboring nanocrystals (site energies, α), on the exchange coupling energy between the nanocrystals (β), and on the Coulomb charging energy of the nanocrystal array (E_c).[71–73] In an ensemble of strongly confined semiconductor nanocrystals, $\Delta\alpha$ is mostly determined by particle size dispersion. The strength of exchange coupling β depends on interparticle spacing δ, the chemical nature of nanocrystals, capping ligands, and in some cases, on the relative orientation of the nanocrystals. β scales approximately as $exp(-\kappa(d + \delta))$, where d is the nanocrystal diameter and κ^{-1} describes the length scale of the wavefunction leakage outside the nanocrystal.[71,72] The charging energy E_c of a spherical nanocrystal of diameter d can be estimated as $E_C = e^2/(4\pi\varepsilon_m\varepsilon_0 d)$, where ε_m is the dielectric constant of the surrounding medium. For efficient charge transport, the dispersion of site energies $\Delta\alpha$ should not exceed β or Anderson localization will dominate the transport properties. If $\beta < E_c$, the nanocrystal array can behave as a Mott insulator.[71]

Numerous early studies revealed low electronic conductivity in semiconductor nanocrystal arrays because of poor exchange coupling and large concentrations of dangling bonds at the nanocrystal surfaces, which introduce carrier trapping mid-gap states.[69,74,75] The electron mobility can be increased by electrochemically charging the nanocrystal solid with several additional electrons per nanocrystal.[70,74] Charge screening by mobile ions present in the electrolyte substantially reduces E_c, facilitating the charging of a nanocrystal solid.[71] The mobility increases because of a combination of trap filling and transport through multiple quantum confined electronic states (e.g., 1S, 1P).[70] However, real technological implementations require solid-state field-effect transistors with insulated gates, where the current is modulated by applying a potential to a gate capacitively coupled to the transistor channel.

Quantum dot solids with high electron and hole mobilities have been assembled from chemically synthesized PbSe and PbTe nanocrystals.[68,76] Figure 10.10c shows small-angle X-ray scattering at grazing incidence (GISAXS) from a self-assembled film of 8-nm PbSe nanocrystals. The appearance of well-resolved reflections confirms both in-plane and vertical particle ordering in the PbSe nanocrystal assemblies. Lead chalcogenide (PbS, PbSe, PbTe) nanocrystal solids offer smaller $\Delta\alpha$ and E_c and larger β values compared with II–VI and III–V nanocrystals. These materials can be synthesized in the form of crystalline nanoparticles with very narrow size distributions (<5% std. dev) (Fig. 10.10a,b), providing a higher density of electronic states available for the charge transport. The large Bohr radius of electrons and holes (e.g., ~23 nm for PbSe and ~46 nm for PbTe) suggests that the electron and hole wavefunctions can effectively spill outside the volume of the nanocrystal, facilitating exchange interactions. Bulk rocksalt PbSe and PbTe have very high static dielectric constants, $\varepsilon \sim 250$ (PbSe) and $\varepsilon \sim 1000$ (PbTe), cf. $\varepsilon \sim 6.2$ for CdSe. This predicts an $E_c < 4$ meV for a 3D array of 8-nm PbSe nanocrystals, which is a small fraction of kT at room temperature.[68]

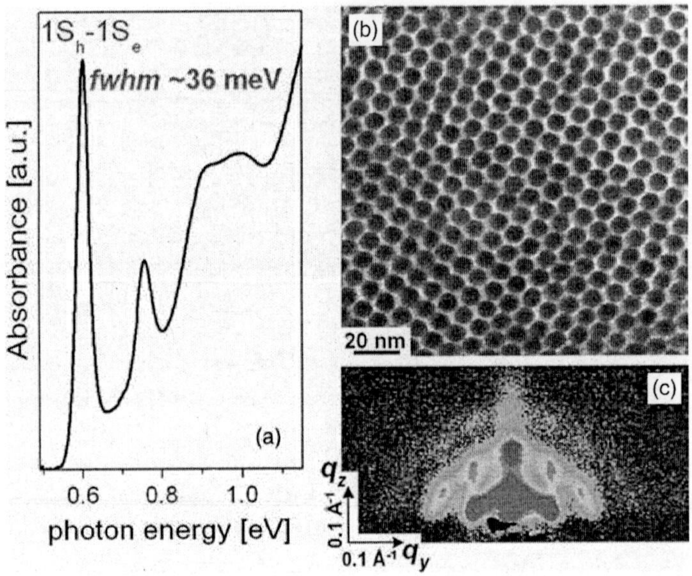

Figure 10.10. (a) Optical absorption spectrum of a colloidal solution of 8-nm PbSe nanocrystals. (b) TEM image of an array of 8-nm PbSe nanocrystals. (c) GISAXS pattern of a PbSe nanocrystal film. The appearance of well-resolved reflections confirms both in-plane and vertical ordering of the PbSe nanocrystals. Reproduced from Ref. 68, Copyright 2005, with permission from the AAAS.

To achieve efficient charge transport through an array of nanocrystals, special attention should be paid to eliminating trapping states at the nanocrystal surface and to enhancing the strength of interparticle exchange coupling. As-synthesized, PbSe and PbTe nanocrystal solids are insulating with a conductance (G) of $<10^{-10}\,\mathrm{S\,cm^{-1}}$ (Fig. 10.11b).[68] The low G can be attributed to poor exchange coupling between the nanocrystals caused by large ~1.1–1.3-nm interparticle spacing (Fig. 10.10b).[77] It was found that the conductance of nanocrystal solids increases by ~10 orders of magnitude after treatment in a dilute solution of hydrazine in acetonitrile (Fig. 10.11c).[68,76] After hydrazine treatment, the current through the nanocrystal film can be modulated by applying a potential to the back gate electrode (Fig. 10.11d). Such behavior is typical for an n-channel field-effect transistor (FET). A negative gate bias depletes electrons in the channel, switching off the device. At low drain-source voltage (V_{DS}), the current between source and drain electrodes (I_D) increases linearly with V_{DS}, and at high V_{DS}, the current saturates as the channel "pinches off" near the drain electrode.[2] Figure 10.12a shows current modulation I_{on}/I_{off} ~2.5×10^3 for an n-channel PbSe nanocrystal FET, with minor hysteresis between V_G scans in the forward and reverse directions. In the ON state, the low-field conductance of the PbSe nanocrystal film is $\sim 0.8\,\mathrm{S\,cm^{-1}}$ ($V_G = 40\,\mathrm{V}$)

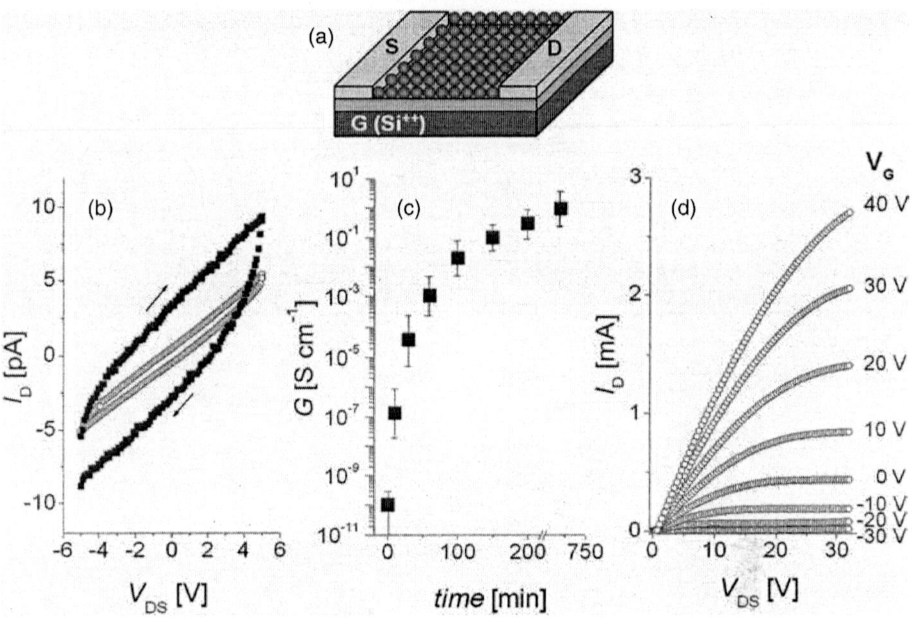

Figure 10.11. (a) Schematic of a nanocrystal TFT where S and D are the gold source and drain electrodes. A highly doped Si substrate (G) was used as the back gate. (b) I–V scans for a film of 8-nm PbSe nanocrystals where the voltage scan rate is $0.5\,V\,s^{-1}$ (black squares) and $0.025\,V\,s^{-1}$ (open circles) and the channel length L and width W are $6\,\mu m$ and $5000\,\mu m$, respectively. (c) Conductance of a PbSe nanocrystal film G versus time of exposure to a 1-M solution of N_2H_4 in acetonitrile. (d) Plot of drain current I_D versus drain-source voltage V_{DS}, as a function of gate voltage V_G for a nanocrystal FET whose channel is composed of 8-nm PbSe nanocrystals treated with hydrazine ($L = 10\,\mu m$, $W = 2000\,\mu m$). Reproduced from Ref. 68, Copyright 2005, with permission from AAAS.

and the current density in the saturation regime approaches $2.7 \times 10^4\,A\,cm^{-2}$. The field-effect electron mobility in hydrazine-treated PbSe and PbTe nanocrystal solids can be as high as $2.8\,cm^2V^{-1}s^{-1}$ for PbSe and above $10\,cm^2V^{-1}s^{-1}$ for PbTe nanocrystals.[78] The hydrazine treatment is generally applicable to lead chalcogenide and some II–VI nanocrystal solids. Conductance of the nanocrystal films can be further enhanced by combining the hydrazine treatment with partial sintering of the nanocrystals. Exposure of the sintered PbSe or PbTe nanocrystal films to a hydrazine solution gives rise to metallic-type conductivity ($G \sim 8.5\,S\,cm^{-2}$) caused by degenerate doping.[68]

Charge transport through an array of semiconductor nanocrystals is strongly affected by the electronic structure of nanocrystal surfaces. It is possible to control the type of conductivity and doping level of quantum dot crystals by adsorbing/desorbing molecular species at the nanocrystal surface. As an

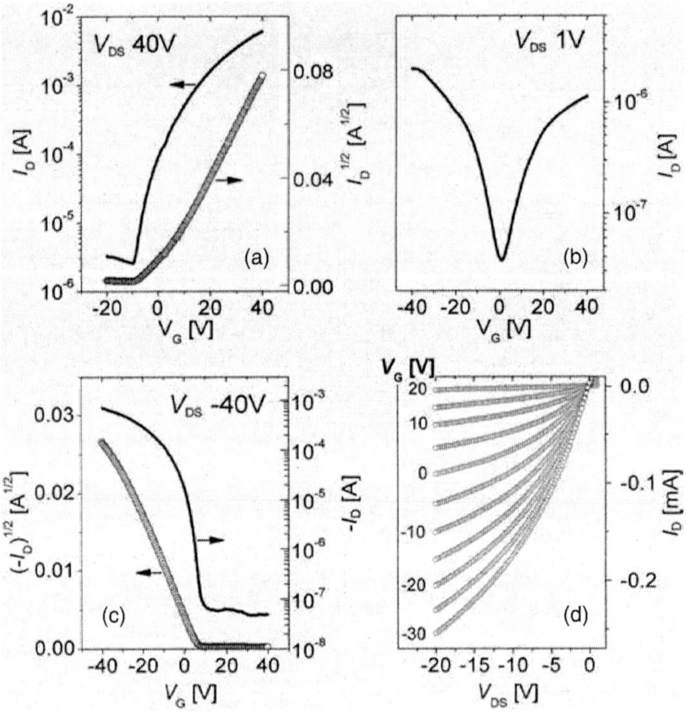

Figure 10.12. Device characteristics of PbSe nanocrystal FETs activated with hydrazine. (a) Plots of I_D and $I_D^{1/2}$ versus V_G at constant V_{DS} = 40 V for an n-channel FET assembled from 8.1-nm PbSe nanocrystals. L = 8 µm, W = 2300 µm. (b) I_D versus V_G plot at constant V_{DS} = 1 V for an ambipolar FET assembled from 8.1-nm PbSe nanocrystals. L = 8 µm, W = 2300 µm. (c) Plots of I_D and $I_D^{1/2}$ versus V_G at constant V_{DS} = −40 V for a p-channel FET assembled from 8.2-nm PbSe nanocrystals. L = 10 µm, W = 3000 µm. (d) Plot of I_D versus V_{DS}, as a function of V_G for a p-channel FET assembled from 8.4-nm PbSe nanocrystals. L = 8 µm, W = 2300 µm. The changes in the transistor polarity were induced by controllable adsorption/desorption of hydrazine molecules from the nanocrystal surface. Reproduced from Ref. [68], Copyright 2005, with permission from the AAAS.

example, vacuum treatment or mild heating (~100 °C) of the hydrazine-activated PbSe nanocrystal films switches their conductivity from n-type (Fig. 10.12a) to ambipolar (Fig. 10.12b) and, finally, to p-type (Fig. 10.12c,d) as the hydrazine desorbs.[68] The resulting p-type FETs show room-temperature hole mobilities μ_{sat} of 0.2 $cm^2 V^{-1} s^{-1}$ and current modulations of above 10^2 (Fig. 10.12d). Switching between electron and hole transport is reversible upon reexposure to hydrazine.

At low temperatures, charge transport in nanocrystal solids primarily occurs through the variable range hopping (VRH) mechanism[79] where a hopping electron finds the lowest activation energy E and the shortest hopping distance.[80] Because of the important role of Coulomb interactions, the low-temperature conductance in nanocrystal solids often follows the $\ln G \sim T^{-1/2}$ law known as Efros–Scklovskii variable range hopping.[79]

Hydrazine treatment does not cause changes of nanocrystal shape and size, but it reduces the interparticle spacing from ~1.1 nm to ~0.3 nm.[68,76] Hydrazine is a Brönsted base and can gently remove the capping ligands by reacting with oleic acid adsorbed on the surface of as-synthesized nanocrystals. Replacement of bulky oleic acid with small hydrazine molecules reduces the interparticle spacing, thereby increasing the exchange coupling energy. The bidentate hydrazine molecules can also link PbSe nanocrystals.[70] Moreover, hydrazine is a strong Lewis base, which enables it to saturate dangling bonds at the nanocrystal surface by donating a lone pair of electrons.[32] Finally, replacement of oleic acid ($\varepsilon \sim 2$) with hydrazine ($\varepsilon \sim 52$) also substantially reduce the charging energy E_c. The increase of β and the decrease of E_c helps to close the Hubbard gap, thus enabling the insulator–metal Mott transition in the nanocrystal solid.[80] Finally, hydrazine behaves as a charge-transfer n-type dopant.[81]

Colloidal synthesis allows for controlling not only the size but also the shape of semiconductor nanocrystals. The anisotropic shape of the nanocrystals can provide certain advantages for their use as the building blocks for electronic and optoelectronic devices. Sun and Sirringhaus observed about an order-of-magnitude improvement of carrier mobility in the films assembled from 65 nm by 10-nm ZnO nanorods (Fig. 10.13a) compared with the films of spherical ZnO nanoparticles.[82,83] The morphology of nanorod films strongly depends on the processing conditions. Colloidal solutions of rod-like nanoparticles can form lyotropic liquid-crystalline phases with nematic and smectic ordering of individual nanorods.[84,85] In the nematic phase, the nanorods are unidirectional aligned, enabling an easy path for charge carriers along the nanorods (Fig. 10.13b). The ability to control orientation of the nanorods and to achieve large domains of aligned nanorod films allows significant improvement of the transport properties. Field-effect transistors with mobilities of up to 1.2–1.4 $cm^2V^{-1}s^{-1}$ and the ON/OFF ratio of 10^5–10^6 were demonstrated in spin-coated, in-plane, aligned ZnO nanorod films.[83]

Colloidal nanocrystals can be used for inexpensive and high-throughput solution fabrication of both n- and p-channel field-effect devices by stamping or ink-jet printing. The field-effect mobilities in PbSe and ZnO nanocrystal FETs are comparable with or better than mobilities in pentacene and other solution-processed organic films.[77] The ease of switching between n- and p-transport in PbSe nanocrystal arrays can enable CMOS FET circuits and p–n junctions for solar cells. Recent observation of carrier multiplication (i.e., generation of several excitons per one absorbed photon) in PbS, PbSe, and PbTe nanocrystals[86] makes these materials promising for thin-film photovoltaics. Sargent et al. recently demonstrated ultrasensitive solution-cast infrared photodetectors using PbS nanocrystal solids.[87] The

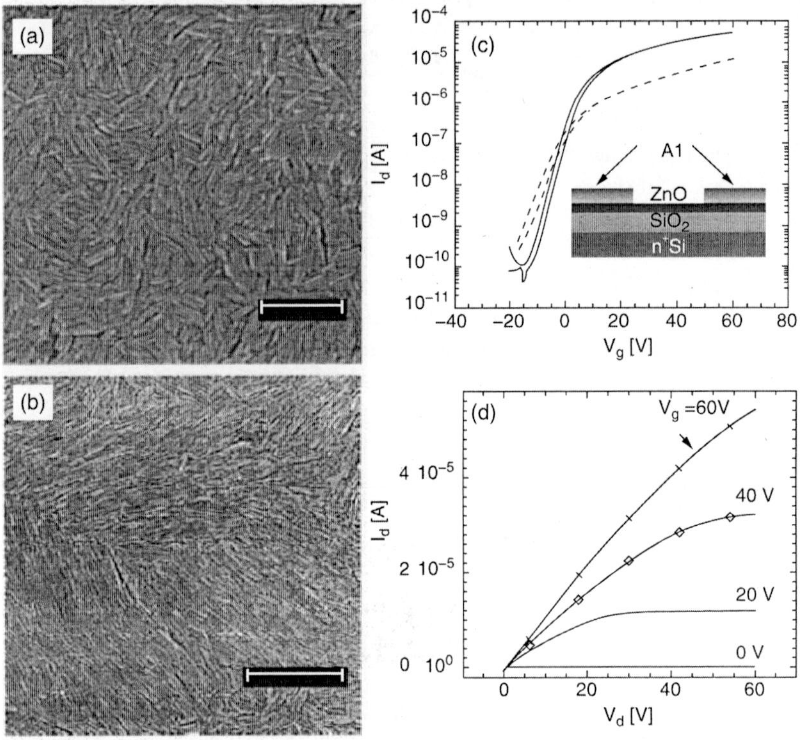

Figure 10.13. (a) SEM image of ZnO nanorods coated with octylamine. Scale bar, 200 nm. (b) Uniform nanorod film fabricated by spin coating of ZnO nanorods. Scale bar, 500 nm. The nanorods assemble into domains with nematic ordering. (c) Saturated transfer characteristics for a thin-film transistor fabricated by spin coating of ZnO nanorods with different ligands: octylamine (solid line), butylamine (dashed line). $V_d = 60$ V. (d) Output characteristics of a spin-coated device made from octylamine-stabilized ZnO nanorods. The device structure is shown in the inset in (c). Reproduced from Ref. 83, Copyright 2006, with permission from the American Chemical Society.

normalized detectivity of their solution-processed infrared (IR) detectors was superior to the best epitaxially grown devices operating at room temperature.

The nanoscale periodicity inherent to conductive nanocrystal solids is promising for thermoelectric applications because enhanced boundary scattering of phonons should lower the thermal conductivity.[88] The large surface-to-volume ratio of nanocrystal FETs might be exploited in chemical sensors. The thin films of aligned ZnO nanorods are a promising candidate for transparent conductive coatings in printable displays, solar cells, and light-emitting devices (LEDs).

10.3.3 Multicomponent Nanoparticle Assemblies

Semiconducting, magnetic, and metallic nanoparticles can self-assemble into complex, multicomponent superlattices (Fig. 10.14).[24,89] A large variety of binary superlattices have been assembled from monodisperse PbS, PbSe, $CoPt_3$, Fe_2O_3, LaF_3, Au, Ag, and Pd nanocrystals by mixing corresponding colloidal solutions and slowly evaporating the carrier solvent. Superlattices with AB, AB_2, AB_3, AB_4, AB_5, AB_6, and AB_{13} stoichiometry with cubic, hexagonal, tetragonal, and orthorhombic symmetries have been identified. Binary nanoparticle arrays isostructural with, for example, NaCl, CuAu, AlB_2, $MgZn_2$, $MgNi_2$, Cu_3Au, Fe_4C, $CaCu_5$, CaB_6, and $NaZn_{13}$, have been synthesized from different nanoparticle combinations.[24,89] Formation of binary nanoparticle superlattices is guided by competing Coulombic, dipolar, and van der Waals interparticle interactions.[24] The ability to mix and match different nanoparticles and to assemble them systematically into ordered binary superlattices, with precisely controlled stoichiometry and symmetry, provides a general path to multifunctional nanocomposites (metamaterials).

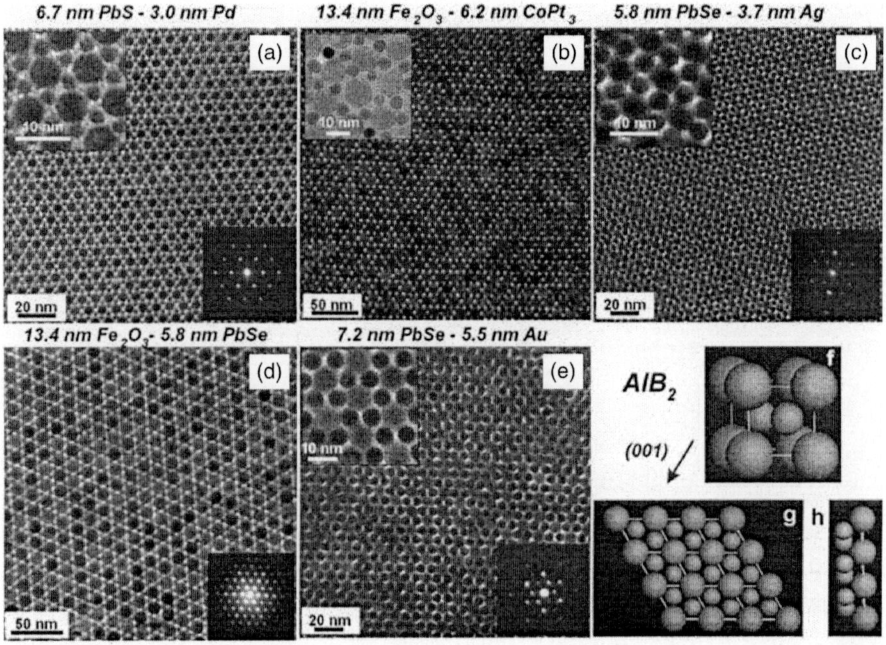

Figure 10.14. Binary superlattices self-assembled from various combinations of magnetic, semiconducting, and metallic nanoparticles. (a–e) TEM micrographs of (001) planes of binary superlattices isostructural with the AlB_2 compound. (f) A sketch of the AlB_2 unit cell. (g,h) Depictions of the front and side views of the superlattice (001) plane, correspondingly. Reproduced from Ref. 24, Copyright 2006, with permission from the American Chemical Society.

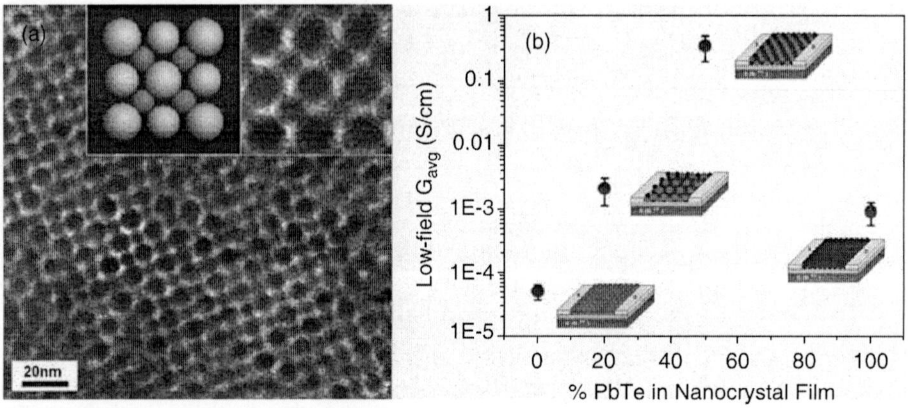

Figure 10.15. (a) TEM micrographs of the PbTe–Ag_2Te binary superlattice, which is isostructural with the CuAu intermetallic compound. (b) Comparison of average low-field conductances measured for pure Ag_2Te films, pure PbTe films, binary 1:1 PbTe–Ag_2Te films, and binary 5:1 PbTe–Ag_2Te films. These data demonstrate the synergistic enhancement of conductivity in the binary nanocrystal solids. Reproduced from Ref. 90, Copyright 2007, with permission from the Nature Publishing Group.

Multicomponent nanoparticle assemblies provide various additional possibilities for designing films with tailorable electronic properties. Electronic measurements on multicomponent nanocrystal solids show that one type of nanocrystal can behave as a dopant with respect to the other component, enabling rational materials design for a targeted application. In the work of Urban et al,[90] the choice of nanocrystal components, PbTe and Ag_2Te, was motivated by the goal of constructing highly conductive, granular, *p*-type nanocrystal composites. This combination of material properties (high conductivity and granularity) was selected to yield a material with high *p*-type electronic conductivity and low thermal conductivity, which are desirable traits for the *p*-type leg of a thermoelectric device. PbTe has the highest ZT of any single-phase bulk solid from 550 K to 700 K, the target range for waste-heat recovery applications. The complimentary component (Ag_2Te) was chosen because it is an effective *p*-type dopant for bulk PbTe.[91] The colloidal monodisperse PbTe and Ag_2Te nanocrystals were coassembled into AB binary nanocrystal superlattices, which are isostructural with the CuAu intermetallic compound (Fig. 10.15a). It was demonstrated that the electrical conductivity of the hydrazine-treated binary superlattices of PbTe–Ag_2Te nanocrystals was approximately 100 times greater than that of similarly treated superlattices of the individual components (Fig. 10.15b), which is consistent with the role of Ag_2Te as a *p*-type dopant in bulk PbTe.[90] This example demonstrates the synergistic effects possible in the multicomponent nanocrystal assemblies and shows the great promise of this approach for rational design of novel solution-processed materials.

10.4 FILMS AND ARRAYS OF INORGANIC NANOWIRES

Semiconductor nanowires are promising materials for many technological applications, such as high-performance nanoscale field-effect transistors,[10] logic gates,[92] memories,[93] sensors for gases and biomolecules,[40] thermoelectric devices,[42] LEDs and lasers,[94,95] as well as photodetectors and solar cells.[96,97] The carrier mobility in high-quality semiconductor nanowires may be comparable with or exceeding that of the single-crystal materials.[6,10] Arrays of single-crystal nanowires fabricated on low-cost, plastic, or glass substrates are expected to find use in applications, such as flexible, wearable, and disposable consumer electronics, sensors, and radio frequency identification (RFID) tags. High temperatures, which are often required for nanowire growth, place limitations on direct fabrication of nanowire-based devices on plastic or glass substrates. To overcome the compatibility problem, the nanowires can be first synthesized using standard high-temperature VLS or vapor-transfer techniques and then transferred onto a low-cost substrate using low-temperature solution processing techniques (see Chapter 11) or dry rubber-stamping techniques (Chapter 13). The film fabrication process should involve both nanowire positioning and alignment because the carrier transport along single-crystal wires is much faster than wire-to-wire hopping in randomly oriented nanowire networks. Ideally, in nanowire devices, the charges should travel directly from source to drain electrodes within single crystal wires. To date, several experimental approaches have been explored: flow-assisted alignment in a microfluidic channel,[98] Langmuir–Blodgett technique,[99] and electric-field–directed alignment.[100,101] The last approach seems to be the most promising because it allows rapid assembly of complex (i.e., multidirectional) networks for nanowire-based electronic circuitry and readily lends itself to very short length scales.[101] Alignment of semiconducting and metallic nanowires has been successfully demonstrated by applying DC or AC electric fields to nanowire suspensions and colloidal solutions (Fig. 10.16). Thus, dielectrophoresis has been used to pull InP, ZnO, and PbSe nanowires toward the electrodes and to align them along the electric field.[100–102] PbSe nanowires could also be grown directly on a chip by the oriented attachment of PbSe nanocrystals, connecting biased source and drain electrodes.[102] The on-chip assembly of semiconducting nanowires may provide a novel concept for growing nanoscale electronic circuitry directly on a chip—in a way, similar to the formation of interconnects between neurons in biological systems.

Because the nanowire synthesis is independent of the device substrate, these processes represent a general approach for exploiting a broad range of materials (including group IV, II–VI, III–V, and IV–VI semiconductors) as the transistor channel materials. Combining different semiconductors allows for addressing the big challenge of the fabrication of CMOS (complimentary metal oxide semiconductor) circuits using solution-processed components.[103] In general, CMOS allows easier circuit design, greater speed, and lower power consumption than n-MOS or p-MOS circuitry alone.[104] Confinement of

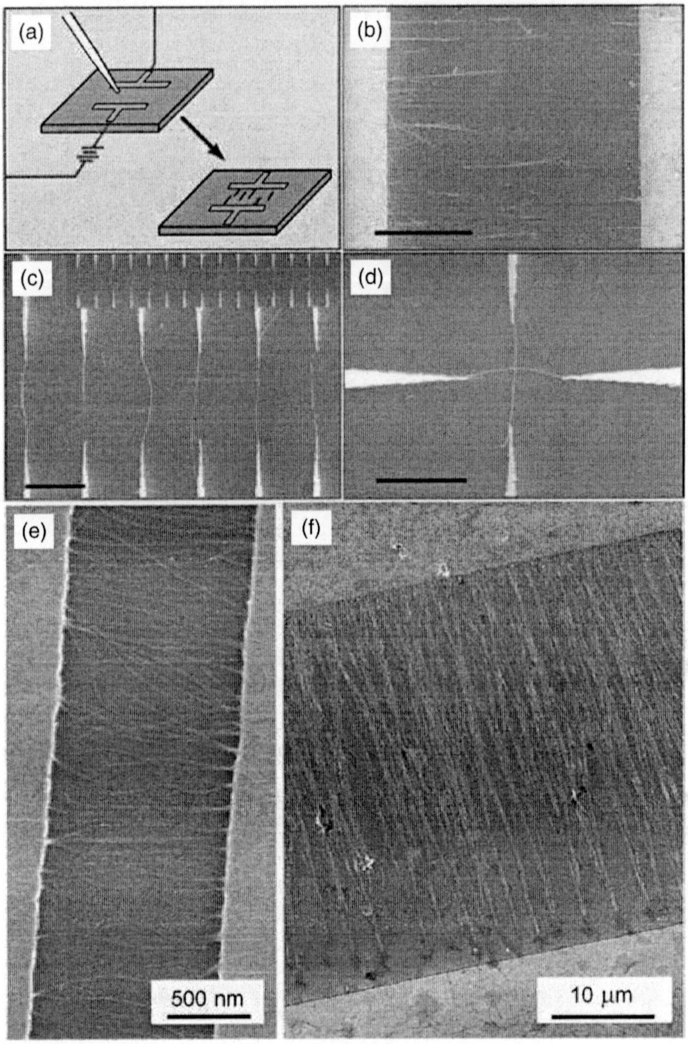

Figure 10.16. Parallel and orthogonal assembly of nanowires with electric fields. (a) Schematic view of alignment by electric field. The electrodes are biased at 50–100V after a drop of nanowire solution is deposited on the substrate. (b) Parallel array of InP nanowires aligned between two parallel electrodes. (c) Spatially positioned parallel array of InP nanowires obtained after electric-field assembly using a bias of 80V. Inset, 15 pairs of parallel electrodes with individual InP nanowires bridging each diametrically opposed electrode pair. (d) Crossed InP nanowire junction obtained using layer-by-layer alignment with the electric field applied in orthogonal directions in the two assembly steps. The applied bias in both steps was 80V. Scale bars in b–d, 10 μm. (e,f) Alignment of PbSe nanowires by a DC electric field. Nanowires aligned across (e) 2-μm and (f) 40-μm channels. Reproduced from Ref. 101, Copyright 2001, with permission from the Nature Publishing Group, and Ref. 102, Copyright 2007, with permission from the American Chemical Society.

acoustic phonons results in a strong suppression of thermal conductivity of semiconductor nanowires.[105,106] Low thermal conductance combined with excellent electrical properties provides the grounds for superior thermoelectric properties of semiconductor nanowires. For example, theoretical calculations on segmented PbTe/PbSe nanowires predict their thermoelectric performance to be at least three times better than that of any known bulk thermoelectric material.[42]

Solution-synthesized semiconductor nanowires have also been employed in dye-sensitized solar cells as an electron-collecting anode material (Fig. 10.17a).[41] The dense array of oriented, crystalline ZnO nanowires attached to a conducting substrate (Fig. 10.17b) was synthesized by a two-step process. First, fluorine-doped tin oxide substrates were coated with a thin film of ZnO nanoparticles, 3–4 nm in diameter, by dip-coating in a concentrated nanoparticle solution. Nanowires were grown by immersing seeded substrates in aqueous solutions containing 25 mM of zinc nitrate hydrate, 25 mM of hexamethylenetetramine, and 5–7 mM of polyethylenimine at 92 °C for 2.5 h. Such mild aqueous chemistry enables inexpensive fabrication, whereas the direct electrical pathways provided by the nanowires ensure the rapid collection of photogenerated carriers. An ordered topology that increases the rate of electron transport, and nanowire electrodes may provide a means to improve the quantum efficiency. Currently, full sun efficiency of 1.5% is demonstrated, which is limited primarily by the surface area of the nanowire array. Raising the efficiency of the nanowire cell to a competitive level depends on achieving higher dye loadings through an increase in surface area. The synthetic strategy should be extended to design of nanowire electrodes with much larger areas available for dye adsorption. The nanowire dye-sensitized solar cell is a promising variant of the excitonic photovoltaic devices. The advantages of the nanowire geometry are even more compelling for inorganic–polymer hybrid

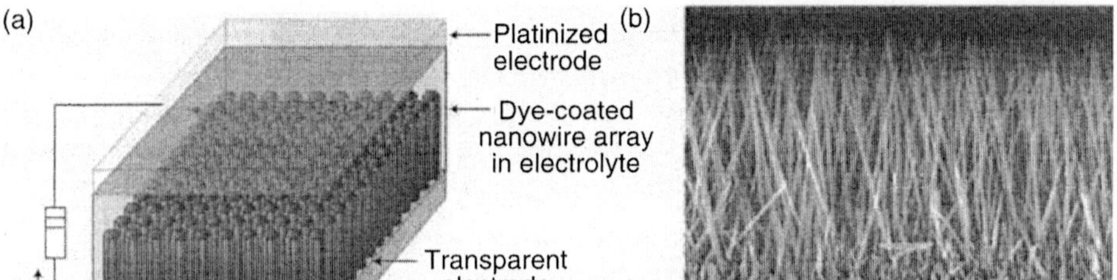

Figure 10.17. (a) Schematic diagram of the nanowire dye-sensitized solar cell. Light is incident through the bottom electrode. (b) SEM cross section of a solution-fabricated ZnO nanowire array on fluorine-doped tin oxide. The wires are in direct contact with the substrate. Scale bar, 5 μm. Reproduced from Ref. 41, Copyright 2005, with permission from the Nature Publishing Group.

devices,[107] in which an oriented, continuous, and crystalline inorganic phase of the proper dimensions could greatly improve the collection of both electrons and holes.

Nanowire films have proven to be an excellent material for gas-sensing applications. Thus, an array of aligned p-doped Si nanowires exhibited parts-per-billion (p.p.b.) sensitivity to NO_2, one of the most dangerous environmental pollutants.[108] The sensing effect is observed as an increase of electrical conductivity in the presence of NO_2. The increase in current was attributed to the strong electron-withdrawing capabilities of NO_2, which have the equivalent effect of hole carrier injections into the p-type Si nanowires. The device could detect concentrations down to 20 p.p.b. of NO_2. This sensitivity metric is comparable with sensors fabricated from metal-oxide nanowires (5–20 p.p.b.).[109] Using nanowire arrays allows for combining sensitivity with a good signal-to-noise ratio resulting from averaging the response signal over multiple wires. Chemical modification of the nanowire surface can impart additional selectivity. In the recent work of Heath et al., the Si nanowire surface was primed with alkane-, aldehyde-, and amino-silanes to create an integrated set of sensors functionalized to provide a fingerprint pattern of response for a given gas.[108] Such "nano-electronic noses" could discriminate low concentrations of acetone and hexane solvent vapors. These inexpensive, low-power gas sensors may find applications that range from real-time pollution regulation to highly portable biological- and chemical-threat detectors.

10.5 APPLICATIONS USING NETWORKS AND ARRAYS OF CARBON NANOTUBES

Single- and multiwall carbon nanotubes represent an important class of solution-processable electronic materials. Their superior properties, including very high carrier mobility, high chemical stability, and impressive mechanical characteristics offer great promise for technological applications. Among a variety of applications envisioned for carbon nanotubes, we will here focus on field-effect transistors and gas sensors, where a high carrier mobility and ultimate surface-to-volume ratio of single-wall CNTs can be used to the full extent. For these applications, CNTs can either be grown directly on a substrate using high-temperature seeded CVD reactions[45,110] or deposited from solution.[111] In both cases, unidirectional alignment of individual nanotubes can be achieved by flow- or field-directed assembly.[112,113] According to Rogers et al., the films of aligned single-wall CNTs contacted by Ti/Pd electrodes (Fig. 10.18) demonstrate hole mobilities of as high as $1000\,cm^2V^{-1}s^{-1}$, which is a higher value than that achieved for nanowires and nanoparticle assemblies.[110] At the same time, the ON–OFF ratio of TFTs assembled from multiple CNTs is typically very poor because of the presence of nanotubes with metallic conductivity.[110,111]

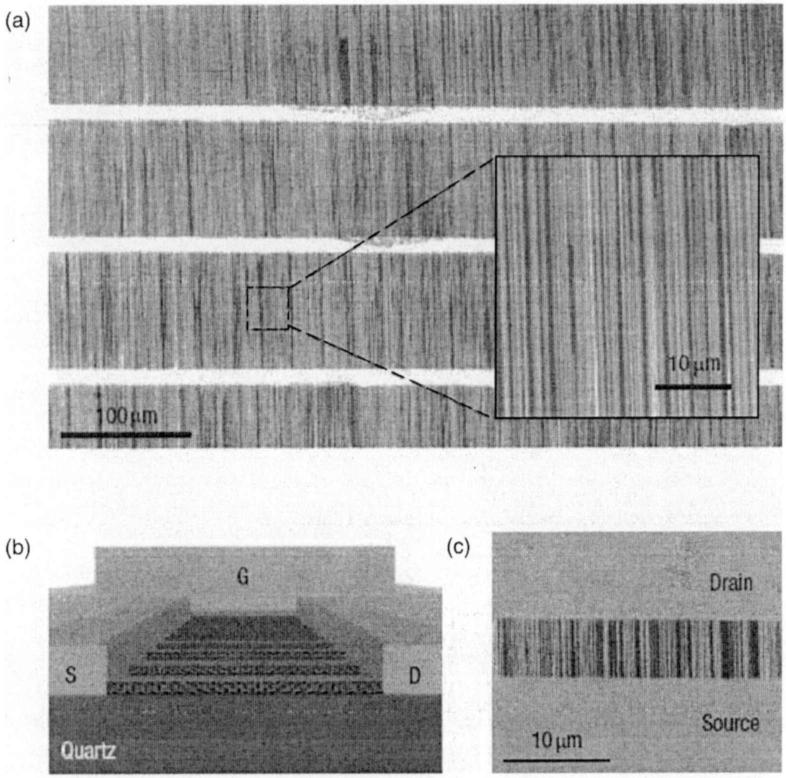

Figure 10.18. (a) SEM image of a pattern of perfectly aligned, linear, single-wall CNTs formed by CVD growth on a quartz substrate. The inset provides a magnified view. (b) Schematic illustration of the layout of a transistor that incorporates aligned CNTs. The device uses source (S), drain (D), and gate (G) electrodes, and a dielectric layer is formed sequentially on top of the nanotubes on quartz. (c) SEM image of the channel region of such a device. Reproduced from Ref. 112, Copyright 2007, with permission from the Nature Publishing Group.

The current modulation in the nanotube TFTs can be significantly improved by controllable electrical breakdown of the metallic nanotubes. This can be achieved by slowly increasing drain voltage, while holding gate voltage at a large positive value to keep the semiconducting nanotubes depleted and insulating.[110,114] Controllable breakdown of the metallic nanotubes allows for fabricating high-performance TFTs on plastic substrates with an ON/OFF ratio exceeding 10^4 and a linear regime field-effect mobility of $480\,cm^2V^{-1}s^{-1}$, the highest value reported for a device on plastic (Fig. 10.19).[110] Using high-k gate dielectrics such as HfO_2 resulted in high peak transconductances ($\sim 800\,S\,m^{-1}$), which is sufficient to employ these nanotube TFTs in high-speed logic circuits. The other approach to fabrication of semiconducting nanotube films is based on separation of semiconducting nanotubes from metallic ones using

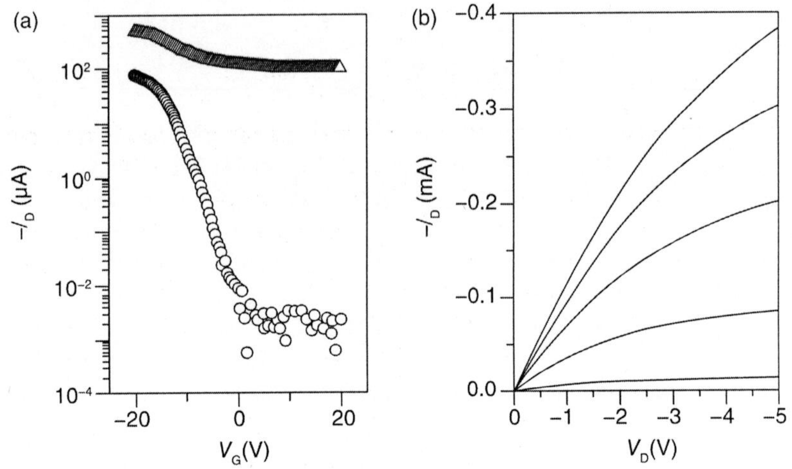

Figure 10.19. (a) Transfer characteristics of a TFT that uses aligned arrays of single-wall CNTs. The data correspond to measurements on the device before (open triangles) and after (open circles) an electrical breakdown process that eliminates metallic nanotubes from the current transport pathway between source and drain electrodes. This process improves the ON/OFF ratio of a nanotube TFT by a factor of more than 10^4. (b) Drain current (I_D) versus drain–source voltage (V_D) at different gate voltages for the nanotube device, measured after breakdown, illustrating good current modulation. The gate voltage varies from −5V to 5V (top to bottom). Reproduced from Ref. 112, Copyright 2007, with permission from the Nature Publishing Group.

density-gradient ultracentrifugation[109] or dielectrophoresis.[115] The devices made of random networks of purified semiconducting nanotubes showed ON/OFF ratios exceeding 20,000 and a hole mobility of 20 $cm^2V^{-1}s^{-1}$.[111]

The films of single-wall carbon nanotubes demonstrated excellent sensing properties. The sensing effect can be based on either change in the nanotube resistance in the presence of electron-donating (electron-withdrawing) species[116,117] or on the changes in the electrical capacitance of nanotubes caused by adsorption of chemical vapors on the nanotube surface.[118] Molecular adsorbates are polarized by the fringing electric fields radiating from the surface of a CNT electrode, which causes an increase in its capacitance.[118] Both approaches demonstrated impressive results, allowing chemical vapors sensing in sub-(part-per-million) range. Coating of the nanotube surface with a chemo-selective material can be used to impart high sensitivity and selectivity to the sensors. Thus, polyethyleneimine coating affords n-type nanotube devices capable of detecting NO_2 at less than 1 p.p.b (parts-per-billion) concentrations while being insensitive to NH_3. Coating Nafion (a polymeric perfluorinated sulfonic acid ionomer) on nanotubes blocks NO_2 and allows for selective sensing of NH_3.[116] Detection of target molecules in a gas mixture was demonstrated with the multiplexed nanotube sensors.[116]

10.6 CONCLUDING REMARKS

During the last two decades, inorganic nanoparticles and nanowires have developed into new classes of materials with their own sets of properties, some of which are unique and not observable in the "molecular" and "macroscopic" worlds. Recent advances in synthesis and self-assembly introduce entirely new ways of controlling the size, shape, arrangement, connectivity, and even topology of crystalline metallic, semiconducting, and oxide nanomaterials. A complete understanding of nucleation and growth in the nanometer regime remains, however, a challenging field for future research activity. Progress in this direction will enable complex multicomponent nanostructures to be created by design in cost-effective ways. These structures will be further used as the building blocks for different electronic, optoelectronic, magnetic, and catalytic materials.

Precise assembly of nanocrystals with desired optical, electronic, and magnetic properties into binary nanoparticle superlattices provides a route to "metamaterials" combining metallic, semiconducting, and magnetic components into ordered superlattices; a diverse collection of these structures have been recently reported, and many novel structures will follow. The modular design of multicomponent solids using nanometer-scale building blocks provides access to unique combinations of properties not available in single-component bulk solids. The controllable assembly of nanoparticles into macroscopic films and structures with a desired arrangement of the building blocks requires fundamental understanding of self-organization phenomena at the nanoscale. Significant progress has been already achieved in this area; however, the lack of fundamental knowledge about particle–particle interactions does not allow for development of the models that can reliably predict the structure and properties of nanoparticle assemblies. Little is known about collective phenomena in the nanoparticle assemblies. Recent studies point to the possibility of miniband formation in properly engineered superlattices of semiconductor nanocrystals. However, the electronic structure, phonon spectra, and other characteristics of single- and multicomponent nanoparticle assemblies have never been a subject of thorough theoretical studies. Because of the absence of theoretical background, the nanocrystal superlattice approaches to materials design are still in the early stages of development. The generation of truly high-performance materials, requiring the optimization of many competing parameters, is still some way off. At the same time, early results represent an important step in the rational design of nanocrystal superlattices for targeted physical applications. Nanocrystal solar cells offer an exciting research direction and serve as a key development toward achieving low-cost solar energy conversion.

Carbon nanotube films possess enormous potential for a variety of applications. The major limitations at present are associated with heterogeneity of as-synthesized nanotubes and with difficulties in separating CNTs with semiconducting and metallic characteristics. If this problem will be solved in a cost-effective manner, carbon nanotubes will find extremely broad use in both high-performance and low-cost (disposable) electronic and sensing devices.

ACKNOWLEDGMENTS

We are deeply indebted to all our colleagues and collaborators, whose names appear in the cited literature. Work at the Molecular Foundry was supported by the Director, Office of Science, Office of Basic Energy Sciences, Division of Materials Sciences and Engineering, of the U.S. Department of Energy under Contract DE-AC02-05CH11231.

REFERENCES

1. International Technology Roadmap for Semiconductors. 2005 Edition. Accessed: http://www.itrs.net/.
2. Kagan, C. R.; Andry, P., editors. 2003. *Thin Film Transistors*. Marcel Dekker, New York.
3. Schmid, G., editor. 1994. *Clusters and Colloids*. VCH, Weinheim, Germany.
4. Murray, C. B.; Kagan, C. R.; Bawendi, M. G. 2000. Synthesis and characterization of monodisperse nanocrystals and close-packed nanocrystal assemblies. *Ann. Rev. Mater. Sci.* 30:545–610.
5. Peng, X.; Manna, L.; Yang, W. D.; Wickham, J.; Scher, E.; Kadavanich, A.; Alivisatos, A. P. 2000. Shape control of CdSe nanocrystals. *Nature* 404:59–61.
6. Cui, Y.; Duan, X.; Huang, Y.; Lieber, C. M. 2003. In *Nanowires and Nanobelts-Materials, Properties, and Devices*, edited by Wang, Z. L. Kluwer Academic/Plenum Publishers, Boston, MA.
7. Dresselhaus, M. S.; Dresselhaus, G.; Eklund, P. C. 1996. In *Science of Fullerenes and Carbon Nanotubes*. Academic Press, San Diego, CA.
8. Vossmeyer, T.; Reck, G.; Schulz, B.; Katsikas, L.; Weller, H. 1995. Double-layer superlattice structure built up of $Cd_{32}S_{14}(SCH_2CH(OH)CH_3)_{36}\cdot 4H_2O$ clusters. *J. Am. Chem. Soc.* 117:12881–12882.
9. Yin, Y.; Alivisatos, A. P. 2005. Colloidal nanocrystal synthesis and the organic-inorganic interface. *Nature* 437:664–670.
10. Cui, Y.; Zhong, Z.; Wang, D.; Wang, W. U.; Lieber, C. M. 2003. High performance silicon nanowire field effect transistors. *Nano Lett.* 3:149–152.
11. Durkop, T.; Getty, S. A.; Cobas, E.; Fuhrer, M. S. 2004. Extraordinary mobility in semiconducting carbon nanotubes. *Nano Lett.* 4:35–39.
12. Mitzi, D. B.; Chondroudis, K.; Kagan, C. R. 2001. Organic-inorganic electronics. *IBM J. Res. & Dev.* 45:29–45.

13. Schmid, G., editor. 2003. *Nanoparticles—from Theory to Applications*. Wiley VCH, Weinheim, Germany.

14. Seo, W. S.; Lee, J. H.; Sun, X.; Suzuki, Y.; Mann, D.; Liu, Z.; Terashima, M.; Yang, P. C.; McConnell, M. V.; Nishimura, D. G.; Dai, H. 2006. FeCo/graphitic-shell nanocrystals as advanced magnetic-resonance-imaging and near-infrared agents. *Nature Mater.* 5:971–976.

15. Cho, K.-S.; Talapin, D. V.; Gaschler, W.; Murray, C. B. 2005. Designing PbSe nanowires and nanorings through oriented attachment of nanoparticles. *J. Am. Chem. Soc.* 127:7140–7147.

16. Rogach, A. L.; Talapin, D. V.; Shevchenko, E. V.; Kornowski, A.; Haase, M.; Weller, H. 2002. Surface derivatization and isolation of semiconductor cluster molecules. *Adv. Funct. Mater.* 12:653–664.

17. Murray, C. B.; Norris, D. J.; Bawendi, M. G. 1993. Synthesis and characterization of nearly monodisperse CdE(E = S, Se, Te) semiconductor nanocrystallites. *J. Am. Chem. Soc.* 115:8706–8715.

18. Steigerwald, M. L.; Alivisatos, A. P.; Gibson, J. M.; Harris, T. D.; Kortan, R.; Muller, A. J.; Thayer, A. M.; Duncan, T. M.; Douglass, D. C.; Brus, L. E. 1988. Surface derivatization and isolation of semiconductor cluster molecules. *J. Am. Chem. Soc.* 110:3046–3050.

19. Mattoussi, H.; Radzilowski, L. H.; Dabbousi, B. O.; Thomas, E. L.; Bawendi, M. G.; Rubner, M. F. 1998. Electroluminescence from heterostructures of poly(phenylene vinylene) and inorganic CdSe nanocrystals. *J. Appl. Phys.* 83:7965–7974.

20. Schubert, U. S.; Tekin, E.; Smith, P. J.; Hoeppener, S.; van den Berg, A. M. J.; Susha, A. S.; Rogach, A. L.; Feldmann, J.; Schubert, U. S. 2007. Inkjet printing of luminescent CdTe nanocrystal-polymer composites. *Adv. Funct. Mater.* 17: 23–28.

21. Sun, Y.; Xia, Y. 2002. Shape-controlled synthesis of gold and silver nanoparticles. *Science* 298:2176–2179.

22. Jana, N. R.; Peng, X. 2003. Single-phase and gram-scale routes toward nearly monodisperse Au and other noble metal nanocrystals. *J. Am. Chem. Soc.* 125: 14280–14281.

23. Brust, M.; Walker, M.; Bethell, D.; Schiffrin, D. J.; Whyman, R. J. 1994. Synthesis of thiol-derivatised gold nanoparticles in a two-phase Liquid–Liquid system. *Chem. Commun.* 7:801–802.

24. Shevchenko, E. V.; Talapin, D. V.; Murray, C. B.; O'Brien, S. 2006. Structural characterization of self-assembled multifunctional binary nanoparticle superlattices. *J. Am. Chem Soc.* 128:3620–3637.

25. Sun, S.; Murray, C. B. 1999. Synthesis of monodisperse cobalt nanocrystals and their assembly into magnetic superlattices. *J. Appl. Phys.* 85:4325–4330.

26. Murray, C. B.; Sun, S.; Doyle, H.; Betley, T. 2001. Monodisperse 3d transition-metal (Co, Ni, Fe) nanoparticles and their assembly into nanoparticle superlattices. *MRS Bulletin* 26:985–991.

27. Park, J.; An, K.; Hwang, Y.; Park, J.-G.; Noh, H.-J.; Kim, J.-Y.; Park, J.-H.; Hwang, N.-M.; Hyeon, T. 2004. Ultra-large-scale syntheses of monodisperse nanocrystals. *Nature Mater.* 3:891–895.

28. Sun, S.; Murray, C. B.; Weller, D.; Folks, L.; Moser, A. 2000. Monodisperse FePt nanoparticles and ferromagnetic FePt nanocrystal superlattices. *Science* 287:1989–1992.

29. Shevchenko, E. V.; Talapin, D. V.; Rogach, A. L.; Kornowski, A.; Haase, M.; Weller, H. 2002. Colloidal synthesis and self-assembly of $CoPt_3$ nanocrystals. *J. Am. Chem. Soc.* 124:11480–11485.

30. Micic, O. I.; Curtis, C. J.; Jones, K. M.; Sprague, J. R.; Nozik, A. J. 1994. Synthesis and characterization of InP quantum dots. *J. Phys. Chem.* 98:4966–4969.

31. Murray, C. B.; Sun, S.; Gaschler, W.; Doyle, H.; Betley, T. A.; Kagan, C. R. 2001. Colloidal synthesis of nanocrystals and nanocrystal superlattices. *IBM J. Res. & Dev.* 45:47–56.

32. Talapin, D. V.; Rogach, A. L.; Kornowski, A.; Haase, M.; Weller, H. 2001. Highly luminescent monodisperse CdSe and CdSe/ZnS nanocrystals synthesized in a hexadecylamine-trioctylphosphine oxide-trioctylphospine mixture. *Nano Lett.* 1:207–211.

33. Manna, L.; Scher, E. C.; Alivisatos, A. P. 2000. Synthesis of soluble and processable rod-, arrow-, teardrop-, and tetrapod-shaped CdSe nanocrystals. *J. Am. Chem. Soc.* 122:12700–12706.

34. Gudiksen, M. S.; Wang, J.; Lieber, C. M. 2001. Synthetic control of the diameter and length of single crystal semiconductor nanowires. *J. Phys. Chem. B* 105: 4062–4064.

35. Law, M.; Goldberger, J.; Yang, P. 2004. Semiconductor nanowires and nanotubes. *Ann. Rev. Mater. Res.* 34:83–122.

36. Pan, Z. W.; Dai, Z. R.; Wang, Z. L. 2001. Nanobelts of semiconducting oxides. *Science* 291:1947–1949.

37. Trentler, T. J.; Hickman, K. M.; Goel, S. C.; Viano, A. M.; Gibbons, P. C.; Buhro, W. M. 1995. Solution-liquid-solid growth of crystalling III-V semiconductors—an analogy to vapor-liquid-solid growth. *Science* 270:1791–1794.

38. Tang, Z.; Kotov, N. A.; Giersig, M. 2002. Spontaneous organization of single CdTe nanoparticles into luminescent nanowires. *Science* 297:237–240.

39. Patolsky, F.; Zheng, G.; Hayden, O.; Lakadamyali, M.; Zhuang, X.; Lieber, C. M. 2004. Electrical detection of single viruses. *Proc. Natl. Acad. Sci. USA* 101:14017–14022.

40. Hahm, J.-i.; Lieber, C. M. 2004. Direct ultrasensitive electrical detection of DNA and DNA sequence variations using nanowire nanosensors. *Nano Lett.* 4:51–54.

41. Law, M.; Greene, L. E.; Johnson, J. C.; Saykally, R.; Yang, P. 2005. Nanowire dye-sensitized solar cells. *Nature Mater.* 4:455–459.

42. Lin, Y.-M.; Dresselhaus, M. S. 2003. Thermoelectric properties of superlattice nanowires. *Phys. Rev. B* 68:075304–075318.

43. Moisala, A.; Nasibulin, A. G.; Kauppinen, E. I. 2003. The role of metal nanoparticles in the catalytic production of single-walled carbon nanotubes—a review. *J. Phys.: Condens. Matter* 15:S3011–3035.

44. Dai, H. 2001. In *Carbon Nanotubes*. Springler, Berlin, Germany.

45. Hata, K.; Futaba, D. N.; Mizuno, K.; Namai, T.; Yumura, M.; Iijima, S. 2004. Water-assisted highly efficient synthesis of impurity-free single-walled, carbon nanotubes. *Science* 306:1362–1364.

46. Brus, L. 1986. Electronic wave functions in semiconductor clusters: Experiment and theory. *J. Phys. Chem.* 90:2555–2560.

47. Efros, Al. L.; Efros, A. L. 1982. Interband absorption of light in a semiconductor sphere. *Sov. Phys. Semicond.* 16:772–775.

48. Coe, S.; Woo, W.-K.; Bawendi, M. B.; Bulovic, V. 2002. Electroluminescence from single monolayers of nanocrystals in molecular organic devices. *Nature* 420:800–803.

49. Bruchez, M. P.; Moronne, M.; Gin, P.; Weiss, S.; Alivisatos, A. P. 1998. Semiconductor nanocrystals as fluorescent biological labels. *Science* 281: 2013–2016.

50. Klajn, R.; Bishop, K. J. M.; Fialkowski, M.; Paszewski, M.; Campbell, C. J.; Gray, T. P.; Grzybowski, B. A. 2007. Plastic and moldable metals by self-assembly of sticky nanoparticle aggregates. *Science* 316:261–264.

51. Buffat, P.; Borel, J.-P. 1976. Size effect on melting temperature of gold particles. *Phys. Rev. A* 13:2287–2298.

52. Goldstein, A. N.; Echer, C. M.; Alivisatos, A. P. 1992. Melting in semiconductor nanocrystals. *Science* 256:1425–1427.

53. Wu, Y.; Li, Y.; Ong, B. S.; Liu, P.; Gardner, S. 2005. High-performance organic thin-film transistors with solution-printed gold contacts. *Adv. Mater.* 17: 184–187.

54. Wu, Y.; Li, Y.; Ong, B. S. 2006. Printed silver ohmic contacts for high-mobility organic thin-film transistors. *J. Am. Chem. Soc.* 128:4202–4203.

55. Ridley, B. R.; Nivi, B.; Jacobson, J. M. 1999. All-inorganic field effect transistors fabricated by printing. *Science* 286:746–749.

56. Kim, H.; Cho, K.; Kim, D.-W.; Lee, H.-R.; Kim, S. 2006. Bottom-and top-gate field-effect thin-film transistors with p channels of sintered HgTe nanocrystals. *Appl. Phys. Lett.* 89:173107–173110.

57. Murray, C. B.; Shevchenko, E.; Talapin, D. V. 2007. *U.S. Patent* 20070057255.

58. Nozik, A. J. 2002. Quantum dot solar cells. *Physica E* 14:115–120.

59. Gur, I.; Fromer, N. A.; Geier, M. L.; Alivisatos, A. P. 2005. Air-stable all-inorganic nanocrystal solar cells processed from solution. *Science* 310:462–465.

60. Talapin, D. V.; Shevchenko, E. V.; Kornowski, A.; Gaponik, N.; Haase, M.; Rogach, A. L.; Weller, H. 2001. A new approach to crystallization of CdSe nanoparticles into ordered three-dimensional superlattices. *Adv. Mater.* 13: 1868–1871.

61. Murray, C. B.; Kagan, C. R.; Bawendi, M. G. 1995. Self-organization of CdSe nanocrystallites into three-dimensional quantum dot superlattices. *Science* 270: 1335–1338.

62. Bolhuis, P. G.; Frenkel, D.; Mau, S.-C.; Huse, D. A. 1997. Entropy difference between crystal phases. *Nature* 388:235–236.

63. Talapin, D. V.; Shevchenko, E. V.; Murray, C. B.; Titov, A. V.; Král, P. 2007. Dipole-dipole interactions in nanoparticle superlattices. *Nano Lett.* 7: 1213–1219.
64. Lazarenkova, O. L.; Balandin, A. A. 2001. Miniband formation in a quantum dot crystal. *J. Appl. Phys.* 89:5509–5515.
65. Jiang, C.-W.; Green, M. A. 2006. Silicon quantum dot superlattices: Modeling of energy bands, densities of states, and mobilities for silicon tandem solar cell applications. *J. Appl. Phys.* 99:114902–114909.
66. Kagan, C. R.; Murray, C. B.; Nirmal, M.; Bawendi, M. G. 1996. Electronic energy transfer in CdSe quantum dot solids. *Phys. Rev. Lett.* 76:1517–1520.
67. Zaitseva, N.; Dai, Z. R.; Leon, F. R.; Krol, D. 2005. Optical properties of CdSe superlattices. *J. Am. Chem. Soc.* 127:10221–10226.
68. Talapin, D. V.; Murray, C. B. 2005. PbSe nanocrystal solids for n- and p-channel thin film field-effect transistors. *Science* 310:86–89.
69. Morgan, N. Y.; Leatherdale, C. A.; Drndic, M.; Jarosz, M. V.; Kastner, M. A.; Bawendi, M. G. 2002. Electronic transport in films of colloidal CdSe nanocrystals. *Phys. Rev. B* 66:075339–075348.
70. Yu, D.; Wang, C.; Guyot-Sionnest, P. 2003. *n*-Type conducting CdSe nanocrystal solids. *Science* 300:1277–1280.
71. Vanmaekelbergh, D.; Liljeroth, P. 2005. Electron-conducting quantum dot solids: novel materials based on colloidal semiconductor nanocrystals. *Chem. Soc. Rev.* 34:299–312.
72. Collier, C. P.; Saykally, R. J.; Shiang, J. J.; Henrichs, S. E.; Heath, J. R. 1997. Reversible tuning of silver quantum dot monolayers through the metal-insulator transition. *Science* 277:1978–1981.
73. Remacle, F.; Beverly, K. C.; Heath, J. R.; Levine, R. D. 2003. Gating the conductivity of arrays of metallic quantum dots. *J. Phys. Chem. B* 107:13892–13901.
74. Wehrenberg, B. L.; Yu, D.; Ma, J.; Guyot-Sionnest, P. 2005. Conduction in charged PbSe nanocrystal films. *J. Phys. Chem. B.* 109:20192–20199.
75. Ginger, D. S.; Greenham, N. C. 2000. Charge injection and transport in films of CdSe nanocrystals. *J. Appl. Phys.* 87:1361–1368.
76. Urban, J. J.; Talapin, D. V.; Shevchenko, E. V.; Murray, C. B. 2006. Self-assembly of PbTe quantum dots into nanocrystal superlattices and glassy films. *J. Am. Chem. Soc.* 128:3248–3255.
77. Chandler, R. E.; Houtepen, A. J.; Nelson, J.; Vanmaekelbergh D. 2007. Electron transport in quantum dot solids: Monte Carlo simulations of the effects of shell filling, Coulomb repulsions, and site disorder. *Phys. Rev. B* 75:085325–085335.
78. Talapin, D. V.; Urban, J. J.; Murray, C. B. Unpublished results.
79. Yu, D.; Wang, C.; Wehrenberg, B. L.; Guyot-Sionnest, P. 2004. Variable range hopping conduction in semiconductor nanocrystal solids. *Phys. Rev. Lett.* 92: 216802–216806.
80. Mott, N. F. 1993. *Conduction in Non-Crystalline Materials*, 2nd edition. Clarendon Press, Oxford, U.K.
81. Klinke, C.; Chen, J.; Afzali, A.; Avouris, P. 2005. Charge transfer induced polarity switching in carbon nanotube transistors. *Nano Lett.* 5:555–558.

82. Sun, B.; Sirringhaus, H. 2005. Solution-processed zinc oxide field-effect transistors based on self-assembly of colloidal nanorods. *Nano Lett.* 5:2408–2413.

83. Sun, B.; Sirringhaus, H. 2006. Surface tension and fluid flow driven self-assembly of ordered ZnO nanorod films for high-performance field effect transistors. *J. Am. Chem. Soc.* 128:16231–16237.

84. Talapin, D. V.; Shevchenko, E. V.; Murray, C. B.; Kornowski, A.; Forster, S.; Weller, H. 2004. CdSe and CdSe/CdS nanorod solids. *J. Am. Chem. Soc.* 126: 12984–12988.

85. Li, L.-S.; Alivisatos, A. P. 2003. Semiconductor nanorod liquid crystals and their assembly on a substrate. *Adv. Mater.* 15:408–411.

86. Klimov, V. I. 2006. Mechanisms for photogeneration and recombination of multiexcitons in semiconductor nanocrystals: Implications for lasing and solar energy conversion. *J. Phys. Chem. B* 110:16827–16845.

87. Konstantatos, G.; Howard, I.; Fischer, A.; Hoogland, S.; Clifford, J.; Klem, E.; Levina, L.; Sargent, E. H. 2006. Ultrasensitive solution-cast quantum dot photodetectors. *Nature* 442:180–183.

88. Harman, T. C.; Taylor, P. J.; Walsh, M. P.; LaForge, B. E. 2002. Quantum dot superlattice thermoelectric materials and devices. *Science* 297:2229–2232.

89. Shevchenko, E. V.; Talapin, D. V.; Kotov, N. A.; O'Brien, S.; Murray, C. B. 2006. Structural diversity in binary nanoparticle superlattices. *Nature* 439: 55–59.

90. Urban, J. J.; Talapin, D. V.; Shevchenko, E. V.; Kagan, C. R.; Murray, C. B. 2007. Synergism in binary nanocrystal superlattices leads to enhanced p-type conductivity in self-assembled PbTe/Ag_2Te thin films. *Nature Mater.* 6:115–121.

91. Orihashi, M.; Noda, Y.; Kaibe, T. H.; Nishida, I. A. 1997. Evaluation of thermodynamic properties of impurity-doped PbTe. *J. Jpn. Inst. Met.* 61:241–246.

92. Huang, Y.; Duan, X.; Cui, Y.; Lauhon, L. J.; Kim, K.-H.; Lieber, C. M. 2001. Logic gates and computation from assembled nanowire building blocks. *Science* 294:1313–1317.

93. Duan, X.; Huang, Y.; Leiber, C. M. 2002. Nonvolatile memory and programmable logic from molecule-gated nanowires. *Nano Lett.* 2:487–490.

94. Qian, F.; Gradecak, S.; Li, Y.; Wen, C.-Y.; Lieber, C. M. 2005. Core/multishell nanowire heterostructures as multicolor, high-efficiency light-emitting diodes. *Nano Lett.* 5:2287–2291.

95. Duan, X.; Huang, Y.; Agarwal, R.; Lieber, C. M. 2003. Single-nanowire electrically driven lasers. *Nature* 421:241–245.

96. Law, M.; Greene, L. E.; Johnson, J. C.; Saykally, R.; Yang, P. 2005. Nanowire dye-sensitized solar cells. *Nature Mater.* 4:455–459.

97. Hayden, O.; Agarwal, R.; Lieber, C. M. 2006. Nanoscale avalanche photodiodes for highly sensitive and spatially resolved photon detection. *Nature Mater.* 5:352–353.

98. Huang, Y.; Duan, X.; Wei, Q.; Lieber, C. M. 2001. Directed assembly of one-dimensional nanostructures into functional networks. *Science* 291:630–633.

99. Jin, S.; Whang, D.; McAlpine, M. C.; Friedman, R. S.; Wu, Y.; Lieber, C. M. 2004. Scalable interconnection and integration of nanowire devices without registration. *Nano Lett.* 4:915–919.

100. Harnack, O.; Pacholski, C.; Weller, H.; Yasuda, A.; Wessels, J. M. 2003. Rectifying behavior of electrically aligned ZnO nanorods. *Nano Lett.* 3: 1097–1101.

101. Duan, X.; Huang, Y.; Cui, Y.; Wang, J.; Lieber, C. M. 2001. Indium phosphide nanowires as building blocks for nanoscale electronic and optoelectronic devices. *Nature* 409:66–69.

102. Talapin, D. V.; Black, C. T.; Kagan, C. R.; Shevchenko, E. V.; Afzali, A.; Murray, C. B. 2007. Alignment, electronic properties, doping, and on-chip growth of colloidal PbSe nanowires. *J. Phys. Chem. C* (in press).

103. Duan, X.; Niu, C.; Sahi, V.; Chen, J.; Parce, J. W.; Empedocles, S.; Goldman, J. L. 2003. High-performance thin-film transistors using semiconductor nanowires and nanoribbons. *Nature* 425:274–278.

104. Dodabalapur, A. 2004. Betting on organic CMOS. *Materials Today* 7:56.

105. Li, D.; Wu, Y.; Kim, P.; Shi, L.; Yang, P.; Majumdar, A. 2003. Thermal conductivity of individual silicon nanowires. *Appl. Phys. Lett.* 83:2934–2936.

106. Li, D.; Wu, Y.; Kim, P.; Shi, L.; Mingo, N.; Liu, Y.; Yang, P.; Majumdar, A. 2003. Thermal conductivity of Si/SiGe superlattice nanowires. *Appl. Phys. Lett.* 83: 3186–3188.

107. Huynh, W. U.; Dittmer, J. J.; Alivisatos, A. P. 2002. Hybrid nanorod-polymer solar cells. *Science* 295:2425–2427.

108. McAlpine, M. C.; Ahmad, H.; Wang, D.; Heath, J. R. 2007. Highly ordered nanowire arrays on plastic substrates for ultrasensitive flexible chemical sensors. *Nature Mater.* 5:379–384.

109. Zhang, D.; Liu, Z.; Li, C.; Tang, T.; Liu, X.; Han, S.; Lei, B.; Zhou, C. 2004. Detection of NO_2 down to ppb levels using individual and multiple In_2O_3 nanowire devices. *Nano Lett.* 4:1919–1924.

110. Kang, S. J.; Kocabas, C.; Ozel, T.; Shim, M.; Pimparkar, N.; Alam, M. A.; Rotkin, S. V.; Rogers, J. A. 2007. High-performance electronics using dense, perfectly aligned arrays of single-walled carbon nanotubes. *Nature Nanotechnol.* 2: 230–236.

111. Arnold, M. S.; Green, A. A.; Hulvat, J. F.; Stupp, S. I.; Hersam, M. C. 2006. Sorting carbon nanotubes by electronic structure using density differentiation. *Nature Nanotechnol.* 1:60–65.

112. Dai, H. 2002. Carbon nanotubes: synthesis, integration, and properties. *Acc. Chem. Res.* 35:1035–1044.

113. Vigolo, B.; Pénicaud, A.; Coulon, C.; Sauder, C.; Pailler, R.; Journet, C.; Bernier, P.; Poulin, P. 2000. Macroscopic fibers and ribbons of oriented carbon nanotubes. *Science* 290:1331–1334.

114. Collins, P. G.; Arnold, M. S.; Avouris, P. 2001. Engineering carbon nanotubes and nanotube circuits using electrical breakdown. *Science* 292:706–709.

115. Krupke, R.; Hennrich, F.; v. Löhneysen, H.; Kappes, M. M. 2003. Separation of metallic from semiconducting single-walled carbon nanotubes. *Science* 301: 344–347.

116. Valentini, L.; Armentano, I.; Kenny, J. M.; Cantalini, C.; Lozzi, L.; Santucci, S. 2003. Sensors for sub-ppm NO_2 gas detection based on carbon nanotube thin films. *Appl. Phys. Lett.* 82:961–963.

117. Qi, P.; Vermesh, O.; Grecu, M.; Javey, A.; Wang, Q.; Dai, H.; Peng, S.; Cho, K. J. 2003. Toward large arrays of multiplex functionalized carbon nanotube sensors for highly sensitive and selective molecular detection. *Nano Lett.* 3:347–351.

118. Snow, E. S.; Perkins, F. K.; Houser, E. J.; Badescu, S. C.; Reinecke, T. L. 2005. Chemical detection with a single-walled carbon nanotube capacitor. *Science* 307:1942–1945.

CHAPTER 11

Functional Structures Assembled from Nanoscale Building Blocks

YU HUANG

Department of Materials Science and Engineering, University of California, Los Angeles, CA

11.1 INTRODUCTION

Ever since the invention of the first transistor 60 years ago, semiconductor electronics has been increasingly permeating into every aspect of our daily lives.[1,2] The rapid miniaturization of semiconductor electronics to the submicron scale has led to remarkable advances in computing power, while reducing cost. These advances have been made possible by many scientific and technological innovations associated with "top-down" manufacturing, in which small features are patterned in bulk semiconductor materials by lithography, deposition, and other processing to form functional devices. However, significant waste is associated with this conventional semiconductor manufacturing process. More than 95% of the bulk of the precious silicon wafers serves no other purpose than to form a mechanical support for the circuitry patterned into its surface for microchip fabrication. The cost associated with such waste is easily absorbed into the price for high-value gadgets, such as computers, digital cameras, portable music players, or cell phones. But for products involving larger areas and higher volumes, such as flat panel displays, electronic paper, or smart clothing, this cost is prohibitively high. Furthermore, the high temperatures required to grow crystalline silicon (>1400 °C) and the difficulties associated with handling large brittle silicon wafers make the traditional silicon microelectronics unsuitable for large-area applications. By using only as much semiconductor material (e.g., amorphous or polycrystalline silicon) as needed for electronic circuitry, on the surface of an inexpensive substrate such as glass, significant reductions in cost can be achieved, thereby enabling applications such as large-area flat-panel displays (e.g., Liquid Crystal Displays). However, the performance of these materials is poor compared with conventional single-crystal semiconductors, thereby limiting them to low-frequency, low-speed applications. Additionally, these materials still require moderate processing temperatures (400–700 °C), which makes them less compatible with flexible plastic substrates.

To go beyond the fundamental and economic limitations of conventional top-down technologies will require new methods.[3–7] A bottom-up approach, in which functional electronic structures are assembled from chemically synthesized, well-defined nanoscale building blocks, much like the way nature uses proteins and other macromolecules to construct complex biological systems, represents a flexible alternative to conventional top-down methods.[5–7] In this approach, high-quality nanoscale building blocks, such as nanocrystals, nanotubes, or nanowires, are first synthesized using various chemical approaches with precisely controlled chemical composition and physical dimension.[8–10] These nanoscale building blocks typically have nearly perfect crystalline structure and can exhibit unique electronic and optical properties. Because of their molecular scale dimension, the presynthesized building blocks can be treated like macromolecules and can be processed in the solution phase. Through a certain low-temperature solution assembly process, these nanoscale building blocks are organized into ordered arrays on various supporting substrates, from which the unique function and properties of nanostructures may be explored for novel applications in functional electronic and optoelectronic systems. This basic principle underpins several combined advantages not readily achievable in traditional technologies, such as follows: (1) high-quality single-crystal building blocks ensure high device performance; (2) low-temperature, low-cost solution processing enables large-area application on flexible substrates such as plastics and glasses; and (3) flexible bottom-up assembly allows for integration of different materials without the concern of chemical or structure incompatibility, therefore enabling multifunction integration. This chapter reviews recent advances in bottom-up, solution-based assembly of electronic and optoelectronic systems from high-quality nanoscale building blocks, with a focus on semiconductor nanowires as a building block.

11.2 BUILDING BLOCKS: SYNTHESIS AND PROPERTIES

A key element in the bottom-up technology is the building blocks. Individual molecules[11–15] and quantum dots,[16–18] which can be classified as zero-dimensional (0D) structures, have been proposed as building blocks for bottom-up assembly of nanoscale electronics. These 0D structures have been intensively pursued over the past decade, because they represent the smallest building blocks with corresponding high potential for massive integration. However, the use of individual molecules or quantum dots in nanoelectronics has been limited by challenges in establishing reliable electrical contacts needed to study their fundamental properties and to interconnect them. It has thus been difficult to elucidate and understand the intrinsic properties of individual devices and, moreover, to develop and demonstrate realistic schemes for scalable interconnection and integration of 0D devices into functional architectures. On the other hand, exploring ensembles of nanocrystals in monolayer or multilayer format is attracting considerable interest for low-temperature solution processing of crystalline thin-film electronics. Recent studies have demonstrated PbSe-based QD TFTs with performance reaching that of amorphous silicon devices (e.g., see Chapter 10).[19]

One-dimensional (1D) nanostructures have also been the focus of extensive studies because of their unique physical properties and potential to revolutionize broad areas of nanotechnology. First, 1D nanostructures represent the smallest dimension structure that can efficiently transport electrical carriers and, thus, are ideally suited for the ubiquitous task of moving and routing charges (information) in nanoscale electronics and optoelectronics. Second, 1D nanostructures can also exhibit a critical device function and thus can be exploited as both the wiring and device elements in architectures for functional nanosystems.[20] In this regard, two material classes, carbon nanotubes[21–31] and semiconductor nanowires,[32–42] have shown particular promise.

Single-walled carbon nanotubes (NTs) can exhibit either metallic or semiconducting behavior depending on diameter and helicity.[21] The unique electronic properties of NTs open up the possibility of creating several different devices that could have potential in nanoelectronics.[20,22–24] For example, single-walled NTs have been used to fabricate room-temperature field-effect transistors (FETs),[25,26] diodes,[27–29] and recently logic circuits.[30,31] However, the inability to control whether NT building blocks are semiconducting or metallic makes specific device fabrication largely a random event, because of the large device-to-device variability.

Semiconductor nanowires represent important and highly versatile nanometer-scale wire structures,[32,33] which are generally synthesized via a metal cluster-catalyzed vapor–liquid–solid (VLS) growth mechanism.[34–36] Here the catalyst is envisioned as a nanocluster or nanodroplet that defines the diameter of the NWs and serves as the site that directs preferentially the addition of reactant to the end of a growing NW, much like a living polymerization catalyst directs the addition of monomers to a growing polymer chain (Fig. 11.1a). Within this framework, a broad range of semiconductor NWs,[34–36] typically with diameters of on the order of 10 nm and lengths extending up to tens of micrometers (Fig. 11.1b and c), can be rationally and predictably synthesized in single-crystal form with all key parameters, including chemical composition, diameter and length, and physical properties controlled.[37–39] Additionally, recent advances have also demonstrated well-controlled synthesis of axial/radial heterostructures, enabling more versatile engineering and improvement of NW properties.[40,41]

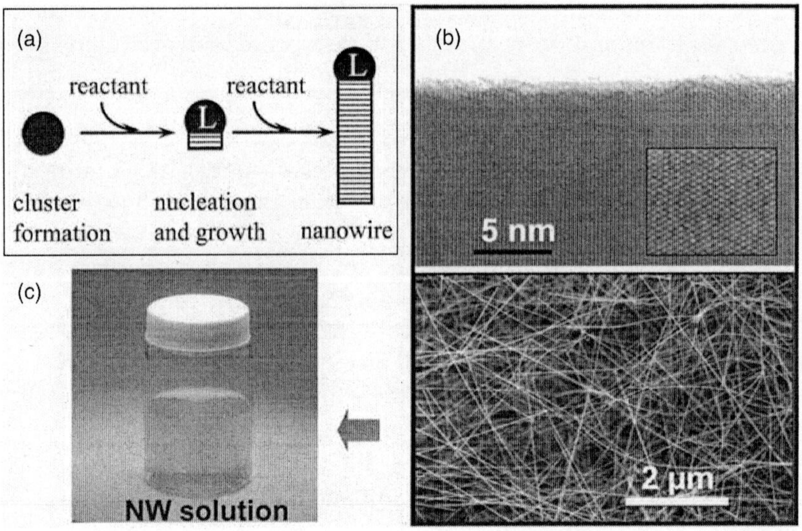

Figure 11.1. Semiconductor NW building blocks. (a) Schematics illustrating the underlying concept for catalytic growth of NWs. Liquid catalytic clusters act as the energetically favored site for localized chemical reaction, absorption of vapor phase reactant, and crystallization of crystalline NWs. (b) Transmission electron microscope (TEM) images (top) and scanning electron microscope (SEM) image (bottom) of as-synthesized GaN semiconductor NWs. High-resolution TEM image illustrates the single crystalline nature of the wires. (c) NWs suspended in appropriate solvent for future processing.

The basic electronic properties of NWs are characterized using electrical transport studies in a single nanowire field-effect transistor configuration (NW-FET) (Fig. 11.2a and b). The NW-FETs are prepared by placing NWs onto the surface of an oxidized silicon substrate, where the underlying conducting silicon is used as a global back gate.[42,43] Source and drain electrodes are defined by electron beam lithography, followed by electron beam evaporation of metal contacts and electrical transport measurements that are taken at room temperature. Current (I_{sd}) versus source-drain voltage (V_{sd}) and I_{sd} versus gate voltage (V_g) are then recorded for an NW-FET to characterize its electrical properties. The gate-sweeping measurement of the NW-FET enables elucidation of important qualitative and quantitative properties of NWs. For example, changes in V_g produce variations in the electrostatic potential of the NW, and hence, they modulate the carrier concentration and conductance of the NW (Fig. 11.2b). Depending on the conductance modulation, it is possible to estimate the carrier mobility in individual NWs using a standard transistor formula: $dI_{sd}/dV_g = \mu(C/L^2)V_{sd}$ and $C \cong 2\pi\varepsilon\varepsilon_0 L/\ln(2h/r)$, where μ is the carrier mobility, C is the capacitance, ε is the dielectric constant, h is the thickness of

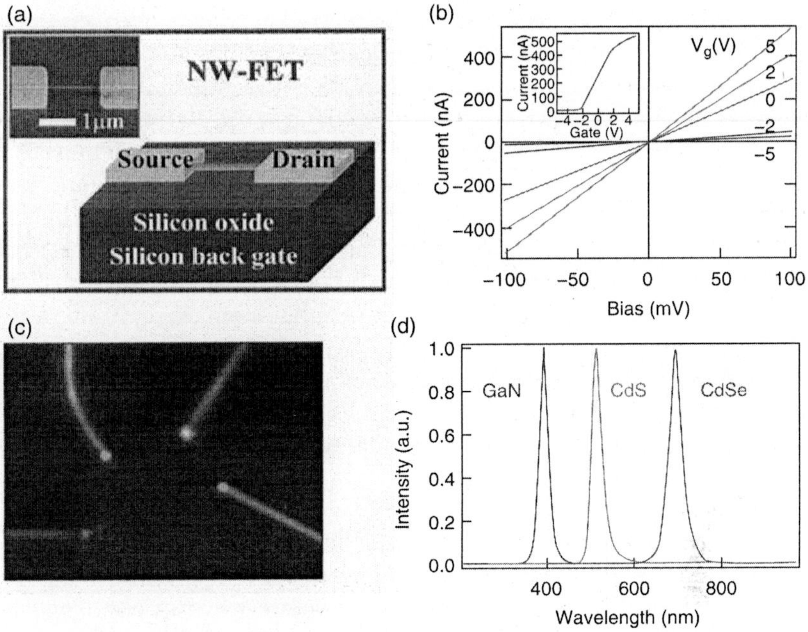

Figure 11.2. Nanowire electronic and optical properties. (a) Schematic of an NW–FET used to characterize electrical transport properties of individual NWs. (inset) SEM image of an NW-FET; two metal electrodes, which correspond to source and drain, are visible at the left and right sides of the image. (b) Current versus voltage for an n-type InP NW-FET. The numbers inside the plot indicate the corresponding gate voltages (V_g). The inset shows current versus V_g for V_{sd} of 0.1 V. (c) Real-color photoluminescence image of various NWs shows different color emissions. (d) Spectra of individual NW photoluminescence. All NW materials show a clean band-edge emission spectrum with narrow FWHM around 20 nm. (*See color insert.*)

TABLE 11.1. Summary of Carrier Mobility in Various NW Materials in Comparison with Their Bulk Material Counterparts of Similar Carrier Concentration

Materials		p-Si	p-GaN	n-GaN	n-CdS	n-InP
Carrier Mobility $cm^2/V{\cdot}s$	NW	40–300	~12	150–650	100–400	400–4000
	Bulk	10–400	~10	100–300	~200	1000–3000

the SiO_2 dielectric, L is the length, and r is the radius of the NW. A broad range of NW materials, including p-type Si, n-type GaN, CdS, and InP NWs show excellent carrier mobility comparable with bulk materials (Table 11.1),[44] which demonstrates the high quality of these materials.

Optical studies on various direct band-gap semiconductor NWs have demonstrated that these NW materials can also exhibit excellent photoluminescence properties, with the emission wavelength controllable either by chemical composition or physical dimension. For example, photoluminescence studies on individual GaN, CdS, and CdSe showed clean luminescence with spectra maxima of ~370, ~510, and ~700 nm, respectively,[44] which is consistent with near-band-edge emission in these materials (Fig. 11.2c and d). Additionally, a photoluminescence study on InP nanowires of various diameters showed that the emission spectra systematically shifted from 900 nm to 700 nm when the NW diameter was reduced from 50 nm to 10 nm in diameter. This systematic blueshift of emission with reduced NW diameter can be attributed to the quantum confinement effect in these molecular-scale nanowires.[45]

The fundamental physical properties of nanowire materials can be improved even more to surpass their bulk counterpart using precisely engineered NW heterostructures. It has been recently demonstrated that Si/Ge/Si core/shell nanowires exhibit electron mobility surpassing that of state-of-the-art technology.[46] Group III–V nitride core/shell NWs of multiple layers of epitaxial structures with atomically sharp interfaces have also been demonstrated with well-controlled and tunable optical and electronic properties.[47,48] Together, the studies demonstrate that semiconductor nanowires represent one of the best-defined nanoscale building block classes, with well-controlled chemical composition, physical size, and superior electronic/optical properties, and therefore, that they are ideally suited for assembly of more complex functional systems.

11.3 HIERARCHICAL ASSEMBLY OF NANOWIRES

To explore in full the potential of these NW building blocks for integrated functional systems would require the development and implementation of efficient and scalable strategies for assembling them into increasingly complex architectures. First, methods are needed to assemble NWs into highly integrated arrays with controlled orientation and spatial location. Second, approaches must be devised to assemble NWs on multiple length scales and to make interconnects among nano-, micro-, and macroscopic worlds. Toward this end, several approaches, including electrical field, fluidic flow, and Langmuir–Blodgett film techniques, have been developed.

11.3.1 Fluidic Flow-Directed Assembly

In this method, NWs can be aligned by passing a suspension of NWs through microfluidic channel structures, for example, formed between a poly(dimethylsiloxane) (PDMS) mold [49] and a flat substrate (Fig. 11.3a). Images of NWs assembled on substrate surfaces (Fig. 11.3b) within microfluidic flows demonstrate that virtually all NWs are aligned along the flow direction. This alignment readily extends over hundreds of micrometers, and

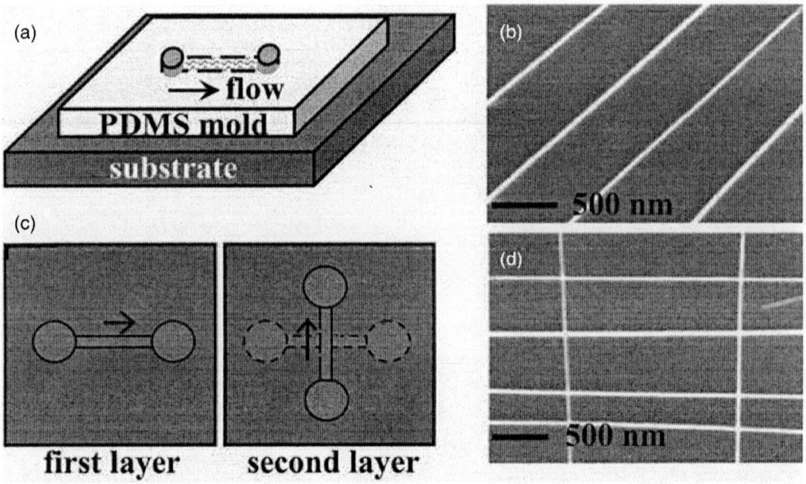

Figure 11.3. Fluidic flow directed assembly of NWs. (a,b) Schematic (a) and SEM image (b) of parallel NW arrays obtained by passing an NW solution through a channel on a substrate; (c,d) Schematic (c) and SEM image (d) of a crossed NW matrix obtained by orthogonally changing the flow direction in a sequential flow alignment process. [Adapted from Ref. 49.]

it is limited only by the size of the fluidic channels used. The alignment of NWs within the channel flow can be understood within the framework of shear flow.[50,51] Specifically, the channel flow near the substrate surface resembles a shear flow, and the linear shear force aligns the NWs in the flow direction before they are immobilized on the substrate.

The fluidic flow assembly approach can be used to organize NWs into more complex crossed NW structures, which are critical for building high-density nanodevice arrays, using a layer-by-layer deposition process (Fig. 11.3c and d). The formation of crossed and more complex structures requires that the nanostructure–substrate interaction be sufficiently strong that sequential flow steps do not affect preceding ones; we find that this condition is readily achieved by modifying the substrate surface with proper functional chemical groups. For example, alternating the flow in orthogonal directions in a two-step assembly process yields crossbar structures in high yield (Fig. 11.3d). Experiments have demonstrated that crossbars extending over hundreds of microns on a substrate with only hundreds of nanometers of separation between individual cross points are obtained through a very straightforward, parallel, low-cost, and fast process.

Fluidic flow-directed assembly of multiple crossed NW arrays offers significant advantages. First, it is intrinsically very parallel and scalable, with the alignment readily extending over very large length scales. Second, this approach is general for virtually any elongated nanostructure, including carbon nanotubes and DNA molecules. Third, it allows for the directed assembly of

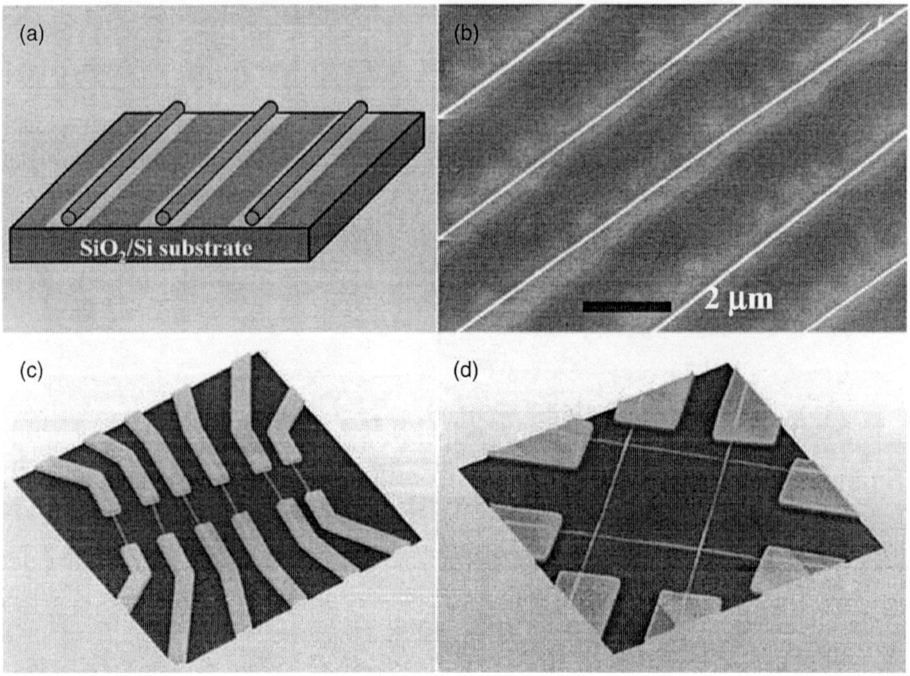

Figure 11.4. Fluidic flow-directed assembly of NWs. (a,b) Schematic (a) and SEM image (b) of regular NW arrays obtained by flowing NW solution over a chemically patterned surface. (c,d) Parallel and crossed NW device arrays obtained with fluidic flow assembly. [Adapted from Refs. 44 and 49.]

geometrically complex structures by simply controlling the angles between flow directions in sequential assembly steps. For example, equilateral triangles have been assembled in a three-layer deposition sequence using 60° angles among the three flow directions.[44] The method of flow alignment thus provides a flexible way to meet the future requirements of many device configurations. An important feature of this sequential assembly scheme is that each layer is independent, and thus, a variety of homo- and hetero-junction configurations can be obtained simply by changing the composition of the NW suspension used for each flow step.

Combining fluidic assembly with conventional lithography, a variety of integrated device structures, including parallel arrays (Fig. 11.4a and b) and crossed arrays (Fig. 11.4c and d),[49] can be explored. Such integrated device arrays offer a variety of opportunities to investigate the fundamental properties of these NW materials and to explore them for various device applications. In particular, the crossed NW matrix represents an important geometrical configuration for nanocircuits, where each crossing point can function as an independently addressable nanoscale device element.

11.3.2 Langmuir–Blodgett Technique-Assisted NW Assembly

The Langmuir–Blodgett (LB) technique has been widely used to prepare monolayers of fatty acids and amphiphilic molecules, and later, it was extended to prepare molecular electronic devices based on a monolayer of molecules or nanocrystals.[52,53] Recently, LB has been exploited for controlled assembly of NWs over large areas[54,55] (Fig. 11.5); LB was applied to compress uniaxially an NW monolayer floating on a water surface via a computer-controlled process, which thereby produces aligned NW arrays parallel to the direction of the trough with controlled spacing (Fig. 11.5a). The aligned NW monolayer was then transferred to any desired planar substrate, such as silicon, plastic, or glass, yielding densely packed NW arrays on the substrate surface (Fig. 11.5b and c). The LB technique can be applied to large areas of NW

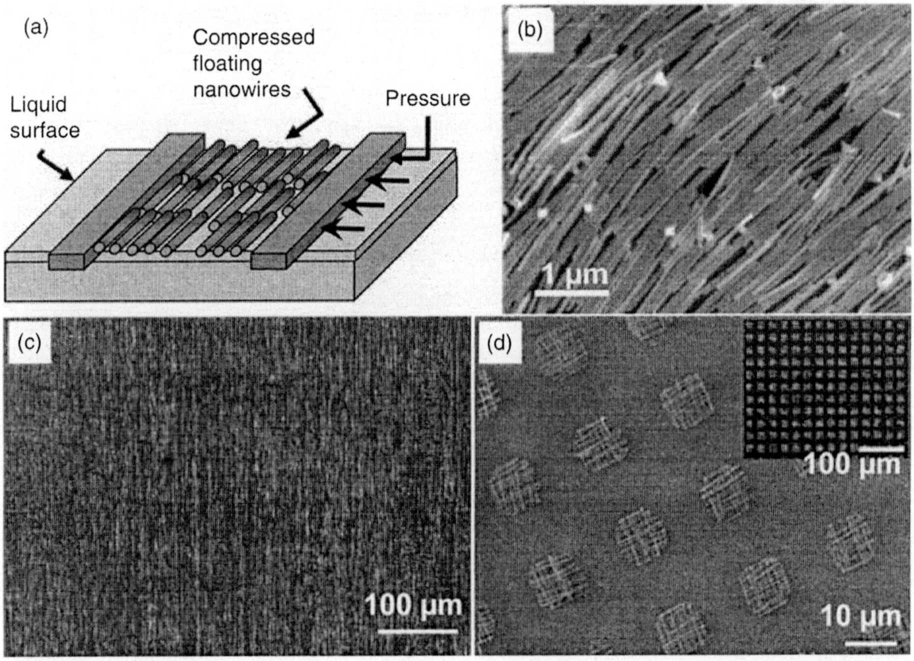

Figure 11.5. Langmuir–Blodgett (LB) technique-assisted NW assembly (a) Schematic of LB-guided assembly. NWs are floated on a water surface and compressed through computer guided pressure. (b) SEM image of silver NWs deposited on Si wafer. Reprinted with permission from Ref. 54. Copyright 2003 American Chemical Society. (c) Large-area image of parallel NW array deposited uniformly on a substrate. Reprinted with permission from Ref. 55. Copyright 2003 American Chemical Society; (d) SEM image of patterned crossed NW arrays; inset, large-area dark-field optical micrograph of the patterned crossed NW arrays. Reprinted with permission from Ref. 55. Copyright 2003 American Chemical Society.

assemblies, limited only by the numbers of NWs dispersed on the water surface. Similar to fluidic flow-directed assembly, this approach also offers the flexibility of assembling various nanowires in a layer-by-layer fashion and to produce both parallel and cross packed NW arrays (Fig. 11.5d). However, to do this usually requires coating of the NW surface with a molecular layer to render a hydrophobic surface, which needs to be removed before continued processing.

11.4 NANOWIRE ELECTRONICS AND OPTOELECTRONICS

The ability to synthesize NWs rationally with controlled electronic properties and to assemble NWs into regular arrays readily enables exploration of these NW structures for a variety of functional device arrays.

11.4.1 Crossed Nanowire Devices

Direct assembly of highly integrated functional electronic circuits based on NWs requires (1) the development of new device concepts with scalable device configuration and (2) high yield assembly of these devices with controllable functional properties. The crossed NW matrix represents an ideal configuration since the critical device dimension is usually defined by the cross point, which can be readily scaled down to the nanometer level, and the crossed NW configuration is naturally a scalable architecture, thus enabling the possibility of massive system integration. Moreover, the crossed NW matrix is a versatile structure and can be configured into a variety of critical device elements, such as diodes and transistors. For example, a p-n diode can be obtained by simply crossing a p- and n-type NW as demonstrated in the cases of p-Si/n-Si, p-InP/n-InP, p-InP/n-CdS and p-Si/n-GaN materials (Fig. 11.6a). Electrical measurements of such crossed junctions show clear current rectification across the junction and linear current behavior in individual NWs, which demonstrates the formation of a p-n diode at the crossing point (Fig. 11.6b). The crossed NW p-n diodes can be assembled with high yield and reproducibility. For example, studies on more than 100 crossed NW devices assembled from p-Si NWs and n-GaN NWs show that over 95% of the junctions exhibit current rectification with turn-on voltages of around 1.0 V.[42,49] Reproducible assembly of crossed NW structures with predictable electrical properties contrasts sharply with results from NT-based devices and readily enabled integrated p-n diode arrays. Significantly, electrical transport measurements made on a typical 4-by-1 crossed p-Si/n-GaN junction array (Fig. 11.6c and d) show that the four nanoscale cross points form independently addressable p-n diodes with clear current rectification and similar turn-on voltages.

Nanoscale FETs can also be achieved in the crossed NW configuration using one NW as the active conducting channel and another crossed NW as

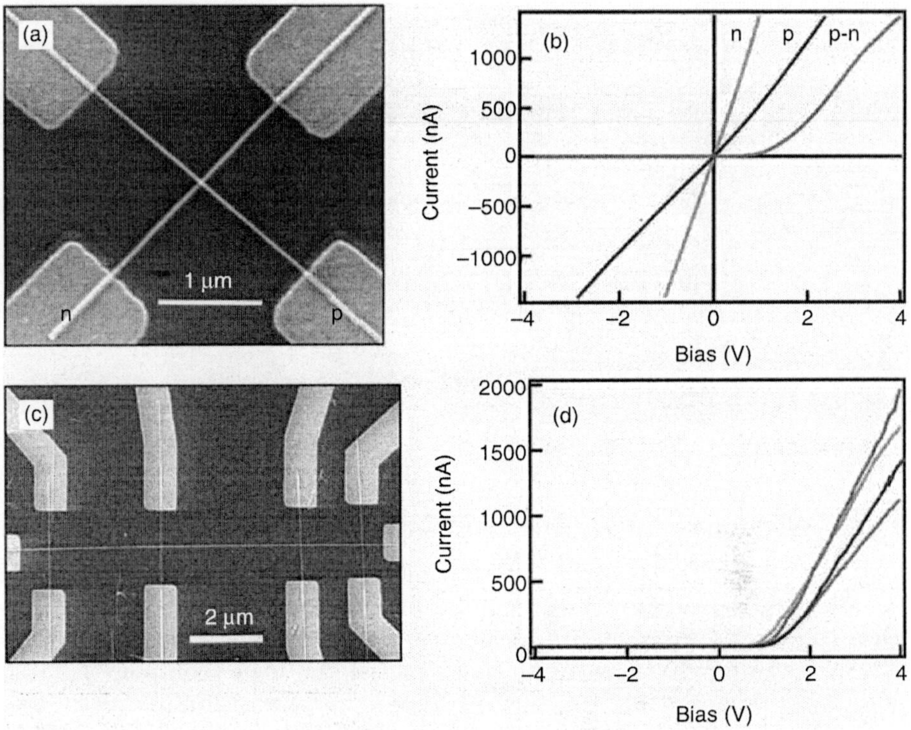

Figure 11.6. Crossed nanowire p-n diode. (a) A typical SEM image of a crossed NW p-n diode. (b) Current-voltage (I–V) relation of the crossed p-n diode. Linear or nearly linear I–V behavior of the p-type and n-type NWs indicates good contact between NWs and metal electrodes. I–V curves across the junction show clear current rectification. (c) An SEM image of an NW p-n diode array. (d) I–V behavior for a 4(p) × 1(n) multiple junction array. [Adapted from Ref. 57.]

the gate electrode (Fig. 11.7a and b). In this configuration, the three critical FET device metrics are naturally defined at the nanometer scale in the assembled crossed NW FETs (cNW-FETs): (1) a nanoscale channel width determined by the diameter of the active NW; (2) a nanoscale channel length defined by the crossed gate NW diameter; and (3) a nanoscale gate dielectric thickness determined by the NW surface oxide. These distinct nanoscale device metrics lead to improved device characteristics such as high gain, high speed, and low-power dissipation. For example, the conductance modulation of an NW FET is much more significant with the NW gate ($>10^5$) than that with a global back gate (<10) (Fig. 11.7c). Moreover, the local NW gate enables independently addressable FET arrays and, thus, enables highly integrated nanocircuits.

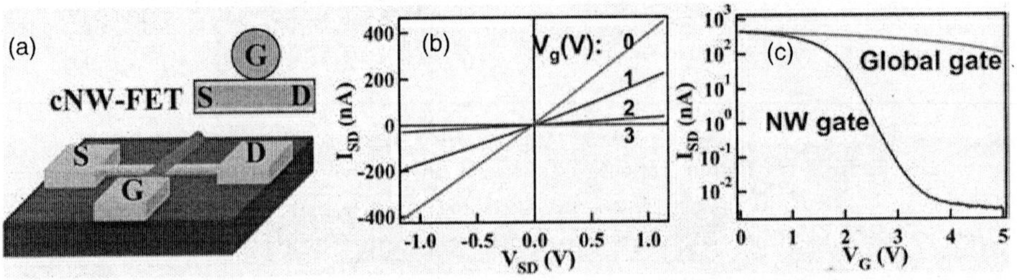

Figure 11.7. Crossed nanowire FET. (a) Schematics illustrating the crossed NW-FET (cNW-FET) concept. (b) Gate-dependent I–V characteristics of a cNW-FET formed using a p-Si NW as the conducting channel and n-GaN NW as the local gate. (c) I_{SD} versus V_G for n-NW and global back gates when V_{SD} is set at 1V. [Adapted from Ref. 57.]

11.4.2 Nanoscale Logic Gates and Computational Circuits

High-yield assembly of crossed NW p-n diodes and cNW-FETs from p-Si and n-GaN materials enables more complex functional electronic circuits, such as logic gates, to be produced. Logic gates are critical blocks of hardware in current computing systems, producing a logic-1 and logic-0 output when the input logic requirements are satisfied. Diodes and transistors represent two basic device elements in logic gates.[56] Transistors are more typically used in current computing systems because they can exhibit voltage gain. Diodes do not usually exhibit voltage gain, although they may also be desirable in some cases[56]; for example, the circuit architecture and constraints on the assembly of nanoelectronics might be simplified using diodes because they are two-terminal devices, in contrast to three-terminal transistors. In addition, by combining diodes and transistors in logic circuits, it is possible to achieve high voltage gain, while maintaining a simplified device architecture. Importantly, the flexibility of bottom-up assembly readily allows integration of diode- and FET-based logic.[57]

For example, a two-input logic OR gate was realized using a 2(p) by 1(n) crossed NW p-n diode array.[44] When either of the inputs to the p-NW is high, a high output is obtained at the n-NW as the p-n diode is forward biased; a low output is only achieved when both inputs are low, thus realizing the same function as a conventional logic "OR" gate (Fig. 11.8a and b). A logic AND gate was also assembled from two p-n diodes and one cNW-FET (Fig. 11.8c and d), and a logic NOR gate with a gain over 5 was assembled from three cNW-FETs in series (Fig. 11.8e and f). Importantly, logic OR, AND, and NOR gates form a complete set of logic elements and enable the organization of virtually any logic circuits. For example, NW logic gates have been interconnected to form an XOR gate and a logic half adder, which were used to carry out digital computations in a way similar to conventional electronics (Fig. 11.8g and h).

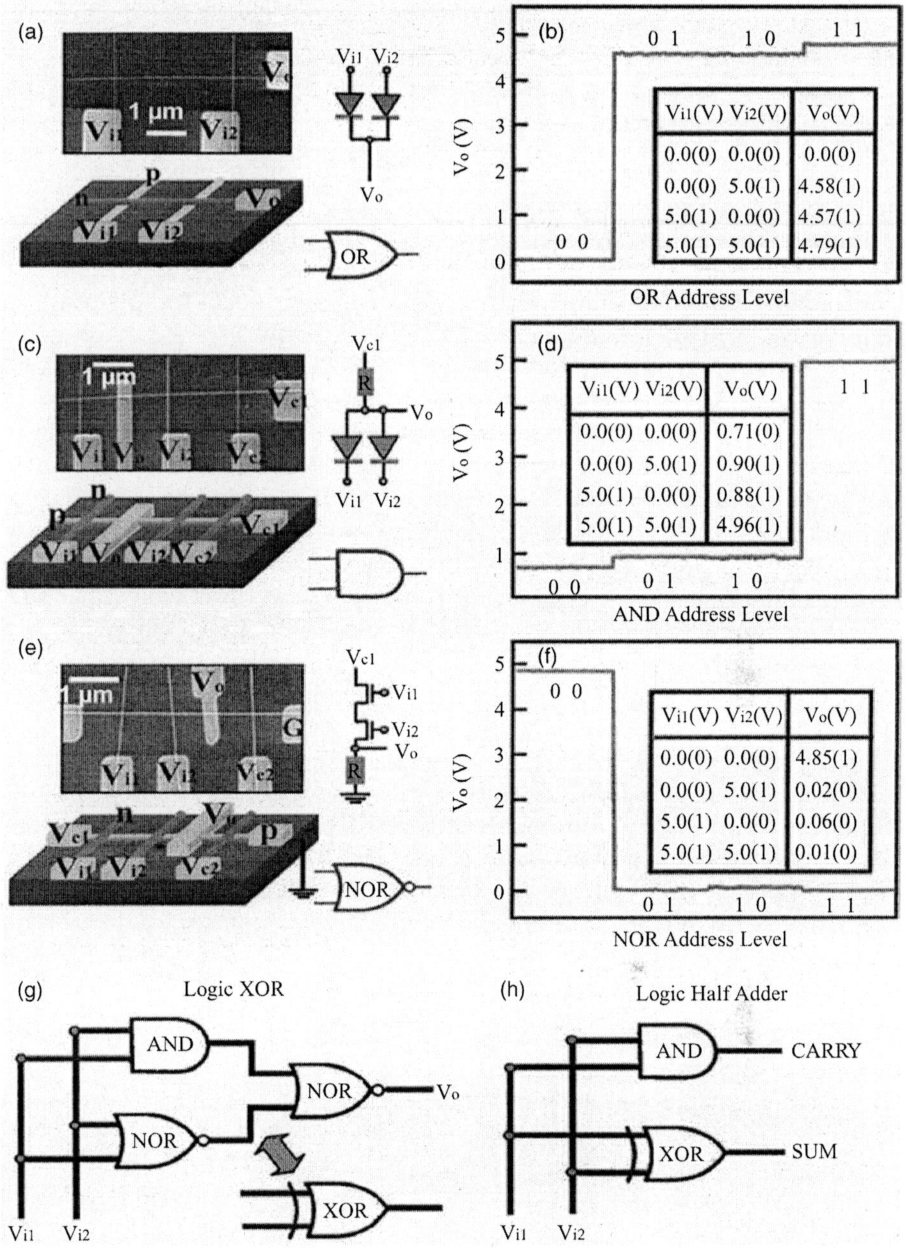

Figure 11.8. Nanoscale logic gates. (a,b) Schematic and measured output versus address level relation for a logic OR gate assembled from two crossed NW p-n diodes. The insets in (a) show the SEM image and equivalent electronic circuit of the device. The inset in (b) shows the experimental truth table for the nano-OR gate. (c,d) Schematic and measured output versus address level relation for a logic AND gate assembled from two diodes and one crossed NW-FET. The insets in (c) show the SEM image and equivalent electronic circuit of the device. The inset in (d) shows the experimental truth table for the nano-AND gate. (e,f) Schematic and measured output versus address level relation for a logic NOR gate assembled from three crossed NW FETs. The insets in (e) show the SEM image and equivalent electronic circuit of the device. The inset in (f) shows the experimental truth table for the nano-NOR gate. (g,h) Schematics for a logic XOR gate and logic half adder realized by interconnecting individual logic gate elements. [Adapted from Ref. 57.]

11.4.3 Nanoscale Optoelectronics

The availability of a broad range of NW materials readily allows one to choose materials with different properties to tailor device functions in a manner that is unique to the bottom-up assembly approach. In addition to nanoscale electronics, the broad range of optically active III–V and II–VI group compound semiconductor NW materials are attractive as building blocks for miniaturized photonic and optoelectronic devices. To this end, a wide range of photonic devices, including nanoscale light-emitting diodes (LEDs), diode arrays, and photodetectors have been demonstrated.[39,42,59–62]

In the previous section, we have discussed p-n diodes, obtained by crossing a p- and n-type NW. In direct band-gap semiconductors like InP, the p-n diode also forms the basis for critical optoelectronics devices, including the LED and the laser diode (LD). To assess whether these p-n diodes might behave similarly, we have studied the electroluminescence (EL) from these nanoscale devices. Significantly, EL can be readily observed from these nanoscale junctions in forward bias. A 3D plot of the EL intensity taken from a typical NW p-n diode at forward bias (Fig. 11.9a) shows that the emission comes from a point-like source, and moreover, comparison of EL and PL images recorded on the same sample (inset, Fig. 11.9a) shows that the position of the EL maximum corresponds to the crossing point in the PL image. These data thus demonstrate that the emitted light indeed comes from the crossed NW p-n junction.[39]

EL spectra (peaked 820 nm) recorded from the crossed NW LEDs exhibit blueshifts relative to the bulk band gap of InP (925 nm). The blueshifts are caused in part by quantum confinement of excitons, although other factors may also contribute. Furthermore, PL studies have demonstrated that the PL peak can be systematically blueshifted as the NW diameter is decreased,[53] and

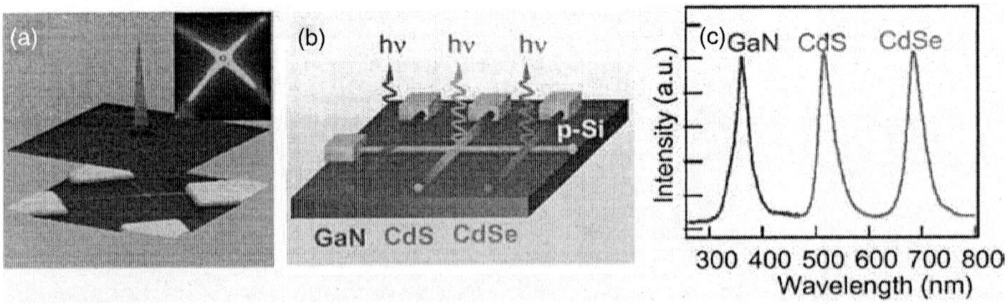

Figure 11.9. NW LED. (a) Crossed InP nanowire LED. (top) Three-dimensional (3D) plot of light intensity of the electroluminescence from a crossed NW LED. Light is only observed around the crossing region. (bottom) 3D atomic force microscope image of a crossed NW LED. (inset) Photoluminescence image of a crossed NW junction. (b–c) Multicolor nanoLED array. (b) Schematic of a tricolor nanoLED array assembled by crossing one n-GaN, n-CdS, and n-CdS NW with a p-Si NW. The array was obtained by fluidic assembly and photolithography with ca. 5-μm separation between NW emitters. (c) Normalized EL spectra obtained from the three elements. [Reprinted with permission from Ref. 59. Copyright 2005 Wiley-VCH Verlag.]

thus, these results provide a means for controlling the color of the LEDs in a well-defined way. Indeed, EL results recorded from p-n junctions assembled from smaller (and larger) diameter NWs show larger (smaller) blueshifts. In addition, the emission color from nanoLEDs can be further varied by using chemically distinct semiconductor NWs with different band gaps. Considering the wide range of group IV, III–V, and II–VI semiconductor NW materials available,[59] it is possible to assemble a variety of NW-based nanoLEDs for different spectral regimes.[59] Indeed, It has been recently demonstrated that crossing p-type Si NWs with n-type GaN, CdS, CdSeS, CdSe, and InP can produce nanoLEDs with emission spectra covering the whole spectral regime from ultraviolet (UV) to near infrared (IR).[59] In such heterostructure devices, p-Si NWs are used as the passive hole-injector, and the n-type compound NWs are used as the electron-injector and the active emitter, which defines the wavelength of emission.

The bottom-up assembly approach allows a flexible combination of chemically distinct nanoscale building blocks that would otherwise be structurally and/or chemically incompatible, in a sequential growth process typical of planar fabrication. This capability should enable assembly of nanostructures with functions not readily obtained by other methods and open up new opportunities. For example, we have exploited the ability to form nanoLEDs with nonemissive Si NW hole injectors to assemble multicolor arrays consisting of n-type GaN, CdS, and CdSe NWs crossing a single p-type SiNW (Fig. 11.9b). Normalized emission spectra recorded from the array demonstrates three spectrally distinct peaks with maxima at 365, 510, and 690 nm (Fig. 11.9c), which is consistent with band-edge emission from GaN, CdS, and CdSe, respectively. The ability to assemble/integrate different materials together seamlessly and to tune the emission wavelength independently from each nanoLED offers substantial potential for producing specific wavelength sources and demonstrates an important step toward integrated nanoscale photonic circuits. Although lithography sets the integration scale of these multicolor arrays, it should be possible to create much denser nanoLED arrays via (1) controlled growth of modulated NW superlattice structures[40] and/or (2) selective assembly of different semiconductor materials.[59]

In addition, the ability to assemble nanoscale electronic and optoelectronics devices based on a common crossed NW architecture allows for assembling integrated optoelectronic circuits consisting of crossed NW LED and FET elements (Fig. 11.10a-b). Specifically, one GaN NW forms a p-n diode with the Si NW, and a second GaN NW functions as a local gate, as described previously. Measurements of current and emission intensity versus gate voltage show that (1) the current decreases rapidly with increasing voltage, as expected for a depletion mode FET; and (2) the intensity of emitted light also decreases with increasing gate voltage (Fig. 11.10c). When the gate voltage is increased from 0 to +3V, the current is reduced from ca. 2200 nA to an off state, where the supply voltage is −6V. The advantages of this integrated approach include switching with much smaller changes in voltage (0 to 3 vs. 0 to 6V) and the

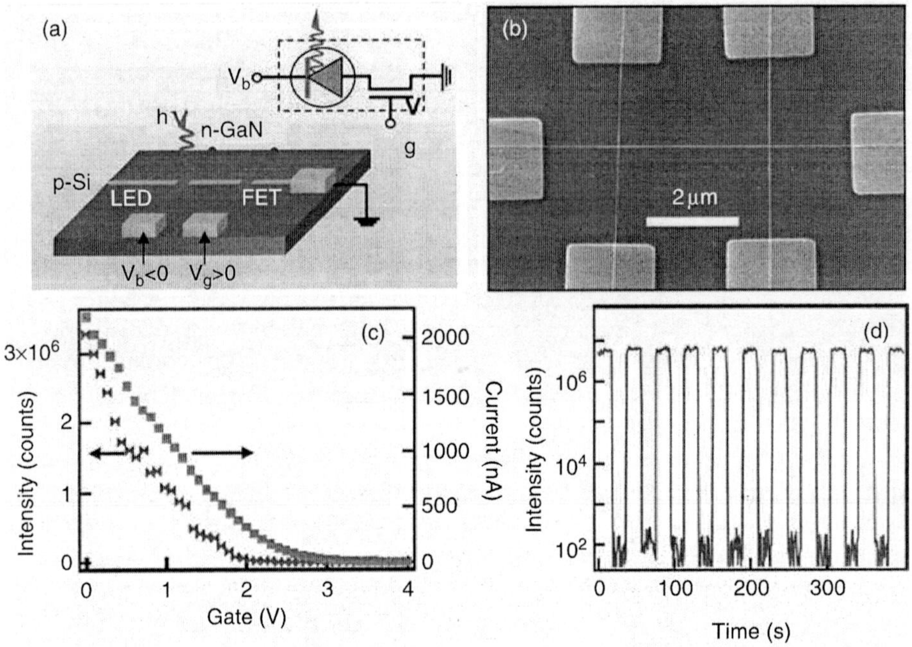

Figure 11.10. NW smart pixels. (a) Schematic of an integrated crossed NW FET and LED and the equivalent circuit. (b) Shows SEM image of a representative device. (c) Plots of current and emission intensity of the nanoLED as a function of voltage applied to the NW gate at a fixed bias of –6V. (d) EL intensity versus time relation when a voltage applied to NW gate is switched between 0 and +4V for a fixed bias of –6V. [Reprinted with permission from Ref. 59. Copyright 2005 Wiley-VCH Verlag.]

potential for much more rapid switching. The ability to use the nanoscale FET to switch reversibly the nanoLED ON and OFF has also been demonstrated (Fig. 11.10d). Recently, this concept of heterojunction NW electronic and photonic circuits has also been realized on plastic substrates because of the flexibility of the assembly approach.[63]

The potential of coupling the bottom-up assembly of nanophotonic devices together with top-down fabricated silicon structures has also been investigated, because this coupling could provide a new approach for introducing efficient photonic capabilities into integrated silicon electronics. To this end, a hybrid top-down/bottom-up approach was employed[59] (1) using lithography to pattern p-type silicon wires on the surface of a silicon-on-insulator (SOI) substrate, and then (2) assembling n-type emissive NWs on top of the silicon structures using a fluidic flow or LB approach to form arrays consisting of p-n junctions at cross points (Fig. 11.11).

Conceptually, this hybrid structure (Fig. 11.12) is virtually the same as the crossed NW structures described above and should produce EL in forward

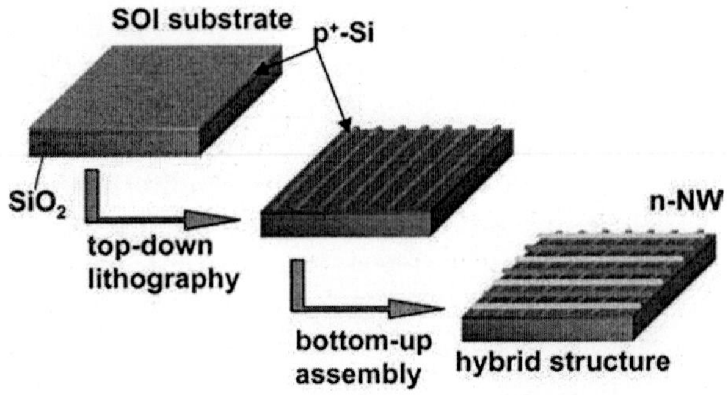

Figure 11.11. Integration of nanowire photonics with silicon electronics. Schematic illustrating fabrication of hybrid structures. A silicon-on-insulator (SOI) substrate is patterned by standard electron-beam or photolithography followed by reactive ion etching. Emissive NWs are then aligned onto the patterned SOI substrate to form photonic sources. [Reprinted with permission from Ref. 59. Copyright 2005 Wiley-VCH Verlag.]

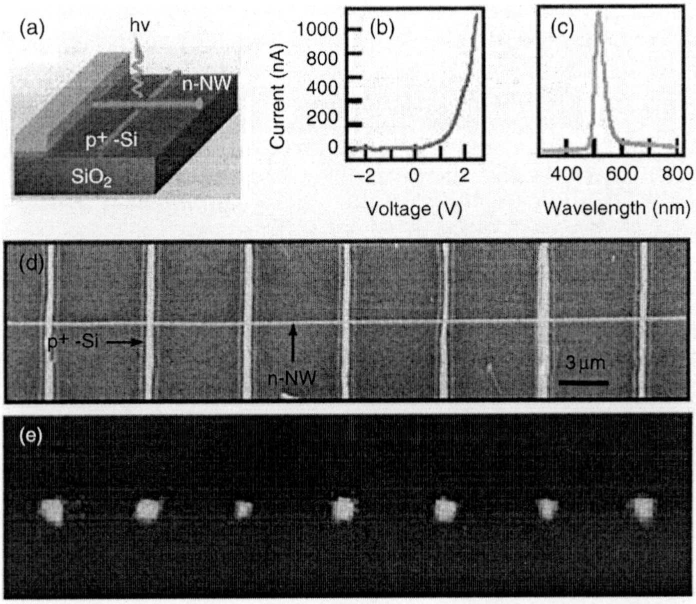

Figure 11.12. Si electrode and NW hybrid device. (a) Schematic of a single LED fabricated by the method outlined in Fig. 11.11. (b) I–V behavior for a crossed p-n junction formed between a fabricated p^+-Si electrode and an n-CdS NW. (c) EL spectrum from the forward biased junction. (d) SEM image of a CdS NW assembled over seven p^+-silicon electrodes on a SOI wafer; (e) EL image recorded from an array consisting of a CdS NW crossing seven p^+-Si electrodes. The image was acquired with +5V applied to each silicon electrode while the CdS NW was grounded. [Reprinted with permission from Ref. 59. Copyright 2005 Wiley-VCH Verlag.]

bias. Notably, I–V data recorded for a hybrid p-n diode formed between the p-Si and an n-CdS NW show clear current rectification (Fig. 11.12b) and a sharp EL spectrum peaked at 510 nm (Fig. 11.12c), which is consistent with CdS band-edge emission. Importantly, the photonic devices produced in this approach are highly reproducible and can be readily implemented in integrated arrays. For example, a 1 × 7 crossed array consisting of a single CdS NW over 7-fabricated p-Si wires (Fig. 11.12d) exhibits well-defined emission from each of the cross points in the array (Fig. 11.12e). Similar results were also obtained for two-dimensional arrays, demonstrating clearly that bottom-up assembly has the potential to introduce photonic function into integrated silicon microelectronics.

11.5 NANOWIRE THIN-FILM ELECTRONICS—CONCEPT AND PERFORMANCE

The assembly of a broad range of nanoscale electronic and optoelectronic devices using NW building blocks demonstrates the great potential of NWs for powering the next generation of electronics. However, significant challenges remain to be solved before these nanoscale devices can be brought into real applications. For example, better assembly schemes are needed to organize NWs in a precisely controlled manner and new circuit architectures must be developed to be compatible with NW device structures. Although a totally new paradigm of nanowire nanoelectronics is further away, more intermediate application of NWs in current technologies may originate by adopting a more conventional device configuration.

For example, a new concept of NW thin-film transistor (NW-TFT) has been developed, in which single-crystal NWs are exploited as conducting channels for routing electrons in large-area thin-film electronics (Fig. 11.13).[64] The concept of NW-TFTs takes a new perspective and shifts the focus of using NWs for nanoelectronics to their applications in higher performance electronics over larger area substrates. In this concept, an oriented NW thin film was assembled to form a novel electronic substrate that can be processed using standard methods to yield high-performance NW-TFTs. Unlike amorphous-Si or polycrystalline-Si TFTs (Fig. 11.13a), in which carriers have to travel across multiple grain boundaries resulting in low mobility,[65,66] NW-TFTs have conducting channels formed by multiple parallel, single-crystal NW paths. Therefore, charges travel from source to drain within single crystals, ensuring high carrier mobility.

11.5.1 p-Si Nanowire Thin-Film Transistors

Duan et al. demonstrated the first NW-TFT based on silicon NWs.[64] Single-crystal Si NWs with controlled diameters around 30 nm are synthesized and then dispersed into solution. A densely oriented NW thin film is produced by

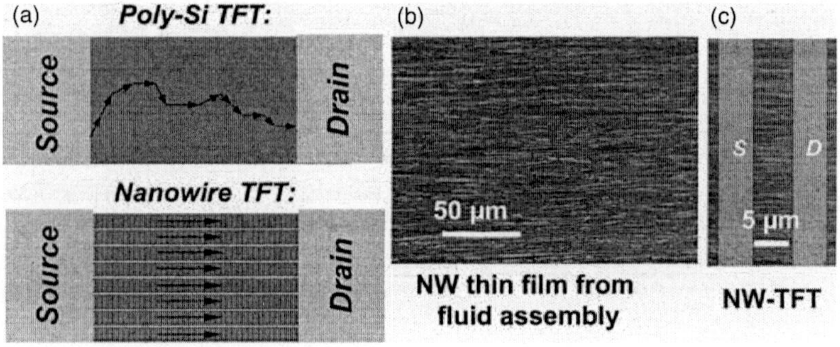

Figure 11.13. Semiconductor nanowire thin-film electronics. (a) Underlying concept of nanowire thin-film transistors (NW-TFTs). Electrical carriers travel across multiple grain boundaries in polycrystalline silicon TFTs (top); while in NW-TFTs (bottom), electrical carriers travel through multiple parallel single-crystal conducting channels, ensuring high carrier mobility. (b–c) Silicon NW-TFT. (b) Large-area optical micrograph of NW thin-film assembled from solution using the fluidic flow-directed assembly approach. (c) Optical image of an NW-TFT with the source-to-drain electrodes bridged by parallel arrays of NWs. [Reprinted by permission from Macmillan Publishers Ltd: Nature (Ref. 64), copyright (2003).]

assembling the NWs onto the surface of a selected substrate using assembly approaches described previously in this article. The NW thin film is then processed using standard lithography and metallization processes to yield TFTs (Fig. 11.13b and c).[64]

Electrical transport studies on NW-TFTs assembled from p-type silicon nanowires show characteristics similar to those of conventional enhancement mode p-channel TFTs with an ON/OFF ratio of nearly 8 orders of magnitude (Fig. 11.14a and b) and a hole mobility of ~119 $cm^2/V{\cdot}s$, which is significantly larger than those of organic, amorphous-Si (<1 $cm^2/V{\cdot}s$)[65] or conventional p-type poly-Si TFTs (~10–40 $cm^2/V{\cdot}s$),[66] and comparable with that of p-type single-crystal silicon material, such as in a silicon-on-insulator (SOI) MOSFET (~180 $cm^2/V{\cdot}s$).[66] Statistical studies also demonstrate that NW-TFTs exhibit highly reproducible and predictable device characteristics, which is critical for practical applications. Figure 11.14c shows a histogram of the threshold voltage distribution of 20 devices having a standard deviation of only 0.22 V, which can be improved even more by optimizing the device fabrication process. All these studies clearly demonstrate that NW-TFTs have an overall performance comparable with single-crystal-based devices.

It is important to note that the entire NW-TFT device fabrication process is performed essentially at room temperature. Therefore, the assembly of high-performance NW-TFTs can be readily applied to low-cost glass and plastic substrates. It has been demonstrated, for example, that NW-TFTs can be fabricated on plastic substrates (Fig. 11.14d) with carrier mobilities comparable with those made on silicon substrates. Moreover, studies demonstrate that

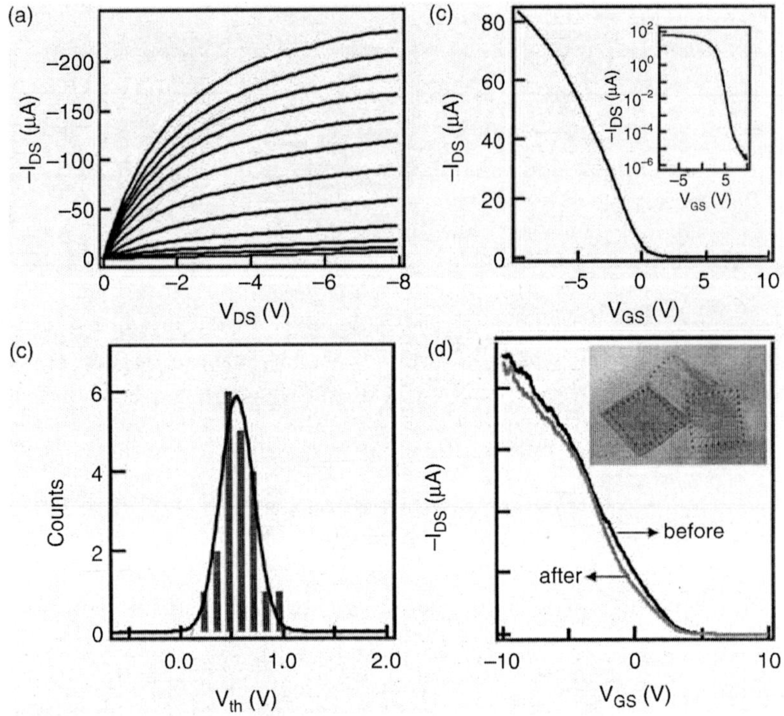

Figure 11.14. p-Si nanowire TFT. (a) Drain-source current (I_{DS}) versus drain-source voltage (V_{DS}) at increasing gate voltages (V_{GS}), in steps of 1 V starting from the top at $V_{GS} = -10$ V. (b) $-I_{DS}$ versus V_{GS} at $V_{DS} = -1$ V. The inset shows $-I_{DS}$ versus V_{GS} at $V_{DS} = -1$ V, using a log scale, highlighting the ON–OFF ratio of nearly 10^8 and subthreshold swing of ~600 mV/decade. (c) Histogram of threshold voltages (V_{th}) from NW-TFT devices shows a high device-to-device reproducibility and a tight distribution. Gaussian fitting shows a standard deviation of only 0.22 V. (d) NW-TFTs on plastics. I_{DS} versus V_{GS} relation at $V_{DS} = -1$ V. The curves show the transfer characteristics of the same device before and after flexing the plastic substrate (radius of curvature ~55 mm), demonstrating the mechanical flexibility of NW-TFTs on plastics. The inset shows a picture of plastic devices with NW-TFTs. [Reprinted by permission from Macmillan Publishers Ltd: Nature (Ref. 64), copyright (2003).]

NW-TFT device performance does not change significantly upon slightly flexing the plastic substrate (Fig. 11.14d).[67] The use of free-standing nanostructures for flexible electronics has also been demonstrated by various groups using nanotubes,[68] nanowires,[63] and nano- or micro-ribbons.[69,70]

11.5.2 High-Speed Integrated Si NW-TFT Circuits

The reproducible assembly of high-performance NW-TFTs in high yield readily enables construction of more complicated high-speed circuits on

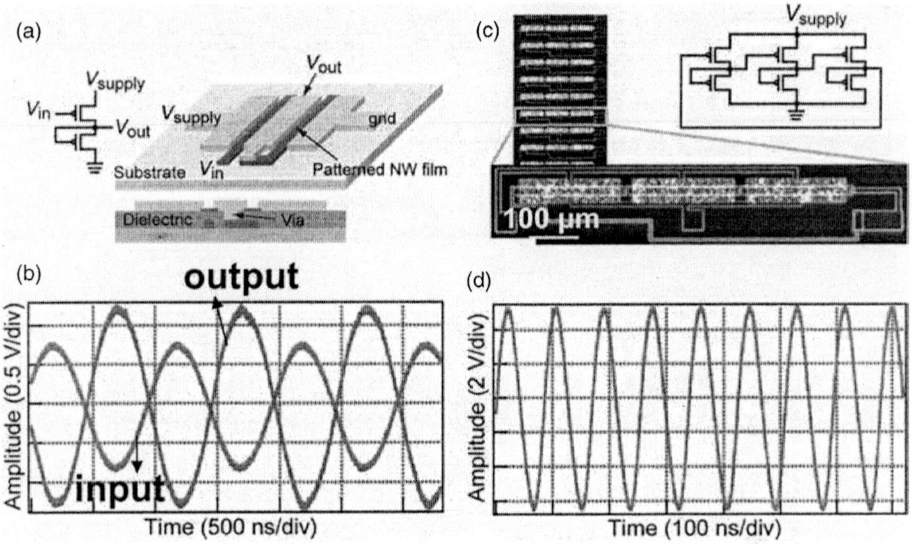

Figure 11.15. High-speed NW circuits on glass. (a) Circuit diagram and schematics of NW-TFT inverters. (b) Output waveform of an inverter fabricated on glass driven by a 1-MHz input sine wave with V_{supply} = 15 V. (c) Optical images and circuit diagram of an NW-TFT ring oscillator. (d) Oscillation of 11.7 MHz in a ring oscillator structure with V_{supply} = 43 V. [Reprinted by permission from Macmillan Publishers Ltd: Nature (Ref. 71), copyright (2005).]

low-cost glass or plastic substrates. For example, the Lieber group has recently integrated NW-TFTs to produce an inverter, which shows greater than unity voltage gain and the expected phase inversion when the inverter is driven by a 1-MHz sine wave (Fig. 11.15a and b).[71] Integrating three such inverters in series, where the input of each inverter is connected to the output of the previous one with a feedback loop, a ring oscillator has been built on a glass substrate. Electrical characterization shows that the NW-TFT oscillator exhibits a self-sustained stable output voltage oscillation, with a maximum oscillation frequency greater than 10 MHz and stage delay on the order of 10 ns (Fig. 11.15c and d),[71] which is significantly better than reported stage delay times of organic ring oscillators (>300 ns)[72] and of the best amorphous silicon ring oscillators (~210 ns).[73] Although this AC performance of current NW-TFT circuits is still inferior to that of the best reported poly-Si TFT circuits (stage delay 100–500 ps), it is believed that continued development in NW-TFT technology, i.e., optimizing NW synthesis, assembly, and the device fabrication process or employing alternative materials, can significantly improve the performance of NW-TFT circuits. For example, Sun et al. have demonstrated gigahertz operation in transistors fabricated from microstructured GaAs wires on plastics.[74]

11.5.3 3D Integrated Functional Electronic System

The power and flexibility of bottom-up assembly can be extended to the fabrication of more complex three-dimensional device structures. A recent report of a 3D multifunctional device based on layer-by-layer assembly of nanowire building blocks represents a great example that couples the flexibility of a bottom-up assembly approach with the breadth of high-quality NWs and takes the field to the next level of complexity.[75] In this work, Lieber's group demonstrated a general approach involving a repeating sequence of (1) contact printing of NWs with optimized properties for targeted function and (2) lithography-based device fabrication on a variety of substrates ranging from silicon to glass and plastic (Fig. 11.16a and b).

In contrast to the flow-directed and Langmuir–Blodgett approaches described above, the printing approach used to deposit the NWs is a dry stamping approach analogous to that described in Chapter 13. Up to 10 addressable vertically stacked layers have been demonstrated. Each layer

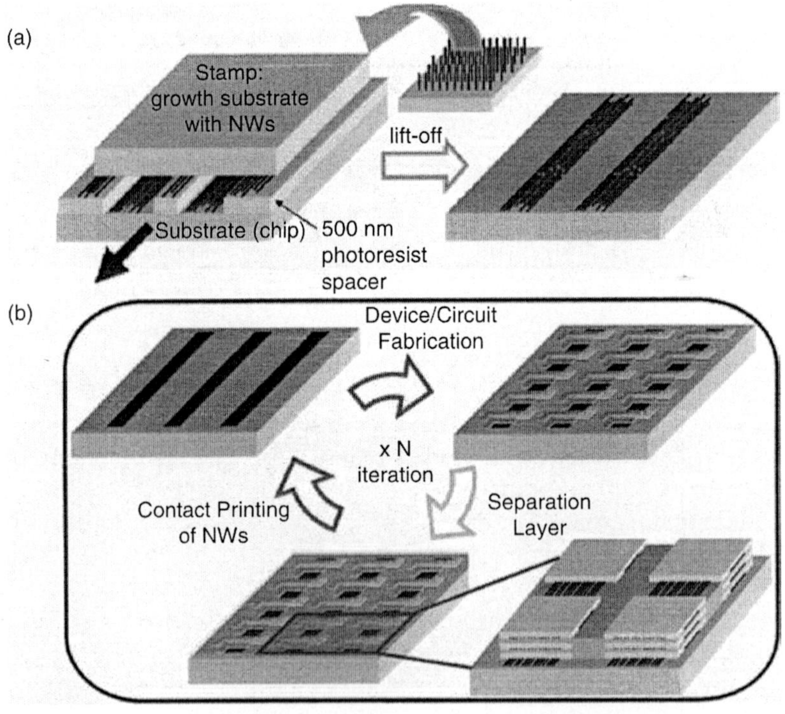

Figure 11.16. Overview of 3D NW circuit integration. (a) Contact printing of NWs from growth substrate to prepatterned substrate. (b) Three-dimensional NW circuit is fabricated by the repetition of the contact printing, device fabrication, and separation layer deposition steps N times. [Reprinted with permission from Ref. [75]. Copyright 2007 American Chemical Society.] (*See color insert.*)

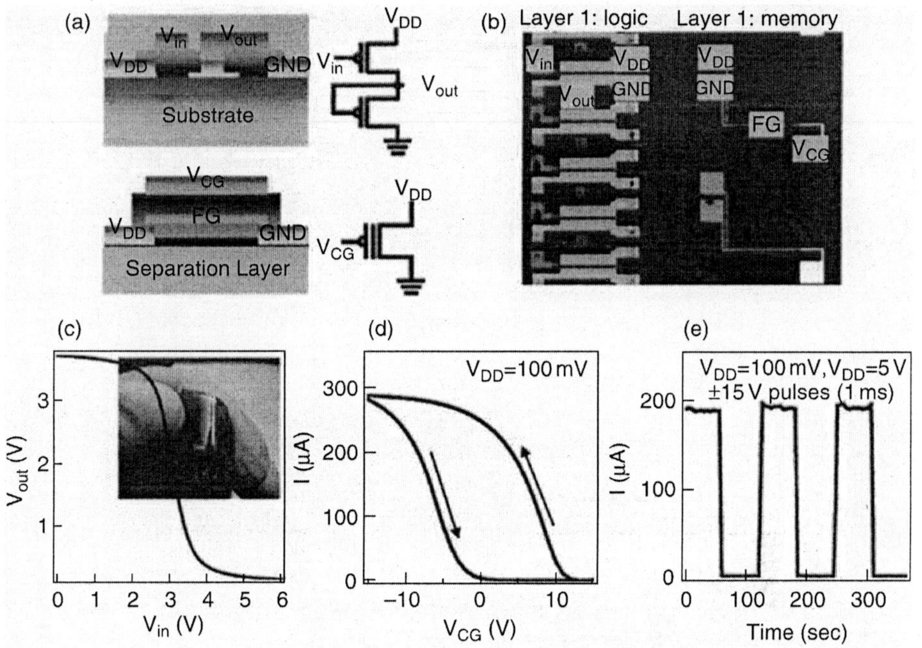

Figure 11.17. Three-dimensional multifunctional circuits on plastics. (a) Schematics and circuit diagrams of inverter (top) and floating gate memory (bottom) elements. (b) Optical image of inverters (layer 1) and floating gate memory (layer 2) on Kapton. (c) DC inverter characteristics. Inset shows functional devices on flexible Kapton substrate. (d) Hysteresis in current-voltage characteristics of a memory layer element. (e) Switching characteristic of the memory device. [Reprinted with permission from Ref. 75. Copyright 2007 American Chemical Society.] (*See color insert.*)

consists of high-performance FETs assembled from Ge core (10 nm)/Si shell (2 nm) single-crystal NWs as the channel material, which has been reported to have better performance than state-of-the-art planar Si devices. They were configured as top-gated devices with channel width and length of 200 and 2 micrometer, respectively, and with a high-k HfO_2 gate dielectric. Key properties of the NW-FETs are reproducible throughout the layers with a general ON current of around 3 mA and a transconductance of around 1 mS. They also designed a 3D inverter-memory structure on a plastic substrate consisting of a lower layer of PMOS inverters and an upper layer of floating gate memory (Fig. 11.17a and b). An electrical measurement demonstrated that the inverter layer showed well-defined inversion with a quasi-DC gain of 3.5 (Fig. 11.17c), it exhibited greater than unity voltage gain, and phase inversion was achieved when the inverter was driven up to a 50-MHz sin wave, outperforming organic and amorphous Si electronics. Measurements on the memory layer showed large and reproducible hysteresis loops (Fig. 11.17d), and it was demonstrated that the writing and erasing cycles can be repeated by applying 1-ms pulses of ±15V to the control gate (Fig. 11.17e). Performance of these NW-film-based devices can be improved even more by using shorter channel lengths and thinner gate dielectrics.

11.6 SUMMARY AND PERSPECTIVE

This chapter provides a brief survey of work on bottom-up assembly of high-quality NW building blocks for various applications, ranging from nanoscale electronics and optoelectronics to large-area thin-film technologies.

NW building blocks are a class of nanoscale materials that can be synthesized with well-controlled chemical compositions, physical structures, and precisely engineered electronic and optical properties. The availability of versatile and high-performance built-in functions during synthesis makes NWs ideal candidates for bottom-up assembled functional systems. As for molecules, the small size of NWs makes them suitable for solution processing. To this end, a few methodologies based on fluids have been already developed to assemble NWs into integrated device arrays using microchannel fluidics or LB technology. A wide range of nanoscale electronic and optoelectronic devices have been assembled from NW building blocks, and several conceptually new nanodevice structures have been realized, such as crossed NW (homo/hetero) p-n junctions and crossed NW-FETs. The critical device dimensions in these structures are defined by chemical synthesis rather than by conventional lithography, and the devices are amenable to high-density integration. The flexibility of bottom-up assembly has also enabled the demonstration of integrated functional electronic circuits, such as logic gates, computational circuits, multicolor LEDs, and smart pixels.

Without dealing with complex device architectures, a new concept of NW TFTs has also been proposed and realized using oriented semiconductor NWs or nanoribbon thin films. Recent studies have demonstrated that high-performance TFTs, high-frequency circuits, and 3D circuits can be readily produced on a variety of substrates, including glass and plastics, with assembled NW structures. The device performance of these solution-assembled NW-TFTs not only greatly surpasses that of solution-processed organic TFTs, but also it is significantly better than conventional amorphous Si or poly-Si TFTs, approaching that of single-crystal, silicon-based transistors. Furthermore, within a similar framework, group III–V or II–VI NWs materials of high intrinsic carrier mobility or optical functionality could be readily built into thin films on flexible substrates to enable new multifunctional electronics/optoelectronics that are not possible with traditional macroelectronics.

There is a very bright future for these electronically well-defined NW building blocks both in fundamental science and in new nanotechnologies. Specifically, we believe that only the very edge of a broad range of ideas has been touched on and that many fascinating fundamental problems remain in these

NW systems, such as the role of finite size and coherent states. Well-defined semiconductor NWs also represent exciting systems to probe fundamental questions about localization or delocalization of electrical carriers and optical excitons in 1D. Future investigations will be not only critical to understanding fundamental issues in 1D systems, but also they will be central to improved device characteristics and possibly conceptually and fundamentally new types of devices, such as single photon emitters and detectors, which could be critical for future quantum communication and computation. Continuing efforts will be required to develop even better control of NW synthesis and increasingly sophisticated assembly approaches that can vary device functionality over multilength scales. Lastly, developing new device concepts and integrated architectures will be increasingly essential as we move closer to highly integrated nanosystems.

In a more immediate term, the new concept of NW thin films opens a new avenue to high-performance macroelectronics and offers opportunities for a paradigm-shift in the area of electronics. Its realization makes it possible to move high-performance microelectronics from single-crystal substrates to glass and plastic substrates—cutting the production cost dramatically while retaining the performance. The flexible bottom-up approach also makes it feasible to integrate macroelectronics, microelectronics, and potentially nanoelectronics at the device level and to integrate different semiconductor materials on a single substrate to achieve multiple functionalities in a way that is not possible with current technologies. Taken together, the NW-TFT concept can impact a broad range of existing applications, from flat panel displays to image sensor arrays, and can enable a whole new range of flexible, wearable, or disposable electronics for computing, storage, and wireless communication.

REFERENCES

1. Accessed: http://www.intel.com/technology/mooreslaw/index.htm.
2. 2007 Executive Summary of International Technology Roadmap for Semiconductors. Accessed: http://www.itrs.net.
3. Meindl, J. D.; Chen, Q.; Davis, J. A. 2001. Limits on silicon nanoelectronics for terascale integration. *Science* 293:2044–2049.
4. Thompson, S. E.; Parthasarathy, S. 2006. Moore's law: The Future of Si microelectronics. *Materials Today* 9:20–25.
5. Lieber, C. M. 2001. The incredible shrinking circuit. *Sci. Am.* 285:50–56.
6. Jacoby, M. 2002. Nanoscale electronics. *Chem. Eng. News Sept.* 30:38–43.
7. Heath, J. R.; Kuekes, P. J.; Snider, G. S.; Williams, R. S. 1998. A defect-tolerant computer architecture: Opportunities for nanotechnology. *Science* 280:1716–1721.
8. Burda, C.; Chen, X.; Narayanan, R.; El-Sayed, M. A. 2005. Chemistry and properties of nanocrystals of different shapes. *Chem. Rev.* 105:1025–1102.
9. Dai, H. 2002. Carbon nanotubes, from synthesis to integration and properties. *Accounts Chem. Research* 35:1035–1044.

10. Duan, X.; Lieber, C. M. 2005. Semiconductor nanowires: Rational synthesis. In *Dekker Encyclopedia of Nanoscience and Nanotechnology*, edited by Schwarz J. A. , Marcel Dekker, Inc., New York.
11. Reed, M. A.; Tour, J. M. 2000. Computing with molecules. *Sci. Am.* 282:86–95.
12. Joachim, C.; Gimzewski, J. K.; Aviram, A. 2000. Electronics using hybrid-molecular and mono-molecular devices. *Nature* 408:541–548.
13. Collier, C. P.; Wong, E. W.; Belohradsky, M. F.; Raymo, M.; Stoddart, J. F.; Kuekes, P. J.; Williams, R. S.; Heath, J. R. 1999. Electronically configurable molecular-based logic gates. *Science* 285:391–394.
14. Collier, C. P.; Mattersteig, G.; Wong, E. W.; Luo, Y.; Beverly, K.; Sampaio, J.; Raymo, F. M.; Stoddart, J. F.; Heath, J. R. 2000. A [2]catenane-based solid state electronically reconfigurable switch. *Science* 289:1172–1175.
15. Reed, M. A.; Chen, J.; Rawlett, A. M.; Price, D. W.; Tour, J. M. 2001. Molecular random access memory cell. *Appl. Phys. Lett.* 78:3735–3737.
16. Alivisatos, A. P. 1996. Semiconductor clusters, nanocrystals, and quantum dots. *Science* 271:933–937.
17. Klein, D. L.; Roth, R.; Lim, A. K. L.; Alivisatos, A. P.; McEuen, P. L. 1997. A single-electron transistor made from a cadmium selenide nanocrystal. *Nature* 389:699–701.
18. Devoret, M. H.; Schoelkopf, R. J. 2000. Amplifying quantum signals with the single-electron transistor. *Nature* 406:1039–1046.
19. Talapin, D. V.; Murray, C. B. 2006. PbSe nanocrystal solids for n- and p-channel thin film field-effect transistors. *Science* 310:86–89.
20. Collins, P. G.; Avouris, P. 2000. Nanotubes for electronics. *Sci. Am.* 283:62–70.
21. Odom, T. W.; Huang, J.-L.; Kim, P.; Lieber, C. M. 1998. Atomic structure and electronic properties of single-walled carbon nanotubes. *Nature* 391:62–64.
22. Odom, T. W.; Huang, J.-L.; Kim, P.; Lieber, C. M. 2000. Structure and electronic properties of carbon nanotubes. *J. Phys. Chem. B* 104:2794–2809.
23. Dekker, C. 1999. Carbon nanotubes as molecular quantum wires. *Phys. Today* 52:22–28.
24. Dai, H.; Kong, J.; Zhou, C.; Franklin, N.; Tmobler, T.; Cassell, A.; Fan, S.; Chapline, M. 1999. Controlled chemical routes to nanotube architectures, physics, and devices. *J. Phys. Chem. B* 103:11246–11255.
25. Tans, S. J.; Verschueren, R. M.; Dekker, C. 1998. Room-temperature transistor based on a single carbon nanotube. *Nature* 393:49–52.
26. Martel, R.; Schmidt, T.; Shea, H. R.; Hertel, T.; Avouris, P. 1998. Single- and multi-wall carbon nanotube field-effect transistors. *Appl. Phys. Lett.* 73:2447–2449.
27. Yao, Z.; Postma, H. W. C.; Balents, L.; Dekker, C. 1999. Carbon nanotube intramolecular junctions. *Nature* 402:273–276.
28. Fuhrer, M. S.; Nygrad, J.; Shih, L.; Forero, M.; Yoon, Y. G.; Mazzoni, M. S. C.; Choi, H. J.; Ihm, J.; Louie, S. G.; Zettl, A.; McEuen, P. L. 2000. Crossed nanotube junctions. *Science* 288:494–497.
29. Derycke, V.; Martel, R.; Appenzeller, J.; Avouris, P. 2001. Carbon nanotube inter- and intramolecular logic gates. *Nano Lett.* 1:453–456.

30. Bachtold, A.; Hadley, P.; Nakanishi, T.; Dekker, C. 2001. Logic circuits with carbon nanotube transistors. *Science* 294:1317–1320.

31. Collins, P. G.; Arnold, M. S.; Avouris, P. 2001. Engineering carbon nanotubes and nanotube circuits using electrical breakdown. *Science* 292:706–709.

32. Xia, Y.; Yang, P.; Sun, Y.; Wu, Y.; Mayers, B.; Gates, B.; Yin, Y.; Kim, F.; Yan, H. 2003. One-dimensional nanostructures: synthesis, characterization and applications. *Adv. Mater.* 15:353–389.

33. Lieber, C. M.; Wang, Z. L. 2007. Functional nanowires. *MRS. Bull.* 32:99–104.

34. Morales, A. M.; Lieber, C. M. 1998. A laser ablation method for the synthesis of crystalline semiconductor nanowires. *Science* 279:208–211.

35. Duan, X.; Lieber, C. M. 2000. General synthesis of compound semiconductor nanowires. *Adv. Mater.* 12:298–302.

36. Duan, X.; Lieber, C. M. 2000. Laser-assisted catalytic growth of single crystal GaN nanowires. *J. Am. Chem. Soc.* 122:188–189.

37. Cui, Y.; Lauhon, L. J.; Gudiksen, M. S.; Wang, J.; Lieber, C. M. 2001. Diameter-controlled synthesis of single-crystal silicon nanowires. *Appl. Phys. Lett.* 78: 2214–2216.

38. Gudiksen, M. S.; Wang, J.; Lieber, C. M. 2001. Synthetic control of the diameter and length of single crystal semiconductor nanowires. *J. Phys. Chem. B* 105:4062–4064.

39. Duan, X.; Huang, Y.; Cui, Y.; Wang, J.; Lieber, C. M. 2001. Indium phosphide nanowires as building blocks for nanoscale electronic and optoelectronic devices. *Nature* 409:66–69.

40. Gudiksen, M. S.; Lauhon, L. J.; Wang, J.; Smith, D.; Lieber, C. M. 2002. Growth of nanowire superlattice structures for nanoscale photonics and electronics. *Nature* 415:617–620.

41. Lauhon, L. J.; Gudiksen, M. S.; Wang, D.; Lieber, C. M. 2002. Epitaxial core-shell and core-multi-shell nanowire heterostructures. *Nature* 420:57–61.

42. Huang, Y.; Duan, X.; Cui, Y.; Lieber, C. M. 2002. Gallium nitride nanowire nanodevices. *Nano Lett.* 2:101–104.

43. Cui, Y.; Duan, X.; Hu, J.; Lieber, C. M. 2000. Doping and electrical transport in silicon nanowires. *J. Phys. Chem. B* 104:5213–5216.

44. Huang, Y., Lieber, C. M. 2004. Integrated nanoscale electronics and optoelectronics: Exploring nanoscale science and technology through semiconductor nanowires. *Pure Appl. Chem.* 76:2051–2068.

45. Gudiksen, M. S.; Wang, J.; Lieber, C. M. 2002. Size dependent photoluminescence from single indium phosphide nanowires. *J. Phys. Chem B* 106:4036–4039.

46. Xiang, J.; Lu, W.; Hu, Y.; Wu, Y.; Yan, H.; Lieber, C. M. 2006. Ge/Si nanowire heterostructures as high-performance field-effect transistors. *Nature* 441:489–493.

47. Li, Y.; Xiang, J.; Qian, F.; Gradecak, S.; Wu, Y.; Yan, H.; Blom, D. A.; Lieber, C. M. 2006. Dopant-free GaN/AlN/AlGaN radial nanowire heterostructures as high electron mobility transistors. *Nano Lett.* 6:1468–1473.

48. Qian, F.; Gradecak, S.; Li, Y.; Wen, C.; Lieber, C. M. 2006. Core/multishell nanowire heterostructures as multicolor, high-efficiency light-emitting diodes. *Nano Lett.* 5:2287–2291.

49. Huang, Y.; Duan, X.; Wei, Q.; Lieber, C. M. 2001. Directed assembly of one-dimensional nanostructures into functional networks. *Science* 291:630–633.
50. Stover, C. A.; Koch; D. L.; Cohen, C. 1992. Observation of fiber orientation in simple shear-flow of semidilute suspensions. *J. Fluid Mech.* 238:277–296.
51. Koch, D. L.; Shaqfeh, E. S. G. 1990. The average rotation rate of a fiber in the linear flow of a semidilute suspension. *Phys. Fluids A* 2:2093–2102.
52. Yang, P. 2003. Nanotechnology: Wires on water. *Nature* 425:243–244.
53. Ulman, A. 1991. *An Introduction to Ultrathin Organic Films: From Langmuir-Blodgett to Self-Assembly*. Academic Publishers, San Diego, CA.
54. Tao, A.; Kim, F.; Hess, C.; Goldberger, J.; He, R.; Sun, Y.; Xia, Y.; Yang, P. 2003. Langmuir-Blodgett silver nanowire monolayers for molecular sensing with high sensitivity and specificity. *Nano. Lett.* 3:1229–1233.
55. Whang, D.; Jin, S.; Wu, Y.; Lieber, C. M. 2003. Large-scale hierarchical organization of nanowire arrays for integrated nanosystems. *Nano Lett.* 3:1255–1259.
56. Horowitz, P.; Hill, W. 1989. *The Art of Electronics*. Cambridge Univ. Press, Cambridge., U.K.
57. Huang, Y.; Duan, X.; Cui, Y.; Lauhon, L.; Kim, K.; Lieber, C. M. 2001. Logic gates and computation from assembled nanowire building blocks. *Science* 294: 1313–1317.
58. Duan, X.; Wang, J.; Lieber, C. M. 2000. Synthesis and optical properties of gallium arsenide nanowires. *Appl. Phys. Lett.* 76:1116–1118.
59. Huang, Y.; Duan, X.; Lieber, C. M. 2005. Nanowires for integrated multicolor nanophotonics. *Small* 1:142–147.
60. Zhong, Z.; Qian, F.; Wang, D.; Lieber, C. M. 2003. Synthesis of p-type gallium nitride nanowires for electronic and photonic nanodevices. *Nano Lett.* 3:343–346.
61. Hayden, O.; Agarwal, R.; Lieber, C. M. 2006. Nanoscale avalanche photodiodes for highly sensitive and spatially resolved photon detection. *Nature Mater.* 5:352–356.
62. Li, Y.; Qian, F.; Xiang, J.; Lieber, C. M. 2006. Nanowire electronic and optoelectronic devices. *Materials Today* 9:18–27.
63. McAlpine, M. C.; Friedman, R. S.; Jin, S.; Lin, K.; Wang, W. U.; Lieber, C. M. 2003. High-performance nanowire electronics and photonics on glass and plastic substrates. *Nano Lett.* 3:1531–1535.
64. Duan, X.; Niu, C.; Sahi, V.; Chen, J.; Parce, J. W.; Empedocles, S.; Goldman, J. L. 2003. High-performance thin-film transistors using semiconductor nanowires and nanoribbons. *Nature* 425:274–278.
65. Street, R. A. 2000. *Technology and Applications of Amorphous Silicon*. Springer, Berlin, Germany.
66. Hara, A.; Takei, M.; Takeuchi, F.; Suga, K.; Yoshino, K.; Chida, M.; Kakehi, T.; Ebiko, Y.; Sano, Y.; Sasaki, N. 2004. High performance low temperature polycrystalline silicon thin film transistors on non-alkaline glass produced using diode pumped solid state continuous wave laser lateral crystallization. *Jpn. J. Appl. Phys.*, Pt. 1 43:1269–1276.
67. Duan, X. 2007. Assembled semiconductor nanowire thin films for high-performance flexible macroelectronics. *MRS Bull.* 32:134–141.

68. Snow, E. S.; Campbell, P. M.; Ancona, M. G.; Novak, J. P. 2005. High-mobility carbon-nanotube thin-film transistors on a polymeric substrate. *Appl. Phys. Lett.* 86:033105–033107.
69. Sun, Y.; Menard, E.; Rogers, J. A.; Kim, H.-S.; Kim, S.; Chen, G.; Adesida, I.; Dettmer, R.; Cortez, R.; Tewksbury, A. 2006. Gigahertz operation in flexible transistors on plastic substrates. *Appl. Phys. Lett.* 88:183509–183511.
70. Menard, E.; Lee, K. J.; Khang, D. Y.; Nuzzo, R. G.; Rogers, J. A. 2004. A printable form of silicon for high performance thin film transistors on plastic substrates. *Appl. Phys. Lett.* 84:5398–5400.
71. Friedman, R. S.; McAlpine, M. C.; Ricketts, D. S.; Ham, D.; Lieber, C. M. 2005. High-speed integrated nanowire circuits. *Nature* 434:1085.
72. Clemens, W.; Fix, W.; Ficker, J.; Knoblock, A.; Ullmann, A. 2004. From polymer transistors toward printed electronics. *J. Mater. Res.* 19:1963–1973.
73. Hiranaka, K.; Yamaguchi, T.; Yanagisawa, S. 1984. Self-alignment processed amorphous-silicon ring oscillators. *IEEE Elec. Dev. Lett.* 7:224–224.
74. Ahn, J. H.; Kim, H. S.; Lee, K. J.; Jeon, S.; Kang, S. J. Y.; Sun, G.; Nuzzo, R. G.; Rogers, J. A. 2006. Heterogeneous three-dimensional electronics by use of printed semiconductor nanomaterials. *Science* 314:1754–1757.
75. Javey, A.; Nam, S.; Friedman, R. S.; Yan, H.; Lieber, C. M. 2007. Layer-by-layer assembly of nanowires for three-dimensional, multifunctional electronics. *Nano Lett.* 7:773–777.

CHAPTER 12

Patterning Techniques for Solution Deposition

PAUL W. BRAZIS,* DANIEL GAMOTA,** JIE ZHANG,** and JOHN SZCZECH**

*Corporate Research, Underwriters Laboratories, Northbrook, IL; **Motorola Physical Realization Research Center, Schaumburg, IL

12.1 INTRODUCTION

The advent of a new class of materials systems based on nanoscale particles dispersed or suspended in carrier and/or binders has captured the attention of the microelectronics technical community. These materials provide the opportunity to use inexpensive solution processing equipment versus expensive vacuum deposition equipment commonly used in the microelectronics manufacturing industry. Experts in the microelectronics industry have suggested that over the course of the next five years, the industry will experience a paradigm shift in manufacturing and, more importantly, will enjoy revenue streams created from never-before-seen products based on printed electronics.

It has long been accepted that scientific and engineering fields will require new theories to explain the phenomena that occur at the nanoscale dimension. Although the term "nanotechnology" has more recently captured the attention of the public, it has existed in high-technology fields such as microelectronics and semiconductor integrated circuits (ICs) for several years. The semiconductor industry has continued to sustain sales growth from the development and qualification of semiconductor processing tools that enable reduction of features on the ICs to offer enhanced computational speed over the years (Fig. 12.1). As the consumer demand for faster, application-rich electronic products grows, the need to fabricate ICs with increased transistor density requires semiconductor processing tools that enable transistor feature resolutions

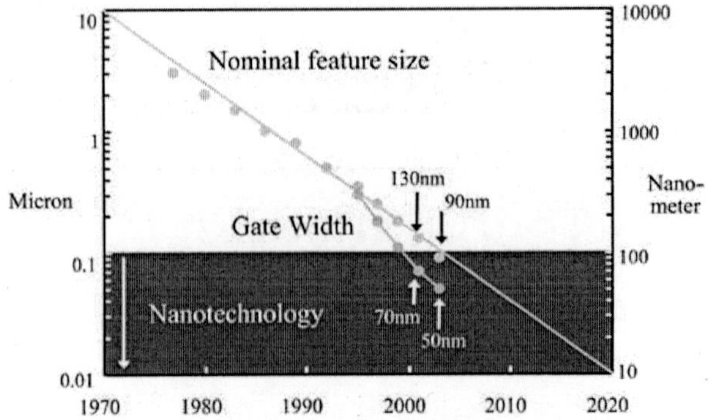

Figure 12.1. Silicon-integrated circuit nominal feature size over the years.

approaching 50 nm and lower. Consumers also are increasingly demanding more pervasive electronics, including mobile and large-area entertainment, enhanced communications, and improved user experiences. Thus, to fulfill these market needs and to continue growth through innovation, the semiconductor industry must either continue to follow the self-fulfilling prophecy of Moore's law by enhancing existing tool sets or embrace an entirely new technological path, which will require the development of novel nanomanufacturing platforms: materials, processing equipment, and device testing systems.

An example of a revolutionary technology platform that will touch the semiconductor industry is printed electronics based on semiconductor inks composed of nanoscale inorganic semiconducting particles. These technologies require the establishment of new value and supply chains, standards, and standard operating procedures that are currently at different stages of development. In the past few years, there has been a substantial increase in government and industrial research to demonstrate manufacturing-acceptable materials and processes to accelerate this development. Specifically, as these printing technology platforms continue to mature and the fundamental knowledge assets are strengthened, the International Technology Roadmap for Semiconductors (ITRS) and semiconductor trade associations will create the foundation for future introduction of these printing technologies in next-generation electronic products.

Although the excitement over the potential to design faster and more powerful computing architectures using nanoscale features is tempered by the time that will be required to launch these products, semiconducting inks composed of nanoscale particles is considered a relatively near-term opportunity. This opportunity—electronically functional inorganic nanoscale inks—leverages the first segment of the nanotechnology value chain (Fig. 12.2). The availability of commercial- and near-commercial-grade nanoscale electronically functional inks has increased in the past few years. Moreover, during the past two years, there have been several public announcements by start-ups and large companies that claim to have made substantial financial commitments to the

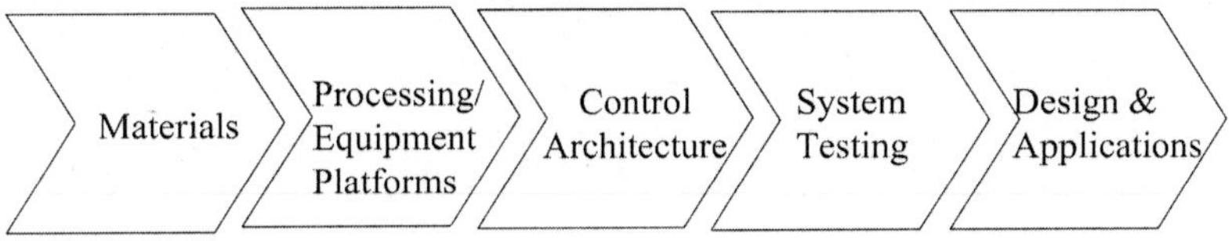

Figure 12.2. Nanotechnology value chain.

development of the enabling printing technologies to fabricate ICs composed of nanotechnology-enabled inks: nanoscale conductive, dielectric, and semiconducting inks. More recently, companies have presented electrical results of the first all-printed integrated circuits prepared using these nano-inks.[1–3]

12.2 OPPORTUNITIES FOR PRINTABLE INORGANIC VERSES ORGANIC MATERIALS SYSTEMS

Historically, the focus of printed electronics has been on organic semiconducting materials, which are generally considered to be more conducive to ink-based deposition techniques than inorganic materials. The commitment to this approach is particularly strong, as exemplified by the heavy investment by the European Union through its PolyApply 6th and 7th Framework initiatives.[4] However, organic semiconducting materials have exhibited far lower carrier mobility than more conventional semiconductors (such as silicon and germanium), appearing to top out on the order of 1 cm^2/Vs (Fig. 12.3) for printable systems. Dimitrakopoulos and Malenfant propose a theoretical limit for organic semiconductor mobility of 10 cm^2/Vs, irrespective of processing method.[5] This limit is about two to three orders of magnitude less than that for silicon and germanium, which have mobility values on the order of 1000 cm^2/Vs. Since mobility is proportional to the switching time of a transistor (see Chapter 1), the use of the best printable organic semiconductors will result in a thousand-fold reduction in circuit switching speed (and further reductions in field-effect mobility are expected from devices produced in volume, as most cited values are laboratory scale). Considering also that current large-volume printing technology cannot create feature sizes much smaller than 10 to 50 μm, printable organic semiconducting devices are not expected to give a suitable performance for most wireless applications. A benchmark of the evolution of organic semiconductor mobility (summarized in Fig. 12.3) strongly suggests that values are not expected to rise significantly in the foreseeable future.

Since inorganic materials may have intrinsic mobility values of 100 to 1000 cm^2/Vs and greater, their use in printed electronics would likely enable far greater device performance than organic devices. Historically, inorganic semiconductor deposition has required both high temperatures and vacuum environments. However, advances in nanotechnology, particularly in particle fabrication and analysis, enable the development of functional inks that can

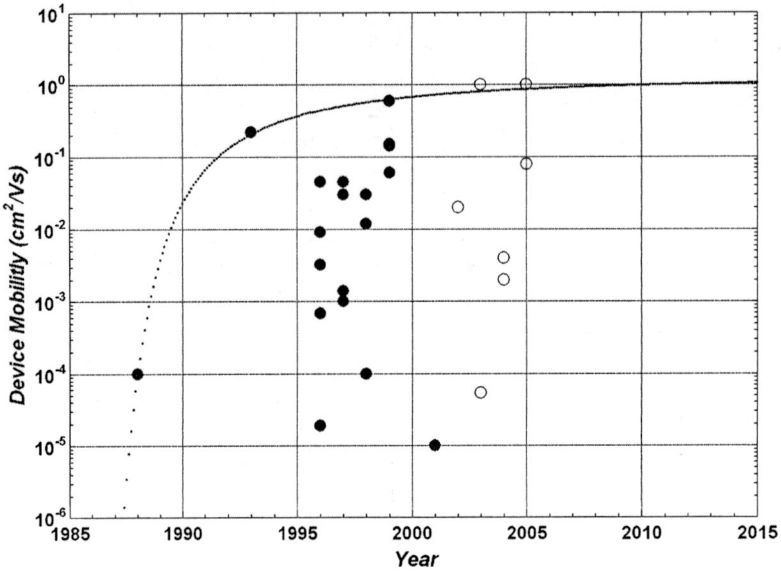

Figure 12.3. Benchmark of peer-reviewed academic reports of organic semiconductor device field-effect mobility versus time of report. All data points are for spin-coated organic semiconducting transistors. Solid points are derived from the benchmark study completed in 2002 by Brazis and Dyrc at Motorola (unpublished). The curve is a calculated estimation, based on these data, of what the expected mobility values will be in the future. The open points are data derived in 2005 from the public journals for verification of the 2002 prediction.[6–38]

be deposited at temperatures compatible with low-cost substrates and using conventional graphic arts printing processes (enabling very high product throughput). The development of an inorganic semiconducting ink, therefore, will likely enable very low-cost production of high-performance wireless or interactive products on flexible substrates. It is conceivable that through this approach to printed electronics, circuit performance several orders of magnitude greater than what is expected from the projected state-of-the art organic-based printed electronics technology can be achieved.

One of the primary advantages cited for organic electronics is that organics are far less costly than inorganic materials. This is based on the general assumption that plastics are cheaper than, for example, silicon and metals. However, it is not clear whether this will be true for volume production. For example, the price of organic semiconducting materials is currently very high (often thousands of U.S. dollars per gram) and available batch quantity is very low. It is possible that increasing production volume will reduce cost, although this has not yet been proven. In contrast, inorganic nanoparticle materials are already produced in huge volumes (several tons) for commodity applications (e.g., milling and paints). Therefore, low cost and scale-up are already demonstrated for nanoparticle dispersions and with greatly lower risk for supply chain development.

Another advantage cited for organic electronics is their perceived low environmental impact and high expected consumer safety. This assumption is generally based on the notion that plastics are easily recycled and are considered safe to humans and animals. However, the materials used are often completely new compositions with poorly understood health and safety attributes. The assumption that all plastics are completely safe for humans is inaccurate, as is exemplified by recent concerns about the toxicity of polyvinyl chloride (PVC).[39] In contrast, most inorganic nanoparticle materials are already on the consumer market and have extensive historical data on their safety in a variety of applications. Some materials, such as zinc oxide, are even considered reasonably safe for ingestion and therefore are commonly used in food and cosmetics. However, the health effects and interactions of nanoparticles on the human body are still a topic of debate.[40]

The use of inorganic materials for printed electronics has been reported in the literature, although a fully commercially viable route toward achieving low-cost, high-performance electronics on low-cost and flexible substrates has not yet been achieved. In the literature, two primary approaches are found: conversion of printable precursor materials and direct deposition of nanoparticles. Work by Shimoda et al. demonstrates the former, with a process of "printing of silicon" (see Chapter 5).[41] A liquid silicon precursor (cyclopentasilane, Si_5H_{10}) material is deposited by a conventional graphic arts technique (here, by inkjet printing) and then thermally decomposed into silicon by heating above 280 °C (although 540 °C for 2 h was cited as the preferred conditions to achieve reasonable device performance). Although silicon was successfully delivered using a graphic arts printing technique, the requirement of relatively high conversion temperatures may focus this process more toward systems with rigid substrates, such as quartz, glass, and metal. Since large-scale vacuum processes for amorphous and poly silicon deposition already exist for these types of substrates,[42] any printing technique for Si must demonstrate a clear manufacturing advantage over these processes to become technologically viable. Similar techniques for chemical conversion of other common inorganic materials, including zinc oxide, have also been reported and require relatively high (>600 °C) conversion temperatures.[43]

As an example of the latter technique, Volkman et al. demonstrated the feasibility of using spin-cast zinc oxide nanoparticles encapsulated in 1-dodecanethiol to fabricate a functional transistor.[44] The zinc oxide was deposited on a thermally grown silicon dioxide layer on a conventional silicon wafer, with thermally evaporated gold source and drain electrodes. As reported, the process requires very small particles (3 nm or less) and a 400 °C forming gas anneal. A similar approach was also reported by Petrat, demonstrating *n*-channel thin-film transistor operation using a nanoparticle solution of zinc oxide dispersed onto a thermally grown silicon dioxide layer on a conventional silicon wafer.[45] In addition to zinc oxide, gallium nitride dispersed onto a silicon wafer has also been attempted, albeit so far with limited success, most likely because of poor matching of the semiconductor to a sufficiently compatible dielectric material.[46] Although these reports support the viability of direct solution deposition of inorganic materials, the use of a thermally grown dielectric does not reveal the full ability of an inorganic semiconducting ink to function in an all-printed device. To date, attempts to deposit inorganic semiconducting materials on printable polymeric dielectrics have resulted in nonfunctional devices, exemplifying the need to develop a complete suite of materials for electrically functional printed materials.

12.3 PRINTING AND THE MICROELECTRONICS INDUSTRY—PRESENT AND FUTURE

For the past 30 years, a variety of noncontact and contact printing technology platforms have been used in the microelectronics industry. In general, the contact platforms are used to process high viscosity materials. As an example of contact printing, screen and stencil printing are used to print lead-based and lead-free solders on IC wafers/chips or on printed wiring boards (PWBs) to form the interconnects (Fig. 12.4). As the size of features on the IC and the PWB have decreased, the solder particle size dimensions have decreased while the dispersion requirements have tightened to ensure the proper rheology of the solder paste.

Noncontact printing is used in the microelectronics industry to print low-viscosity electronic-grade materials: solder fluxes, dielectric films, anti-static

Figure 12.4. Left—Scanning electron microscope image of the input/output pads of a silicon IC bumped with eutectic lead/tin solder after solder reflow. Right—Photograph of printed wiring board interconnect pads that have been printed with lead/tin solder prior to reflow.

coatings, and environmental barriers (Fig. 12.5). The most common form of noncontact printing is ink jet. These materials systems can be water- or solvent-based and may require thermal treatment to remove carrier liquids after deposition, or they can be thermosets requiring thermal or ultraviolet radiation exposure for processing.

These contact and noncontact printing platforms are examples of technologies that are in use today for high-volume assembly of electronic products, such as wireless phones and electronic engine control units. As manufacturing engineers have developed the enabling hardware/software architectures to integrate these technologies, a plethora of additional manufacturing opportunities have been identified and are at different stages of development.

Since the late 1990s and early 2000s, engineers have investigated the potential use of printing platforms to fabricate photovoltaic arrays, primary batteries, solid-state lighting, flexible displays, and active devices/circuitry. These activities have fueled the creation of start-ups, consortia, and joint ventures, with the mission to commercialize printed electronics-based products. Currently, there are ongoing efforts for printed electroluminescent (EL) displays and solid-state lighting systems, as well as for printed organic light-emitting diode (OLED/PLED) displays. Examples of printed energy sources include printed photovoltaic/solar cells and printed primary batteries. Ongoing efforts for printed circuits consider both printed passive and printed active circuits. Passive circuits include interconnect traces, antennae, resistors, capacitors, and inductors. Active circuits under development consist of transistors, as

Figure 12.5. Photograph of microdispensing system depositing an inorganic dielectric dispersion onto a patterned, metallized polyester film. The pattern is of a transistor gate electrode array. The microdispensing head atomizes the dispersion, generating a liquid spray much like a dual orifice atomizer found on an airbrush.

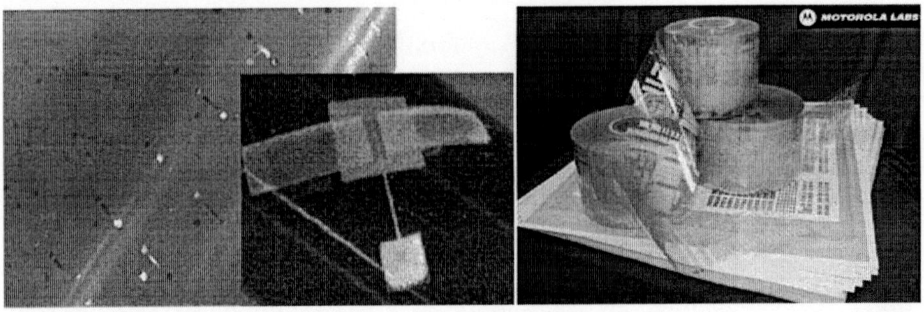

Figure 12.6. Left—Magnified photograph of an all-printed transistor consisting of silver nanoparticle electrodes, a ceramic nanoparticle dielectric and organic semiconductor layer. Right—Photograph depicting three 3000-ft-long, 6-in-wide rolls of all-printed transistors fabricated using graphic arts printing technologies from Motorola.

shown in Fig. 12.6, and passive elements that, in combination, form printed integrated circuits (pICs) providing logic operation functionality.

Printed EL products are used for backlighting of automobile instrument panels and cell phone key pads. OLED/PLED technologies will provide large and small high-resolution displays that will compete with liquid crystal displays and solid-state lighting. Printed battery technology has been in commercial use for instant cameras for over 20 years and more recently has the ability to provide power for electronics and cosmetic/pharmaceutical products with novel form factors.[47] Printed passive circuitry fabricated on ceramic substrates has been in commercial use for more than two decades. Within the last five years, the microelectronics industry has observed the commercial launch of printed passive elements embedded in organic rigid and flexible PWBs to reduce the electronic component part count and PWB real estate; this technology has enabled the launch of ultra-small Bluetooth ear modules.[49] The most recent development in printed circuits technology is the printing of active circuitry for the design of radio frequency identification (RFID) tags, display backplanes, and sensors.

12.4 PRINTED ELECTRONICS VALUE CHAIN

The printed electronics value chain is similar to the nanotechnology value chain in that it has five major segments: materials, processing equipment (printing manufacturing equipment) platforms, control architectures (hardware/software), system testing tools, and design and applications. Each segment can be further subdivided into discrete elements, each of which creates a revenue stream. The materials segment, for example, is divided into functional inks, substrates, barrier coatings, and self-assembly materials. As mentioned, the development of functional inks is viewed as the enabler for the printed electronics market.

12.5 ELECTRICALLY FUNCTIONAL INKS

In general, inks are composed of four fundamental components: functional element, binder, solvent, and additive. The main differentiator between graphic arts and printed electronics inks is the functional element: a colorant material for graphic arts and electronic functional materials for printed electronics. "Functional ink" is a term used to define inks consisting of an electrically functional element. The binder, solvent, and additives form the carrier and are chemically functionalized for a specific printing technology by modifying ink rheology. The function of the carrier in the ink is identical for both graphic arts and printed electronics inks: (1) functional elements dispersion, (2) viscosity, and (3) surface tension modification. Also functional inks are formulated with specific rheological properties to enable printing using various contact and noncontact printing platforms and are categorized based on their fundamental electrical materials properties.

The three classes of functional inks necessary to print ICs are (1) semiconducting, (2) conducting, and (3) dielectric. Semiconducting inks are composed of intrinsically semiconducting organic polymers in solvents, inorganic nanoscale particles (Group III–V, II–VI, and IV semiconductors and carbon nanotubes) suspended in carrier fluids, or organic–inorganic blends; conducting inks are composed of intrinsically conductive organic polymers in solvents, metallic particles suspended in binders, or organic–metallic blends; dielectric inks are composed of organic polymers in solvents, organic polymer thermosets, or ceramic-filled organic polymers.

Figure 12.7 shows the transmission electron microscopy (TEM) image of one nano-enabled ink that is designed for printing ICs using high-resolution printing technologies. The nanoparticle interaction in this ink impacts ink rheological properties and, therefore, ink printability. The dispersability of nanoparticles is improved by nanoparticle surface passivation/coating to prevent nanoparticle agglomeration and the use of surfactants. Also, adequately controlling the nanoparticle geometry simplifies the necessary post-processing conditions (lower curing temperature) and assists to achieve the desired printing resolution (agglomeration can dramatically reduce ink printability by clogging and branch-out ink flow pathways).

Ink printability is a function of the physical and chemical properties of the four components (functional element, binder, solvent, and additive) in the ink formulation. Rheological properties and surface energy are used to determine an ink's printability for a selected printing process.[50] Viscosity, a commonly reported property of rheology, is the deformation resistance of an ink for given environmental conditions (e.g., shearing force and shearing speed, interaction of surface and surface geometry, and temperature), and thus, each printing technology requires a different ink rheology. Surface tension, a measurement of surface energy, defines the interaction between the ink and a substrate. It is a critical parameter and is a measure of ink wetting to a selected substrate, resulting in its final deposition geometry or dot-gain.

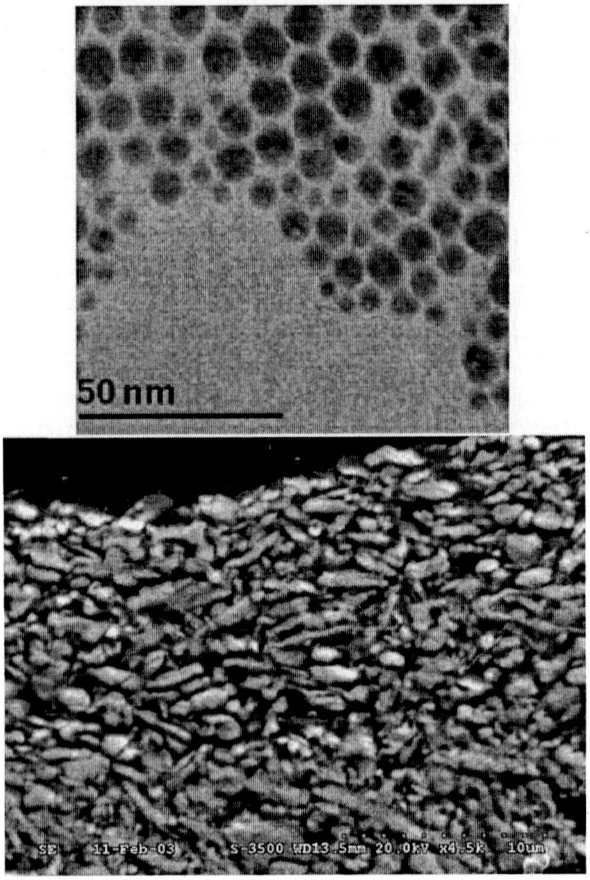

Figure 12.7. Top—Micrograph of a suspension containing nanosized silver particles. Bottom—Micrograph of a micron-sized silver composite ink.

In addition to the ink electrical property testing, these inks are subjected to shelf/pot life storage testing, in-flight processing, impact/post-impact processing, and in-process characterization to ensure high-volume manufacturing compatibility. Intrinsic fluid properties, such as viscosity, surface tension, and density, have a pronounced effect on the printing process window and general printing parameter settings. On-press fluid property modification during printing is common practice and in general is based on the printing press age and preventive maintenance schedule, operator experience, printing quality requirements, and factory/facility environmental conditions (e.g., temperature and humidity). Solution stability/shelf-life is directly attributed to ink printability and, therefore, printing quality and, in the case of printed electronics, it will affect electronic device performance and overall functionality. A short shelf-life of the ink will affect its rheological properties and will require frequent operational parameter changes to maintain optimal printing quality and/or printability; it also leads to a variation in electronic properties resulting from particle settling and functional material degradation.

An additional constraint placed on ink formulation is ink deposition and subsequent drying/curing conditions on flexible substrates. The deposition of an ink on a given substrate is a function of substrate wetting (surface energy compatibility between substrate and ink) and deposition parameters. Wetting of the substrate is a combination of mechanical and chemical phenomena, and it can be quantified by surface tension and/or wetting angle for a specific ink and substrate combination. The effects of wetting directly influence printing quality, film formation, edge definition, and dot gain/loss (dots are printed larger/smaller than intended). After ink deposition, the curing/sintering processing, as well as the postprocessing conditions, are critical for each selected printing process to ensure optimal printed device functionality. In general, ink curing conditions are designed to match the selected printing technology and press capabilities. However, for printed electronics, electrically functional inks are still in development and, therefore, the ink processing parameters are being optimized. Most functional inks require longer curing conditions to ensure optimal ink electrical functionality and to avoid ink-to-ink contamination during subsequent layer-on-layer printing. Therefore, appreciating the requirements of the postprocessing conditions is critical to develop and obtain optimally functioning printed electronics devices and circuits.

12.6 PRINTING TECHNOLOGIES

In recent years, the term "printed electronics" has been used to refer to a variety of deposition technologies. In many instances, printed electronic structures can be fabricated using vacuum and nonvacuum technologies (Table 12.1). In this chapter, printed electronics will be used in reference solely to deposition technologies used in traditional graphic arts printing, an example of which is given in Figure 12.8. These printing technologies have the advantage of depositing electronically functional inks in volumes significantly higher than what is achievable with more traditional microelectronics processes (Figs. 12.9 and 12.10).

12.6.1 Contact Printing

The most commonly used graphic arts contact printing technologies for high-volume manufacturing are screen, flexo (i.e., flexographic printing), litho (i.e., lithographic printing), and gravure.[49] Each technology has its advantages, which define its marketplace in the printing industry. Table 12.2 provides a general review of these printing technologies and the required material rheological properties. Screen printing, a technique that creates a sharp-edged image using a stencil (Fig. 12.11), is commonly used for thick film printing, where the thickness of the as-printed material is greater than 25 μm. It has

TABLE 12.1. Examples of Fabrication Technologies for Semiconductors

Thin-Film Deposition	Physical Vapor Deposition (sputtering, pulsed laser) Chemical Vapor Deposition Electrochemical Deposition Molecular Beam Epitaxy Atomic Layer Deposition Thermal Oxidation Spin Coating
Pattern Transfer	Photolithography Nanolithography Soft Lithography Liquid Imaging
Implantation	Ion Implantation Diffusion Furnace
Removal	Reactive Ion Etch Dry/Wet Etch Plasma Ashing Chemical Mechanical Planarization
Graphic Arts Printing	Gravure Flexography Screen Jetting Embossing

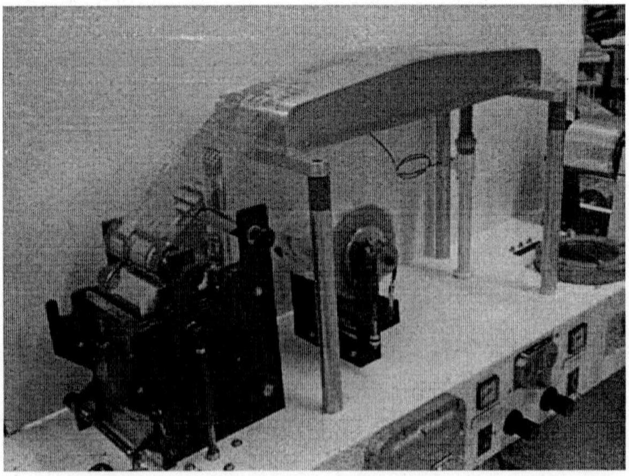

Figure 12.8. A photograph of a traditional roto-gravure label press that has a web width of 4 in. The press consists of a single gravure head that can be used to apply conductive inks, such as nanoparticle silver suspensions, onto surface-treated polymer substrates to generate patterned interconnects and electrodes.

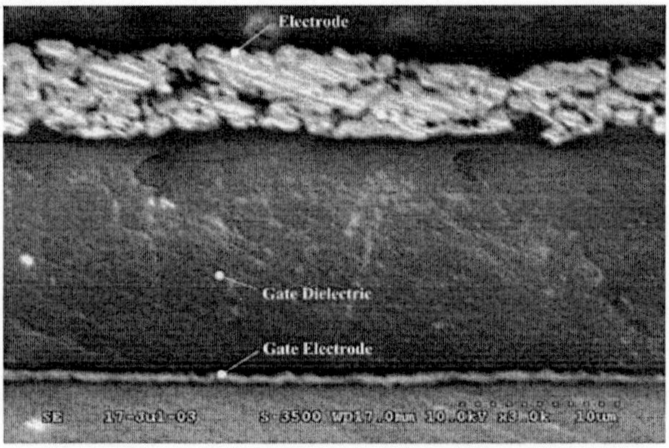

Figure 12.9. Micrographs of a cross section of a three-layer construction (source/dielectric/gate) for a transistor fabricated by graphic arts printing technologies.

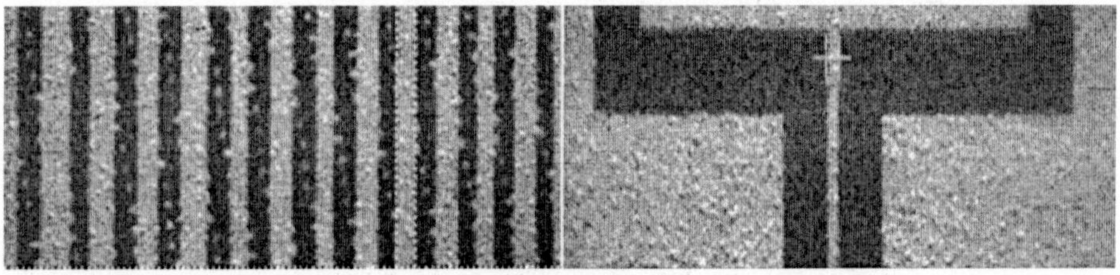

Figure 12.10. Micrographs of devices fabricated using gravure printing technology. Left—shows the interdigitated transistor gate fabricated by an ink composed of nanoscale metallic particles. Right—Channel fabricated when using a silver-filled adhesive to print the transistor source and drain.

TABLE 12.2. Common Material Rheological Properties and Resulting Film Thickness for Printing Technologies

	Viscosity (cPs)	Thickness (μm)	Resolution (μm)	Application
Screen	10K–100K	25	50–100	Membrane switch, label printing
Flexo	1K–5K	5	40	High throughput, label printing
Offset Litho	2K–10K	3–10	25	High resolution, high throughput, newspaper/magazine printing
Gravure	600–1K	3	15	High resolution, high throughput, low coloring graphic arts

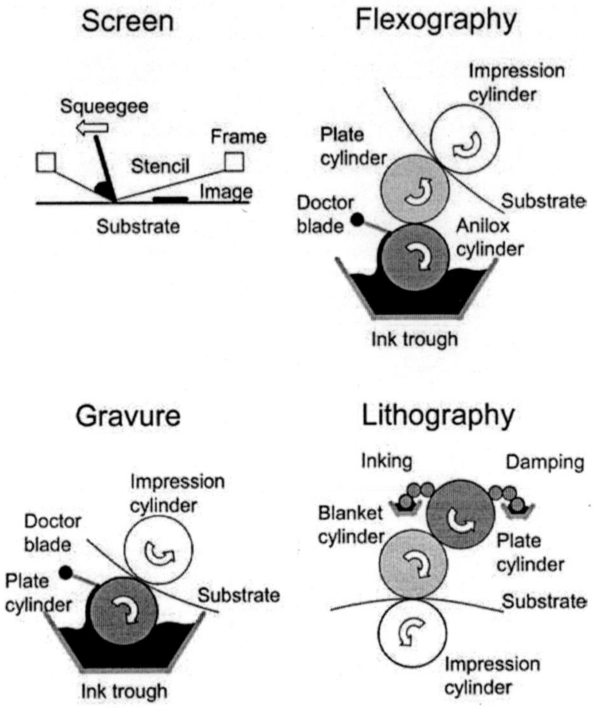

Figure 12.11. A simple representation of the process and components associated with four common graphic arts printing technologies: screen, flexographic, gravure, and lithographic printing.

been used for membrane switch printing where the conductivity and reliability performance of the electrical traces are achieved by printing thicker lines. The printing resolution achieved by screen printing is usually >100 μm and is directly related to the screen mesh and wire dimensions of the stencil. A recent application of screen printing is RFID antenna fabrication, which requires greater film thickness to achieve the necessary electrical conductivity.[50] Flexographic printing is also being considered for RFID antenna printing because of the potential for high throughput. Flexo typically uses a raised image on a printing plate cylinder (typically made of a soft rubber-like material) to transfer images to a substrate, as shown in Fig. 12.11. In general, flexo ink has a much lower viscosity compared with screen-printable inks and the thickness of flexo printed structures is lower, <5 μm for one pass with a typical lateral resolution of 40 μm.[49] Higher conductivity is achieved by using multiple printing passes and by using higher metallic loaded inks. An additional advantage of flexography printing is the availability of printing presses that can operate under extreme curing/sintering conditions as required by the inks—i.e., the extended in-line thermal curing necessary for the electrically conductive inks.

Lithographic printing is a higher resolution (25 μm) printing technology whose resolution is achieved by precisely controlling and balancing the plate surface energy for ink (imaging) and water/oil (masking) wetting (Fig. 12.11).[50] For example, the positive region of the image would be inked with a hydrophobic material, whereas the negative portion of the image would be inked with water. When a compatible ink/water mixture is introduced, the ink will only adhere to the positive region of the image. The as-printed film thickness using lithographic printing is approximately 5 μm and depends on the critical processing parameters. Functional inks for lithographic printing must be further developed to meet the requirements of printability and electrical conductivity for printed electronics. Gravure printing is commonly used for currency and security printing because of the high resolution and high definition it can achieve.[49] Gravure or, more specifically, roto-gravure printing is a direct printing process that consists of an engraved metal image cylinder, a doctor blade, and a rubberized impression roll (Fig. 12.11). The imaged cylinder is partially submersed in a trough of ink, and as it rotates, the ink wets the surface of the carrier and fills the engraved areas of the cylinder. The purpose of the doctor blade is to wipe excess ink from the cylinder surface while the impression roll presses the substrate against the cylinder, allowing the ink that remains in the engraved areas to transfer to the substrate. High-resolution printing is a result of the precision image definition on the chromium-coated gravure cylinder (hard surface for little distortion) and thin as-printed film thickness (<3 μm), which reduces ink spreading (i.e., the dot gain phenomenon).

12.6.2 Noncontact Printing—Ink Jet

In an effort to decrease the cost of fabricating large-area, thin-film electronic devices, electronics manufacturers have been developing technologies capable of flexible manufacturing, reduced changeover time, and improved product yield and throughput. This growing attention is propelling electronic system tooling vendors toward rapid prototyping fabrication platforms that make use of nontraditional electronic manufacturing tools and materials and are, instead, replacing them with graphic-art–based printing platforms that print electrically functional inks. These technologies offer the potential for low-cost, small-batch manufacturing of large-area, thin-film electronics, such as display drivers and RFIDs.[49]

Over the years, ink jet printing has been used to dispense a variety of materials, other than traditional pigmented inks used in photo and text imaging, in an effort to improve existing electronics manufacturing techniques. In fact, ink jet systems have been successful in fabricating electroluminescent displays using doped polymers for OLEDs.[51–54] Even commercial ink jet manufacturers, such as, Hewlett-Packard and Seiko Epson, have started investigating ink jet as a means to print organic polymers to manufacture flexible, low-cost OLED displays.[55] Meanwhile, others have realized the benefits of using ink jet to enhance the rapid prototype capabilities of traditional electronics fabrication techniques by dispensing photoresist materials to fabricate micron-sized circuit electrode structures.[56,57]

One of the basic building blocks of an electronic device is the conductor or electrode, whose primary function is to provide a path for electrons to flow. Traditionally these structures are formed using subtractive, chemical etching techniques. The process involves selective masking and chemical etching to create regions of metallization on nonconducting substrates.[58] To create conductive patterns on electrically insulating surfaces, single- or double-sided, metal laminated substrates are masked to prevent areas of wanted metallization from being chemically etched away. The masked substrates are subsequently placed into a chemical bath that removes unwanted metal from the substrate surface. The substrate is then cleaned, and what remains is the desired metal pattern. Such practice not only wastes precious metal material, but also it generates substantial amounts of chemical pollutants. Furthermore, this method of producing metallization does not lend itself to rapid product translation because of the need for machining new masks and lengthy substrate processing steps.

However, for printed circuit electrodes or interconnects to be effective, they must allow electrons to flow freely with the least amount of resistance, in order to minimize power consumption through resistive heating. From an expansion of Ohm's law, it is apparent that the resistance Z is proportional to the resistivity δ_R of the conductor material.[59]

$$\mathrm{Z} = \delta_R \frac{L}{A_c} \tag{1}$$

where L and A_c represent the length and cross-sectional area of the deposited material film, respectively. Since thin films typically have very large length L to cross-sectional area A_c ratios, it is necessary that the resistivity of the material used to print the metallization layers have significantly low value to minimize power losses.

Researchers have already been using ink jet systems to print several types of novel materials, some of which have been used to fabricate circuit electrodes and interconnects. The four main types of materials used are metallo-organics, conductive polymers, molten metals, and metallic nanoparticle suspensions.

12.6.3 Functional Inks for Ink Jet

12.6.3.1 Metallo-Organics. One of the first reported attempts to create patterned metallization structures using ink-jet printing was demonstrated by the NASA Jet Propulsion Laboratory in the mid-to-late 1980s, where a metallo-organic, low-firing silver precursor ink solution was used to print metallization for solar panels.[60,61] At the time, Purdue's Turner Laboratory pioneered the use of low-firing, metallo-organic inks and developed three formulations consisting of copper 2-ethylhexoate, silver neodecanoate, and gold amine 2-ethylhexoate.[62–69] Several years later, similar attempts to make metallization layers using gold and copper precursor solutions via ink-jet printers were also pursued by other researchers using very similar and, in some cases, identical ink formulations.[70–73]

However, using such metallo-organic solutions typically yields prints that are susceptible to discontinuity from crystallization, as depicted in Fig. 12.12, if the thermal annealing processes are not stringently controlled. One of the main difficulties with metallo-organic inks is impurities in the solution, which can increase film resistivity by several orders of magnitude.[74] In addition, precursor solutions also have a relatively short shelf-life because of the formation of precipitates during storage, as shown in Fig. 12.13.

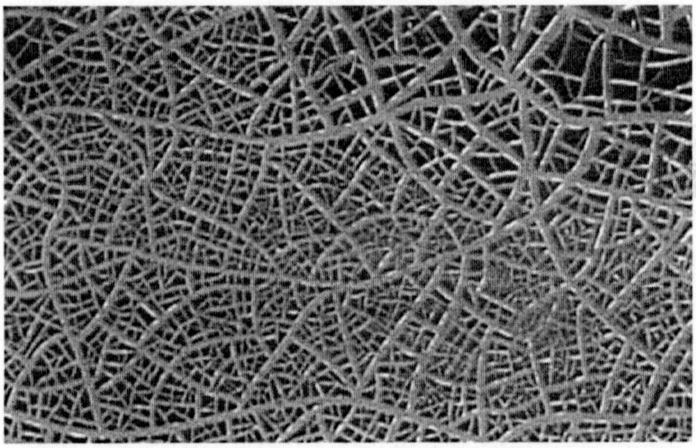

Figure 12.12 Image of an organo-metallic palladium precursor solution that was deposited by the author on a flexible polyimide substrate via spin coating and thermally processed at 200 °C for 30 seconds.

Figure 12.13. Image depicting the formation of precipitates, seen as small dark spots, in a palladium precursor solution after 1 day of storage. The image shown was taken of a drop of precursor solution that was dispensed onto a microscope slide and covered with a glass cover slide.

12.6.3.2 Conducting Polymers. The area of conductive polymer inks has been the focus of much interest in recent years; yet it seems to be the most scarcely reported area with regard to its direct use with ink-jet printing. It is also an area that can cause confusion, since most of the formulated polymer inks are versatile and can take on multiple roles, such as performing as a light-emitting polymer for displays and/or acting as a semiconducting polymer for thin-film, transistor-based electronics.[49] A recent survey of the use of ink-jet technology to print nontraditional materials cited one specific study, which involved patterning polymer solutions onto a rigid substrate and then doping the films into a conducting form with iodine vapor.[75–78]

12.6.3.3 Molten Metals. Other researchers resorted to printing molten metals using specially made ink jet printheads, which can temporarily withstand the relatively high melting temperatures of the metals. One such approach involved the printing of PbSn solder.[79–84] Most of the structures printed were specifically targeted for fabricating solder onto wafer boards or etched printed wiring boards to provide vertical interconnects for surface mount components. One of the problems associated with this technique is that the performance of the printhead would deteriorate over time because of prolonged exposure to the ~220 °C melting temperature required to keep the solder in liquid form. Another problem involved the oxidation of solder, which would increase the surface tension of the solder and essentially clog the printhead and cease solder jetting.[85] Typically, when this occurred, solder jetting could not be reinitiated and the printhead could not be salvaged and required replacement. Finally, the bumpy surface profiles from the solidification of the solder and large height-to-width aspect of the deposits are not well suited for the fabrication of metallization electrodes for thin-film electronics.

Another process that was developed to print metals used a molten aluminum alloy 2024 and ink-jet technology to form three-dimensional mechanical structures.[86,87] The approximate melting temperature of 2024 aluminum is between 510 °C and 638 °C, more than double that of eutectic solder.[88] Although this method was not specifically targeted for printing thin-film metallization layers, it showed that ink-jet technology could be used to dispense materials at extreme temperatures. However, the process suffers from the same complications as that of printed solder, in that an oxygen-free environment was required to prevent surface oxidation to achieve success in printing. In addition, the high melting point of aluminum makes use of currently implemented electronic substrates impractical.

12.6.3.4 Nanoscale Metallics. Many of the metallization materials previously discussed, such as gold, can be developed into fine particles through either chemical reduction from a metal salt or from vaporization of the bulk metal.[89] In fact, metals, such as gold, silver, palladium, and iron, can be harvested in elemental form from vegetation and grains that have been planted in metal-enhanced soils.[90] Researchers at the University of Texas in El Paso have germinated and grown alfalfa plants in gold-rich agar media and upon analysis of the plant tissue found gold nanoparticle clusters with particle sizes on the order of 5 nm.[91,92] One of the challenges, however, is dispersing these particles into jettable, stable inks containing high metal concentrations.[93]

Ulvac, located in Tokyo, Japan, has been able to develop dispersed, ultra-fine particle suspensions of gold and silver. As a result, researchers began using these inks, primarily targeting the gold suspensions.[94] Researchers at the Massachusetts Institute of Technology were successful in printing low concentrations (10 %wt) of gold and silver nanoparticle suspensions on heated, rigid substrates to create microelectromechanical systems (MEMS).[95] This method of printing yielded rough metallization surfaces and is not well suited for high throughput manufacturing, since it requires the substrate to reach a steady-state temperature of ~300 °C for the nanoparticle suspension to conduct electrically. Similarly, researchers at the Swiss Federal Institute of Technology demonstrated a technique that patterned a 30% (by weight) gold nanoparticle suspension, using an ink-jet printhead, onto rigid substrates.[96,97] The deposited nanosuspension was then sintered using single and multiple lasers of spot sizes between 10 and 100 μm at power levels of 50–600 mW. As a result, conductive lines were created. However, the fabricated lines suffered from poor edge definition, as a result of rapid heating of the deposited nanoparticle suspension produced by the laser. In addition, the temperature at which the laser locally heated the nanoparticle suspension was never mentioned and may be not be compatible with conventional, flexible electronic substrates.

12.7 STRUCTURE OF A PRINTED TRANSISTOR

A typical electronic component that is targeted during a printing deposition is the thin-film transistor (TFT), which is composed of four main elements, as shown in Fig. 12.14: (1) the gate, (2) the gate dielectric, (3) the source and drain, and (4) the semiconductor layer. The TFT structure has the advantage of requiring only one type of semiconductor material (for example, requiring only p-type material—this is in contrast to other structures, such as bipolar and most JFET devices, which generally require both n- and p-type semiconductor material). Thin-film transistors also tend to be more tolerant of impurities within the semiconductor material, which may be expected from the processing and handling methods used for printed electronics manufacture. It is noted that large amounts of impurities can have a detrimental effect on device performance. Through circuit and device design modifications, however, some mitigation of these effects can be achieved.

Each printed TFT element is composed of nanoscale conducting, dielectric, or semiconducting particles. The electrical performance of the printed transistors and printed integrated circuits is dependent on the uniformity of the

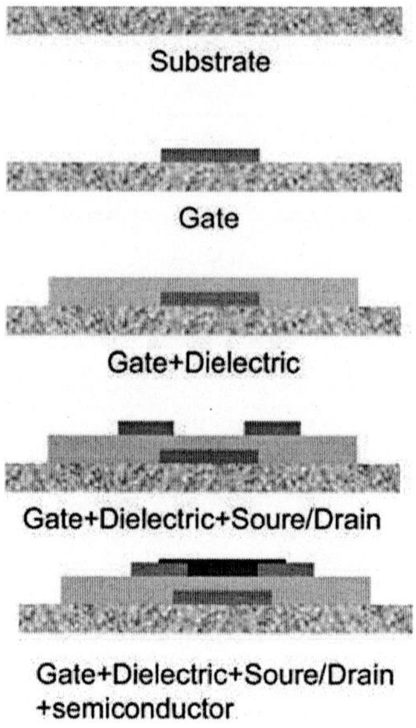

Figure 12.14. Structure and fabrication process flow of a printed bottom-gated thin-film transistor (TFT).

fundamental materials properties of the nanoscale particles. Figure 12.14 presents the flow process for fabricating a printed device. Since the processes are additive, each subsequent process effects the final structure. As an example, if the nanoparticle size distribution of the conductive nanoscale particles in the gate is not uniform and greater than the desired final gate dielectric thickness, the device will not function properly because of the shorting between the gate and the source/drain. In addition, if the particle size distribution is not controlled in the conductive ink, the channel length (L) between the source and the drain can only be designed based on the largest particles (D_{max}) in the ink, $L > 10D_{max}$.

12.8 PATTERNING TECHNIQUES FOR SOLUTION DEPOSITION: TECHNOLOGY DIFFUSION

Aside from market conditions, supply, and demand, four elements (Fig. 12.15) strongly impact the successful commercialization and diffusion of a printed (or other) technology platform into the market: (1) standards, (2) awareness, (3) supply chain, and (4) quality control/assurance. Each of these elements is typically addressed prior to commercialization, although these elements are not a prerequisite for success.

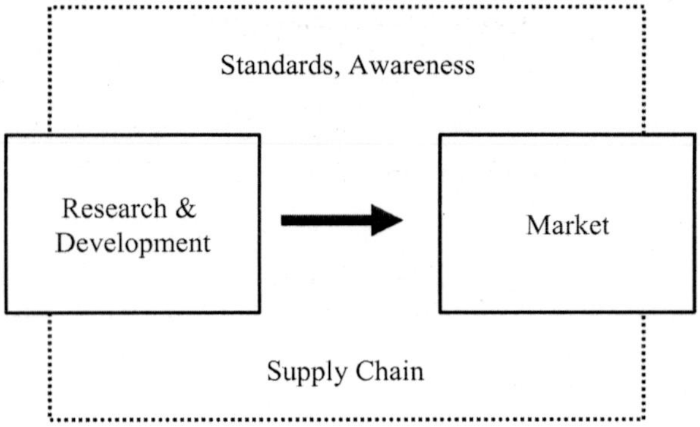

Figure 12.15. Technology diffusion elements for success.

12.8.1 Standards

The development of standards is important to establish a common framework by which scientists and engineers are able to maintain technical dialogue prior to and after commercialization. Standards can be categorized into two main classes: (1) anticipatory and (2) enabling. An example of an anticipatory standards portfolio is IEEE 802.*X*, which was established over 20 years ago by a group of engineers that teamed together to create a portfolio of standards based on historical perspectives and trends in communications technology. The second category is populated by standards that were established to bring structure and discipline to an important technical field that was inundated with a multiplicity of technology platforms, thereby creating a fragmented environment. In this case, a standard was required to reach consensus on a mechanism by which the different technology solutions could thrive without providing a significant competitive advantage to any one solution. It is important that the community adopts standards early during the development period to ensure the commercial acceptance of the patterning techniques for solution deposition of inorganic materials. More specifically, a family of standards encompassing, for example, materials, device, circuitry, manufacturing and processing, and environmental testing, must be established.

12.8.2 Awareness

The utility and value of the ability to solution deposit inorganic materials must be promoted and discussed in a variety of forums, both technical and nontechnical. In general, the technical forums are required to share observations during experimentation as well as product development. Forums such as workshops, symposia, ad hoc consortia, and study groups are necessary to provide an environment for engineers and scientists to disclose and discuss problems and solutions. The nontechnical forums are important to excite the mass market with compelling visions for a simplified, exciting, enriched future enabled by the nanotechnology platforms.

12.8.3 Roadmapping for Supply Chain Development

A well-established supply chain composed of companies that have demonstrated an ability to supply nanoscale or other printing-based materials, devices, manufacturing platforms, components, modules, and testing platforms is critical to maintain and accelerate the diffusion process. As the market acceptance rate increases, the incumbents will be faced with increasing competition as the early adoption stage is exited, the mass market adoption rate increases, and the barrier to implementation is reduced. The supply chain can have raw materials suppliers, converting and processing services suppliers, and systems integrating services suppliers, to name a few general classes of supply chain members. It is important that the different supply chain members work in unison to support the timely delivery of the product.

12.8.4 Quality Control/Assurance

Mass market acceptance is typically dependent on a product providing a level of quality. As the technology/product platform "Crosses the Chasm" from early adoption, it is necessary that the product deliver on its perceived value. Of critical importance to achieving acceptance by the mass market is to establish rigorous quality control/assurance methodologies that can be easily deployed to various manufacturing sites. The advent of a revolutionary technology platform, solution-processable inorganic materials, and the preferred patterning techniques of these materials, must be quickly followed by the establishment of standard operating procedures that can be maintained and updated as necessary to provide the greatest level of quality control. At present, there are some efforts at both government labs and industrial labs to codify these procedures.

12.9 CONCLUSIONS

The trend in the technical community toward developing patterning techniques for solution-deposited inorganic materials is growing. Through the development of solution-processable materials and the leveraging of graphic arts printing processes, a sustainable platform for printed electronics manufacture will be realized. Although organic materials have been used historically for printed electronics, it is likely that solution-processable materials based on inorganic materials will be necessary for the realization of very high-performance electronics, especially electronics suitable for printed wireless applications. Along with these technical advances, a sustainable value chain also requires well-defined standards for uniform reporting of data, electrical and materials characterization, and health/safety reporting. Together these efforts will initiate a renaissance in consumer electronics, bringing to reality a wide and pervasive range of new, exciting products, such as flexible displays, active labels, novel and interactive signage, and likely other applications not yet envisioned.

REFERENCES

1. Reuss, R.; Hopper, D.; Park, J. 2005. Macroelectronics. *MRS Bull.* 31:447–454.
2. InterTECH October 11–12th Meeting. 2004. Durham, England. Accessed: http://www.intertech2004.com/.
3. Gabriel, J. P. 2004. Carbon nanotube field effect transistors and sensors based on nanotube networks. *Mater. Res. Soc. Proc.*, Session HH 14.5.
4. More information can be obtained at http://www.polyapply.org/.
5. Dimitrakopoulos, C.; Malenfant, P. 2002. Organic thin film transistors for large area electronics. *Adv. Mater.* 14:99–117.
6. Paloheimo, J.; Kuivalainen, P.; Stubb, H.; Vuorimaa, E.; Yli-Lahti, P. 1990. Molecular field-effect transistors using conducting polymer Langmuir–Blodgett films. *Appl. Phys. Lett.* 56:1157–1159.
7. Yoshino, K.; Takahashi, H.; Muro, K.; Ohmori, Y.; Sugimoto, R. 1991. Optically controlled characteristics of Schottky gated poly(3-alkylthiophene) field effect transistor. *J. Appl. Phys.* 70:5035–5039.
8. Ohmori, Y.; Takahashi, H.; Muro, K.; Uchida, M.; Kawai, T.; Yoshino, K. 1991. Fabrication and characteristics of Schottky gated poly(3-alkylthiophene) field effect transistors. *Jpn. J. Appl. Phys.* 30:L610–L611.
9. Assadi, A.; Svensson, C.; Willander, M.; Inganäs, O. 1988. Field-effect mobility of poly(3-hexylthiophene). *Appl. Phys. Lett.* 53:195–197.
10. Fuchigami, H.; Tsumara, A.; Kezuka, H. 1993. Polythienylenevinylene thin-film transistor with high carrier mobility. *Appl. Phys. Lett.* 63:1372–1374.
11. Bao, Z.; Dodabalapur, A.; Lovinger, A. 1996. Soluble and processable regioregular poly (3-hexylthiophene) for thin film field-effect transistor applications with high mobility. *Appl. Phys. Lett.* 69:4108–4110.
12. Dimitrakopoulos, C. D.; Afzali-Ardakani, A.; Furman, B.; Kymissis, J.; Purushothaman, S. 1997. Trans-trans-2,5-bis-[2-{5-(2,2′-bithienyl)}ethenyl] thiophene: Synthesis, characterization, thin film deposition and fabrication of organic field-effect transistors. *Synth. Met.* 89:193–197.
13. Garnier, F.; Hajlaoui, R.; El Kassmi, A.; Horowitz, G.; Laigre, L.; Porzio, W.; Armanini, M.; Provasoli, F. 1998. Dihexylquaterthiophene, a two-dimensional liquid crystal-like organic semiconductor with high transport properties. *Chem. Mater.* 10:3334–3339.
14. Kobashi, M.; Takeuchi, H. 1998. Inhomogeneity of spin-coated and cast non-regioregular poly(3-hexylthiophene) films. Structures and electrical and photophysical properties. *Macromolecules* 31:7273–7278.

15. Wang, J.; Lin, Y.; Qian, W.; Jackson, T. 1998. Thin-film transistors with high carrier mobility based on solution-cast poly(3-alkyl thiophene). *1998 Mater. Res. Soc. Fall Meeting: Symposium R*. pp. 372.
16. Narasimhan, M.; Hagler, M.; Cammarata, V.; Thakur, M. 1998. Junction devices based on sulfonated polyaniline. *Appl. Phys. Lett.* 72:1063–1065.
17. Katz, H.; Laquindanum, J. 1999. Thin film transistor and organic semiconductor material thereof. *U. S. Patent 5,936,259.*
18. Kagan, C. R.; Mitzi, D. B.; Dimitrakopoulos, C. D. 1999. Organic-inorganic hybrid materials as semiconducting channels in thin-film field-effect transistors. *Science* 286:945–947.
19. Geens, W.; Tsamokuras, D.; Poortmans, J.; Hadziioannou, G. 2001. Field-effect mobilities in spin-cast and vacuum-deposited PPV-type pentamers. *Synth. Met.* 122:191–194.
20. Bao, Z.; Feng, Y.; Dodabalapur, A.; Raju, V.; Lovinger, A. 1997. High-performance plastic transistors fabricated by printing techniques. *Chem. Mater.* 9:1299–1301.
21. Bao, Z.; Lovinger, A.; Cherniavskaya, O. 2000. Material issues for construction of organic and polymeric driving circuits for display and electronic applications. *Macromolecular Symposia* 154:199–207.
22. Wang, J.; Gundlach, D. 1999. High mobility polymer thin film transistors based on copolymers of thiophene and 3-hexyl thiophene. *41st Electronic Materials Conference Digest*. pp. 16.
23. Bao, Z.; Lovinger, A. 1999. Soluble regioregular polythiophene derivatives as semiconducting materials for thin film field-effect transistors. *Chem. Mater.* 11:2607–2612.
24. Rogers, J.; Bao, Z.; Makhija, A.; Braun, P. 1999. Printing process suitable for reel-to-reel production of high-performance organic transistors and circuits. *Adv. Mater.* 11:741–745.
25. Hamilton, M.; Martin, S.; Kanicki, J. 2004. Thin-film organic polymer phototransistors. *IEEE Trans. Electron Dev.* 51:855–877.
26. Deen, M.; Kazemeini, M.; Haddara, Y.; Yu, J.; Vamvounis, G.; Holdcroft, S.; Woods, W. 2004. Electrical characterization of polymer-based FETs fabricated by spin-coating poly(3-alkylthiophene)s. *IEEE Trans. Electronic Dev.* 51:1892–1901.
27. Cui, T.; Liang, G.; Varahramyan, K. 2003. An organic poly(3,4-ethylenedioxythiophene) field-effect transistor fabricated by spin coating and reactive ion etching. *IEEE Trans. Electron Dev.* 50:1419–1422.
28. Meixner, R.; Yildirim, F.; Schliewe, R.; Goebel, H.; Bauhofer, W.; Krautschneider, W. 2005. Low-temperature process for manufacturing all polymer thin-film transistors. *Polytronic 2005—5th International Conference on Polymers and Adhesives in Microelectronics and Photonics*. pp. 195–197.
29. Deen, M.; Kazemeini, M. 2005. Photosensitive polymer thin-film FETs based on poly(3-octylthiophene). *Proc. IEEE* 93:1312–1320.
30. Subramanian, V.; Frechet, J.; Chang, P.; Huang, D.; Lee, J.; Molesa, S.; Murphy, A.; Redinger, D.; Volkman, S. 2005. Progress toward development of all-printed RFID tags: Materials, processes, and devices. *Proc. IEEE* 93:1330–1338.

31. Ong, B.; Wu, Y.; Liu, P. 2005. Design of high-performance regioregular polythiophenes for organic thin-film transistors. *Proc. IEEE* 93:1412–1419.

32. Kuo, C.; Payne, M.; Anthony, J.; Jackson, T. 2003. Solution processed OTFTs with 1 cm^2/V-s mobility. *62nd DRC Device Research Conference. Conference Digest.* 2:8–9.

33. Wagner, V.; Wobkenberg, P.; Hoppe, A.; Seekamp, J. 2006. MHz organic field effect transistors. *Proceedings of OEC-06: Peer Reviewed Papers*. pp. 7.

34. Kempa, H.; Fugmann, U.; Hahn, U.; Schmidt, G.; Meier, B.; Bartzsch, M.; Fisher, T.; Stanerl, M.; Reuter, K.; Preibler, K.; Huebler, A. 2006. On the applicability of different mass printing methods for the deposition of organic functional materials. *Proceedings of OEC-06: Peer Reviewed Papers*. pp. 39.

35. Ohe, T.; Nomoto, K.; Wada, M.; Kasahara, J. 2006. Solution-processed naphthacene-derivative organic TFT. *Proceedings of OEC-06: Peer Reviewed Papers*. pp. 27.

36. Bock, K. 2005. Organic electronics-Towards a cost-efficient heterointegration platform for multi-functional systems. *Polymers and Adhesives in Microelectronics and Photonics, Polytronic 2005*. pp. 228–232.

37. Cui, T.; Liang, G.; Varahramyan, K. 2003. An organic poly(3,4-ethylenedioxythiophene) field-effect transistor fabricated by spin coating and reactive ion etching. *IEEE Trans. Electron Dev.* 50:1419–1422.

38. Manuelli, A.; Knobloch, A.; Bernds, A.; Clemens, W. 2002. Applicability of coating techniques for the production of organic field effect transistors. *2nd International IEEE Conference on Polymers and Adhesives in Microelectronics and Photonics, POLYTRONIC 2002*. pp. 201–204.

39. Belliveau, M.; Lester, S. 2004. PVC: Bad news comes in threes. *Center of Health, Environment, and Justice Report*. December 7.

40. Warheit, D. 2004. Nanoparticles health impacts?. *Materials Today* 7:32–35.

41. Shimoda, T.; Matsuki, Y.; Furusawa, M.; Aoki, T.; Yudasaka, I.; Tanaka, H.; Iwasawa, H.; Wang, D.; Miyasaka, M.; Takeuchi, Y. 2006. Solution-processed silicon films and transistors. *Nature* 440:783–786.

42. For example, the thin-film amorphous silicon manufacturing technology in operation at Universal Solar Ovonic (Auburn Hills, MI; Accessed: http://www.ovonic.com/).

43. Norris, B.; Anderson, J.; Wager, J.; Keszler, D. 2003. Spin-coated zinc oxide transparent transistors. *J. Phys. D: Appl. Phys.* 36:L105–L107.

44. Volkman, S.; Mattis, B.; Molesa, S.; Lee, J.; Vombrock, A.; Bakhishev, T.; Subramanian, V. 2004. A novel transparent air-stable printable n-type semiconductor technology using ZnO nanoparticles. *Proceedings from IEDM'04 (San Francisco, CA)*. pp. 32.1.1–32.1.4.

45. Petrat, F. 2006. Printed electronics—Opportunities for the chemical industry. *Presented at the Organic Electronics Conference and Exhibition 2006, Frankfurt-am-Main, Germany; Sept. 27, 2006.*

46. Wu, H.; Liu, J.; Shi, J.; Spencer, M. 2006. Thin film transistor using GaN nanoparticles. *Presented at the 2006 Lester Eastman Conference on High-Performance Devices, Ithaca, NY; Aug. 3, 2006.*

47. Webb, W. 2007. Printed electronics: Ink on the brink. *Electronics Weekly*. September 24. Accessed: http://www.electronicsweekly.com/Articles/2007/09/24/42254/printed+electronics+ink+on+the+brink.htm.
48. Knickerbocker, J.; Andry, P.; Buchwalter, L.; Deutsch, A.; Horton, R.; Jenkins, K.; Kwark, Y.; McVicker, G.; Patel, C.; Polastre, R.; Schuster, C.; Sharma, A.; Sri-Jayantha, S.; Surovic, C.; Tsang, C.; Webb, B.; Wright, S.; McKnight, S.; Sprogis, E.; Dang, B. 2005. Development of next-generation system-on-package (SOP) technology based on silicon carriers with fine-pitch chip interconnection. *IBM J. Res. and Dev.* 49:725–753.
49. Gamota, D.; Brazis, P.; Kalyanasundaram, K.; Zhang, J. 2004. *Printed Organic and Molecular Electronics*. Kluwer Academic Publishers, Boston, MA.
50. Swedberg, C. 2005. Precisia launches flexwing tag antenna. *RFID Journal*. September 9. Accessed: http://www.rfidjournal.com/article/articleview/1859/1/1/.
51. Shibusawa, M.; Kobayashi, M.; Hanari, J.; Sunohara, K.; Ibaraki, N. 2003. A 17-inch WXGA full-color OLED display using the polymer ink jet technology. *IEICE Trans. Electron.*, E86-C(11):2269–2274.
52. Shimoda, T.; Kanbo, S.; Kobayashi, H.; Seki, S.; Kiguchi, H.; Yudasaka, I.; Kimura, M.; Miyashita, S.; Friend, R.; Burroughes, J.; Towns, C. 1999. Multicolor pixel patterning of light-emitting polymers by ink-jet printing. *SID Symp. Dig. Tech. Papers* 30:376–379.
53. Bharathan, J.; Yang, Y. 1998. Polymer electroluminescent devices processed by inkjet printing: I. Polymer light-emitting logo. *Appl. Phys. Lett.* 72:2660–2662.
54. Hebner, T.; Wu, C.; Marcy, D.; Lu, M.; Sturm, J. 1998. Ink-jet printing of doped polymers for organic light emitting devices. *Appl. Phys. Lett.* 72:519–521.
55. Rogers, J. 2002. Move over, silicon. *The Economist.* Dec. 12:20–21.
56. Percin, G.; Levin, L.; Khuri-Yakub, B. 1997. Piezoelectrically actuated droplet ejector. *Rev. Sci. Instr.* 68:4561–4563.
57. Percin, G.; Lundgren, T.; Khuri-Yakub, B. 1998. Controlled ink-jet printing and deposition of organic polymers and solid particles. *Appl. Phys. Lett.* 73:2375–2377.
58. Jawitz, M. 1997. *Printed Circuit Board Materials Handbook*. In Electronic Packaging and Interconnection Series. McGraw-Hill, New York.
59. Serway, R. 1992. *Physics for Scientists & Engineers*. 3rd, updated edition. Saunders Golden Sunburst Series, Vol. 1169. Saunders College Pub., Philadelphia, PA.
60. Vest, G. 1983. Metallization with generic metallo-organic inks. *JPL Proceedings of the Flate-Plate Solar Array Research: Forum on Photovoltaic Metallization Systems* (NASA Center for AeroSpace Information; Pasadena, CA). (Document ID. 19831115).
61. Vest, G.; Vest, R. 1984. Metallo-organic decomposition (MOD) silver metallization for photovoltaics. *JPL Proceedings of the 23rd Project Integration Meeting* (NASA Center for AeroSpace Information; Pasadena, CA). (Document ID. 19850006969).
62. Vest, R. 1986. Ink jet printing of silver metallization for photovoltaics. NASA Center for AeroSpace Information, Pasadena, CA. (Document ID. 19860067973).

63. Vest, R. 1985. Ink jet printing of silver metallization for photovoltaics. NASA Center for AeroSpace Information, Pasadena, CA. (Document ID. 19860019057).

64. Vest, R. 1985. Ink jet printing of silver metallization for photovoltaics. NASA Center for AeroSpace Information, Pasadena, CA. (Document ID. 19850025256).

65. Vest, R.; Vest, G. 1985. MOD silver metallization: screen printing and ink jet printing. *JPL Proceedings of the 25th Project Integration Meeting*, NASA Center for AeroSpace Information; Pasadena, CA. (Document ID. 19850601).

66. Vest, R.; Vest, G. 1986. Ink jet printing of silver metallization for photovoltaics. *JPL Proceedings of the 26th Project Integration Meeting*. NASA Center for AeroSpace Information, Pasadena, CA. (Document ID. 19870006988).

67. Vest, R.; Singaram, S. 1989. Silver ink for jet printing. *NASA Tech Brief* 13(8): i,1–2.

68. Vest, R. 1985. Ink jet printing of silver metallization for photovoltaics. NASA Center for AeroSpace Information, Pasadena, CA. (Document ID. 19850022407).

69. Vest, R. 1985. Ink jet printing of silver metallization for photovoltaics. NASA Center for AeroSpace Information, Pasadena, CA. (Document ID. 19850017632).

70. Sharma, S.; Bhasin, K.; Narang, S. 1998. Direct deposition of a gold layer. *U.S. Patent 5,846,615*.

71. Cuk, T; Troian, S. M.; Hong, C. M.; Wagner, S. 2000. Using convective flow splitting for the direct printing of fine copper lines. *Appl. Phys. Lett.* 77:2063–2065.

72. Hong, C.; Wagner, S. 2000. Inkjet printed copper source/drain metallization for amorphous silicon thin-film transistors. *IEEE Electron Dev. Lett.* 21:384–386.

73. Hong, C. 2000. Jet printed copper metallization. In *Electrical Engineering*. Princeton University, Princeton, NJ.

74. Calvert, P. 2001. Inkjet printing for materials and devices. *Chem. Mater.* 13:3299–3305.

75. Calvert, P.; Yoshioka, Y.; Jabbour, G. 2002. Inkjet printing of biopolymers, conductors and dielectrics. *Abstracts of Papers of the American Chemical Society* 224:U56–U57.

76. Calvert, P.; Yoshioka, Y.; Jabbour, G. 2002. Inkjet printing of self-assembling polymers. *Abstracts of Papers of the American Chemical Society* 224:U398–U398.

77. Calvert, P.; Ghaemi, M. 2001. Inkjet printing and ionic self-assembly of polymers. *Abstracts of Papers of the American Chemical Society* 221:U372–U372.

78. Liu, Q.; Orme, M. 2001. High precision solder droplet printing technology and the state-of-the-art. *J. Materials Processing Technology* 115:271–283.

79. Baggerman, A.; Schwarzbach, D. 1998. Solder-jetted eutectic PbSn bumps for flip-chip. *IEEE Transactions on Components Packaging and Manufacturing Technology Part B-Adv. Packaging* 21:371–381.

80. Hayes, D.; Wallace, D. 1998. Solder jet printing: Wafer bumping and CSP applications. *Chip Scale Review* 2:75–80.

81. Wallace, D.; Hayes, D. 1998. Solder jet technology update. *International J. Microcircuits Electron. Packaging* 21:1–4.

82. Waldvogel, J.; Diversiev, G.; Poulikakos, D.; Megaridis, C.; Attinger, D.; Xiong, B.; Wallace, D. 1998. Impact and solidification of molten-metal droplets on electronic substrates. *J. Heat Transfer* 120:539.

83. Hayes, D. 1993. Picoliter solder droplet dispersion. *International J. Microcircuits Electron. Packaging* 16:173–179.

84. Wallace, D. 1988. *Automated Electronic Circuit Manufacturing using Ink-Jet Technology*, Vol. 3. Division of Industrial Science and Technological Innovation (ISTI), Directorate for Scientific, Technological, and International Affairs (STIA), National Science, Washington, DC. pp. 69.

85. Howell, E. 2001. Dynamic surface tension measurements of liquid solder using oscillating jets of elliptical cross section. *Mechanical and Industrial Engineering*. University of Illinois at Chicago, Chicago, IL. pp. 75.

86. Orme, M.; Liu, Q.; Smith, R. 2000. Molten aluminum micro-droplet formation and deposition for advanced manufacturing applications. *Aluminum Transactions J.* 3:95–103.

87. Orme, M.; Smith, R. 2000. Enhanced aluminum properties by means of precise droplet deposition. *J. Manufacturing Sci. Engineer-Trans. ASME* 122:484–493.

88. 1979. *Properties and Selection: Nonferrous Alloys and Pure Metals, Metals Handbook*, 9th edition, Vol. 2. ASM International, Metals Park, OH.

89. Whyman, R. 1996. Gold nanoparticles. A renaissance in gold chemistry. *Gold Bulletin* 29:11–15.

90. Spotts, P. 2002. No fairy tale: Researchers spin straw into gold. *The Christian Science Monitor*. August 29.

91. Gardea-Torresdey, J. 2003. Use of XAS and TEM to determine the uptake of gold and silver and nanoparticle formation by living alfalfa plants. *Abstracts of Papers of the American Chemical Society* 225:U837–U837.

92. Gardea-Torresdey, J. L.; Parsons, J. G.; Gomez, E.; Peralta-Videa, J.; Troiani, H. E.; Santiago, P.; Jose Yacaman, M. 2002. Formation and growth of Au nanoparticles inside live alfalfa plants. *Nano Lett.* 2:397–401.

93. Tay, B.; Edirisinghe, M. 2002. Dispersion and stability of silver inks. *J. Mater. Sci.* 37:4653–4661.

94. Szczech, J.; Megaridis, C.; Gamota, D.; Zhang, J. 2002. Fine-line conductor manufacturing using drop-on demand PZT printing technology. *IEEE Trans. Electron. Packaging Manufacturing* 25:26–33.

95. Fuller, S.; Wilhelm, E.; Jacobson, J. 2002. Ink-jet printed nanoparticle microelectromechanical systems. *J. Microelectromechanical Systems* 11:54–60.

96. Bieri, N.; Chung, J.; Haferl, S.; Poulikakos, D.; Grigoropoulos, C. 2003. Microstructuring by printing and laser curing of nanoparticle solutions. *Appl. Phys. Lett.* 82:3529–3531.

97. Chung, J.; Ko, S.; Bieri, N.; Grigoropoulous, C.; Poulikakos, D. 2004. Conductor microstructures by laser curing of printed gold nanoparticle ink. *Appl. Phys. Lett.* 84:801–803.

CHAPTER 13

Transfer Printing Techniques and Inorganic Single-Crystalline Materials for Flexible and Stretchable Electronics

JONG-HYUN AHN, MATTHEW A. MEITL, ALFRED J. BACA, DAHL-YOUNG KHANG, HOON-SIK KIM, and JOHN A. ROGERS
University of Illinois at Urbana-Champaign, Urbana, IL

13.1 INTRODUCTION

Large-area electronics, sometimes known as "macroelectronics," in mechanically flexible formats are attractive for diverse applications such as flexible displays, large-area solar cells, conformable X-ray imagers, and structural and personal health monitors.[1–3] To reduce the cost and to increase the performance of such systems, new or improved electronic materials and fabrication methods are required. The dominant semiconductor for use in rigid macroelectronic systems is amorphous silicon (a-Si), which, for example, forms the active component of transistors in flat panel displays. Deposition of this material onto flexible plastic substrates can be accomplished at relatively low temperatures using vacuum processing techniques. A disadvantage of a-Si is its relatively low mobility, which limits the performance of the resulting transistors and circuits. New laser annealing approaches for *in situ* recrystallization of a-Si can produce large-grained polycrystalline silicon that exhibits mobilities, as determined from measurements in transistor devices, approaching the levels observed for single-crystalline silicon.[4] The tight control over the laser exposures, and the need for thick buffer layers to avoid thermal damage of plastic substrates, make reliable implementation of this approach on large-scale flexible devices very challenging. Solution-processable materials avoid some of these challenges and provide potentially low-cost means for implementing circuits on large-area or unusual substrates.[5–11] The other chapters in this book describe the many possible choices for such materials, ranging from chalcogenides to hybrid organic/inorganics and others, with advantages and disadvantages for each. In some cases, impressive devices and circuits have been demonstrated. Achieving levels of performance comparable, for example, to single crystal bulk or traditional thin-film inorganic semiconductors remains, however, challenging.

Recent work explores, as an alternative to a-Si, laser-annealed a-Si and solution-processable materials, as well as the use of single-crystalline inorganic semiconductor nanomaterials for high-performance flexible macroelectronics (e.g., see Chapters 10 and 11).[12–14] These approaches use structural elements—i.e., micro/nanoscale wires, platelets, ribbons, and so on—that are mechanically flexible by virtue of at least one small dimension (e.g., thickness, diameter, etc.). There are two general techniques for generating the printable elements: one that uses direct growth by chemical vapor deposition and another that forms micro/nanoscale objects from conventional wafers using photolithography and special etching procedures.[12–14] The latter approach is attractive because it relies on well-developed, commodity wafer technology, and it facilitates the use of selected high-temperature processing steps (e.g., formation of ohmic contacts) to be performed on the starting wafer, separate from the plastic substrate. Following these steps, a room-temperature dry printing technique, with excellent registration capability, transfers organized collections of high-quality inorganic semiconductor elements and devices from the starting wafer onto plastic substrates.[13,14] This type of approach, which forms the focus of the current chapter, relates to some of the solution-based stamping and nanocomponent techniques described in earlier chapters (e.g., see Chapters 10–12), has successfully been used to demonstrate bendable device arrays and circuits on thin plastic substrates, and is scalable to large areas.

This chapter briefly reviews the methods for fabricating the semiconductor elements and the rubber stamping techniques (i.e., dry transfer printing) for transferring these elements onto plastic, glass, or elastomeric substrates. The electrical and mechanical properties of typical TFTs and circuits built in this fashion demonstrate some of the capabilities. The chapter content is organized into five main sections. The first summarizes the fabrication of micro/nanostructures of single-crystalline semiconductors from conventional high-quality wafers of these materials through "top-down" approaches, in which photolithographic patterning and etching create the desired structures on the top surface of the wafers. The second section describes a dry printing method to transfer ordered wires/ribbons produced in this manner, as well as other semiconductor nanomaterials such as single-walled carbon nanotubes, onto substrates, including plastic sheets, using elastomeric polydimethylsiloxane

(PDMS) stamps as transfer elements. The third section presents the effective use of thin films of these wafer-based materials on plastic substrates for high-performance transistors. Several examples of devices and their electronic and mechanical characteristics demonstrate the current state of the technology. The fourth section presents strategies for heterogeneous integration of dissimilar classes of semiconductors into single systems, in either two- or three-dimensional (2D or 3D) layouts. The fifth section describes stretchable electronics that use nanoribbons of high-quality, single-crystalline inorganic semiconductor materials engineered into "wavy" geometrical configurations and bonded to PDMS substrates, for electronics that offer not only bendability but also fully reversible stretchability. The last section summarizes the results included in previous sections and concludes with perspectives on the trends for future work related to flexible macroelectronics.

13.2 INORGANIC SINGLE-CRYSTALLINE SEMICONDUCTOR MATERIALS FOR FLEXIBLE ELECTRONICS

Macroelectronics circuits represent a new class of electronics where the sizes of the integrated circuits are much larger compared with those of conventional semiconductor wafer-based microelectronics (see Chapter 1 for details).[1,3] Flexible macroelectronics uses, as basic building blocks, thin-film type transistors formed over large areas on bendable substrates for applications ranging from switching elements in liquid crystal displays to electronics in structural health monitors. Rather than shrinking the size of an individual circuit component as in conventional wafer-scale electronics, macroelectronics focuses on increasing the size of the overall system. This size, then, represents the primary scaling metric.

Macroelectronics requires semiconductor materials that are highly flexible (and sometimes stretchable), lightweight, rugged, low in cost, and that can be implemented in circuits that display both high performance and reliable behavior. Various forms of "thin films" ranging from amorphous and polysilicon to thin films of chalcogenides quantum dots have shown great promise.[4,9,10,15,16] However, as is the case for silicon films of various degrees of crystallinity, the performance of these materials is highly dependent on the locations and sizes of grain boundaries, which tend to hinder electrical transport and lead to poor performance and device reliability. In this section, we discuss the fabrication of micro/nanostructures of inorganic semiconductor materials, which we refer to as microstructured semiconductors (μs-Sc), derived from high-quality wafers via a top-down approach. Such elements provide single-crystalline electrical transport pathways as a way to achieve high performance in macroelectronic systems.

The fabrication method generates functional elements via anisotropic etching of high-quality semiconductor wafers, referred to as "mother" wafers.[13,17–25] The process begins with photolithographic definition of patterns of resist followed by wet or dry chemical etching to produce the desired structures from the near-surface regions of the mother wafers. Micro/nanostructures formed in this way exhibit well-controlled doping levels, carrier concentrations, dimensions, and crystallinity.[13,22] Equally important, with appropriately defined anchor structures, they can remain attached to the mother wafer at certain points, thereby retaining their lithographically defined spatial layouts even after complete undercut etching. These two features represent the main advantages of this top-down approach over the bottom-up growth methods.

The earliest demonstrations of this top-down approach used layered substrates such as silicon-on-insulator (SOI) wafers, where the etching of the buried SiO_2 layer produces silicon ribbons, wires, or bars. Micro/nanostructures of silicon (μ*s*-Si) produced using this approach are extremely smooth and show excellent mechanical flexibility.[13,22] Thin-film transistors and circuits built using these structures as building blocks exhibit high performance,[17,22,26] which can satisfy the requirements even for demanding macroelectronic systems in mechanically bendable as well as fully stretchable configurations[19] needed for conformal "wrapping" of electronics on curvilinear surfaces. A main disadvantage of the SOI or layered wafer approach is that the cost of the semiconductor can be high because of the relatively high cost of such wafers. As a result, simple routes for generating micro/nanostructures from lower cost, bulk wafers are of interest.

We recently developed a simple route to generating single-crystalline silicon ribbons, platelets, and bars from bulk silicon (111) wafers.[18,21,24,27] Figure 13.1 depicts the fabrication sequence, which involves a trench etching step followed by an anisotropic wet chemical etch to slice off structures from the top surface of the wafer. For ribbons with rectangular profiles, the process starts with a Si wafer with a (111) orientation. Aligning photoresist lines perpendicular to the $\mathrm{Si}(1\bar{1}0)$ planes and isotropically etching trenches into the exposed silicon by reactive ion etching leads to structures depicted in the first frame of Fig. 13.1a.

The next step involves the formation of a layer of resist on the top surface and sidewalls of the trenches. We found that a bilayer of SiO_2 (~60 nm) and Si_3N_4 (~300 nm) forms an effective resist for the solution that we use for the anisotropic etch.[18] Selectively protecting the top and sidewall surfaces of the rectangular relief features by depositing titanium and gold metal via directional deposition using the collimated flux from a electron-beam evaporator at an oblique angle with respect to the wafer surface leaves the fast etching planes [i.e., $\mathrm{Si}(1\bar{1}0)$] exposed, as shown in the middle frame in Fig. 13.1a. The final step exploits these faster etching planes by anisotropic wet chemical etching of the unpassivated regions in either inorganic or organic anisotropic chemical etchants. The $\mathrm{Si}(1\bar{1}0)$ etch fronts proceed in a horizontal fashion until they meet, thereby creating freestanding single-crystalline silicon micro/nanostructures with rectangular cross sections. Figure 13.1b shows 500-nm-thick, 200-μm-long, and 7-μm-wide silicon ribbons. Figure 13.1c shows ultra-

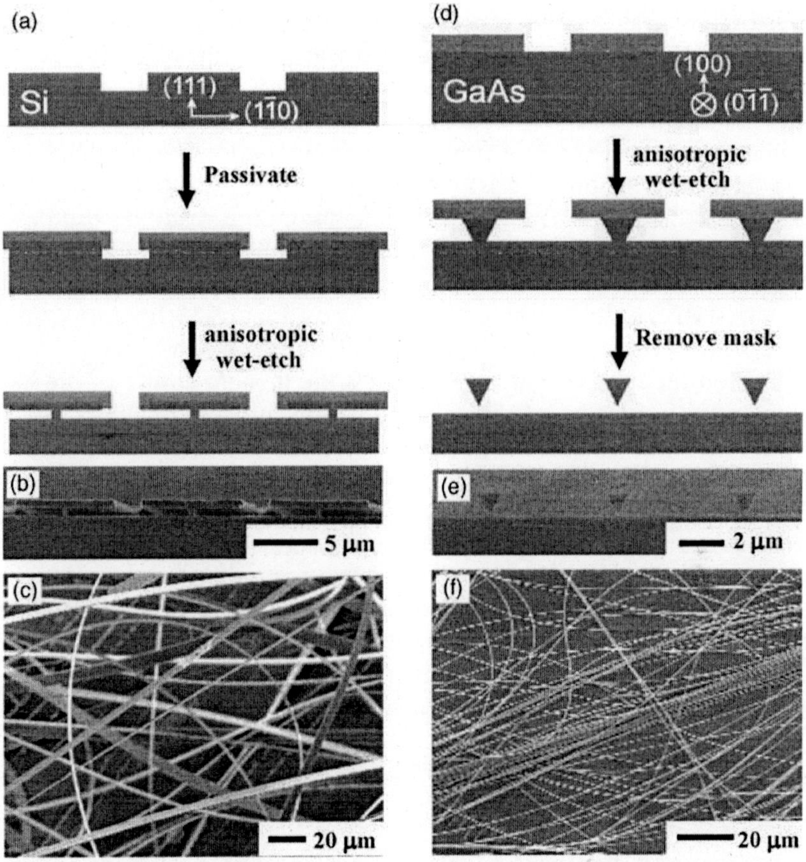

Figure 13.1. Schematic illustration of processes for fabricating semiconducting wire/ribbons from bulk wafers.

long (up to 6 cm) silicon ribbons with smooth surface morphologies and good mechanical flexibility characteristic of ribbons produced by this approach. By carefully controlling the processing parameters, ribbons with dimensions ranging from less than 100 nm (thickness) to as large as 6 cm (length) can be produced using this approach, in both flexible and stretchable configurations.[18] This procedure can be repeated to consume the entire thickness of the wafer and thereby to yield large quantities of ribbons and related elements. A limitation of this approach is the large number of processing steps. Large quantities of μ*s*-Si can be produced in a single set of steps by sculpting ripple patterns into the sidewalls of the etched trenches by use of programmed etching sequences in an inductively coupled plasma reactive ion etching (ICPRIE) system. Using a process similar to that depicted in Fig. 13.1a with wafers etched in this fashion leads to the generation of bulk quantities of single-crystalline semiconducting ribbons.[20]

In a simpler but related process, anisotropic chemical etching of III–V compound wafers, comprising a (100) top surface and zinc blend face-centered cubic crystal lattices, can produce freestanding wires with triangular cross sections.[14,24] Figure 13.1d shows representative fabricating steps for generating triangular wires from GaAs wafers (μs-GaAs). The top frame illustrates the formation of lines of photoresist (PR) aligned along the $(0\bar{1}\bar{1})$ direction on the GaAs (100) surface. The middle frame shows the resulting profiles after wet etching of the patterned GaAs wafer in a solution of phosphoric acid and hydrogen peroxide. Complete etching followed by removal of the PR yields GaAs wires (the bottom frame in Fig. 13.1d). Figure 13.1e shows partially undercut GaAs wires. Figure 13.1f shows fully undercut GaAs wires randomly assembled on the mother wafer and depicts the flexible nature of the nanowires generated using this method.[24] As with the process of Fig. 13.1a, these steps can be repeated to consume the entire thickness of the GaAs and to yield large quantities of wires.

Micro/nanostructures generated using these and related top-down approaches are geometrically and electrically homogeneous, with layouts that can be controlled over a wide range to realize not only ribbons and wires but also bars, platelets, membranes, and other structures. The main limitations of the top-down approach are as follows: (1) The composition of the fabricated objects is limited to materials that are readily available in wafer or thin-film forms, (2) the etching processes can lead to some level of roughness on the surfaces of the structures, and (3) dimensions of less than 20 nm, for other than the thickness, are difficult to obtain reliably.

13.3 TRANSFER PRINTING USING AN ELASTOMER STAMP

Useful integration of the micro/nanoscale inorganic material elements described in the previous section requires methods that can deliver them in desired layouts to target substrates. A challenge is that the mother substrates often exhibit properties and characteristics (mechanical fragility, high cost per area/weight, temperature requirements, opacity, etc.) that are greatly different than those desired for flexible macroelectronics. In addition, certain applications require the ability to integrate several different types of inorganic materials elements on a single substrate. System-on-a-chip applications,[28] for example, may benefit greatly from a reliable micro/nanoscale heterogeneous integration strategy. Such a strategy should demonstrate the following characteristics: high throughput in number of elements transferred per unit time, high placement accuracy (few microns or better), high (near perfect) yield or low defect density, low-temperature operation (< about 120 °C), chemical noninvasiveness, and the ability to handle small, fragile objects. Modern pick-and-place tools can transfer objects at truly impressive rates (~26,000 per hr at \$0.007 per ea),[29] but they cannot achieve, in existing embodiments, throughputs required for most macroelectronic applications (e.g., millions of discrete components per product). In addition, many transferable elements/components are thin, mechanically flexible/floppy, fragile, small (<200 μm), and therefore, impossible to handle using conventional machines. Instruments equipped with micromanipulators can transfer some small, fragile components[30–32] but only at the expense of throughput and cost. These constraints leave as two viable options self-assembly in fluids and rubber stamping or dry transfer approaches. Figure 13.2 illustrates both options. Each begins with the fabrication and release of small elements from a mother wafer. In fluidic self-assembly, an excess of these micro/nanoscale elements dispersed randomly in an agitated liquid (e.g., ethylene glycol) drift and fall until they encounter receptor sites on a substrate. These sites bind the elements/components via shape-recognition and/or capillary forces from molten solder. One example[29] of self-assembly of microscale components generated from an SOI wafer demonstrates potentially very high throughput and simultaneous integration of several different types of components on a plastic substrate. Currently, however, several challenges must be addressed before such approaches can operate at sufficiently high yields for demanding applications (e.g.,

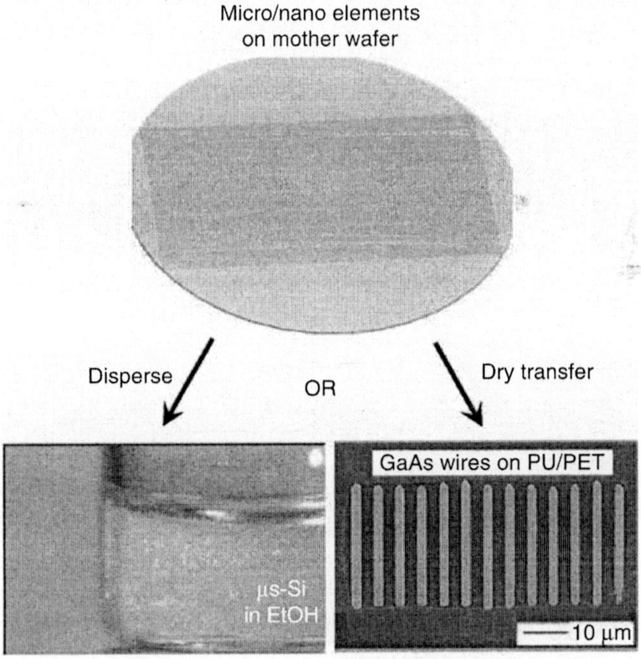

Figure 13.2. Inorganic micro/nanoscale elements derived from wafers can be used to form solution suspensions for delivery to another substrate (e.g., plastic sheet) by casting processes (left frame), or they can be dry transfer printed in a way that preserves their lithographically defined spatial organizations (right frame).

displays). Complex or high-aspect ratio shapes, for example, do not bind to receptor sites as easily as highly symmetric shapes, and it is unclear that the self-assembly of arbitrarily shaped objects can be accomplished without arbitrarily complicated procedures. In addition, the necessity for great excesses of micro-objects (10 to 100 times the number of receptor sites) might make the self-assembly approach impractical except in very large-volume production lines. Additional research on these approaches, however, may soon prove self-assembly to be the ultimate strategy in terms of throughput in microscale manufacturing, at least in specific cases. Although previous chapters describe self-assembly approaches, this section focuses primarily on the stamping approach to transfer.

This printing strategy is one of a class of microfabrication techniques referred to as "soft lithography" that uses molded stamps composed of soft materials.[33] One of the most commonly employed materials for the production of high-resolution stamps in soft lithography is polydimethylsiloxane (PDMS) rubber. PDMS is selected for its very high-resolution molding capabilities,[34] chemical stability, and low mechanical stiffness (~1–10 MPa) that facilitates fully conformal contact on reasonably smooth (e.g., coarsely polished) surfaces without exhibiting excessive tackiness, which allows easy handling and manageable levels of distortion. Soft lithography using molded PDMS stamps offers several interesting microfabrication capabilities: nanofabrication and patterning resolutions below 100 nm,[34–36] patterning on nonplanar surfaces,[37–40] and potentially low-cost, large-area operation.

In this context, stamp-based or dry transfer printing methods represent a deterministic alternative for the transfer of inorganic elements onto new substrates that, in contrast to self-assembly, maintains solid mechanical contact to the elements via stamps throughout the entire transfer process. As a result, the elements maintain their organization, thereby minimizing misplacement of the elements and other yield-reducing errors. The array in the lower right of Fig. 13.2 represents a precisely ordered set of inorganic micro-elements transferred in this way to a polyethyleneterephtalate (PET) sheet coated with a thin film of polyurethane (PU). The transfer printing process begins with the release of micro-elements from the mother wafer, as described above. Organization of the released elements is maintained by structures that anchor or tether them to the mother substrate. Without these structures, the elements may lose organization partially or completely, but interconnecting the elements into sufficiently large (~millimeter) networks can afford organization of elements relative to each other and negligible motion of the networks relative to the mother wafer in some cases. The printing proceeds with the application of a stamp to elements on the mother substrate. The stamp contacts the elements either uniformly (i.e., using a flat stamp) or with molded protruding relief features. After contact, the separation of the stamp from the mother substrate retrieves the contacted elements onto the stamp. The stamp then contacts another substrate (i.e., the receiving or "target" substrate) and delivers the elements. Separation of the stamp from the target completes the process. In most instances, the stamps are unaffected by the process and remain usable for thousands of repetitions or more, but some variations described in the following sections (e.g., chemical modification of the stamp surface; Section 13.3.1) could possibly require frequent replacement of the stamps. If the process uses molded stamps that contact only a selected fraction of elements on the mother substrate, then the process may be repeated to distribute the elements from that substrate in a well-controlled manner across a larger target substrate or onto multiple target substrates.[41] This strategy is known as "selective transfer" or "area-multiplication," and it represents one of the key aspects of transfer printing that makes it valuable for large-area and certain other applications. Conversely, transfer printing may be used to condense elements or to combine different types of elements onto a single substrate, for example, by multilayer, quasi-3D processing. The printing procedure is generally a purely additive process, and as a result, the delivery of multiple layers of elements to a substrate is relatively straightforward and uncomplicated[14,40,42] (see Figs. 13.4, 13.7, 13.13, and 13.15 as well as Section 13.5.2).

These capabilities of transfer printing make it a powerful solution for deterministic assembly of inorganic micro/nanoscale elements on dissimilar substrates. The rest of Section 13.3 is devoted to the science that governs transfer of objects from one substrate to a stamp and from there to a new substrate and the current state of the technology. Section 13.3.1 discusses surface chemistries that may be used to guide transfer, and Section 13.3.2 discusses thin-film adhesives or glues for adhering arrays of elements to target substrates. Section 13.3.3 describes kinetic effects in transfer printing that originate from the viscoelastic properties of elastomer stamps and that allow reversible switching in the strength of adhesion between micro/nanoscale elements and the stamps. Section 13.3.4 explores the importance of crack-like features for the separation of elements from the mother substrate, and Section 13.3.5 discusses the use of carrier films to manipulate arrays of elements (e.g., molecules) that are too small to define by top-down methods. Section 13.3.6 concludes with a brief description of a printing press that may be used for this variety of transfer printing.

13.3.1 Surface Chemistry

Successful implementation of dry transfer printing requires a well-designed strategy of adhesion such that the elements to be printed adhere to the stamp more strongly than to the mother substrate and then adhere to the target more strongly than to the stamp. The strength of adhesion begins with the interface interactions between the bodies involved in the process, i.e., the stamps, the elements, and the substrates.

Several notable interfacial chemistry schemes can strengthen the adhesion between the bodies involved in transfer printing. Figure 13.3 illustrates one such reaction in which silanol groups on the surfaces of adjoined bodies

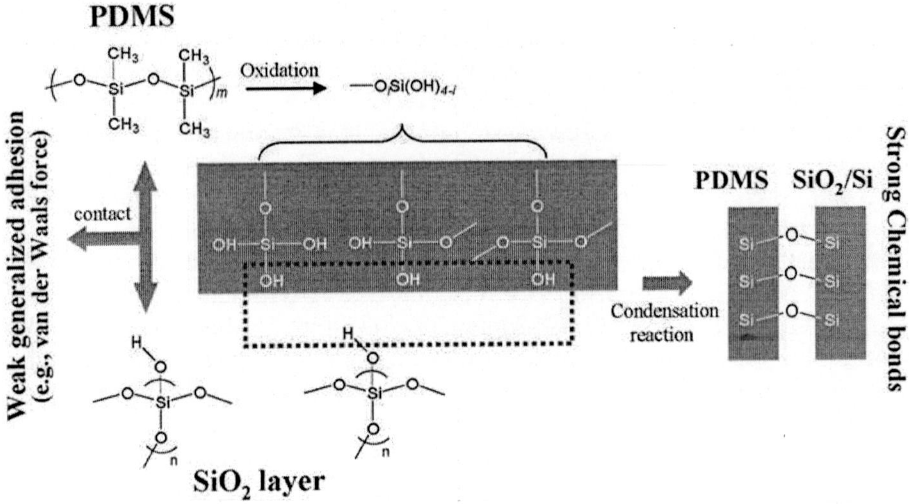

Figure 13.3. Schematic illustration of the surface chemistry of PDMS and semiconductor ribbons coated with thin layers of SiO_2.

produce water and strong covalent Si—O—Si bonds that bridge the two bodies. This reaction can improve the reliability of retrieving elements (e.g., GaAs nanowires) that are anchored to a mother wafer via a chemically treated PDMS stamp.[14,24] To employ this reaction, mild oxygen plasma and subsequent exposure to atmospheric moisture produce –OH groups on the surface of a PDMS stamp. This stamp contacts an array of elements coated with a film of SiO_2 that supports some concentration of –OH groups on its surface. Si—O—Si bridges form as the pair remains connected for some minutes. Similarly strong adhesion originates from reactions between silanol and other metal—OH groups on the adjoined body, for example, titanol.[43] These interfacial bonding schemes lead to strong adhesion, easily strong enough to produce cohesive failure in PDMS, and therefore, the density of these bonds should be carefully controlled and limited to low values to allow removal of the elements from the stamps during printing.[14] Another notable interfacial interaction relevant to transfer printing occurs between thiol surface groups and gold or gallium arsenide.[44,45] Thiol functionality is usually supplied in the form of self-assembled monolayers of alkanethiols or alkanedithiols that are deposited on a surface by vapor[46] or solution[45] methods. Yet another interface interaction useful for guiding transfer in these printing methods occurs during the fusion of two contacting thin layers of noble metals in a process called cold-welding.[35,45,47–49]

In the absence of these or other specific surface interactions between the bodies involved in transfer printing, nonspecific surface interactions can guide the transfer. Although they are relatively weak, short-ranged interactions,

Van der Waals forces and hydrogen bonding can lead to significant levels of adhesion from the atomic-scale conformal contact that occurs between the bodies. PDMS exhibits chemical inertness, and it usually interacts with inorganic bodies in contact via Van der Waals forces.[50–52] Despite the relatively low surface energy of PDMS (20 mJ/m^2),[53] these interactions can be strong enough to pull elements off of a mother substrate.[13,21,22,26,40,54] During printing, stronger nonspecific forces (Van der Waals or others) can then preferentially bond the elements to a target substrate. When both the elements and the target substrate have rigid, smooth surfaces, the interactions between the bodies are the same as those in direct bonding of semiconductor wafers.[55,56] Figures 13.4a through 13.4c show several examples of these kinds of printed systems, directly bonded to smooth, rigid substrates. Figure 13.4a shows n-type silicon chiplets printed onto a p-type silicon wafer, and Figure 13.4b shows nanoscale printed silicon beams spanning and directly bonded to microscale silicon beams printed onto a silicon wafer. This direct-bond transfer strategy can work on curved surfaces too, as demonstrated by the silicon beams bonded directly to a cylindrical glass lens (radius = 5 mm; Fig. 13.4c).

13.3.2 Thin-Film Adhesives

Regardless of the nature of the surface interactions between contacted bodies in a transfer printing system (e.g., specific covalent bonds and Van der Waals), reliable adhesion between them requires complete, atomic-scale contact that spans a significant fraction, if not the entirety, of at least one face of an element to be printed. This kind of contact originates in systems with shape complementarity (in the simplest case, contact between two flat, very smooth surfaces, e.g., as in wafer-bonding[55,56]) or with at least one low-modulus body that can conform to relief in the other. PDMS provides this conformability during the retrieval stage of transfer printing, conforming to topographical features in the printable elements upon contact. During printing, thin films of soft adhesive materials can provide conformability between the elements and the target substrate and greatly enhance the fidelity of transfer.[14,22,42,46] These adhesive films are critically important when the target substrate has even modest roughness. Figures 13.4d and 13.4e show examples, with a thin film of PDMS guiding transfer of silicon structures to a glass cylinder (Fig. 13.4d) and a thin film of polyurethane guiding transfer to a PET sheet (Fig. 13.4e). The procedure for using thin-film adhesives is straightforward and is described below. A stamp loaded with inorganic elements contacts a substrate onto which a thin-film adhesive has been deposited, usually by spin coating. At the point of contact, the thin-film adhesive is typically uncured or only partially cured, enabling it to flow or conform to the printable elements during printing. Mild heating while the loaded stamp is in contact with the target substrate can assist this flow. The thin-film adhesive then cures and the stamp is removed, completing the printing process. Optionally, full curing of the adhesive layer may occur after printing. Examples of materials used for these kinds of

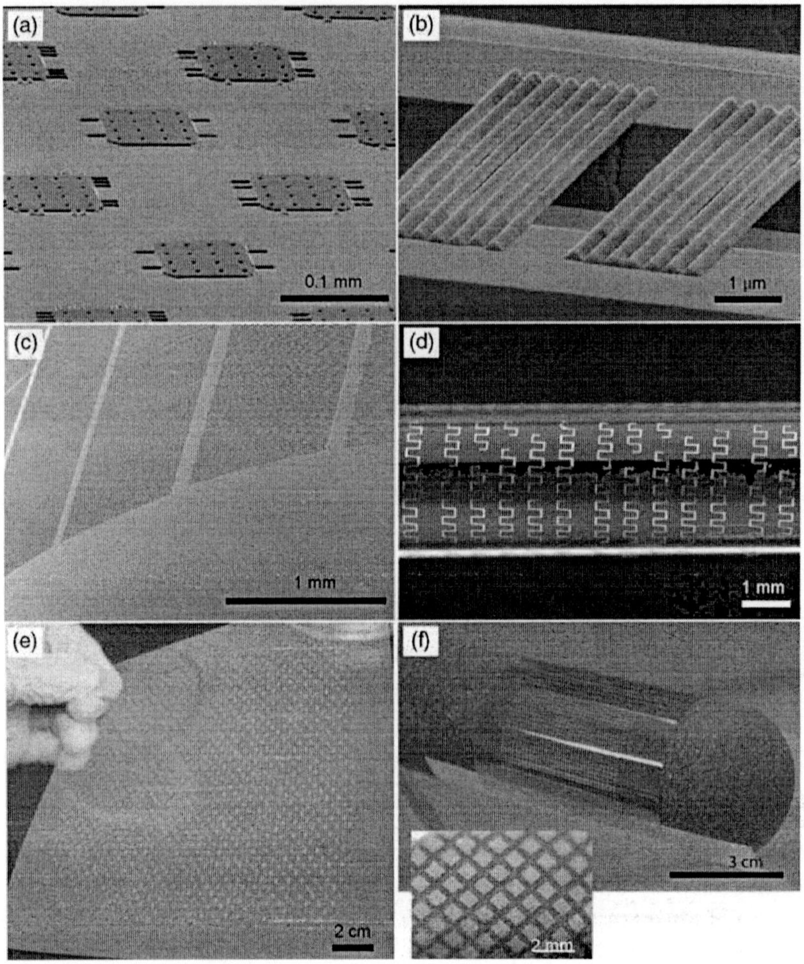

Figure 13.4. Single-crystalline silicon structures transfer printed on various rigid and flexible substrates with and without the use of adhesives. (a) Silicon chiplets directly bonded to a silicon wafer. (Reprinted with permission from Ref. 40. Copyright 2006 Nature Publishing Group.) (b) printed silicon nanobeams bridging printed silicon microbeams. (c) Silicon beams on a glass lens. (Reprinted with permission from Ref. 40. Copyright 2006 Nature Publishing Group.) (d) Silicon structures on a glass rod using a thin-film (PDMS) adhesive. (e) Large-area (15 × 15 cm) transfer of silicon onto a PET sheet coated with PU. (Reprinted with permission from Ref. 41. Copyright 2005 Wiley-VCH Verlag.) (f) μs-silicon ribbons on a cylindrical roller stamp.

adhesives include polyimide[42] and benzocyclobutene-containing siloxane polymers (BCB[57]). Photocurable materials such as epoxies (e.g., Su-8[22]) and polyurethanes (e.g., Norland optical adhesives[14,41]) also serve as effective adhesive layers for transfer printing, curing by exposure to light while the loaded stamp contacts the adhesive coating on the target substrate.

For many applications, it is desirable that the adhesive layer accept printable elements readily in its fully cured state. This characteristic usually requires the layer to be soft in its cured form. Adhesive thin films composed of low-modulus PDMS elastomer meet this requirement well[18] and can guide transfer of elements to a target quickly (without exposure to heat or light). Surprisingly, the direction of transfer can be well defined even when the composition of the adhesive is identical to that of the stamp. Successful transfer is thus determined by several factors: surface chemistry, conformability (modulus), geometrical/mechanical factors (e.g., adhesive film thickness), and others.

13.3.3 Kinetic Effects

The elastomer stamps used for transfer printing are viscoelastic materials, exhibiting mechanical properties (e.g., elastic modulus and loss modulus) that depend on strain rate; as a result, the fidelity or even the direction of transfer can also be determined by kinetic effects. The strength of adhesion between a viscoelastic body (e.g., a PDMS stamp) and an adjoined rigid elastic body (e.g., an inorganic micro-element) varies strongly with the rate at which the two separate.[40,58–60] Slow separation, measured by the speed at which the separation front propagates, can lead to relatively weak levels of adhesion, whereas fast separation requires more energy. A roll test[40,59] represents one simple set of experiments that can measure this rate dependence. As an elastic cylinder rolls down a flat viscoelastic slab, the cylinder assumes a constant velocity, at which the loss in gravitational potential energy balances the energy absorbed by viscoelastic separation processes (minus a negligible amount[61] of mechanical energy evolved at the advancing line of contact). Measurements of stainless steel cylinders rolling down PDMS slabs show that for speeds ranging from 0.02 to 4.8 cm/s, the gravitational potential energy loss or "energy release rate" varies from under 0.4 J/m^2 to over 4 J/m^2. The energy release rate is typically given by the equation $G(v) = G_0(1 + \varphi(v))$, where G_0 represents the energy release rate in the limit of slow delamination speed v. G_0 is determined by the thermodynamic work of adhesion between the two substrates and by irreversible processes that occur at the interface between the two bodies at the separation front or "crack tip" (e.g., extension of polymer chains[62]). Surface modification of the PDMS can therefore tailor $G(v)$. $\varphi(v)$ represents a kinetic amplification of G_0 caused by viscoelastic dissipation in the PDMS near the crack tip.[58]

In the context of transfer printing,[40] this rate-dependent adhesion effect has important consequences, as illustrated in Fig. 13.5. During the retrieval step, fast separation of a stamp from the mother or "donor" substrate can greatly increase the tendency of elements originally housed on the donor to adhere preferentially to the stamp. Scanning electron micrographs in Fig. 13.5 show the results of slow and fast delamination of a stamp from an array of freestanding silicon beams anchored at their ends to a donor SOI wafer. Slow delamination leaves the beams intact, whereas the stronger adhesion caused by fast

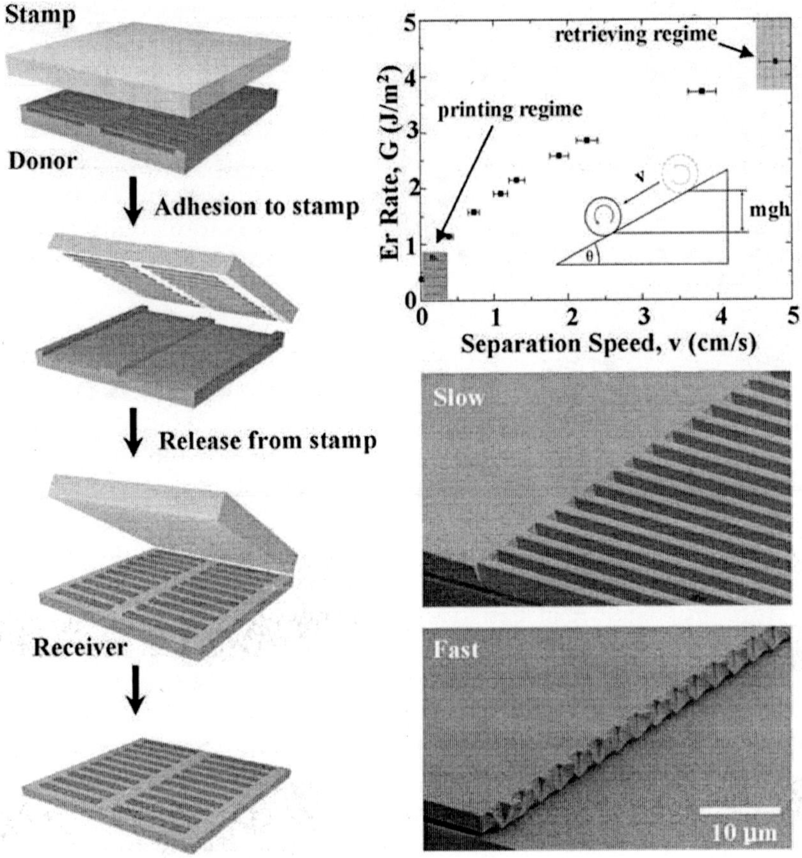

Figure 13.5. Schematic illustration of the generic process flow for transfer printing solid objects (left) and the rate dependence of stamp adhesion (right). The printing strategy illustrated here relies on the relatively strong adhesion of rubber to solids at fast peel rates, v, and significantly weaker adhesion at slower rates. Er rate = Energy release rate. (Reprinted with permission from Ref. 40. Copyright 2006 Nature Publishing group.)

delamination is enough to break the beams off of the wafer for subsequent printing. During the printing step, the relatively weak adhesion caused by slow delamination can be helpful for guiding the elements off of the stamp and onto the target. One important caveat is that viscoelasticity in thin-film adhesives can lead to kinetically variable adhesion between the elements and the target. In these instances, successful rate-dependent printing strategies require additional considerations.

13.3.4 Stress Concentration and Fracture

Printable elements, especially those prepared by top-down methods, are often prepared such that they are attached to the mother substrate by tethers and anchoring structures that keep the otherwise freestanding elements in place. To separate elements from the mother substrate in the retrieval step, stamps must successfully break these tethers, which are typically composed of the same inorganic material as the elements. This process often requires careful design of the elements and tethers such that the elements separate by well-controlled and reliable fracture. Simple fracture theory[63] points to several important factors that lead to easy fracture, notably the presence of cracks in the inorganic material and the geometry of those cracks.

Early fracture theory by Inglis[64] suggests that fracture occurs because of the presence of high stresses generated at the tips of cracks. Looking at an elliptical crack in a plate under tension perpendicular to the long axis of the crack, he found that the maximum stress occurs at the crack tips and has a value σ_m given by $\sigma_m = \sigma_0\left(1+\sqrt{(a/\rho)}\right)$, where σ_0 is the tension applied to the plate, a is one half the crack length, and ρ is the radius of curvature of the crack at the tip. Clearly, as the crack tip sharpens, ρ drops and the ratio of the maximum stress to the applied stress increases strongly. When the maximum stress exceeds the strength of the bonds that hold together the material, the bonds break, the crack propagates, and the fracture occurs. Fracture in plates that contain sharp cracks, thus, occurs more easily than for plates that contain similarly sized, rounded cracks. Fracture of printable elements from anchoring structures occurs according to the same mechanics even though their geometries are more complex than the idealized cracked plate and the mode of fracture is nonobvious. A series of model systems composed of robust (2.5-micron-thick) silicon beams machined from an SOI wafer demonstrate the effects of crack sharpness on their ability to break free from the wafer.[54] Figure 13.6 illustrates these systems that differ primarily by the sharpness of the features in the beams. Figure 13.6a shows beams with round corners produced by nearly isotropic SF_6 plasma etching of the top silicon. Figure 13.6b shows beams with sharper corners produced by etching in aqueous KOH/isopropanol, and the beams in Fig. 13.6c, etched using aqueous KOH without isopropanol, have yet sharper corners. The characteristic radii that describe the corners in the three systems are 1700 nm for SF_6-etched beams, 120 nm for KOH/isopropanol, and <10 nm for KOH. A standardized stamping procedure attempted to retrieve beams from each of these systems, taking care to control the rate (see Section 13.3.3) using automated mechanical stages (see Section 13.3.6). The results of the retrieval attempts show several clear trends across the range of sharpness. Most notably, the ease of fracture is much greater for sharp-cornered beams than for round-cornered ones. The bottom frames in Fig. 13.6 show the results of the retrieval attempts, with no retrieval of the round-cornered beams, partial (about 75%) retrieval of the intermediately

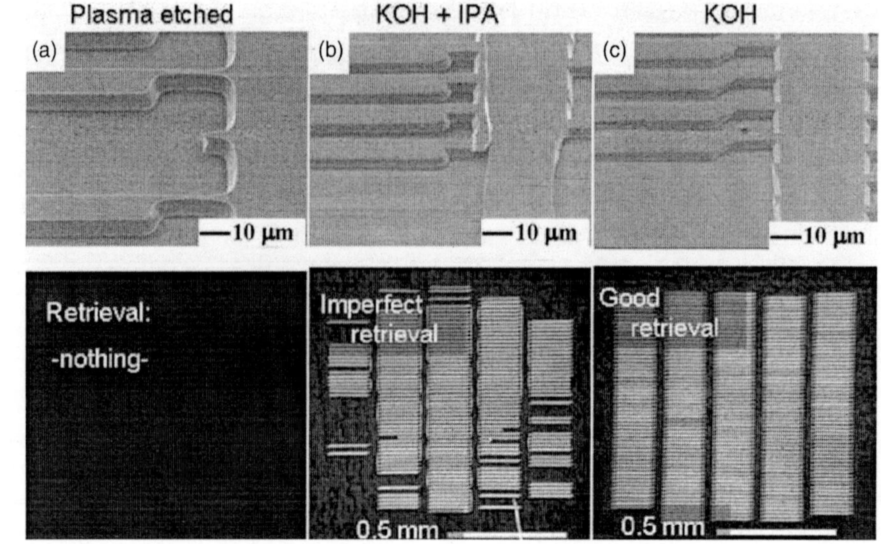

Figure 13.6. Effects of stress concentration on breakability. Scanning electron micrographs (top frames) showing breaks of anchored silicon beams etched with (a) plasma, (b) KOH and isopropyl alcohol, and (c) KOH and optical micrographs (bottom frames) of printing results that demonstrate the relative ease of breakability, or the ability for a stamp to separate elements from the anchoring structures in each system. (Reprinted with permission from Ref. 54. Copyright 2007 American Institute of Physics.)

sharp-cornered beams, and near-perfect retrieval of the beams with the sharpest corners. Additionally, the fracture in the sharpest system (Fig. 13.6c) occurs in a well-defined way, with beams that break cleanly at the anchors. The rounder-cornered system in Fig. 13.6b, on the other hand, forms cracks that propagate into the anchoring structures or into the beam, resulting in broken and unintentionally interconnected beams.

Finite element analysis (FEA) using ABAQUS software shows that the results above may be explained by the stresses that are generated as the structures are lifted from the mother substrate. The analysis simulates a mild upward stress (0.1 MPa) applied to models that closely resemble the beams described above. The maximum tensile stresses observed in these analyses occur where fracture of the beams is observed in experiment, at the intersection of the beam and anchoring structure. The magnitude of this maximum stress is greater for more sharply cornered systems, reaching more than four orders of magnitude greater than the applied stress and leading to the ease of fracture observed for these systems. The distribution of stress fields in these analyses also shows that the spatial extent (full-width at half-maximum) of these stress fields generated in the elements follows the characteristic radius of the features in the model to within a factor of about two. FEA shows that the functional dependence of the maximum stresses generated in these systems on the sharpness of the corners is the same as that generated in cracked plates, linear in the square root of the sharpness of the cracks.

Fracture mechanics, thus, plays an important role during the transfer printing of inorganic materials elements. In the most general sense, every step of transfer printing as it is described here is a competition of at least two competing fractures: fracture between the stamp and the elements and fracture between the elements and the substrate, mother or target. For this reason, the presence and properties of cracks and crack-like features at the interfaces involved influence, to a great extent, the direction of transfer.

13.3.5 Carrier Films and Carbon Nanotubes

The smallest printable elements have at least one (usually two) dimensions that are comparable with the reach of short-range surface interactions, hydrogen bonding, and chemical reactions described in Section 13.3.1. The physics that governs the transfer of these truly nanoscale objects are dominated by surface interactions and chemistry and are less influenced by the mechanical/geometrical considerations described in Sections 13.3.2 through 13.3.4. As a result, many of the strategies that are necessary for reliable transfer of discrete micrometer-scale to many-nanometer-scale elements are ineffective for the transfer (especially retrieval onto a PDMS stamp) of discrete elements a few nanometers in size. A sacrificial thin film deposited on top of these kinds of elements may then aid in the transfer process. Such a carrier film allows an assembly of the nanoscale elements to move from substrate to stamp to substrate as a larger unit in the ways that are described in the sections above (e.g., crack-design, kinetically switchable adhesion, and thin-film adhesives). One particularly interesting example of this kind of printable nanoelement is single-walled carbon nanotubes (SWNTs).

SWNTs share many important characteristics with high-performance (single-crystal) inorganic electronic materials and materials elements. Composed of a single, seamlessly rolled graphene tube, SWNTs often exhibit perfect long-range atomic ordering and compositional purity. Accordingly, they serve as very high-performance conductors[65] and semiconductors[66] depending on the chirality of the tube. These characteristics, both for inorganic semiconductor wafers and for SWNTs, require high-temperature growth, and thus, methods for transfer to low-temperature, large-area, and/or low-cost substrates are desirable. Several methods exist for bulk-synthesis of carbon nanotubes (laser oven, arc-discharge, etc.), which may then be delivered to a surface in thin-film form (e.g., by solution methods).[39] Generally speaking, SWNT growth on a substrate can more reliably produce high-quality SWNT films (pure, with good electrical properties) and new thin-film surface growth techniques[67,68] can generate arrays of SWNTs on the surface of a quartz substrate that, in contrast to those produced by bulk methods, have lengths approaching millimeters or more and very good alignment. These arrays

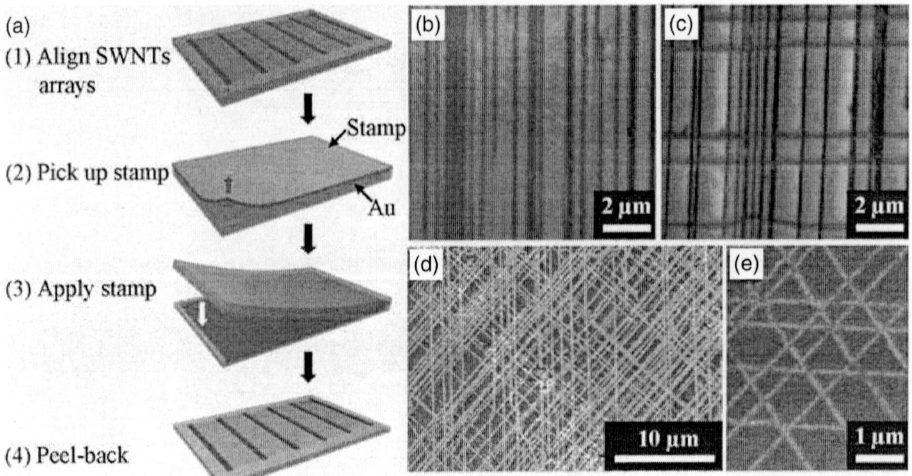

Figure 13.7. (a) Schematic illustration of the dry transfer of CVD-grown single-walled carbon nanotubes onto plastic substrates. (b)–(e) Scanning electron micrographs of SWNT arrays transferred to plastic substrates, with repetitive transfer for crossed arrays.

represent in some ways the ideal assemblages for SWNT-based thin-film electronics. Such SWNT films may be transferred from the growth substrate to some target substrate using carrier films.[69–74] Figure 13.7a illustrates the process. To begin, vacuum metallization techniques deposit gold on the SWNT growth substrate. Optionally, photolithography and etching perforates the gold thin film and some wet etchant selectively removes substrate material directly beneath the SWNT film for easy separation. A PDMS stamp contacts this film and retrieves it for printing as described above. Also optionally, instead of retrieval with a PDMS stamp, spin coating deposits a film of polymer (photoresist, polyimide, polyvinylalcohol, etc.) over the gold to produce a thin polymer "stamp" that may then be retrieved directly along with the gold carrier film and the SWNTs. The stamp then prints the SWNTs and carrier film onto a target that is usually treated with a thin-film adhesive. After printing, wet etching removes the carrier film to complete the transfer process. Figures 13.7b–e display images of SWNT arrays transferred to plastic substrates, with repetitive transfer for crossed arrays (Fig. 13.7c–e).

The use of carrier films is thus an effective way to enable the transfer printing of assemblages of nanoscale materials elements. In addition to SWNTs, this carrier film approach may aid in the transfer of arrays of small inorganic nanoparticles or perhaps even small molecules that for whatever reason must be synthesized or processed on the surface of a substrate that is not suitable for the end-use application.

13.3.6 Machines for Transfer Printing

Commercially viable transfer printing technology of the sort described in these sections for most applications will require great positional control of stamps relative to mother substrates and receiving substrates (few micrometers or better), a method for reliable contact and separation with good speed control, and the capability to perform transfers many times (tens to hundreds per minute). These requirements necessitate automation and the development of new tools (and stamps) designed specifically for the printing process.

Figure 13.8 shows a printing tool built to demonstrate automation of this process. The tool consists of three linear stages, a tilt stage, and a rotation stage to manipulate the stamp between mother substrates and a target substrate that are attached to a vacuum-chuck substrate holder. The print head is equipped with a vision system that checks the registration of the stamp to the printable elements on the mother substrate and to receptor sites on the target. This vision system also monitors the results of a printing event, giving a real-time indication of any failed transfers or defects. Load cells on the print head allow careful control of the pressure during printing. The print head can hold stamps as large as about 75 mm. These stamps may be composed of a bulk piece of PDMS that is laminated against a glass plate on the bottom of the print head, but the low modulus and high thermal expansion coefficient of PDMS can lead to distortions of tens of microns or more in such a stamp.[75] Instead, using stamps composed of a thin PDMS layer backed with a thin, high-modulus (e.g., glass or kapton) material can reduce distortions to a few micron levels.[75,76] The backing material should be rigid to avoid mechanical distortion, but it should be thin enough that separation from substrates occurs readily. One additional consideration is the optical transparency of both the backing and the elastomer (PDMS or otherwise) that allows visual alignment and inspection. Such composite stamps when loaded into the tool may be

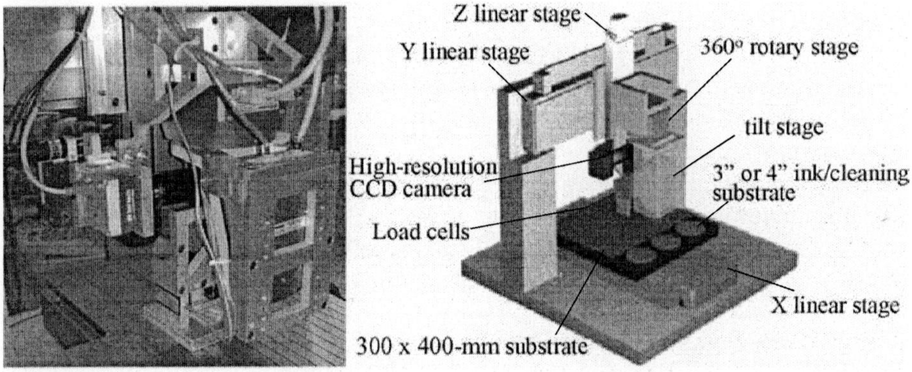

Figure 13.8. Images of an automated mechanical stage used for transfer printing.

backed by a pressurizable bladder that can inflate to control contact and printing pressure carefully.

The tool shown in Fig. 13.8 represents an early effort toward the automation of the transfer printing approach. Other designs for high-resolution contact printing tools also show promise, notably the "wave-printer" built by Philips,[77–79] which uses a series of pneumatic valves to drive contact and separation of a composite glass-backed stamp to some substrate. Later efforts may assume different forms, for example, by using stamps wrapped around drums for wafer-to-reel manufacture. Future printing tools may exhibit unique mechanisms more well suited to such a manufacturing technology, based on key components composed of soft materials. For example, future alignment strategies may seek to take advantage of the ability to deform soft stamps mechanically in a sort of "lock-and-key" approach,[80] in which relief features on the substrate guide the positioning of the stamp as it contacts a substrate. Transfer printing of inorganic materials elements, as described in these sections, continues to show promising results in laboratory settings, although the path to widespread commercial applicability has currently met the engineering issues associated with automation. The development of tools for high-resolution contact printing is, therefore, of great importance.

13.4 FLEXIBLE THIN-FILM TRANSISTORS THAT USE μs-Sc ON PLASTIC

Figure 13.9 illustrates the processing steps for fabricating high-performance, single-crystalline silicon metal oxide semiconductor field-effect transistors using silicon ribbons printed on a thin (~25 μm) polyimide substrate and the dry transfer-printing method described in the previous section.[17] We refer to such devices as μ*s*-Si MOSFETs. In the first step, a conventional, high-temperature doping process defines regions to which ohmic contacts can be formed between metal electrodes and the silicon at the source and drain regions of the transistors. Because this process requires high temperatures (>950 °C), it is carried out on the mother wafer before transfer to plastic substrate (even in optimized polyimide materials, thermal degradation occurs at 350 °C). These doped μ*s*-Si ribbons can then be transferred to a plastic substrate, with or without an adhesive layer, using an elastomeric stamp as the transfer element. For the case described here, a liquid polyamic acid (which becomes polyimide upon a thermal curing) is used as an adhesive layer on a thin (~25 μm) polyimide (PI) substrate. Plasma-enhanced chemical vapor deposition (PECVD) using SiH_4 and N_2O at 250 °C forms a uniform film, for the gate dielectric, on the μ*s*-Si/PI structure. The PECVD SiO_2 on the source/drain contact regions is then removed by reactive ion etching (RIE) through openings in a layer of photoresist patterned by photolithography. A Cr (5 nm)/Au(100 nm) layer deposited by physical vapor deposition forms a gate elec-

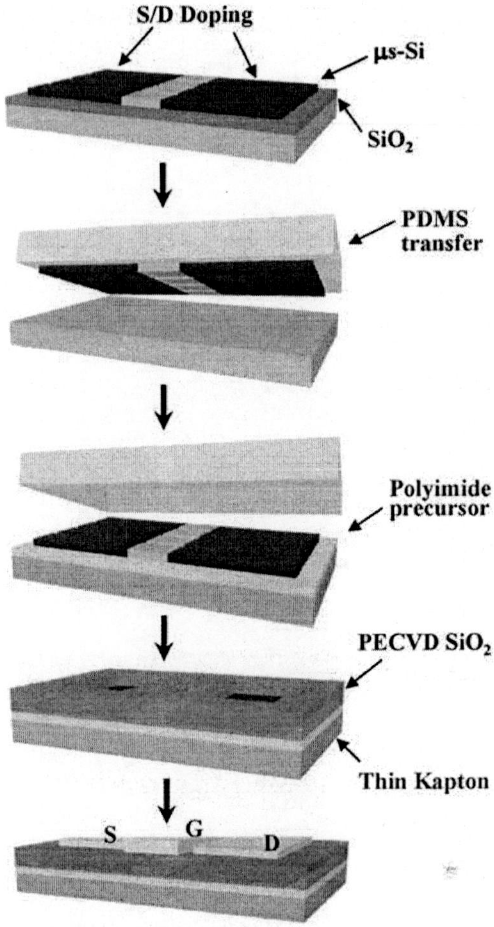

Figure 13.9. Schematic illustration of process steps for fabricating μ*s*-Si MOSFET devices on plastic substrates.

trode on top of the SiO_2. Finally, connecting electrode pads to source, drain, and gate are formed by photolithography and liftoff.

Similar processing steps can be implemented with other semiconductor materials. Figures 13.10a through c show optical microscope images and dc electrical characterizations of ribbon- and wire-based μ*s*-Si MOSFET, and μ*s*-GaN HEMT, and μs-GaAs MESFET, respectively. As with the μ*s*-Si MOSFETs, the ohmic contacts for μ*s*-GaN HEMTs and μ*s*-GaAs MESFETs are formed prior to integration with the plastic substrate. For the μ*s*-GaN HEMTs, the ohmic contacts consist of Ti(15 nm)/Al(60 nm)/Mo(35 nm)/Au(50 nm) on a GaN on Si wafer, thermally annealed at 850 °C for 30 s in a N_2 environment.[81] For the μ*s*-GaAs MESFETs, ohmic contacts consist of AuGe(120 nm)/Ni(20 nm)/Au(120 nm) annealed at 450 °C for 60 s.[82]

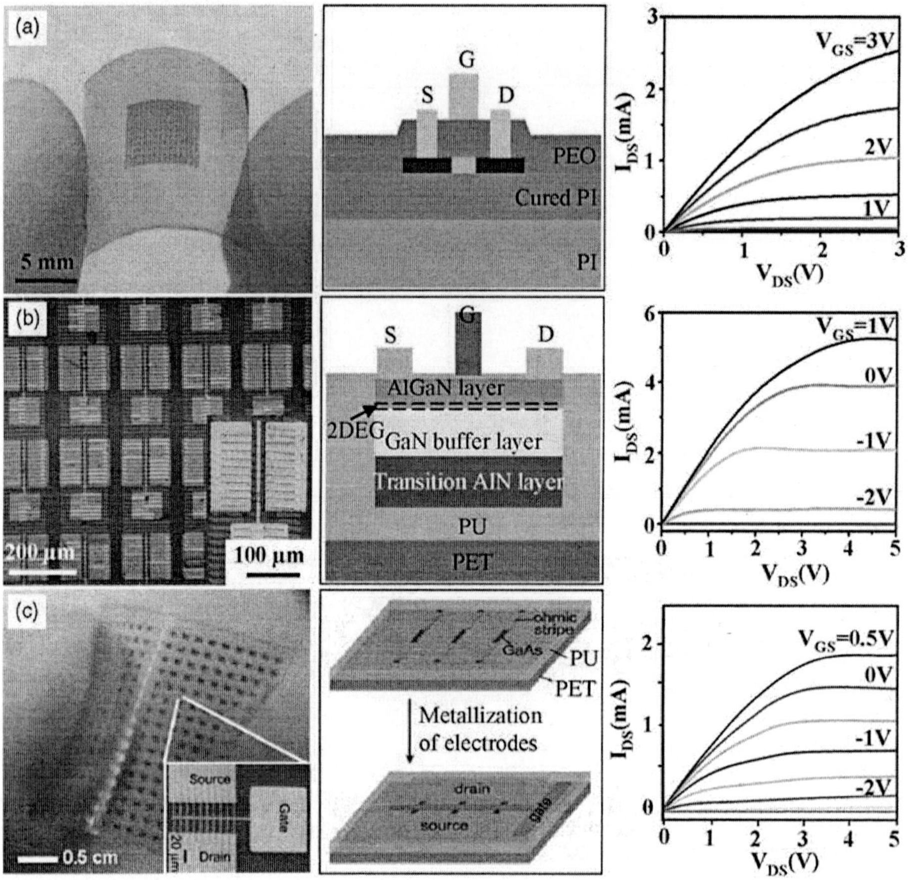

Figure 13.10. Images, cross sections, and electrical characteristics of flexible devices that use semiconducting wires/ribbons and plastic substrates. (a) μ*s*-Si MOSFETs (Reprinted with permission from Ref. 17. Copyright 2006 IEEE.) (b) μ*s*-GaN HEMTs (Reprinted with permission from Ref. 81. Copyright 2006 IEEE.) (c) μ*s*-GaAs MESFETs.

Electrical measurements on these devices show excellent performance. For example, effective mobilities in μ*s*-Si n-channel MOSFETs are ~600 cm^2/Vs in the linear regime and ~530 cm^2/Vs in the saturation regime, as computed with standard models (Fig. 13.10a).[83] These values approach those of similarly scaled single-crystalline silicon MOSFETs on Si wafers, and they exceed slightly those observed in the most sophisticated form of laser-annealed polycrystalline silicon transistors on glass.[84] They exhibit substantially higher mobilities than similar laser-annealed devices on plastic.[85] The transconductance of an average μ*s*-GaN HEMT is 1.6 mS, with a channel length, channel width, and gate length of 20 μm, 170 μm, and 5 μm, respectively (Fig. 13.10b).[77] This transconductance is lower (by about 40%) than those observed in similar wafer-based devices, primarily because of some degradation of the ohmic contacts caused by etchants during ribbon delineation. Process improvements can likely eliminate these effects. The dc characteristics of μs-GaAs MESFETs on plastic with gate lengths of 2 μm and channel lengths of 50 μm exhibit a pinch-off voltage at V_{DS} = 0.1 V (i.e., linear region), an ON/OFF current ratio, and a maximum transconductance of −2.7 V, ~10^6, and ~800 μS, respectively (Fig. 13.10c).[82,86]

Figures 13.11a–c show radio frequency (RF) measurements on μs-Si MOSFETs and μs-GaAs MESFETs. In the case of the μs-Si MOSFET, the unity-current gain frequency (f_T) was measured in the common-source configuration. The small-signal current gain (H_{21}) was extracted from measured S-parameters using a vector network analyzer (V_{NA}). The unity current gain frequency is defined as the frequency at which the short-circuited current gain becomes unity or 0 dB. These f_T values of a typical μs-Si MOSFET and μs-GaAs MESFET determined in this manner are 515 MHz and 1.55 GHz, respectively (Figs. 13.11a and b). In the case of the μs-Si MOSFET, the channel length (L_c) and the overlap gate length (L_o) are 2 μm and 1.5 μm, respectively. In the case of the μs-GaAs MESFET, the gate length and the channel length are 2 μm and 50 μm, respectively, without a recessed gate region. The RF responses of both classes of devices are roughly consistent with dc electrical characteristics and measured capacitances.

13.5 INTEGRATED CIRCUITS ON PLASTIC

13.5.1 Two-Dimensional Integration

Electronic systems require the integration of transistors as well as other components, including capacitors, diodes, inductors, and resistors. Printable ribbon and wire-based μs-Si MOSFETs and μs-GaAs MESFETs, respectively, can be integrated into basic examples of building blocks for digital applications with particular functions (e.g., ring oscillator, differential amplifier, and NAND logic gates).[86,87] Figure 13.12a shows, as an example, optical images, schematic illustrations and electrical measurements of a five-stage ring oscillator. The load and drive devices have channel widths of 30 μm and 200 μm, respectively. The values for L_c and L_o in both cases are 4 μm and 2.5 μm, respectively. This device exhibits a well-defined waveform at a supply voltage, V_{DD} = 4 V. The response shows a frequency of 8.1 MHz, which corresponds to a stage delay of 12 ns. The operating voltages are much lower than those reported for ring oscillators fabricated using organic transistors, polycrystalline Si, and nanowire transistors.[4,88–90] Additional improvements in device design, such as reduction of contact overlap and channel length, should lead to higher oscillation frequencies. Furthermore, the circuit of Fig. 13.12a is supported by a thin PI substrate, which provides bendability to radii of curvature of 0.6 cm (corresponding to surface strains of ~0.2%) without causing the oscillator to fail.

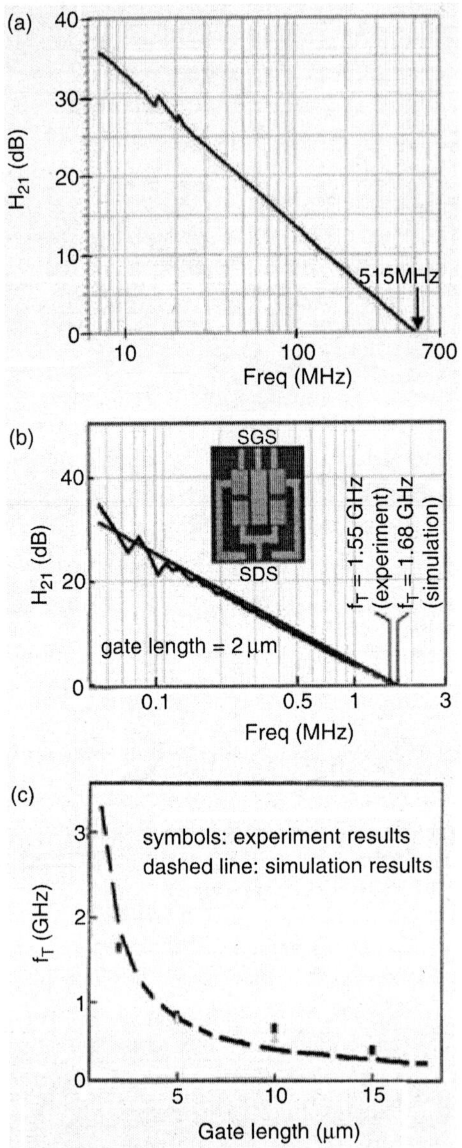

Figure 13.11. High-frequency behavior of TFTs built with semiconductor wires and ribbons. (a) μs-Si MOSFETs on PI substrates. (Reprinted with permission from Ref. 17. Copyright 2006 IEEE) (b) μs-GaAs MESFETs on PET substrates. (c) Dependence of f_T on gate length of μs-GaAs MESFETs. The different symbols represent measurements on different devices; the dashed line corresponds to calculation. (Reprinted with permission from Ref. 82. Copyright 2006 American Institute of Physics.)

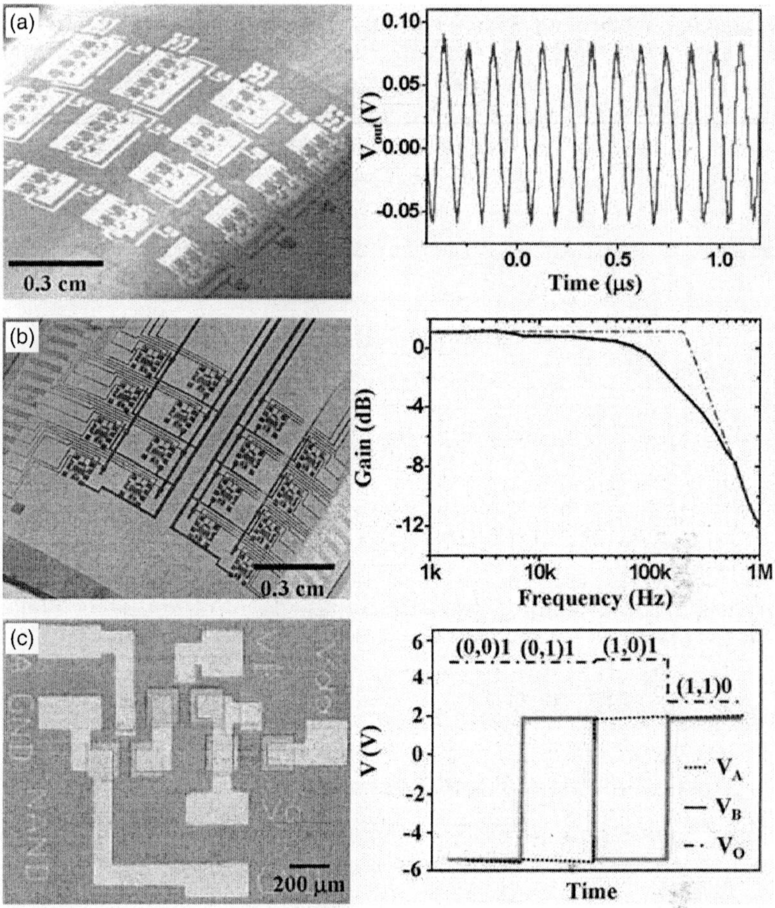

Figure 13.12. Images and electrical measurements of integrated circuits that use semiconductor wires and ribbons on plastic. (a) Silicon ribbon five-stage ring oscillator and (b) differential amplifier. (Reprinted with permission from Ref. 87. Copyright 2007 American Institute of Physics.) (c) GaAs logic gates. (Reprinted with permission from Ref. 86. Copyright 2006 Wiley-VCH Verlag.)

Differential amplifiers are useful building blocks for structural health monitor systems, in which the circuit is wrapped around a curved rigid structure where it remains for monitoring purposes, as shown in Fig. 13.12b.[87] Such devices can be built with μ*s*-Si n-channel MOSFETs on flexible PI substrates. This differential amplifier consists of four different components, which include a current source (three TFTs with $L_c = 30\,\mu m$ and $W = 80\,\mu m$), current mirror (two TFTs with $L_c = 40\,\mu m$, $W = 120\,\mu m$ and $L_c = 20\,\mu m$, $W = 120\,\mu m$), differential pair (two TFTs with $L_c = 30\,\mu m$ and $W_c = 180\,\mu m$), and load (two TFTs with $L_c = 40\,\mu m$ and $W = 80\,\mu m$). The L_o values of these transistors are $15\,\mu m$. This amplifier exhibits an output voltage gain of ~1.3 for an input signal with amplitude 0.5V peak-to-peak (VPP), consistent with the P-SPICE simulations used to design the circuit, and can be operated in a bent configuration as shown in Fig. 13.12b (right). These circuits were specifically designed for preconditioning of low-amplitude analog signals in flexible strain sensors arrays. Such large-area sensor arrays can be used to perform real-time, in-flight monitoring of the deformation of aircraft panels. In this application, the differential to single-ended conversion and high-impedance-to-low-impedance conversions need to be performed on the sensor substrate in order to avoid the use of long metal lines (i.e., interconnects), which can act as antennas that pick up noise.

As another example of circuit integration, three μs-GaAs MESFETs with different channel lengths can be combined to form logic elements such as NAND gates (Fig. 13.12c).[82,86] One MESFET serves as the load, and two identical MESFETs serve as the switching transistors, which have channel lengths of 100 and 50 μm, respectively. All of these devices have channel widths of 150 μm and gate lengths of 5 μm. In the configuration of a NAND gate, when both switching transistors (V_A and V_B) are turned on by applying a high voltage (logic 1), a large current flows through the drain (V_{DD}) of the load transistor to ground (GND), resulting in a low output voltage (V_O) (logic 0). A high output (logic 1) results if one or both of the inputs to the gate are low. The dependence of the output on the inputs of the NAND gate is shown in Fig. 13.12c (right).

13.5.2 Three-Dimensional and Heterogeneous Integration

Many emerging electronic systems benefit from the heterogeneous, three-dimensional integration (3D-HGI) of a variety of disparate materials into a single system that is capable of performing several different functions.[28,91–96] For example, communication devices and infrared (IR) imaging cameras need high-performance compound semiconductors such as GaAs to handle high-frequency radio signals or IR detection, but they might exploit much less-expensive silicon CMOS for data processing. Two primary modes of manufacturing exist for such systems—wafer bonding[55,97] or epitaxial growth[98,99]—but neither is suitable for the types of flexible macroelectronic systems that form the focus of the current chapter. As described in the previous section, the etching and printing process can be used to achieve the goal of constructing 3D-HGI on plastic.[42] This fabrication approach has several attractive features. As a result of the low-temperature processing, it avoids differential thermal expansion and shrinkage effects that can result in multilayer stacked systems; it is applicable to broad classes of semiconductor nanomaterials, including emerging materials such as SWNTs; the integration is nondestructive because of the use of soft elastomeric transfer stamps; layer-to-layer electrical interconnects can be accomplished by direct metallization because of the ultra-thin device geometries and interlayers.

The process begins with the synthesis of different semiconductor nanomaterials (e.g., single-walled carbon nanotubes and single-crystalline nanowires/

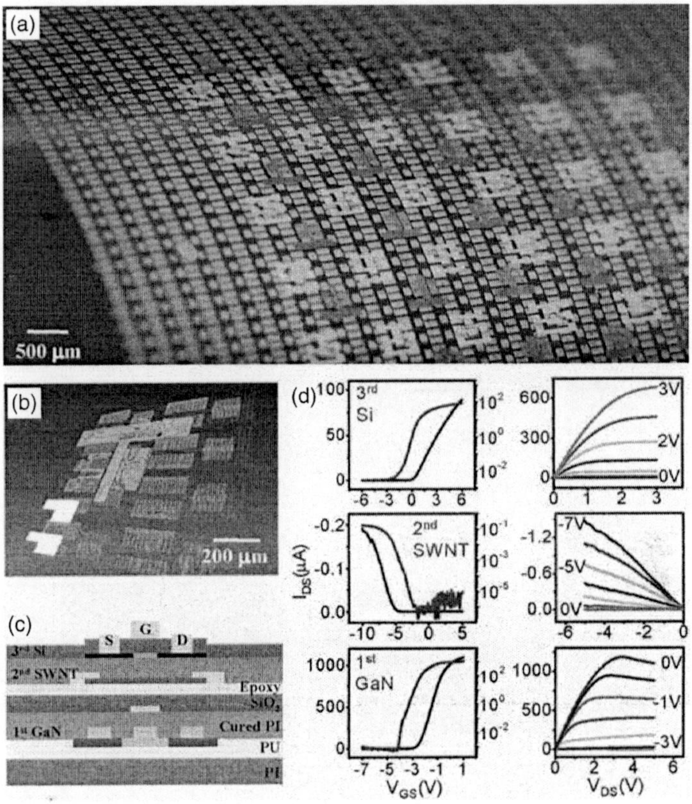

Figure 13.13. 3D heterogeneous integrated devices on plastic, including μs-GaN HEMTs, ms-Si MOSFET, and SWNT TFTs, in a three-layer stack. (a) Large-area image of the 3D circuit. (b) 3D image collected by confocal microscopy. (Reprinted with permission from Ref. 42. Copyright 2006 American Association for the Advancement of Science.) (c) Schematic cross-sectional view. (d) Full transfer and current-voltage characteristics of the three layers. (Reprinted with permission from Ref. 42. Copyright 2006 American Association for the Advancement of Science.)

ribbons of GaN, Si, and GaAs) on separate substrates using procedures described in previous sections.[17,71,81,82] Printing, using approaches outlined previously, delivers these materials to a target substrate, where interconnects and/or additional layers are formed. This process is then repeated to create three-dimensional, functional devices. Because the polymer adhesive layers are thin, small via holes can easily be etched to allow connections between selected devices in different layers. Figure 13.13 shows high magnification optical (a) and confocal (b) images, schematic illustration of cross-sectional view (c), and electrical measurements (d) of heterogeneous multiple layers on PI, with a bottom layer of μs-GaN HEMTs followed by layers of SWNT TFTs and μs-Si MOSFETs. The μs-GaN HEMTs have channel lengths, widths, and gate widths of 20, 170, and 5 μm, respectively. The SWNT TFTs have Cr/Au (2/20 nm) electrodes and a SiO_2/epoxy (300/150 nm) gate dielectric, with channel lengths and widths of 50 and 200 μm, respectively. The μs-Si MOSFETs use the same design as those shown in Fig. 13.10. These devices have similar characteristics to those fabricated on the source wafers. The μs-GaN HEMTs have threshold voltages (V_{th}) of −2.4 ± 0.2 V, ON/OFF ratios of $>10^6$, and transconductances of 0.6 ± 0.5 mS; the SWNT TFTs have V_{th} of −5.3 ± 1.5 V, ON/OFF ratios of $>10^5$, and linear mobilities of 5.9 ± 2.0 cm^2/Vs; and the μs-Si MOSFETs have V_{th} of 0.2 ± 0.3 V, ON/OFF ratios of $>10^4$, and linear mobilities of 500 ± 30 cm^2/Vs.

Good mechanical bendability is critically important for applications in flexible macroelectronics. To examine durability, bending fatigue tests were performed for each device layer using the apparatus shown in Fig. 13.14a. Figure 13.14b presents transconductances (g_m) normalized to the transconductance

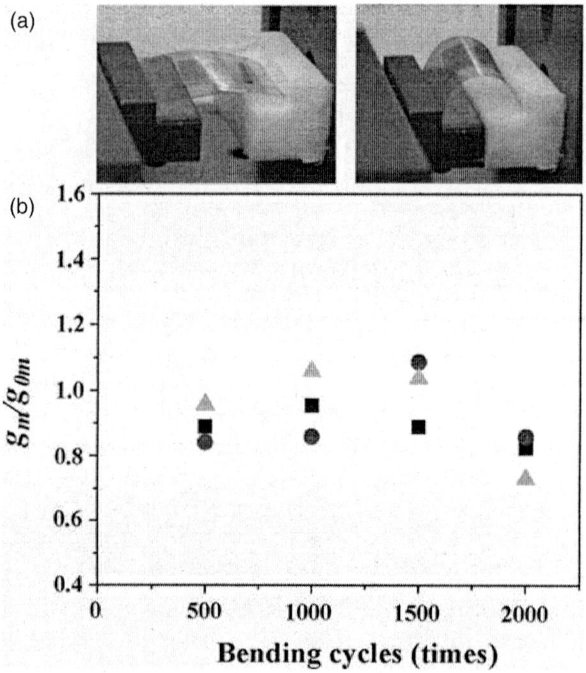

Figure 13.14. Bending tests on three-layer stacks of Si, GaN, and SWNT devices supported by a plastic substrate. (a) Images of the automated mechanical stages used to bend the system. (b) Normalized transconductances (g_m/g_{0m}) of devices in each layer after bending (to a 3.7-mm radius) and unbending (to a flat state) the devices several thousand times. (squares: μs-Si MOSFETs; circles: SWNT TFTs; triangles: μs-GaN HEMTs.) (Reprinted with permission from Ref. 42. Copyright 2006 American Association for the Advancement of Science.)

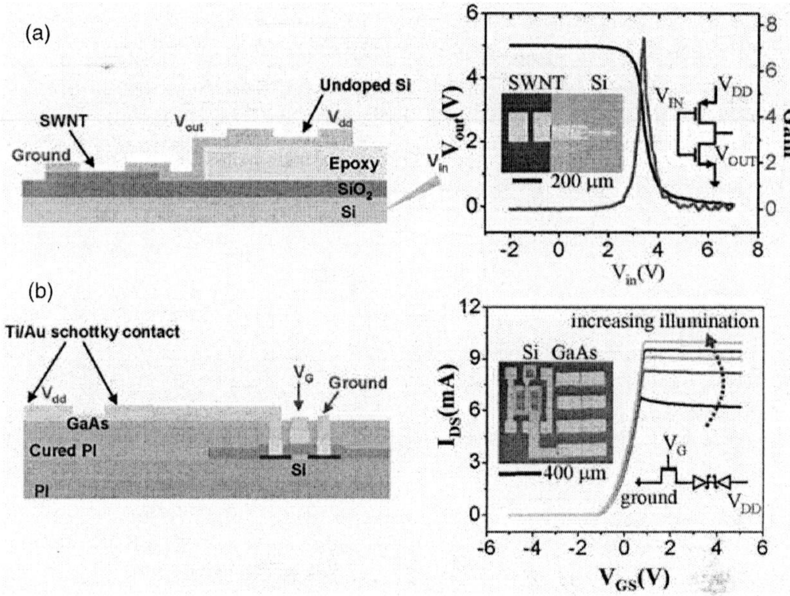

Figure 13.15. (a) Schematic illustration of the cross section of a hybrid SWNT-Si CMOS inverter built on a silicon wafer substrate (left) and transfer characteristic, optical micrograph, and circuit schematic (right). (Reprinted with permission from Ref. 42. Copyright 2006 American Association for the Advancement of Science.) (b) Schematic illustration of the cross section of a GaAs MSM-Si TFT IR detector built on a polyimide substrate (left) and current-voltage response at different levels of illumination with an infrared light source at 850 nm, optical image, and circuit schematic (right). (Reprinted with permission from Ref. 42. Copyright 2006 American Association for the Advancement of Science.)

in the unbent state (g_{0m}), as a function of bending radius. No major changes in device properties are observed for bend radii down to 3.7 mm.

Electrical interconnections formed between different levels in these 3D-HGI devices can create interesting circuit capabilities.[42] The thin polymer interlayers allow these interconnects to be formed easily by evaporating metal lines over and into lithographically defined openings. Figure 13.15 presents some examples. The first example, shown in Fig. 13.15a, is an inverter with a complementary design (CMOS) by use of integrated n-channel μ*s*-Si MOSFETs and p-channel SWNT TFTs, using layouts that equalize the current-driving capability in both pull-up and pull-down directions. The dependence of V_{out} on V_{in} in a Si–SWNT-based TFT inverter with a 5-V supply voltage is shown in Fig. 13.15a. The voltage gain of this inverter is ~7, which is qualitatively consistent with numerical circuit simulations. Another example is GaAs metal–semiconductor–metal (MSM) IR detectors integrated with μ*s*-Si

MOSFETs on PI substrates, which demonstrates a capability for fabricating unit cells that could be implemented in active IR imagers on flexible surfaces. Electrodes of Ti/Au (5/70 nm) deposited on the ends of the GaAs nanoribbons (thicknesses, widths, and lengths of 270 nm, 100 μm, and 400 μm, respectively) form back-to-back Schottky diodes with separations of 10 μm. The resulting detector cells exhibit current enhancement as the intensity of IR illumination increases, in good agreement with P-SPICE simulation. A responsivity of about 0.30 A/W at a wavelength of 850 nm is observed in the range from 1 to 5 V. The bendability of this system, which is comparable with that of the devices in Fig. 13.14, could be useful for advanced systems such as curved focal plane arrays for wide-angle IR night vision imagers.

13.6 μs-Sc ELECTRONICS ON RUBBER

Mechanical stretchability is an important characteristic for extreme bendability in flexible macroelectronics. In addition, this feature is required for systems that require conformal integration with complex curvilinear surfaces such as aircraft wings for structural health monitors or hemispheres for electronic eye type imagers. Bendability can be obtained by using "thin" electronic materials (e.g., μs-Sc) on "thin" substrates (e.g., plastic sheets). Such systems can be rolled into cylinders or cone shapes, neither of which requires significant stretching. A strategy to achieve stretchability, which represents a much more challenging mechanical characteristic than bendability, is to use semiconductor materials preformed into "wavy" shapes that can evolve, similar to an accordion bellows, in response to applied strains. This approach provides an effective stretchability that does not require the materials themselves to experience significant strains. Such "wavy" layouts can be obtained using the well-known "mechanical buckling" phenomena. This section provides a brief summary of stretchable inorganic semiconductor ribbons and their implementation in stretchable electronic devices.

Buckling (or wrinkling) instabilities represent a well-known phenomenon in mechanics, and it has been extensively studied.[100] When a stiff, thin layer of material bonded to a compliant substrate is compressed, the layer relieves the strain by buckling to form periodic, out-of-plane undulations of the layer and the near-surface regions of substrate. Controlled buckling of thin metal films on elastomeric substrates can be used to achieve stretchable interconnects for rigid device islands, as a route to stretchable electronics.[101] Another approach relies on devices that themselves, in addition to the interconnects, are stretchable. Here, "wavy" semiconductor materials and devices are required. The process for fabricating such elements starts with thin strips, or ribbons, of single-crystalline Si or GaAs, or integrated devices, formed using the methods described previously, followed by bonding to prestrained elastomeric (PDMS) supports. Relaxing the prestrain forms buckled, "wavy" device elements that can accommodate externally applied strains by changing their

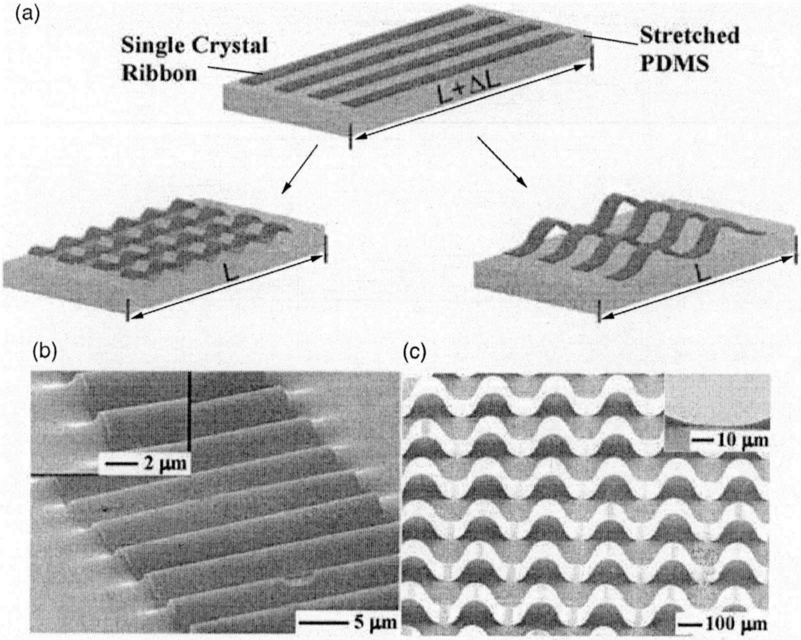

Figure 13.16. (a) Schematic illustration of procedures for fabricating "wavy" and "buckled" semiconductor nanoribbons on elastomeric PDMS substrates. Scanning electron micrographs of wavy Si (b) and buckled GaAs ribbons (c) on elastomeric substrates. (Reprinted with permission from Ref. 23. Copyright 2006 Nature Publishing Group.)

shapes, mainly their wavelengths and amplitudes.[19,23] The bonding in this case derives from the —OH chemistries described in the printing section.[102]

Figure 13.16a shows a schematic illustration of steps for generating a wavy and a buckled structure. Single-crystalline Si or GaAs ribbons can be derived from SOI or epitaxial GaAs wafers using procedures described previously. Before making contact between the released ribbons (which rest on the mother substrates) and the PDMS, the PDMS is stretched, by either thermal or mechanical means. Bonding the ribbons to the PDMS upon contact, followed by release of prestrain, leads the stiff ribbons [E(Young's modulus, Si or GaAs) ~ 100 GPa, cf. E(PDMS) ~ 2 MPa] to buckle into very regular, sinusoidal "wavy" shapes to relieve the compressive strain acting on them by the contracting PDMS substrate, as shown in Fig. 13.16b. In this implementation, the mechanics define the wavelengths, and the ribbons are intimately bonded to the PDMS along their entire lengths. Structures optimized for extreme levels of stretchability can be prepared by selectively activating the PDMS for strong bonding, such that relaxing the prestrain causes the ribbons to delaminate from the PDMS in a pattern consistent with the geometry of the bonding

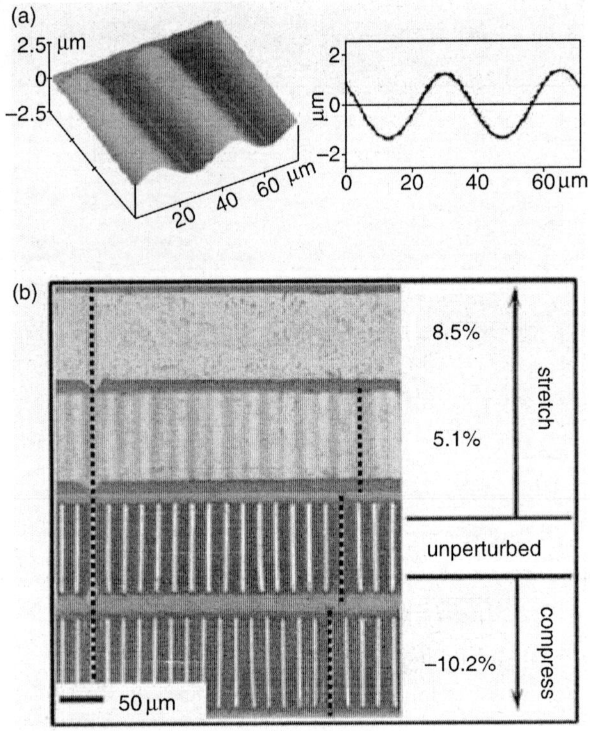

Figure 13.17. (a) AFM images of wavy GaAs ribbons formed on a PDMS substrate prestretched by 1.9%. (Reprinted with permission from Ref. 25. Copyright 2006 Wiley-VCH Verlag.) (b) Optical micrographs of wavy GaAs ribbons formed with a prestrain of 7.8%, collected at different applied strains. The dotted line on the left and right highlight certain peaks in the structure; the variation in the distance between these lines indicates the dependence of the wavelength on applied strain. (Reprinted with permission from Ref. 25. Copyright 2006 Wiley-VCH Verlag.)

sites. The result is structures that involve large-amplitude, bridge-like, air-gap structures (Fig. 13.16c) that are capable of accommodating strains of up to 100%. In practical embodiments, such structures are encapsulated on top with additional PDMS to eliminate the air gaps and to provide fully reversible stretching behaviors.

Both classes of these "wavy" semiconductor structures on PDMS can be reversibly stretched or compressed. Figures 13.17a and b show 3D AFM images of wavy GaAs ribbons formed on a PDMS substrate prestretched by 1.9% and optical images of wavy GaAs ribbons embedded in PDMS. When the initially wavy GaAs ribbons are stretched, their amplitude decreases and wavelength increases. The opposite is true for the compression case.

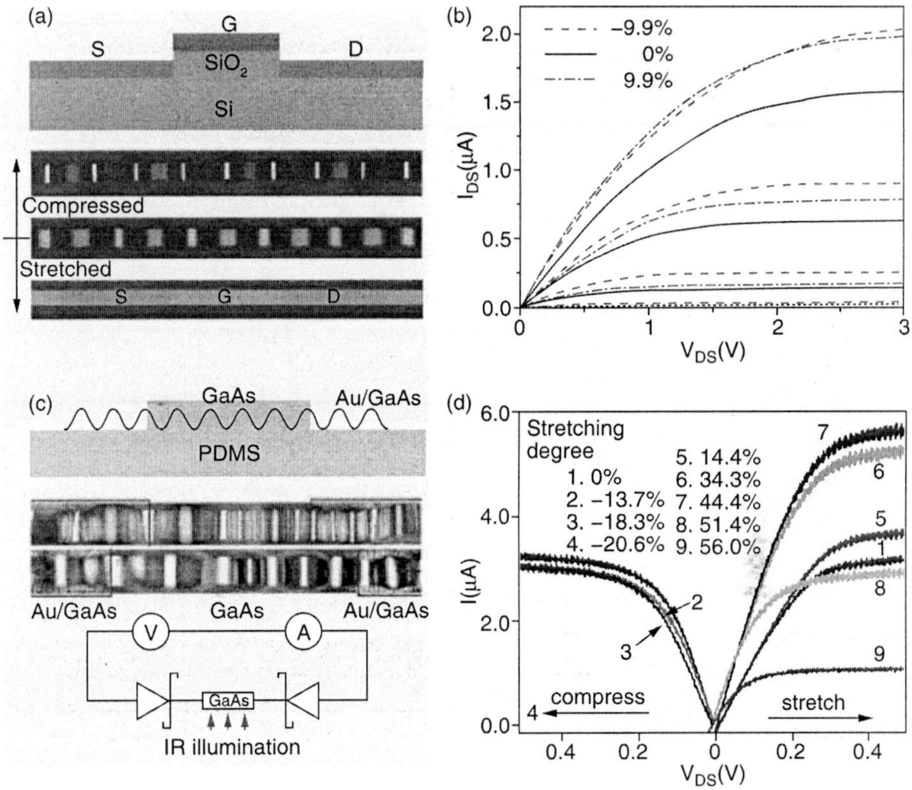

Figure 13.18. (a) Schematic illustration (top) and optical images (bottom) and (b) current-voltage characteristics of a stretchable Si ribbon MOSFET on a PDMS substrate stretched at different levels. The gate voltage varied from 0V to −5V, with a 1-V step. (Reprinted with permission from Ref. 19. Copyright 2006 American Association for the Advancement of Science.) (c) Schematic illustration (top), optical images before and during stretching (middle), and equivalent circuit (bottom) corresponding to a buckled GaAs diode. (Reprinted with permission from Ref. 23. Copyright 2006 Nature Publishing Group.) (d) Current/voltage curves recorded from a buckled photodetector stretched or compressed to different degrees with constant luminance.

The response is entirely consistent with the physics of an accordion bellows.

For functional, stretchable electronic devices on PDMS, all device processing steps, especially high-temperature processes such as doping and contact metallization, are performed on the mother substrate. Subsequently, ribbons with integrated device layers are made into wavy geometries using the same processes mentioned above. As an example, Fig. 13.18 presents a stretchable MOSFET device. Figure 13.18a shows the schematic cross-sectional geometry

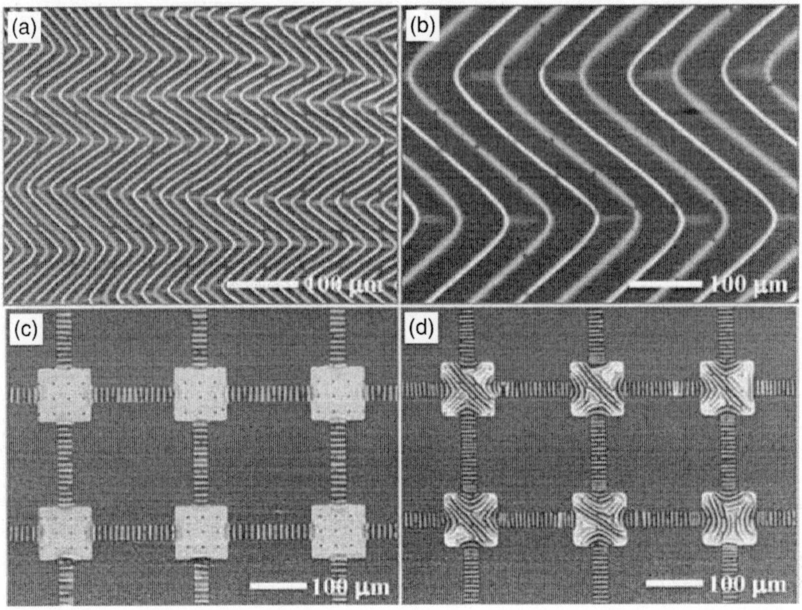

Figure 13.19. Optical images of 2D wavy uniform Si nanomembranes with thickness of 100 nm (a) and 320 nm (b) on PDMS, formed with a thermal prestrain of 3.8%, and structured Si nanomembranes formed with a thermal prestrain of 2.3% (c) and 15% (d). (Reprinted with permission from Ref. 103. Copyright 2007 American Chemical Society.)

(top) and optical images (bottom) and (b) shows the output characteristics. Upon applying strains between −10% and 10%, the transistors continue to operate well. Stretchable GaAs photodiodes are also shown in Figs. 13.18c and d.

Complete 2D stretchability can also be achieved using 2D nanomembranes of semiconductor materials and the same type of buckling mechanics.[103] Figure 13.19 shows some representative images of two-dimensionally buckled, wavy Si nanomembranes derived from SOI wafers. The regular array of small holes (~2 μm in diameter and ~25 μm spaced apart) provides access for the wet chemical etchant for the undercut. When bonded to a biaxially prestretched PDMS (by heating at high temperature, typically 70 °C ~ 150 °C), release of the strain causes the Si nanomembrane to undergo a two-dimensional buckling process. The buckled shape varies with location across the sample; simple "wavy" buckling patterns are near the edges, herringbone (or zigzag) structures are further toward the center part of the sample, and labyrinth (or random) buckling patterns are in the middle. Figures 13.19a and b show the herringbone layout. The edge effects can be exploited to confine buckling at some locations while avoiding buckling entirely in other regions. To demonstrate this possibility, a membrane with an array of square islands (100 μm ×

100 μm) connected by 30-μm × 150-μm ribbons was fabricated and buckled. When the compressive strain is small, the buckling occurs only along the connecting ribbons, not on the square islands (Fig. 13.19c; strain ~2.3%). At sufficiently large strains (~15%), buckling begins to appear on the square island surfaces, as shown in Fig. 13.19d.

13.7 CONCLUSION

In this chapter, we reviewed materials and methods that enable high-performance semiconductor devices on flexible and even stretchable substrates, in two- or three-dimensional layouts. The methods use specialized rubber "stamps" with functional "inks" consisting of high-performance semiconductor materials in the form of micro/nanoscale ribbons, sheets, wires, tubes, and bars. A printing operation delivers these materials to virtually any type of substrate, including lightweight, flexible plastic sheets. This stamping method has several advantages, including low-temperature operation that avoids differential thermal expansion and degradation effects, applicability to broad classes of semiconductor nanomaterials, and an ability to form nondestructive contacts with underlying device layers. These features lead to devices and circuit systems that overcome many of the disadvantages of conventional approaches. Successful commercial implementation of such techniques represents a significant engineering challenge, but one that would enable wide-ranging new applications for electronics, including not only those that are traditionally associated with flexible macroelectronics but also others, such as hemispherical electronic eye imagers, smart sensory surgical gloves, and personal health monitors, that could represent important technologies for the future.

REFERENCES

1. Reuss, R. H.; et al. 2005. Macroelectronics: Perspectives on technology and applications. *Proc. IEEE* 97:1239–1256.
2. Service, R. F. 2006. Inorganic electronics begin to flex their muscle. *Science* 312:1593–1594.
3. Reuss, R. H.; Hopper, D. G.; Park, J.-G. 2006. Macroelectronics. *MRS Bull.* 31: 447–450.
4. Kane, M. G.; Goodman, L.; Firester, A. H.; van der Wilt, P. C.; Limanov, A. B.; Im, J. S. 2005. 100 MHz CMOS circuits using sequential laterally solidified silicon thin-film transistors on plastic. *IEDM Tech. Dig. Int. Electron Dev. Meet.* art.no.1609514:939–941.
5. Peumans, P.; Uchida, S.; Forrest, S. R. 2003. Efficient bulk heterojunction photovoltaic cells using small-molecular-weight organic thin films. *Nature* 425: 158–162.

6. Gelinck, G. H.; et al. 2004. Flexible active-matrix displays and shift registers based on solution-processed organic transistors. *Nature Mater.* 3:106–110.
7. Afzali, A.; Dimitrakopoulos, C. D.; Breen, T. L. 2002. High-performance, solution processed organic thin film transistors from a novel pentacene precursor. *J. Amer. Chem. Soc.* 124:8812–8813.
8. Ridley, B. A.; Nivi, B.; Jacobson, J. M. 1999. All-inorganic field effect transistors fabricated by printing. *Science* 286:746–749.
9. Mitzi, D. B.; Kosbar, L. L.; Murray, C. E.; Copel, M.; Afzali, A. 2004. High-mobility ultrathin semiconducting films prepared by spin coating. *Nature* 428: 299–303.
10. Mitzi, D. B.; Copel, M.; Chey, S. J. 2005. Low-voltage transistor employing a high-mobility spin-coated chalcogenide semiconductor. *Adv. Mater.* 17: 1285–1289.
11. Kagan, C. R.; Mitzi, D. B.; Dimitrakopoulos, C. D. 1999. Organic-inorganic hybrid materials as semiconducting channels in thin-film field-effect transistors. *Science* 286:945–947.
12. McAlpine, M. C.; Friedman, R. S.; Lieber, C. M. 2005. High-performance nanowire electronics and photonics and nanoscale patterning on flexible plastic substrates. *Proc. IEEE* 93:1357–1363.
13. Menard, E.; Lee, K. J.; Khang, D. Y.; Nuzzo, R. G.; Rogers, J. A. 2004. A printable form of silicon for high performance thin film transistors on plastic. *Appl. Phys. Lett.* 84:5398–5400.
14. Sun, Y.; Rogers, J. A. 2004. Fabricating semiconductor nano/microwires and transfer printing ordered arrays of them onto plastic substrates. *Nano Lett.* 4:1953–1959.
15. van der Wilt, P. C.; Kane, M. G.; Limanov, A. B.; Firester, A. H.; Goodman, L.; Lee, J.; Abelson, J. R.; Chitu, A. M.; Im, J. S. 2006. Low-temperature polycrystalline silicon thin-film transistors and circuits on flexible substrates. *MRS Bull.* 31:461–465.
16. Voutsas, A. T. 2007. The role of structural defects and texture variability in the performance of poly-Si thin film transistors. *Thin Solid Films*. 515:7406–7412.
17. Ahn, J.-H.; Kim, H.-S.; Lee, K. J.; Zhu, Z.-T.; Menard, E.; Nuzzo, R. G.; Rogers, J. A. 2006. High-speed mechanically flexible single-crystal silicon thin-film transistors on plastic substrates. *IEEE Electron Dev. Lett.* 27:460–462.
18. Baca, A. J.; Meitl, M. A.; Ko, H. C.; Mack, S.; Kim, H.-S.; Dong, J.; Ferreira, P. M.; Rogers, J. A. 2007. Printable single-crystal silicon micro/nanoscale ribbons, platelets and bars generated from bulk wafers. *Adv. Funct. Mater.* 17:3051–3062.
19. Khang, D.-Y.; Jiang, H.; Huang, Y.; Rogers, J. A. 2006. A stretchable form of single-crystal silicon for high-performance electronics on rubber substrates. *Science* 311:208–212.
20. Ko, H. C.; Baca, A. J.; Rogers, J. A. 2006. Bulk quantities of single-crystal silicon micro-/nanoribbons generated from bulk wafers. *Nano Lett.* 6:2318–2324.
21. Mack, S.; Meitl, M. A.; Baca, A. J.; Zhu, Z.-T.; Rogers, J. A. 2006. Mechanically flexible thin-film transistors that use ultrathin ribbons of silicon derived from bulk wafers. *Appl. Phys. Lett.* 88:213101.

22. Menard, E.; Nuzzo, R. G.; Rogers, J. A. 2005. Bendable single crystal silicon thin film transistors formed by printing on plastic substrates. *Appl. Phys. Lett.* 86: 093507.

23. Sun, Y.; Choi, W.-M.; Jiang, H.; Huang, Y. Y.; Rogers, J. A. 2006. Controlled buckling of semiconductor nanoribbons for stretchable electronics. *Nature Nanotechnol.* 1:201–207.

24. Sun, Y.; Khang, D.-Y.; Hua, F.; Hurley, K.; Nuzzo, R. G.; Rogers, J. A. 2005. Photolithographic route to the fabrication of micro/nanowires of III-V semiconductors. *Adv. Funct. Mater.* 15:30–40.

25. Sun, Y.; Kumar, V.; Adesida, I.; Rogers, J. A. 2006. Buckled and wavy ribbons of GaAs for high-performance electronics on elastomeric substrates. *Adv. Mater.* 18:2857–2862.

26. Zhu, Z.-T.; Menard, E.; Nuzzo, R. G.; Rogers, J. A. 2005. Spin on dopants for high-performance single-crystal silicon transistors on flexible plastic substrates. *Appl. Phys. Lett.* 86:133507.

27. Kuhn, M.; Silversmith, D. 1971. Ionic contamination and transport of mobile ions in MOS structures. *J. Electrochem. Soc.* 118:966–970.

28. Banerjee, K.; Souri, S. J.; Kapur, P.; Saraswat, K. C. 2001. 3-D ICs: A novel chip design for improving deep-submicrometer interconnect performance and systems-on-chip integration. *Proc. IEEE* 89:602–633.

29. Stauth, S. A.; Parviz, B. A. 2006. Self-assembled single-crystal silicon circuits on plastic. *Proc. Natl. Acad. Sci. USA* 103:13922–13927.

30. Aoki, K.; et al. 2003. Microassembly of semiconductor three-dimensional photonic crystals. *Nature Mater.* 2:117–121.

31. Dechev, N.; Cleghorn, W. L.; Mills, J. K. 2004. Microassembly of 3-D microstructures using a compliant, passive microgripper. *J. Microelectromechan. Sys.* 13:176–189.

32. Miyazaki, H. T.; Miyazaki, H.; Ohtaka, K.; Sato, T. 2000. Photonic band in two-dimensional lattices of micrometer-sized spheres mechanically arranged under a scanning electron microscope. *J. Appl. Phys.* 87:7152–7158.

33. Xia, Y. N.; Whitesides, G. M. 1998. Soft lithography. *Ann. Rev. Mater. Sci.* 28:153–184.

34. Hua, F.; et al. 2004. Polymer imprint lithography with molecular-scale resolution. *Nano Lett.* 4:2467–2471.

35. Loo, Y. L.; Willett, R. L.; Baldwin, K. W.; Rogers, J. A. 2002. Additive, nanoscale patterning of metal films with a stamp and a surface chemistry mediated transfer process: Applications in plastic electronics. *Appl. Phys. Lett.* 81:562–564.

36. Odom, T. W.; Thalladi, V. R.; Love, J. C.; Whitesides, G. M. 2002. Generation of 30–50 nm structures using easily fabricated, composite PDMS masks. *J. Am. Chem. Soc.* 124:12112–12113.

37. Rogers, J. A.; Jackman, R. J.; Whitesides, G. M. 1997. Constructing single- and multiple-helical microcoils and characterizing their performance as components of microinductors and microelectromagnets. *J. Microelectromechan. Sys.* 6: 184–192.

38. Rogers, J. A.; Jackman, R. J.; Whitesides, G. M. 1997. Microcontact printing and electroplating on curved substrates: Production of free-standing three-dimensional metallic microstructures. *Adv. Mater.* 9:475–477.
39. Meitl, M. A.; et al. 2004. Solution casting and transfer printing single-walled carbon nanotube films. *Nano Lett.* 4:1643–1647.
40. Meitl, M. A.; et al. 2006. Transfer printing by kinetic control of adhesion to an elastomeric stamp. *Nature Mater.* 5:33–38.
41. Lee, K. J.; et al. 2005. Large-area, selective transfer of microstructured silicon: A printing-based approach to high-performance thin-film transistors supported on flexible substrates. *Adv. Mater.* 17:2332–2336.
42. Ahn, J.H.; et al. 2006. Heterogeneous three-dimensional electronics by use of printed semiconductor nanomaterials. *Science* 314:1754–1757.
43. Loo, Y. L.; Willett, R. L.; Baldwin, K. W.; Rogers, J. A. 2002. Interfacial chemistries for nanoscale transfer printing. *J. Amer. Chem. Soc.* 124:7654–7655.
44. Loo, Y. L.; et al. 2002. High-resolution transfer printing on GaAs surfaces using alkane dithiol monolayers. *J. Vac. Sci. & Technol. B* 20:2853–2856.
45. Zaumseil, J.; et al. 2003. Three-dimensional and multilayer nanostructures formed by nanotransfer printing. *Nano Lett.* 3:1223–1227.
46. Lee, K. J.; Tosser, K. A.; Nuzzo, R. G. 2005. Fabrication of stable metallic patterns embedded in poly(dimethylsiloxane) and model applications in non-planar electronic and lab-on-a-chip device patterning. *Adv. Funct. Mater.* 15:557–566.
47. Ferguson, G. S.; Chaudhury, M. K.; Sigal, G. B.; Whitesides, G. M. 1991. Contact adhesion of thin gold-films on elastomeric supports—Cold welding under ambient conditions. *Science* 253:776–778.
48. Kim, C.; Burrows, P. E.; Forrest, S. R. 2000. Micropatterning of organic electronic devices by cold-welding. *Science* 288:831–833.
49. Kim, C.; Forrest, S. R. 2003. Fabrication of organic light-emitting devices by low-pressure cold welding. *Adv. Mater.* 15:541–545.
50. Huang, Y. G. Y.; et al. 2005. Stamp collapse in soft lithography. *Langmuir* 21:8058–8068.
51. Zhou, W.; et al. 2005. Mechanism for stamp collapse in soft lithography. *Appl. Phys. Lett.* 87:251925.
52. Hsia, K. J.; et al. 2005. Collapse of stamps for soft lithography due to interfacial adhesion. *Appl. Phys. Lett.* 86:154106.
53. Hur, S. H.; Khang, D. Y.; Kocabas, C.; Rogers, J. A. 2004. Nanotransfer printing by use of noncovalent surface forces: Applications to thin-film transistors that use single-walled carbon nanotube networks and semiconducting polymers. *Appl. Phys. Lett.* 85:5730–5732.
54. Meitl, M. A.; et al. 2007. Stress focusing for controlled fracture in microelectromechanical systems. *Appl. Phys. Lett.* 90:083110.
55. Tong, Q.-Y.; Gosele, U. 1999. *Semiconductor Wafer Bonding: Science and Technology*. John Wiley, New York.
56. Haisma, J.; Spierings G. A. C. M. 2002. Contact bonding, including direct-bonding in a historical and recent context of materials science and technology, physics and chemistry—Historical review in a broader scope and comparative outlook. *Mater. Sci. & Eng. R-Reports* 37:1–60.

57. Vitale, S. A.; Chae, H.; Sawin, H. H. 2000. Etching chemistry of benzocyclobutene (BCB) low-k dielectric films in F-2+O-2 and Cl-2+O-2 high density plasmas. *J. Vac. Sci. & Technol. A* 18:2770–2778.

58. Saulnier, F.; Ondarcuhu, T.; Aradian, A.; Raphael, E. 2004. Adhesion between a viscoelastic material and a solid surface. *Macromolecules* 37:1067–1075.

59. Roberts, A. D. 1979. Looking at rubber adhesion. *Rubber Chem. Technol.* 52:23–42.

60. Barquins, M. 1992. Adherence, friction and wear of rubber-like materials. *Wear* 158:87–117.

61. Shull, K. R.; Ahn, D.; Chen, W. L.; Flanigan, C. M.; Crosby, A. J. 1998. Axisymmetric adhesion tests of soft materials. *Macromol. Chem. Phys.* 199:489–511.

62. Degennes, P. G. 1995. Spontaneous adhesion between rubbers and bare solids. *Comptes Rendus Acad. Sci. Serie Ii* 320:193–197.

63. Anderson, T. L. 1995. *Fracture Mechanics: Fundamentals and Applications*. CRC Press, New York.

64. Inglis, C. E. 1913. Stresses in a plate due to the presence of cracks and sharp corners. *Trans. Instit. Naval Architects* 55:219–241.

65. Yao, Z.; Kane, C. L.; Dekker, C. 2000. High-field electrical transport in single-wall carbon nanotubes. *Phys. Rev. Lett.* 84:2941–2944.

66. Javey, A.; et al. 2002. High-kappa dielectrics for advanced carbon-nanotube transistors and logic gates. *Nature Mater.* 1:241–246.

67. Kocabas, C.; et al. 2005. Guided growth of large-scale, horizontally aligned arrays of single-walled carbon nanotubes and their use in thin-film transistors. *Small* 1:1110–1116.

68. Kocabas, C.; Shim, M.; Rogers, J. A. 2006. Spatially selective guided growth of high-coverage arrays and random networks of single-walled carbon nanotubes and their integration into electronic devices. *J. Amer. Chem. Soc.* 128: 4540–4541.

69. Cao, Q.; et al. 2006. Highly bendable, transparent thin-film transistors that use carbon-nanotube-based conductors and semiconductors with elastomeric dielectrics. *Adv. Mater.* 18:304–309.

70. Cao, Q.; et al. 2006. Transparent flexible organic thin-film transistors that use printed single-walled carbon nanotube electrodes. *Appl. Phys. Lett.* 88: 113511.

71. Hur, S. H.; et al. 2005. Printed thin-film transistors and complementary logic gates that use polymer-coated single-walled carbon nanotube networks. *J. Appl. Phys.* 98:114302.

72. Hur, S. H.; Park, O. O.; Rogers, J. A. 2005. Extreme bendability of single-walled carbon nanotube networks transferred from high-temperature growth substrates to plastic and their use in thin-film transistors. *Appl. Phys. Lett.* 86:243502.

73. Hur, S. H.; et al. 2005. Organic nanodielectrics for low voltage carbon nanotube thin film transistors and complementary logic gates. *J. Amer. Chem. Soc.* 127:13808–13809.

74. Kang, S. J.; et al. 2007. High-performance electronics using dense, perfectly aligned arrays of single-walled carbon nanotubes. *Nature Nanotechnol.* 2:230–236.

75. Rogers, J. A.; Paul, K. E.; Whitesides, G. M. 1998. Quantifying distortions in soft lithography. *J. Vac. Sci. & Technol. B* 16:88–97.

76. Menard, E.; Bilhaut, L.; Zaumseil, J.; Rogers, J. A. 2004. Improved surface chemistries, thin film deposition techniques, and stamp designs for nanotransfer printing. *Langmuir* 20:6871–6878.

77. Burdinski, D.; Brans, H. J. A.; Decre, M. M. J. 2005. Single etch patterning of stacked silver and molybdenum alloy layers on glass using microcontact wave printing. *J. Amer. Chem. Soc.* 127:10786–10787.

78. Decre, M. M. J.; Schneider, R.; Burdinski, D.; Schellekens, J.; Saalmink, M.; Dona, R. 2004. Wave printing (I): Toward large-arrea, multilayer microcontact printing. *Mater. Res. Soc. Symp. Proc.* EXS-2:M4.9.1–M4.9.3.

79. Schellekens, J.; Burdinski, D.; Saalmink, M.; Beenhakkers, M.; Gelinck, G.; Decre, M. M. J. 2004. Wave printing (II): Polymer MISFET's using microcontact printing. *Mater. Res. Soc. Symp. Proc.* EXS-2:M2.9.1–M2.9.3.

80. Biebuyck, H. A.; Larsen, N. B.; Delamarche, E.; Michel, B. 1997. Lithography beyond light: Microcontact printing with monolayer resists. *IBM J. Res. & Dev.* 41:159–170.

81. Lee, K. J.; Meitl, M. A.; Ahn, J.-H.; Rogers, J. A.; Nuzzo, R. G.; Kumar, V.; Adesida, I. 2006. Bendable GaN high electron mobility transistors on plastic substrates. *J. Appl. Phys.* 100:124507.

82. Sun, Y.; Menard, E.; Rogers, J. A.; Kim, H.-S.; Kim, S.; Chen, G.; Adesida, I.; Dettmer, R.; Cortez, R.; Tewksbury, A. 2006. Gigahertz operation in mechanically flexible transistors on plastic substrates. *Appl. Phys. Lett.* 88:183509.

83. Sze, S. M. 1981. *Physics of Semiconductor Devices*. Wiley, New York.

84. Sameshima, T.; Usui, S.; Sekiya, M. 1986. XeCl eximer laser annealing used in the fabrication of poly-Si TFT's. *IEEE Electron Dev. Lett.* EDL-7:276–278.

85. Gosain, D. P. 2002. Excimer laser crystallized poly-Si TFT's on plastic substrates. *Proc. SPIE* 4426:394–400.

86. Sun, Y.; Kim, H.-S.; Menard, E.; Kim, S.; Adesida, I.; Rogers, J. A. 2006. Printed arrays of aligned GaAs wires for flexible transistors, diodes and circuits on plastic substrates. *Small* 2:1330–1334.

87. Ahn, J. H.; et al. 2007. Bendable integrated circuits on plastic substrates by use of printed ribbons of single-crystalline silicon. *Appl. Phys. Lett.* 90:213501.

88. Stingelin-Stutzmann, N.; Smits, E.; Wondergem, H.; Tanase, C.; Blom, P.; Smith P.; De Leeuw, D. 2005. Organic thin-film electronics from vitreous solution-processed rubrene hypereutectics. *Nature Mater.* 4:601–606.

89. Afentakis, T.; Hatalis, M.; Voutsas A. T.; Hartzell, J. 2006. Design and fabrication of high-performance polycrystalline silicon thin-film transistor circuits on flexible steel foils. *IEEE Trans. Electr. Dev.* 53:815–822.

90. Friedman, F. S.; McAlpine, M. C.; Ricketts, D. S.; Ham, D.; Lieber, C. M. 2005. Nanotechnology: High-speed integrated nanowire circuits. *Nature* 434:1085.

91. Fike, S. M.; Buchanan, B.; Jokerst, N. M.; Brooke, M. A.; Morris, T. G.; DeWeerth, S. P. 1995. 8 × 8 array of thin-film photodetectors vertically electrically interconnected to silicon circuitry. *Photonics Technol. Lett., IEEE* 7:1168–1170.

92. Tseng, Y.-C.; Xuan, P.; Javey, A.; Malloy, R.; Wang, Q.; Bokor, J.; Dai, H. 2004. Monolithic integration of carbon nanotube devices with silicon MOS technology. *Nano Lett.* 4:123–127.

93. Joachim, C.; Gimzewski, J. K.; Aviram, A. 2000. Electronics using hybrid-molecular and mono-molecular devices. *Nature* 408:541–548.

94. Roelkens, G.; et al. 2005. Integration of InP/InGaAsP photodetectors onto silicon-on-insulator waveguide circuits. *Opt. Express* 13:10102–10108.

95. Strukov, D. B.; Likharev, K. K. 2005. CMOL FPGA: A reconfigurable architecture for hybrid digital circuits with two-terminal nanodevices. *Nanotechnology* 16:888–900.

96. Javey, A.; Nam, S.-W.; Friedman, R. S.; Yan, H.; Lieber, C. M. 2007. Layer-by-layer assembly of nanowires for three-dimensional, multifunctional electronics. *Nano Lett.* 7:773–777.

97. Schmidt, M. A. 1998. Wafer-to-wafer bonding for microstructure formation. *Proc. IEEE* 86:1575–1585.

98. Amano, H.; Sawaki, N.; Akasaki, I.; Toyoda, Y. 1986. Metalorganic vapor phase epitaxial growth of a high quality GaN film using AlN buffer layer. *Appl. Phys. Lett.* 48:353–355.

99. Kuykendall, T.; Pauzauskie, P. J.; Zhang, Y.; Goldberger, J.; Sirbuly, D.; Denlinger, J.; Yang, P. 2004. Crystallographic alignment of high-density gallium nitride nanowire arrays. *Nature Mater.* 3:524–528.

100. Allen, H. G. 1969. *Analysis and design of structural sandwich panels*. Pergamon Press, Tarrytown, NY.

101. Bowden, N.; Brittain, S.; Evans, A. G.; Hutchinson, J. W.; Whitesides, G. M. 1998. Spontaneous formation of ordered structures in thin films of metals supported on an elastomeric polymer. *Nature* 393:146–149.

102. Childs, W. R.; Nuzzo, R. G. 2002. Decal transfer microlithography: A new soft-lithographic patterning method. *J. Amer. Chem. Soc.* 124:13583–13596.

103. Choi, W. M.; et al. 2007. Biaxially stretchable "wavy" silicon nanomembranes. *Nano Lett.* 7:1655–1663.

CHAPTER 14

Future Directions for Solution-Based Processing of Inorganic Materials

M. F. A. M. VAN HEST and D. S. GINLEY
National Renewable Energy Laboratory, Golden, CO

14.1 INTRODUCTION

This book presents a detailed snapshot of the current state of the art in solution-processable inorganic materials and has provided a well-crafted discussion of the directions in which the field is moving. Some compelling themes emerged in the book as to why solution-based deposition is an attractive and growing area of materials science. Initially these themes are the simple comparative advantages of solution deposition versus standard physical vapor deposition (PVD) approaches, including atmospheric pressure processing (thereby requiring much less capital than vacuum-based approaches), increased applicability to large-area and flexible (topologically interesting) substrates, potentially environmentally "greener" processing, with a combination of more efficient materials usage (especially in direct write processes), and lower temperature processing. These advantages taken together point to a new and increasingly important approach to materials and device deposition with significantly reduced costs for capitalization and broad applicability. This is especially important for emerging areas, such as large-area systems, including displays, photovoltaics, and next-generation buildings, as well as other technologies, such as flexible electronics and hybrid inorganic–organic devices.

Although solution-based approaches can be used as a direct replacement for vacuum-based processes, one of the key aspects of atmospheric solution-based processing is that, by virtue of it being non vacuum and using conformal solution-based precursors, it can lead to a next generation of hybrid and nanostructured materials and devices by enabling unique composites and structures. Increasingly, there is interest in combining inorganic, organic, and even biological materials into self-organizing structures. This process can only be done through low-temperature, solution-based processing. In addition, the drive to flexible and 3D-structures lends itself especially to printing technologies.

Some of the key topics discussed in the other chapters of the book center on materials-related issues, deposition approaches, and next-generation approaches to devices and structures. In this chapter, we do not try to be inclusive with respect to the contents of the book, but rather we try to look at some of the important areas of commonality and then look forward to some exciting future directions.

14.2 MATERIALS

As can be seen in the proceeding chapters, an increasing diversity of inorganic materials can be grown and/or deposited using solution-based approaches. These materials include compound semiconductors, oxides, and metals. The availability of new precursors commercially and a growing wealth of synthetic tools have made development possible for many cations previously thought to be synthetically very difficult. Aside from nanoparticle precursors, also metal organic molecules have been designed containing one, two, or even three cations that decompose at low temperatures to produce the desired inorganic materials. Coupled to this has been a set of new deposition tools for solution-based precursors that are capable of deposition from the meter to the nanometer scale.

14.2.1 Semiconductors

The diversity of solution-deposited semiconductors is rapidly expanding. A few years ago only a very limited set of materials could be deposited by solution approaches. This set has expanded dramatically, and not only are thin films being considered, but also now growth of controlled nanostructures and even core shell structures is becoming routine. Some examples observed in this book are CdS (see Fig. 14.1 and Chapter 7), $CuInSe_2$ (see Fig. 14.2 and Chapters 3, 6, and 7), and Si (see Chapter 5). A key point is that materials such as Si, which have historically been very difficult to solution deposit because of oxidation potential, are now possible because of new precursors and synthetic approaches. One driver for this approach and choice of materials is the increasing interest in using them for photovoltaics, where large-area deposition of high-quality materials at low cost is a necessity. There is also a next generation of devices employing quantum dots (QDs), including inorganic/organic hybrids, which are becoming of more interest for a broad range of applications, including third-generation photovoltaic devices.[1–5] Solution-processed materials have tremendous potential in this respect. Key to the

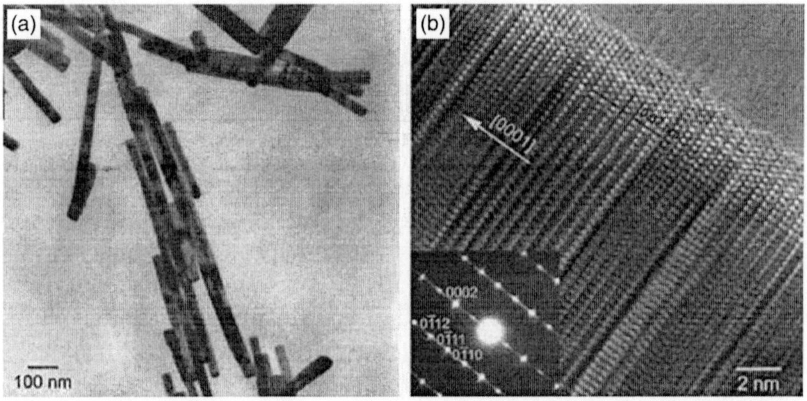

Figure 14.1. Transmission election microscope (TEM) (left) and HRTEM (right) image of CdS nanorods grown by a solvothermal route with ethylenediamine and aqueous solution. The inset shows the electron diffraction (ED) pattern of CdS with the growth axis along the [0001] direction. [Reproduced with permission from Ref. 5. Copyright 2005 Springer.]

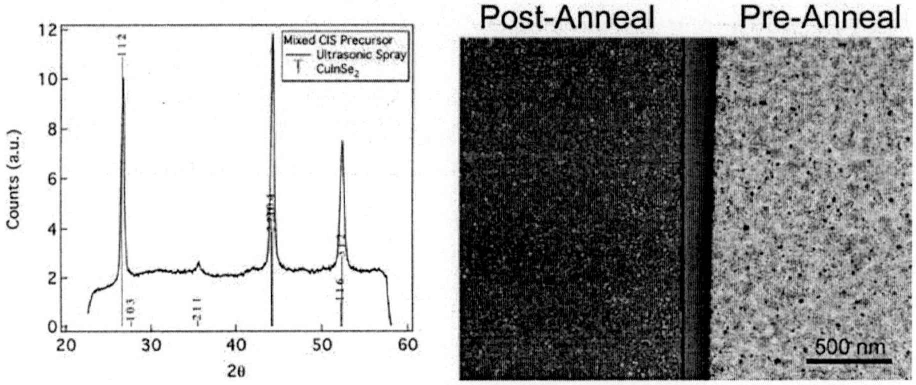

Figure 14.2. Ultrasonic spray deposited $CuInSe_2$ film for photovoltaic applications using mixed-metal organic precursors. XRD shows (left) the growth of nearly phase pure $CuInSe_2$, and optical micrographs show reasonable morphologies (right).

success of this approach is the attainment of high-quality electronic materials (e.g., developing control of grain size, grain boundaries, defects, and doping within the materials). To optimize this complex parameter space simultaneously requires a careful selection of precursor materials, solvents, and additives tailored to the subsequent processing. This area is rapidly evolving, and the coupling of custom synthesized precursors and the use of rapid thermal or laser processing is leading to improved materials through greater control. As indicated, even the "holy grail" of solution processing, Si, is beginning to see progress by solution and nanoparticle routes (see Chapter 5). Although much remains to be done to get large grain and appropriately doped materials, this work represents a tantalizing approach that could revolutionize photovoltaics, lighting, thin-film transistors (TFTs), and other large-area devices. Solution processing also leads rapidly to the ability to prototype new materials in a facile way that is difficult by vacuum-based approaches. The ability to characterize on the nanoscale means that extremely small samples can be used to explore new materials or to validate synthesis. Increasingly there is also an interest in nanostructured semiconductors, from nanorods to complex core shell structures, that can create or enhance functionality and that are much easier to grow by solution means than by any other means.

14.2.2 Oxides

This book covers a variety of solution-processed metal oxide materials for a wide range of applications, including superconductors, transparent TFTs, batteries, tunable dielectrics, fuel cells, and transparent conductive oxides (TCOs) (e.g., see Chapters 2, 4, and 7–9). Oxides are emerging as some of the key materials for a very broad range of applications, including, for example, electronic, optical, mechanical, tribological, and catalytic systems. Historically, these electronic materials have been deposited by either PVD approaches or classic ceramic processing. Solution-based precursors combined with appropriate processing are opening the door for high-quality materials deposition in a non vacuum approach.[6] In many cases, rapid thermal or laser processing allows for the potential of minimizing the thermal load on novel and flexible substrates. This is especially true for laser processing where the process temperature can be high but the thermal load to the substrate can be small.[7,8]

Traditionally oxide materials are predominately crystalline. However, there has been much recent interest in amorphous oxide systems that can be processed at temperatures near room temperature. This approach has been highlighted by the recent emergence of amorphous TCOs such as InZnO[9] and InZnGaO[10] and their application in transparent TFTs.[11,12] Since their mobilities are about 10 times that of amorphous or microcrystalline silicon and their transparency is around 90% in the visible spectral range, this enables transparent electronics, flexible electronics, and related applications. The amorphous-based systems may be ideal candidates for solution processing, possibly requiring low thermal energy, since reordering to achieve crystallinity is often endothermic. The ability to use solution or nanoparticle precursors to reduce processing temperature low enough to work on flexible substrates is potentially a unique advantage of solution-based processing, although precursors need to be specifically tailored for every application. Another unique advantage of solution-based processing is that processing can also lead to unique 3D-structures that would be much more difficult to attain by other nonsolution means. Nanostructured oxide materials have already shown uses in organic photovoltaics (see Fig. 14.3 and Chapter 9), where nanostructured zinc oxide

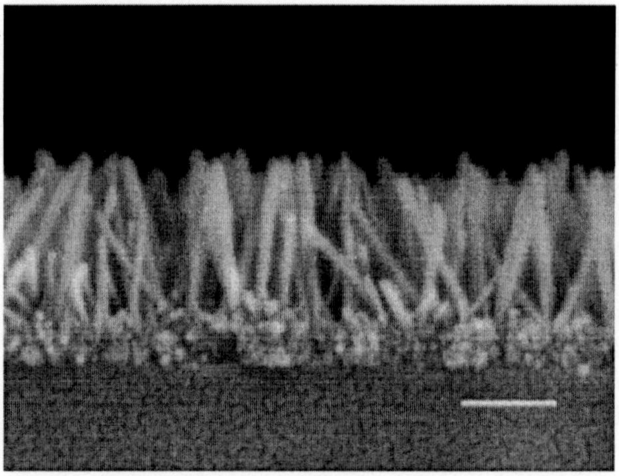

Figure 14.3. Cross-sectional SEM of a ZnO nanorod carpet grown on fluorine-doped SnO_2 by solution growth.[16] Scale bar is 100 nm.

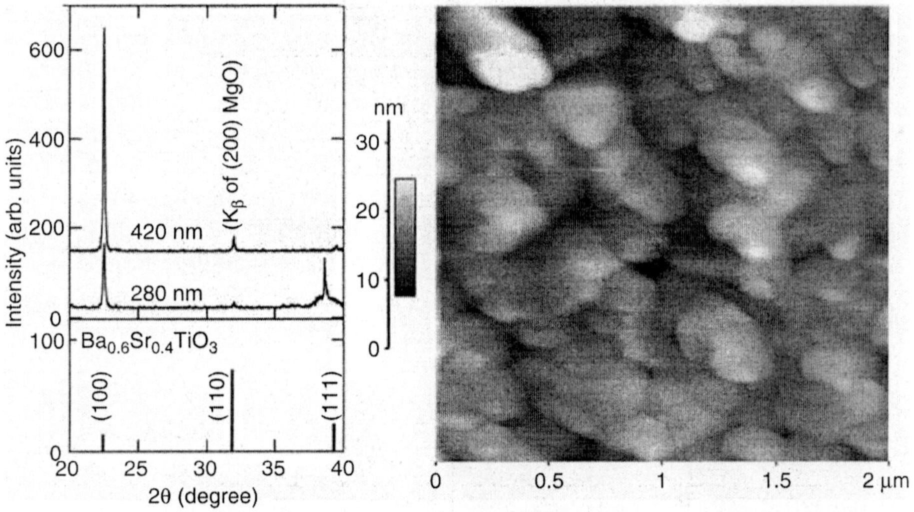

Figure 14.4. XRD (left) and AFM (right) images of a barium strontium titanate film grown by solution processing.[17]

materials potentially result in more stable and lower cost cells.[13,14] Also nano-structured oxides are being used as battery materials for faster charge/discharge rates and increased cycle stability. Solution-processed dielectrics, e.g., barium strontium titanate (BST, see Fig. 14.4), can be easily applied for use in frequency agile electronics and high-K dielectrics.[15]

14.2.3 Metals

The final link in the development of solution-deposited electronics and other devices, such as fuel cells and sensors, is that of metal deposition. Both solution and nanoparticle precursors have been used for the formation of metals with good electronic properties.[18–20] This approach is potentially enabling for contacts for large-area devices, photovoltaics, touch screens, displays, and flexible electronics. New generations of inks have demonstrated the potential to produce high-quality metal grid lines (see Fig. 14.5) with properties comparable with their vacuum-deposited equivalent. Combined with ink-jet printing, this enables better resolution than more conventionally used screen printing. It also enables rapid patterning and prototyping without photolithography. In addition, ink-jet printing provides the possibility to deposit multilayer metallic contacts and even wraparound or 3D-contacts, something that is nearly impossible using conventional methods.

In the future, we expect that the variety of contact materials will expand to include a greater diversity of metals and that development of new interfacial modifiers will aid contact formation, dope at the interface, and stabilize the structure. It is also possible to anticipate organic surface agents that work with flexible substrates and that processing methodologies will be improved to produce high-quality electronic materials, comparable with those prepared by conventional approaches, allowing for new functionality at significantly reduced cost and process temperatures. Currently, features on the order of 20 μm are possible by ink jet printing (see Fig. 14.5). However, with the rapid development of next-generation ink-jet printing and other liquid precursor dispensing techniques, feature sizes smaller than a couple of micrometers will be possible in the not-too-distant future. We note also that stamping techniques, such as those initially developed by Whitesides, Stellacci, and others, can move printing into the nanometer regime.[21,22]

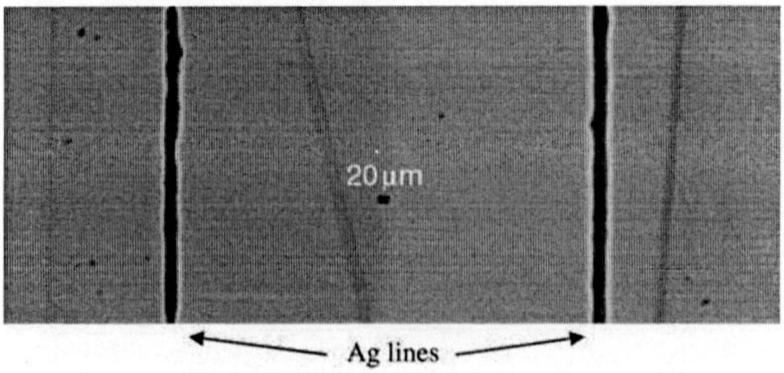

Figure 14.5. Micrograph image of ink-jet-printed silver lines using a metal-organic decomposition precursor. Lines printed on SiN_x-coated ribbon silicon. Line width is ~40 μm.

14.3 DEPOSITION APPROACHES

Coupled intimately to the development of the precursors and processing is the development of suitable deposition technologies capable of meeting the complex needs of uniformity, thickness, resolution, and rate. For some applications, such as semiconductors, the focus may be on large-area uniformity, whereas for other applications, such as non linear optics, thickness control may be key. As a specific example, contact printing for silicon photovoltaics requires high-resolution lines (<50 μm wide) on 6-inch wafers at a processing rate of one wafer per second. To achieve these deposition goals, we have seen a variety of techniques in this book, including dip coating, chemical bath deposition, electrodeposition, spray deposition, and printing approaches, such as ink jet, gravure, and offset (e.g., see Chapters 6–8 and 12). A next generation of techniques includes nano-pen deposition and stamping approaches.[23–28] Each of these deposition techniques has pros and cons. The best deposition approach in the long run will be application specific and depend critically on such issues as value added versus cost and resolution and throughput. Spray deposition, chemical bath deposition, and dip coating will likely be most useful in areas where uniform large-area coverage is needed—dip coating and chemical bath deposition preferably for thin films and spray deposition for thicker films. Ink-jet printing will have application in fields where patterning 2D or 3D-structures is needed, although it can also be used for deposition of thin homogenous films. Electrodeposition can be used for deposition of thin films and to thicken patterned materials deposited by ink-jet printing. Finally dip pen and stamping approaching can move solution deposition into the nanoscale regime.

Overall, the key element in many of these approaches is the complex needs of the "ink" to meet the needs of the deposition technique, to facilitate the formation of the desired material with appropriate properties and to facilitate the desired thickness, roughness, line resolution, and uniformity. This set of constraints will in most cases require an appropriate combination of precursors, solvents, surfactants, adhesion promoters, and rheology control agents (e.g., see Chapter 12 for more discussion).[29] For new metal organic and nanoparticle precursors, the fabrication of inks is still an emerging science.

14.4 NEXT GENERATION OF APPLICATIONS

As the development of precursors and the fabrication of inks evolves and is coupled to a better understanding of how to process them to make very high-quality inorganic materials, this will enable a whole new generation of applications that begin to integrate materials at the nanoscale to obtain new chemical, optical, electronic, and magnetic functionality. In addition, many new devices, sensors, solar cells, and other concepts propose the integration of inorganic, organic, and biologically active materials to allow new functionality (e.g., see

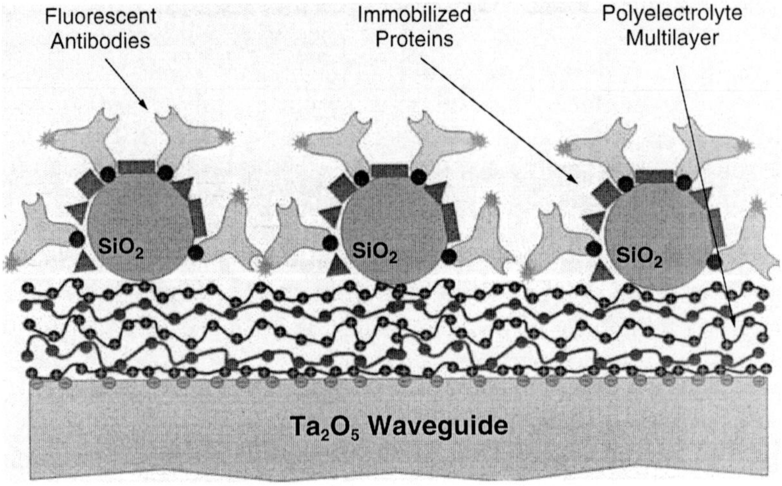

Figure 14.6. Schematic drawing of the working principle of nanoparticle modified planar optical waveguides (PWGs). The amount of immobilized proteins is enhanced because of the larger available surface area provided by immobilized nanoparticles.[33] The SiO_x and Ta_2O_5 structure creates a non linear structure with very high surface area. The bound proteins and antibodies couple optically into this structure providing very high sensitivity to monitor biological changes. [Figure kindly provided by Prof. Janos Vörös, ETH Zurich, Switzerland.]

Fig. 14.6). A key is that in many cases solution processing can lead to new structures that are difficult or impossible to attain by other means. This can include, for example, nanofiber arrays, core-shell structures, nanopods, and nanoribbons.[30–32] These structures can lead to a variety of new functionalities—from 3D prototyping, to third-generation PV structures, to electronic paper, to a new class of non linear optics, to the ability to order nanostructures at very small length scales and maybe even to the holy grail of the energy field, artificial photosynthesis. Below we briefly discuss how some of these concepts are beginning to be realized.

14.4.1 New Solar Cells: Quantum Dot (QD) Structures and Multiple Exciton Generation (MEG)

There is an increasing interest in the possibility of developing higher efficiency solar cell devices that extend beyond the typical single junction Shockley–Queisser efficiency limit of ~30%.[34] One way of doing this is, in bulk materials, through multi-junction devices, such as the evolving multi-junction III–V devices. The other is to operate on the nanoscale by employing quantum dots (QDs). Potentially, structures can be built with variable band gaps through different compositions and sizes (e.g., see Chapter 10).[35] The quantum dots

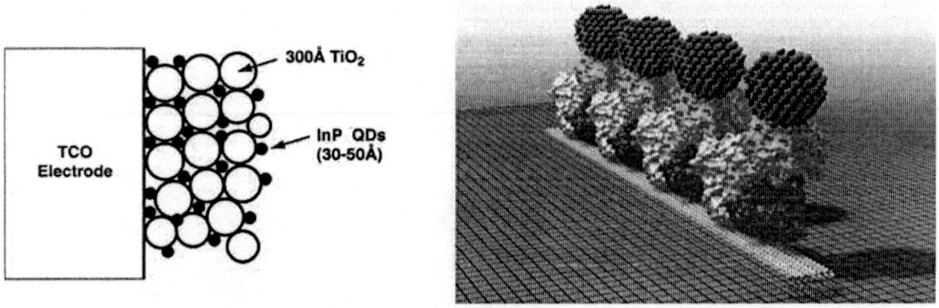

Figure 14.7. Left: Solution-processed quantum dot structure of InP nanoparticles (black dots) in a TiO_2 nanoparticle array (open circles). The QDs could sensitize an optical response similar to the dye in a Gratzel cell. Right: A schematic of very small QDs (balls) bonded to a controlling surfactant that is bonded to a surface specific element, such as a TCO surface.

then need to be integrated into a hybrid device structure to get efficient charge transfer into and out of the structure (see Fig. 14.7).

The QD's absorption can be used directly or it can be used to get multiple exciton generation. The latter has recently been shown in suitably chosen QDs, such as PbSe and Si.[36,37] This discovery allows for the potential of a variety of devices employing both upconversion and downconversion.

In hybrid structures, the QDs can be monodisperse, polydisperse, or even of different compositions to get a broadband response. The backbone can be inorganic or organic. The capping agents on the QDs are critical for coupling the structures and, in some cases, inducing order to form an array. Combining solution-processable QDs with solution-processable organics/polymers to form hybrid structures is a key platform for third-generation photovoltaic technologies. Figure 14.8 shows another approach to producing such nanostructures. In this case, the authors have decorated the surface of a pre-grown semiconductor stack with Au-cluster-containing micelles. A reduction then produces gold dots on the surface, which then act as a mask during reactive ion beam etching, thereby producing nanopillars. These nanopillars can be multifunctional in third-generation PV. They can serve as contacts for carrier transport, quantum-confined structures for upconversion or downconversion, and as antennas in a rectenna (i.e., a rectifying antenna) for solar energy conversion.[38]

14.4.2 Organic–Inorganic Hybrids

Besides the use of solution-based processing for the deposition of inorganic compounds, major developments are taking place in the field of deposition of organic–inorganic hybrid structures. Already indicated was the use of a combination of structured inorganic QDs and organics in the field of photovoltaics,

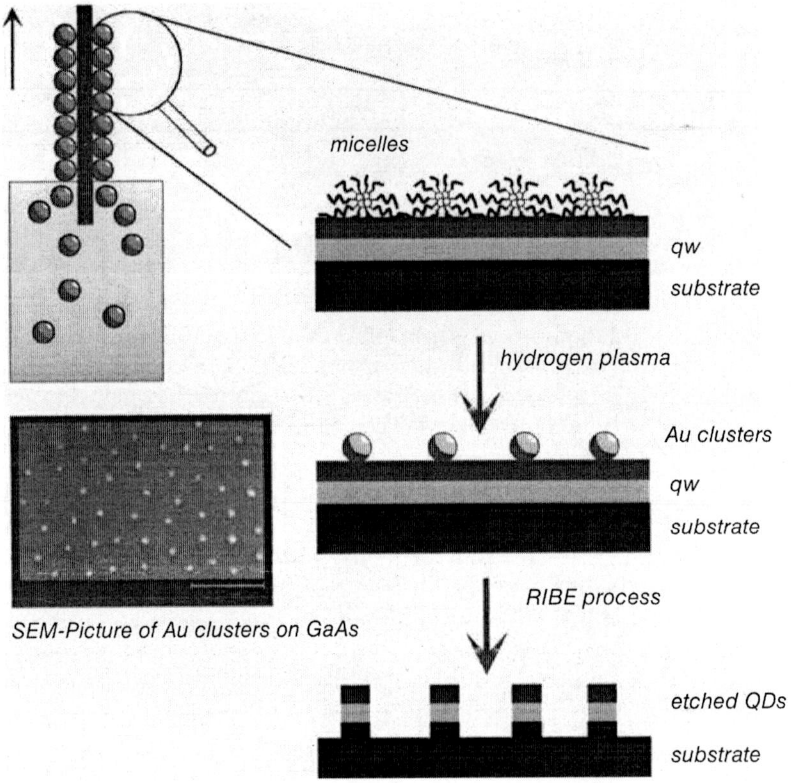

Figure 14.8. Vertically aligned nanopillar, with embedded QDs. Created by using a mask produced by solution processing coupled Au clusters from micelles. [Reproduced with permission from Ref. 39. Copyright 2002 American Institute of Physics.]

but obviously the use of hybrids, especially electronic materials, is not limited to that field. The tailorability of polymers is leading to new generations of organic light-emitting diodes and non linear optical materials among others. In addition, their ease of solution processing and deposition is nearly ideal for large-area technologies such as displays, lighting, photovoltaics, and flexible electronics. Nearly all of these structures integrate inorganic components with the organic components. Solution processing of all components is highly desirable. A good example is in the photovoltaics area, where there is an increasing interest in the development of nanostructured solution-processed inorganic structures, such as the ZnO nanorod array illustrated schematically in Fig. 14.9, which can stabilize and enhance the efficiency of bulk hetero-junction organic photovoltaic (OPV) devices. A cross-sectional SEM of a solution-grown nanorod array was shown earlier in Fig. 14.3. These structures have attained nearly 3% light conversion efficiency to date when filled with the bulk hetero-junction polymer mixture. Although these photovoltaic structures are

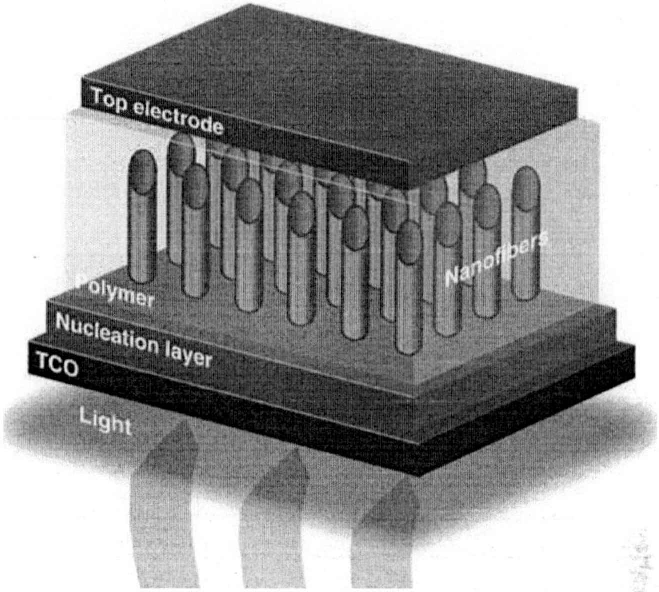

Figure 14.9. Schematic of a solution-grown nanorod/nanotube array (e.g., ZnO), which can be coupled with a light absorbing polymer to produce a functional organic–inorganic photovoltaic device.

not optimized and ZnO may not ultimately be the material of choice, the ability to solution grow regular nanocarpets is interesting in several areas. Interestingly, some of these devices, when made in an inverted configuration, can be made in the air to produce reasonable efficiencies and, thus, may lead to more stable inorganic/polymer structures.[39,40]

The combination of nanoparticles and inorganic nanostructures with polymers and organic molecules provides a solution-based tool kit for producing a wide range of new and novel electronic structures, such as the development of nanoparticle or organic-based displays using OLED-based technology. Recently ink-jet printing of such displays is becoming a viable alternative to conventional processing, which could have a real importance for white light displays for indoor lighting, where large area and low cost are critical.[41]

Another area where solution processing and inorganic nanoparticle incorporation is already beginning to make an impact is for electronic paper. This area was suggested several years ago,[42–44] but it was not until E-Ink Corporation was formed in 1997 that serious development began. Already the technology has produced products, including Sony's Reader Digital Book, Polymer Vision's Readius, and Amazon's Kindel, which when linked to the internet enable downloading of your morning newspaper.[42,45] Figure 14.10 illustrates a couple of the emerging technologies. Part of the reason for the rapid move to market is the ability to prototype solution processes rapidly and the ease of

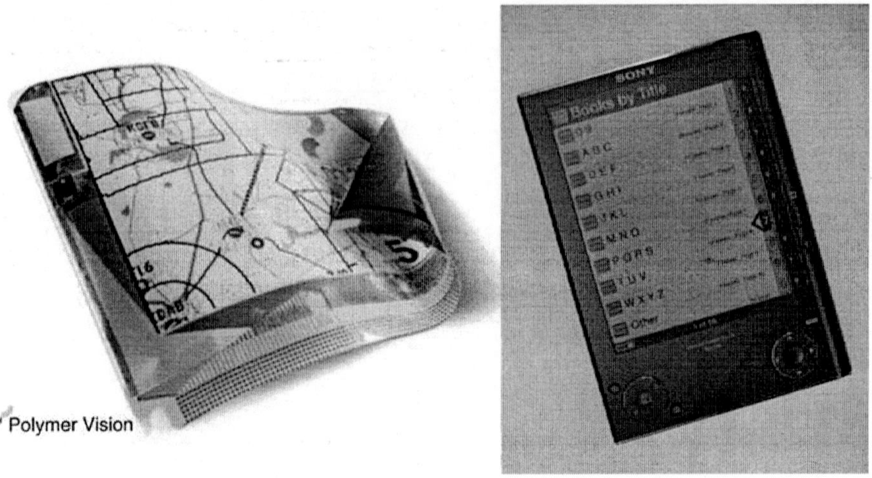

Figure 14.10. Examples of flexible electronic paper are shown with a Polymer Vision flexible map on the left and an E-Ink-based Sony Digital Book on the right.

moving to printing technologies.[46] These devices are being explored in organic, inorganic, and hybrid structures.

There is an increasing interest in the continued development of flexible and wearable displays and rollable electronic paper. To achieve a true electronic newspaper requires a very low-cost approach to flexible displays, which will likely require low-cost, solution-based processing, and more advanced devices will ideally require very high-speed pixel ON/OFF switching (e.g., see Chapter 1). Importantly, mobilities will need to be higher than those in the amorphous Si TFTs used now (0.1–1 cm^2/Vs), and research is looking at nanowire-based transistors where mobilities can be much higher (30–100 cm^2/Vs).[47] The key is that nearly all of these devices can be solution deposited, most likely by ink-jet printing, in a very large format. For these evolving flexible electronics applications, processing temperatures also need to be low (<200 °C, preferably <150 °C). Currently there is a mixture of materials for large-area, low-resolution devices. These materials are well suited to solution-based direct write processing. An important aspect to the future success of these devices is to improve resolution as well as the development of full-color electronic paper.[48]

14.4.3 Non Linear Optics

Since the original observation of the Pockels' and Kerr effects, considerable work has been done on non linear optical effects and devices. Some effects that can be used are second harmonic generation, third harmonic generation, stimulated Raman scattering, and self-focusing.[49] Of key importance is the effect the light generates when traveling through a media. Non linear optical materials can be crystals or thin films. Recently, much attention has been

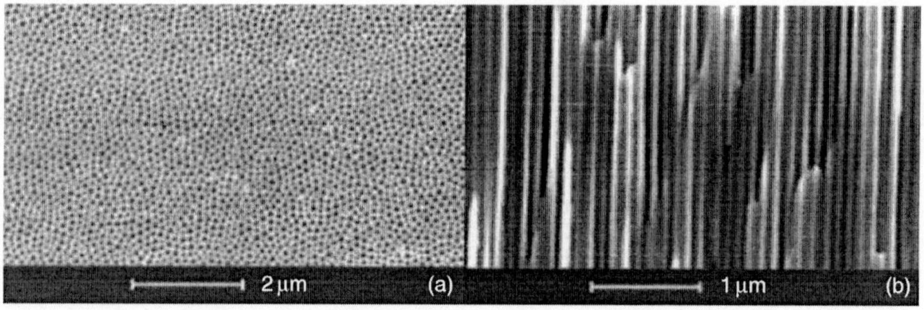

Figure 14.11. A non linear optical material formed from nanopores in an InP membrane. The picture on the left is the top view, whereas that on the right shows the cross section. [Reproduced with permission from Ref. 60. Copyright 2003 Wiley-VCH.]

directed to the development of mesoscale waveguides that can potentially be implemented by solution means.[50] In addition, there are non linear optical effects observed in composite materials, such as those discussed above. The composites are typically organic/polymer or inorganic/polymer materials, and many have the advantage that the matrix defines the gross optical properties, whereas the loading material defines the non linear optical properties. Recent work has opened up this field dramatically and has obtained non linear optical properties from sol-gel and solution-deposited films, organic/inorganic hybrids, nanotube–inorganic hybrids, and solution-deposited inorganic waveguides.[51–59]

An example of such a solution-prepared, non linear optical structure from InP is shown in Fig. 14.11, which produces a nanoporous structure. Such pores can be filled with organics to change the dielectric constant or to help develop the structure in a sol-gel or templated approach. A wide variety of printing approaches, including ink-jet printing, spin coating, dip coating, and spray deposition, can be used to develop these structures, which could involve inorganic, inorganic/organic, or purely organic materials. Deposition techniques such as chemical bath deposition and electrodeposition might not directly seem feasible. However, even these techniques can be used, for example, for electropolymerization to create the polymer *in situ*. One limiting step for solution-based organic/inorganic hybrids is the maximum deposition and processing temperatures that can be used, which are obviously much lower than for deposition of strictly inorganic materials. The emerging abilities and capabilities in the organic area complement and integrate with the inorganic materials area and enable new materials, structures, and devices.[60,61]

Another example of recent work is the demonstration of non linear optical properties of Cu nanoparticles in an ITO matrix.[62] This is an example of a structure that can be obtained fully by solution processing. Coupled with this is the observation of enhanced surface plasmons in some nanoparticles, which potentially produces tailorable, non linear optical properties.[63] This effect can be observed as above with either the presence or the absence of QDs, i.e., nanoholes, leaving an opening for tailored aerogels.

Increasingly, there is an understanding of how to get these self-organized structures to form using solution processing and how to build intentional nanostructures and to enable a broad range of new devices, which can have enhanced optical, electronic, and catalytic properties (e.g., see Chapter 9). The use of solution-based inorganic and organic precursors are enabling in general for these applications; i.e., a metal organic precursor for an inorganic material coupled to a monomer for a polymer can produce reaction unique structures that cannot be obtained by other means.

14.4.4 3D-Structures

Beyond the use of nanoporous arrays as described above is the use of solution processing to produce controlled 3D-structures, including not only the use of self-organized structures but also the use of solution-based precursors to build macroscopic 3D-structures. We illustrate the former in Fig. 14.12 for a set of nanorod arrays grown by solution means. These arrays are increasingly of interest for filtering water and air, catalyst beds, and new opto-electronic devices, such as solar cells. Again, these can only be grown by solution means and, by varying the solution pH and other modifiers in solution, the morphology can be changed substantially.

One of the fields in which solution processing will most likely expand rapidly is in the fabrication of 3D-structures and objects. Already shown are nanorod and nanotube arrays (see Figs. 14.3, 14.11, and 14.12). For 3D-structures, the ability to control the structure, morphology, and composition is very difficult by normal PVD approaches, whereas by controlling processing conditions, it can be done by solution means. On a larger scale, solution-based processing is used to create 3D objects such as prototypes, parts for various machinery, and pieces of equipment.[65–67] For this application, startup companies are investing in creating printer-like equipment to produce spare parts at home (e.g., http://www.desktopfactory.com/ and http://www.2objet.com/). An example is shown in Fig. 14.13 where a prototype engine block has been printed. Some believe that even larger scales of solution-based processing are possible by "printing" a complete house (see Fig. 14.14, http://www.contour-crafting.org/), where everything is scaled and nanoparticulates are replaced by small rocks and the solvent is a mixture of sand, mortar, and water.[68] Even though this last example might seem out of proportion, it is a clear indication that solution-based processing does not really have any size nor application limit.

14.4.5 Catalysis/Artificial Photosynthesis

Finally if you combine many of the concepts above, you can begin to think about developing controlled reaction centers that combine inorganic, organic,

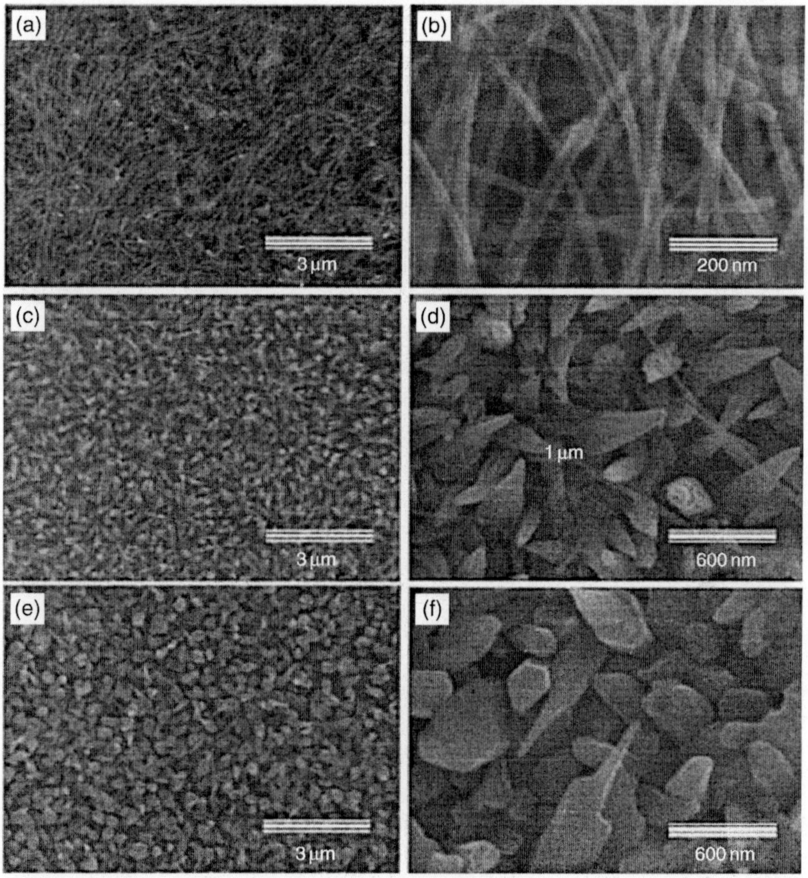

Figure 14.12. SEM images of the ZnO materials formed under various conditions: (a, b) at a zinc nitrate/HMTA concentration of 0.005 M; (c, d) at a zinc nitrate/HMTA concentration of 0.005 M; and (e, f) at a zinc nitrate/HMTA concentration of 0.02 M. All are on an indium tin oxide seed layer. The seed layer used for (c, d) was different from the seed layer used for (a, b, e, and f). [Reproduced with permission from Ref. 64. Copyright 2006 American Chemical Society.][64]

and biological moieties to mimic artificial photosynthesis.[69–71] Figure 14.15 illustrates a scheme for the translation from natural to artificial photosynthesis. Many of these use the active solution-processed species on a nanostructured TiO_2 or other oxide platform.[72,73] Natural photosynthesis makes use of an antenna array feeding excited states to an inorganic reaction center where the actual redox chemistry occurs. It has been a long-standing goal to replicate these functionalities in an inorganic or hybrid system. Accomplishing this may be the "holy grail" of energy conversion and will require hybrid nanostructures designed to transport excitons and do high-energy redox chemistry.

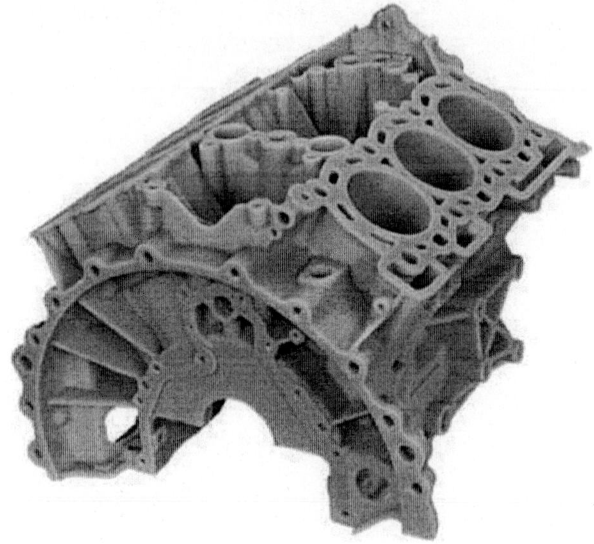

Figure 14.13. An engine block printed by processing photopolymers. (http://www.2objet.com/Misc/Gallery/tabid/70/AlbumID/393-3/Default.aspx).

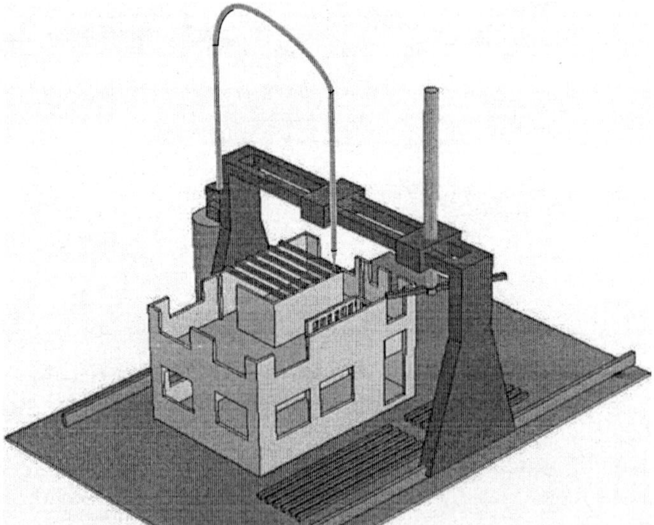

Figure 14.14. Large-scale printing: Design concept for a house "printer". [Reproduced with permission from Ref. 68. Copyright 2004 Elsevier.]

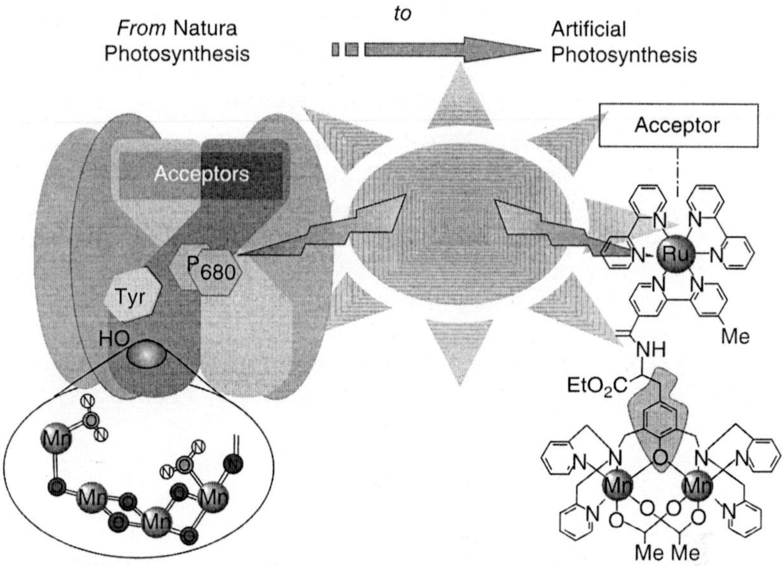

Figure 14.15. Schematic for Artificial photosynthesis.[74] (http://www.fotomol.uu.se/Forskning/Biomimetics/fotosyntes/index.shtm) [Figure kindly provided by Prof. Stenbjörn Styring, Uppsala University, Sweden.]

14.5 CONCLUSIONS

As indicated in the various chapters of this book, the application fields of inorganic solution-based processing are numerous. It is clear that development of new inks will make it possible to transfer applications, for which currently other deposition techniques are used, to full solution-based processing. It is also clear that the possibilities with solution-based deposition for materials outnumber those of conventional deposition, and therefore, it will be enabling for new structures, devices, and science. This is not only true for deposition of inorganics, but also for organic, hybrid, and biological materials. As solution-based techniques are developed, they will lead to new generations of composites, self-organizing structures, and the integration of complex functionalities. Finally, solution-based processing is enabling for 3D-structures as well as for flexible electronics. Solution-based processing is still in its early stages, and clearly, a prosperous and diverse future is on the horizon.

REFERENCES

1. Lira-Cantu, M.; Norrman, K.; Andreasen, J. W.; Casan-Pastor, N.; Krebs, F. C. 2007. Semiconductor oxides as electron acceptors in hybrid organic-inorganic solar cells. *ECS Trans.* 3:1–9.

2. Reale, A.; Brown, T. M.; Di Carlo, A.; Giannini, F.; Brunetti, F.; Leonardi, E.; Lucci, M.; Terranova, M. L.; Orlanducci, S.; Tamburri, E.; Toschi, F.; Sessa, V. 2006. Nanocomposites for organic and hybrid organic-inorganic solar cells. *Proc. SPIE-Int. Soc. Opt. Eng.* 6334:63340Y/63341–63340Y/63348.
3. Lira-Cantu, M.; Krebs, F. C. 2005. Polymer photovoltaics: From conjugated polymers to hybrid organic-inorganic solar cells. *Recent Res. Dev. Appl. Phys.* 8:71–98.
4. Nguyen Cong, H.; Dieng, M.; Sene, C.; Chartier, P. 2000. Hybrid organic-inorganic solar cells: Case of the all thin film PMeT(Y)/CdS(X) junctions. *Sol. Energy Mater. Sol. Cells* 63:23–35.
5. Li, F.; Zhang, Z. 2005. Synthesis of high quality CdS nanorods by solvothermal process, and their photoluminescence. *J. Nanoparticle Research* 7:685–689.
6. Rivkin, T.; Curtis, C.; Miedaner, A.; Perkins, J. D.; Alleman, J.; Ginley, D. S. 2002. Direct-write processing for photovoltaic cells. *Conference Record of the 29th IEEE Photovoltaic Specialist Conference*. pp. 1326–1329.
7. Steen, W. M. 2003. Laser material processing: An overview. *J. Opt. A: Pure Appl. Opt.* 5:S3–S7.
8. Young, R. T.; Wood, R. F. 1982. Laser processing of semiconductor materials. *Ann. Rev. Mater. Sci.* 12:323–350.
9. Fortunato, E.; Ginley, D.; Hosono, H.; Paine, D. C. 2007. Transparent conducting oxides for photovoltaics. *MRS Bull.* 32:242–247.
10. Hosono, H. 2006. Ionic amorphous oxide semiconductors: Material design, carrier transport, and device application. *J. Non-Cryst. Solids* 352:851–858.
11. Boyd, J.; Moore, S. K. 2005. Transparent transistors. *IEEE Spectrum* March 2005: 18–19.
12. Wager, J. F. 2003. Transparent electronics. *Science* 300:1245–1246.
13. White, M. S.; Olson, D. C.; Shaheen, S. E.; Kopidakis, N.; Ginley, D. S. 2006. Inverted bulk-heterojunction organic photovoltaic device using a solution-derived ZnO underlayer. *Appl. Phys. Lett.* 89:143517/143511–143517/143513.
14. Olson, D. C.; Shaheen, S. E.; Collins, R. T.; Ginley, D. S. 2007. The effect of atmosphere and ZnO morphology on the performance of hybrid poly(3-hexylthiophene)/ZnO nanofiber photovoltaic devices. *J. Phys. Chem. C* 111:16670–16678.
15. Kaydanova, T.; Miedaner, A.; Perkins, J. D.; Curtis, C.; Alleman, J. L.; Ginley, D. S. 2007. Direct-write inkjet printing for fabrication of barium strontium titanate-based tunable circuits. *Thin Solid Films* 515:3820–3824.
16. Olson, D. C.; Piris, J.; Collins, R. T.; Shaheen, S. E.; Ginley, D. S. 2006. Hybrid photovoltaic devices of polymer and ZnO nanofiber composites. *Thin Solid Films* 496:26–29.
17. Kaydanova, T.; Miedaner, A.; Perkins, J. D.; Curtis, C.; Alleman, J. L.; Ginley, D. S. 2007. Direct-write inkjet printing for fabrication of strontium titanate-based tunable circuits. *Thin Solid Films* 515:3820–3824.
18. Curtis, C. J.; Rivkin, T.; Miedaner, A.; Alleman, J.; Perkins, J.; Smith, L.; Ginley, D. S. 2002. Direct-write printing of silver metallizations on silicon solar cells. *Mater. Res. Soc. Symp. Proc.* 730:79–84.
19. Ginley, D. S. 1999. Nanoparticle-derived contacts for photovoltaic cells *Proc.—Electrochem. Soc.* 99–11:103–109.

20. Dearden, A. L.; Smith, P. J.; Shin, D.-Y.; Reis, N.; Derby, B.; O'Brien, P. 2005. A low curing temperature silver ink for use in ink-jet printing and subsequent production of conductive tracks. *Macromol. Rapid Commun.* 26:315–318.

21. Xia, Y. N.; Whitesides, G. M. 1998. Soft lithography. *Angew. Chem. Int. Ed.* 37:550–575.

22. Yu, A. A.; Savas, T. A.; Taylor, G. S.; Guiseppe-Elie, A.; Smith, H. I.; Stellacci, F. 2005. Supramolecular nanostamping: Using DNA as movable type. *Nano Lett.* 5:1061–1064.

23. Groeger, R. 2007. Nano contact printing with afm-controlled phase-separated block copolymer layers. *Institut fur Nanotechnologie, Fakultat fur Physik, Universitat Karlsruhe, Karlsruhe, Germany*. Report. i–vii, pp. 1–130.

24. Park, J. D.; Lee, K. D. 2007. Method for manufacturing stamp for nanoimprint lithography having two-dimensional pattern of uniform size and shape, and method for photonic crystal using the same. LG Electronics Inc., S. Korea.

25. Chen, L. Q.; Chan-Park, M. B.; Yan, Y. H.; Zhang, Q.; Li, C. M.; Zhang, J. 2007. High aspect ratio silicon nanomoulds for UV embossing fabricated by directional thermal oxidation using an oxidation mask. *Nanotechnology* 18:355307/355301–355307/355307.

26. Shestopalov, A. A.; Clark, R. L.; Toone, E. J. 2007. Inkless microcontact printing on self-assembled monolayers of Fmoc-protected aminothiols. *J. Am. Chem. Soc.* 129:13818–13819.

27. Kim, Y.-K.; Kim, G. T.; Ha, J. S. 2007. Simple patterning via adhesion between a buffered-oxide etchant-treated PDMS stamp and a SiO_2 substrate *Adv. Funct. Mater.* 17:2125–2132.

28. Huang, C.; Lu, N.; Chi, L. 2007. Surface patterning approaches based on Langmuir-Blodgett technique. *Huaxue Jinzhan* 19:852–859.

29. Ash, M.; Ash, I. 2006. *Handbook of Rheology Modifiers*. Synapse Information Resources, Endicot, NY. pp. 1–1154.

30. Pauzauzkie, P.; Yang, P. 2006. Nanowire photonics. *Materials Today* 9:36.

31. Jie, J. S.; Zhang, W. J.; Jiang, Y.; Lee, S. T. 2006. Single-crystal CdSe nanoribbon field-effect transistors and photoelectric applications. *Appl. Phys. Lett.* 89:133118.

32. Moore, D.; Wang, Z. L. 2006. Growth of anisotropic one-dimensional ZnS nanostructures. *J. Mater. Chem.* 16:3898–3905.

33. Vörös, J.; Ramsden, J. J.; Csucs, G.; Szendro, I.; De Paul, S. M.; Textor, M.; Spencer, N. D. 2002. Optical grating coupler biosensors. *Biomaterials* 23:3699–3710.

34. Shockley, W.; Queisser, H. J. 1961. Detailed balance limit of efficiency of p-n junction solar cells. *J. Appl. Phys.* 32:510–519.

35. Luque, A.; Martí, A.; Nozik, A. J. 2007. Quantum dots: Multiple exciton generation and intermediate bands. *MRS Bull.* 32:236–241.

36. Beard, M. C.; Knutsen, K. P.; Yu, P.; Luther, J. M.; Song, Q.; Metzger, W. K.; Ellingson, R. J.; Nozik, A. J. 2007. Multiple exciton generation in colloidal silicon nanocrystals. *Nano Lett.* 7:2506–2512.

37. Nozik, A. J. 2007. Third generation solar photon conversion. *Abstracts of Papers, 233rd ACS National Meeting* (*Chicago, IL; March 25–29*). PHYS-082.

38. Green, M. A. 2003. *Third Generation Photovoltaics: Advanced Solar Energy Conversion*. Springer-Verlag, Berlin. pp. 1–160.

39. Haupt, M.; Miller, S.; Ladenburger, A.; Sauer, R.; Thonke, K.; Spatz, J. P.; Rietmüller, S.; Möller, M. 2002. Semiconductor nanostructures defined with self-organizing polymers. *J. Appl. Phys.* 91:6057–6059.

40. White, M. S.; Olson, D. C.; Shaheen, S. E.; Kopidakis, N.; Ginley, D. S. 2006. Inverted bulk-heterojunction organic photovoltaic device using a solution-derived ZnO underlayer. *Appl. Phys. Lett.* 89:143517/143511–143517/143513.

41. Van der Vaart, N. C.; Lifka, H.; Budzelaar, F. P. M.; Rubingh, J. E. J. M.; Hoppenbrouwers, J. J. L.; Dijksman, J. F.; Verbeek, R. G. F. A.; Van Woudenberg, R.; Vossen, F. J.; Hiddink, M. G. H.; Rosink, J. J. W. M.; Bernards, T. N. M.; Giraldo, A.; Young, N. D.; Fish, D. A.; Childs, M. J.; Steer, W. A.; Lee, D.; George, D. S. 2005. Distinguished paper: Towards large-area full-color active-matrix polymer OLED television. *J. Soc. Inf. Display* 13:9–16.

42. Kitamura, T. 2006. Electronic paper for human communication and environmental protection. *Oyo Butsuri* 75:995–999.

43. Li, L.; He, J.; Zhang, S.; Li, H.; Chen, C.; Yang, J.; Zou, J. 2004. Microcapsule process and application in electronic paper. *Gongnent Cailiao* 35:407–409, 413.

44. Sung, J. H.; Jang, I. B.; Choi, H. J.; Chen, I.-J. 2004. Titania-coated polystyrene spheres for electronic ink. *Polym. Prepr. (Am. Chem. Soc., Div. Polym. Chem.)* 45:970–971.

45. Omodani, M. 2003. Trends of the study and development of electronic paper. *Oyo Butsuri* 72:176–180.

46. Rogers, J. A.; Baldwin, K.; Bao, Z.; Dodabalapur, A.; Raju, V. R.; Ewing, J.; Amundson, K. 2001. Rubber stamped plastic circuits for electronic paper. *Proceedings—International Symposium on Advanced Packaging Materials: Processes, Properties, and Interfaces*. pp. 98–103.

47. Ju, S.; Facchetti, A.; Xuan, Y.; Liu, J.; Ishikawa, F.; Ye, P.; Zhou, C.; Marks, T.; Janes, D. B. 2007. Fabrication of fully transparent nanowire transistors for transparent and flexible electronics. *Nature Nanotechnol.* 2:378–384.

48. Sunohara, S.; Kitamura, T. 2007. Investigation of color electrophoretic display utilizing electrophoretic colored particles. *Nippon Gazo Gakkaishi* 46:247–253.

49. Günter, P. 2000. Non linear optical effects and materials. *Springer Ser. Opt. Sci.* 72:540 pp.

50. Fan, H.; Lopez, G. P.; Brinker, C. J. 2000. Rapid prototyping of patterned multifunctional nanostructurs. *Mater. Res. Soc. Symp. Proc.* 625:141–150.

51. Sigaev, V. N.; Lotarev, S. V.; Orlova, E. V.; Stefanovich, S. Y.; Pernice, P.; Aronne, A.; Fanelli, E.; Gregora, I. 2007. Lanthanum borogermanate glass-based active dielectrics. *J. Non-Cryst. Solids* 353:1956–1960.

52. Lin, A.; Kim, B. H.; Moon, D. S.; Chung, Y.; Han, W.-T. 2007. Cu^{2+}-doped germano-silicate glass fiber with high resonant non linearity. *Opt. Express* 15:3665–3672.

53. Lin, A.; Kim, B. H.; Ju, S.; Han, W.-T. 2007. Fabrication and third-order non linearity of germano-silicate glass fiber incorporated with Au nanoparticles. *Proc. SPIE-Int. Soc. Opt. Eng.* 6481:64810M/64811–64810M/64817.

54. Kang, E.-H.; Bu, T.; Jin, P.; Sun, J.; Yang, Y.; Shen, J. 2007. Layer-by-layer deposited organic/inorganic hybrid multilayer films containing noncentrosymmetrically orientated azobenzene chromophores. *Langmuir* 23:7594–7601.

55. Kobayashi, T.; Ogawa, R.; Miyazawa, K. I.; Kuwabara, M. 2002. Fabrication of beta -BaB_2O_4 thin films with (00l) preferred orientation through the chemical solution deposition technique. *J. Mater. Res.* 17:844–851.

56. Zhao, Y.; Chen, Y.; Zhang, X.-C.; Raravikar, N. R.; Ajayan, P. M.; Lu, T.-M.; Wang, G.-C.; Schadler Feist, L. S. 2002. Ultrafast all-optical switch using carbon nanotube polymer composites. Rensselaer Polytechnic Institute. U.S. Patent 6,782,154, pp. 1–15.

57. Saji, T.; Hiraga, T.; Moritani, T.; Tanaka, N.; Yanagimoto, H.; Ueno, I.; Tsujita, K. 2001. Production method of organic thin film and optical component. Agency of Industrial Sciences and Technology, Japan; Dainichi Seika Kogyo K. K.; Victor Co. of Japan. Japanese Patent 2001-003195, pp. 1–20.

58. Huang, Y.-Q.; Liu, M.-D.; Zeng, Y.-K.; Liu, S.-B. 2001. Progress of study on ZnO thin film and its properties. *Wuji Cailiao Xuebao* 16:391–397.

59. Watkins, A. N.; Ligler, F. S. 2000. Patterned deposition of sol-gel thin films using an airbrush. *Anal. Lett.* 33:1899–1912.

60. Tiginyanu, I. M.; Kravetsky, I. V.; Langa, S.; Marowsky, G.; Moneckke, J.; Föll, H. 2003. Porous III-V compounds as non linear optical materials. *Phys. Stat. Sol. (a)* 197:549–555.

61. Aakeröy, C. B.; Azoz, N. E.; Kadim, M.; Seddon, K. R.; Trowbridge, L. 1993. Novel colorless composite materials for non linear optics. *Adv. Mater.* 5:364–367.

62. Ryasnyansky, A. I.; Palpant, B.; Debrus, S.; Khaibullin, R. I.; Stepanov, A. L. 2006. Non linear optical properties of copper nanoparticles synthesized in indium tin oxide matrix by ion implantation. *J. Opt. Soc. Am. B* 23:1348–1353.

63. Brolo, A. G.; Kwok, S. C.; Cooper, M. D.; Moffitt, M. G.; Wang, C.-W.; Gordon, R.; Riordon, J.; Kavanagh, K. L. 2006. Surface plasmon-quantum dot coupling from arrays of nanoholes. *J. Phys. Chem. B* 110:8307–8313.

64. Ku, C.-H.; Wu, J.-J. 2006. Aqueous solution route to high-aspect-ratio zinc oxide nanostructures on indium tin oxide substrates. *J. Phys. Chem. B* 110:12981–12985.

65. Chovancova-Lovell, V.; Pekarovicova, A.; Fleming, P. 2006. Novel phase change inks for printing three-dimensional structures. *J. Imaging Sci. Technol.* 50:550–555.

66. Napadensky, E. 2005. Ink-jet 3d printing of photopolymers materials: An emerging rapid prototyping technology. *Proceedings—RadTech Europe 05: UV/EB* 2:193–197.

67. Travitzky, N.; Zimmerman, K.; Melcher, R.; Greil, P. 2006. From polysaccharides to SiSiC composites by 3D printing. *Ceramic Transactions* 175:37–45.

68. Khoshnevis, B. 2004. Automated construction by contour crafting—related robotic and information technologies. *J. Automation in Construction* 13:5–19.

69. Ishida, H. 2002. From artificial photosynthesis to nano-science. *Kokagaku* 32:136–143.

70. Ganguly, T.; Pal, S. K. 2006. Photoinduced electron transfer research to build model compounds of artificial photosynthesis and solar energy conversion. *J. Chinese Chem. Soc.* 53:219–226.

71. Brune, A.; Jeong, G.; Liddell, P. A.; Sotomura, T.; Moore, T. A.; Moore, A. L.; Gust, D. 2004. Porphyrin-sensitized nanoparticulate TiO_2 as the photoanode of a hybrid photoelectrochemical biofeul cell. *Langmuir* 20:8366–8371.

72. Sun, L.; Hammarstrom, L.; Akermark, B.; Styring, S. 2001. Towards artificial photosythesis: Ruthenium-manganese chemistry for energy production. *Chem. Soc. Rev.* 30:36–49.

73. Tributsch, H.; Pohlmann, L. 1995. Synergetic molecular approaches towards artificial and photosynthetic water photoelectrolysis. *J. Electroanal. Chem.* 396:53–61.

74. Huang, P.; Kurz, P.; Styring, S. 2007. EPR investigations of synthetic manganese complexes as bio-mimics of the water oxidation complex in photosystem II. *Appl. Magn. Reson.* 31:301–320.

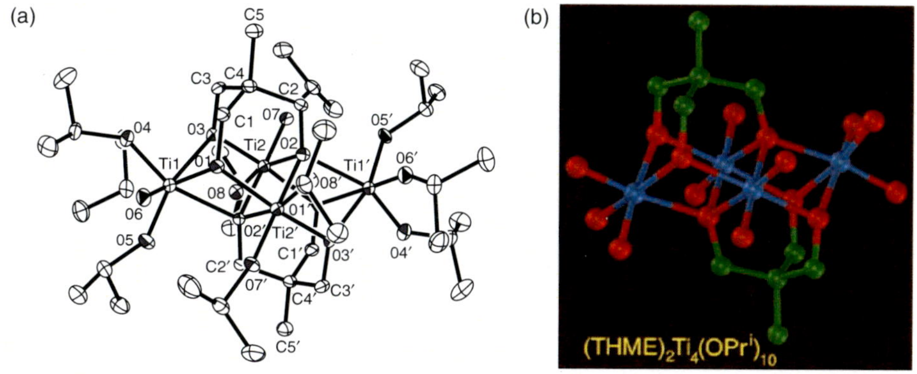

Figure 2.15. (a) ORTEP structure of TiO_2 precursor [(THME)$_2$Ti$_4$(OPr^i)$_{10}$] synthesized for use in thin-film preparation and (b) molecular dynamics simulation of the same precursor. [Reprinted with permission from Ref. 82. Copyright 1995 American Chemical Society.]

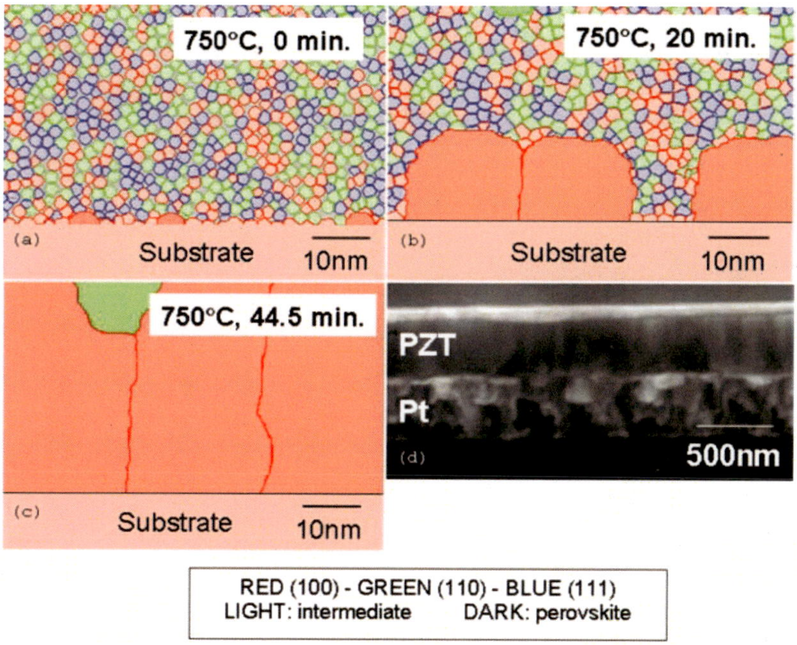

Figure 2.16. (a–c) Simulations of film structural evolution for PZT thin films at various times during heat treatment.[15] (d) A representative SEM photomicrograph illustrating the columnar microstructure of PZT.[48] The lower layer is the lower Pt electrode, the middle layer is the PZT, and the upper layer is the top Pt electrode. [(a)–(c) Reprinted with permission from Ref. 15. (d) Reprinted with permission from Ref. 9. Copyright 1997 American Chemical Society.]

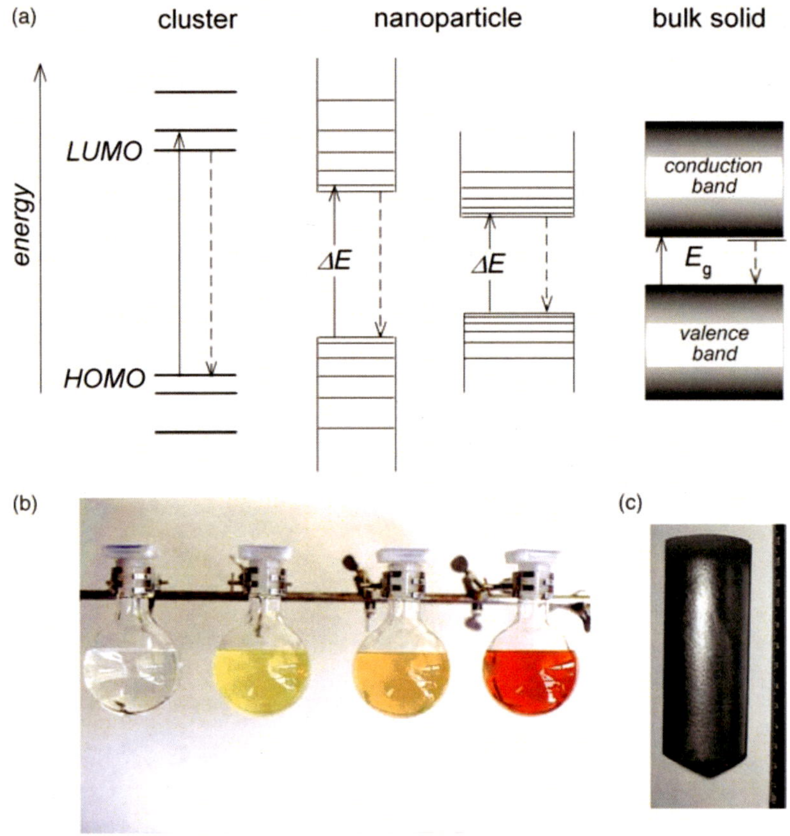

Figure 10.4. (a) Schematic energy diagrams of clusters, nanoparticles, and bulk semiconductors. (b) Manifestation of the size quantization effect as a color change of aqueous colloidal solutions of CdSe nanoparticles (courtesy of A. Rogach). The particle size changes from left to right from ~1.5 to ~4.5 nm. (c) Bulk CdSe crystal.

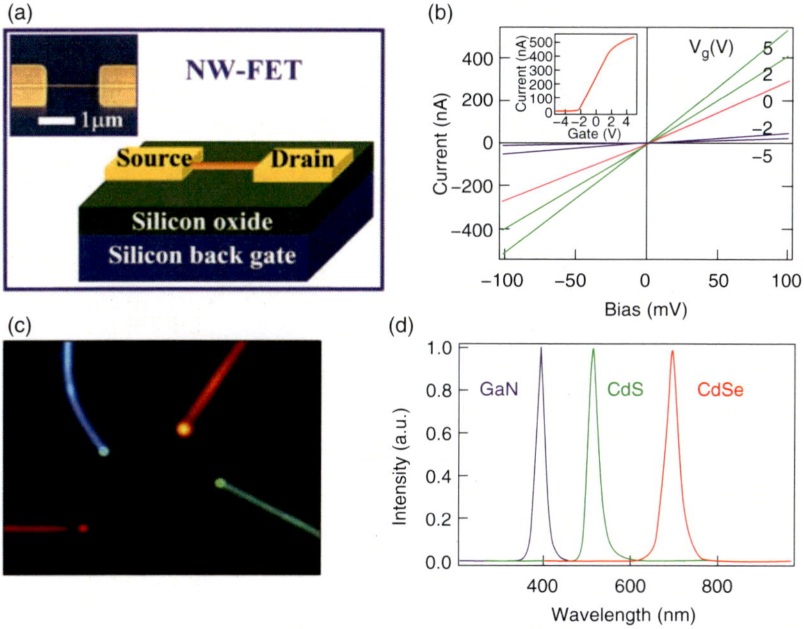

Figure 11.2. Nanowire electronic and optical properties. (a) Schematic of an NW–FET used to characterize electrical transport properties of individual NWs. (inset) SEM image of an NW-FET; two metal electrodes, which correspond to source and drain, are visible at the left and right sides of the image. (b) Current versus voltage for an n-type InP NW-FET. The numbers inside the plot indicate the corresponding gate voltages (V_g). The inset shows current versus V_g for V_{sd} of 0.1 V. (c) Real-color photoluminescence image of various NWs shows different color emissions. (d) Spectra of individual NW photoluminescence. All NW materials show a clean band-edge emission spectrum with narrow FWHM around 20 nm.

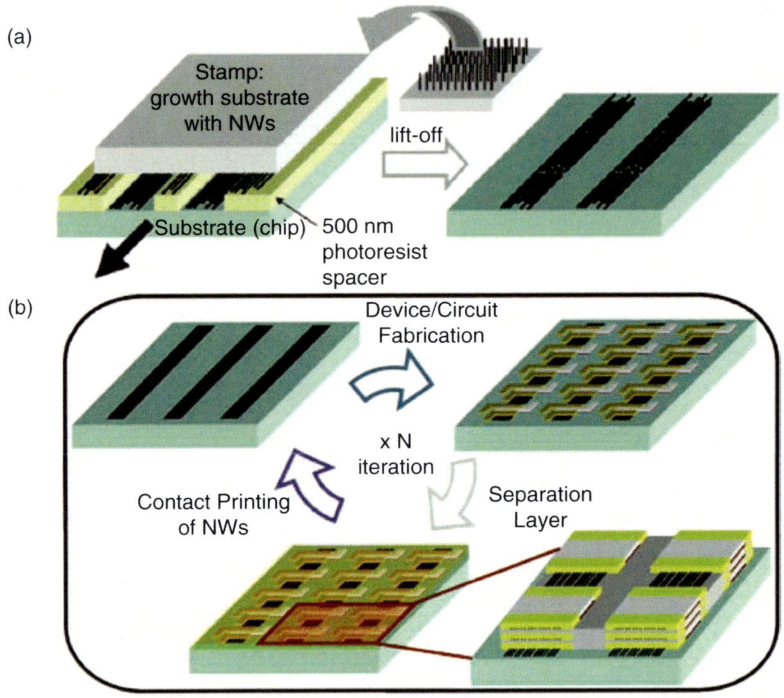

Figure 11.16. Overview of 3D NW circuit integration. (See text for full caption.)

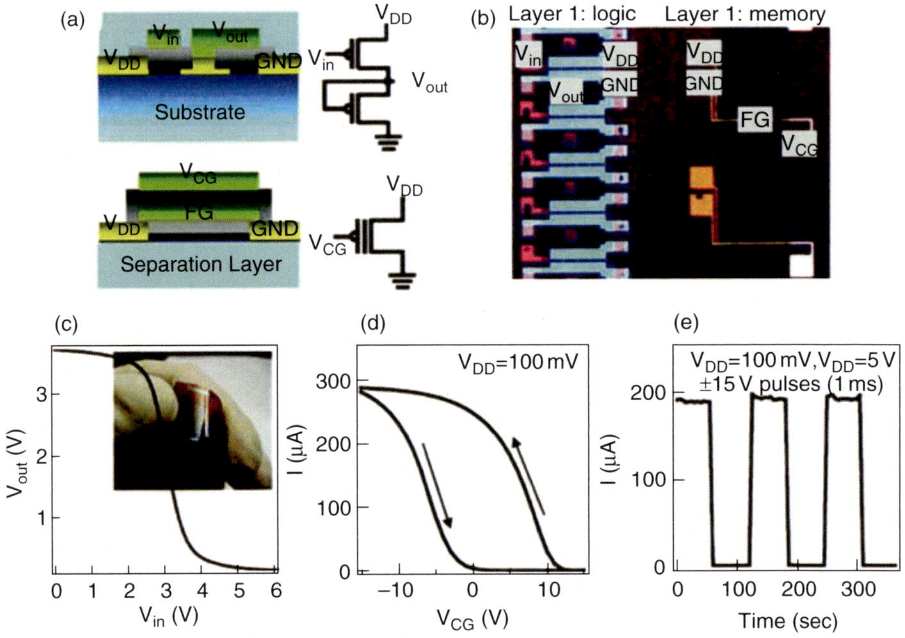

Figure 11.17. Three-dimensional multifunctional circuits on plastics. (See text for full caption.)

INDEX